Solid State Theory

METHODS AND APPLICATIONS

Solid State Theory

METHODS AND APPLICATIONS

Edited by

P. T. LANDSBERG

Professor of Applied Mathematics and Mathematical Physics,
University College, Cardiff, Great Britain

Contributors

D. A. EVANS, P. T. LANDSBERG, T. LUKES, D. J. MORGAN (*University College, Cardiff*) and J. E. PARROTT (*University of Wales Institute of Science and Technology, Cardiff*)

1969

WILEY — INTERSCIENCE

a division of John Wiley & Sons Ltd.

London - New York - Sydney - Toronto

Library of Congress catalog card No. 78–82285.

SBN 471 51383 0

Set on Monophoto Filmsetter and printed by
J. W. Arrowsmith Ltd., Bristol, England

Preface

During the last twenty years solid state devices such as transistors, solar cells and lasers have been invented and produced, and this development shows little sign of slackening. There is consequently no need to argue here the case for a widespread study by theoretically inclined students of the theory of solids. This book is intended for this growing group and should be suitable for their first post-graduate year. But even if a student does not continue for long in this line of work, we believe that the subject has still something important to offer. For it is a kind of laboratory in which the student can develop a first contact with a considerable variety of mathematical techniques. These will remain useful to him even if the technological challenges of tomorrow lead him to different fields of application. In order to display the subject in this light we have had to restrict the extent to which solid state theories themselves could be covered. (Many-body theories, for example, though they are sometimes included in our lecture course, are omitted. Detailed techniques of energy band calculations are not described, and the theory of mechanical effects and dislocations had also to be sacrificed.) But we believe that the areas with which we have dealt are important both as parts of solid state theory and as illustrating useful mathematical techniques. For this reason we have included many problems which we have found useful over the six years that we have given the course of lectures on which this book is based.

We have avoided most of the duplication which occurs when some of the courses, which are represented by the different parts of this book, are running concurrently in our post-graduate lectures, but complete avoidance of such duplication would have been artificial and was not attempted.

We hope that the reader will derive some pleasure from the perusal of this book and from our refusal to invite prolonged confusion by the phrase "it can readily be shown that".

A brief summary follows of what the reader will find in this book, together with an indication of how it differs from other expositions. We start with a chapter which, using simple formulae based on the assumption of constant relaxation times, introduces some of the important experimental methods which have yielded insight into the properties of various solids. Part B describes basic ideas about lattices, the use of Fourier expansions, equations of motion and effective mass theory. The approach here (sections 15 to 17) is somewhat novel, both as regards the formalism of band theory and the treatment of effective mass theory. It utilises the work done in Cardiff on sum rules. Some new material will also be found in the treatment of the *n*-electron Bloch theorem and in section 8. Part C deals with group theory using an approach which emphasizes the role of wave functions as basic functions for the irreducible representations of the group of symmetry transformations of the Hamiltonian. For example the irreducible representations of the space groups are found via Bloch functions, thus obviating the use of factor groups, which some beginners find difficult. Special attention may perhaps be drawn also to the use of projection operators and to the perturbation calculation of the spin–orbit splitting which illustrates the importance of double groups. Part D deals with electron statistics and displays its usefulness in the elucidation of some key properties of *p–n* junctions and of devices. The approach here is based on work which has been done in Cardiff during the last five years or so, and is, we believe, theoretically pleasing. Sections 3, 6, 7 and 10 of Part D represent the first systematic account of this approach. The calculation of an avalanche multiplication factor (section 12, problem 1) and the account of our recent work on thermodynamic efficiencies (section 14) should also be noted. In Part E, after the derivation of the necessary standard results in Chapter IX, there is in Chapter X an account of the calculation of the phonon spectrum for a non-trivial model which makes use of perturbation theory, group theory and critical point analysis in a manner intelligible to beginners, but not over-simplified to the extent of being incapable of extension to more realistic problems. Lastly, Part F gives what is perhaps the first systematic account devoted entirely to one-electron Green's functions. Chapter XI brings together the general properties which are needed, and the applications to periodic lattices are given in Chapter XII which contains an account of recent work as well as some new proofs. The last chapter is devoted to fairly recent work on the important subject of disordered systems.

P.T.L.

Contents

Chapter X Applications of Lattice Dynamics

PART F ONE-ELECTRON GREEN'S FUNCTIONS IN SOLID STATE PHYSICS (*T. Lukes*)

Chapter XI General Theory

PART A

TYPICAL EXPERIMENTS AND THEIR INTERPRETATION

D. J. Morgan

CHAPTER I

Typical Experiments and their Interpretation

A.1 INTRODUCTION

In this chapter we consider some effects which arise from the motion of the charge carriers in semiconductors under the influence of electric and magnetic fields. These effects have become powerful tools for studying the electronic properties of solids, enabling one to obtain values for various parameters (effective mass, mobility, lifetime, etc.) used in the phenomenological description of the motion of the charge carriers. It is possible with the use of these parameters to interpret in a consistent way the results of experiments on many solids, and we now introduce the parameters of interest in this chapter by summarizing the relevant features of the one-electron model of a solid.

In the one-electron approximation the valence electrons in a crystal are regarded as a collection of noninteracting particles moving in a potential which has the periodicity of the crystal lattice. This potential may be thought of as arising from the periodic charge distribution associated with the ion cores plus some averaged potential contribution due to the other electrons. The solution of the Schrödinger equation for a single electron in this potential then provides a set of one-electron states, amongst which the electrons of the crystal are distributed in accordance with the rules of Fermi–Dirac statistics.

For a perfectly periodic potential the one-electron energies occur in bands of permitted states separated by regions of energy for which no electron energy states are allowed. In a semiconductor at zero temperature all occupied energy bands are completely filled with electrons, the highest of these bands (the valence band) being separated from the nearest empty band (the conduction band) by the forbidden energy gap. Electrons in a completely filled energy band cannot contribute to the current. At higher temperatures, electrons are thermally excited across the forbidden gap into the conduction band and become conduction electrons, leaving empty states (holes) in the valence band. Conduction is then possible in both bands.

The dynamical behaviour of the electrons is in many ways similar to that of free particles. An electron responds to applied electric and magnetic fields as if it were free but with an effective mass (§A.3 and §A.4) which is usually different from the free-electron mass. In many cases the effective mass of an electron is a scalar quantity, as is the free-electron mass, but in other very important cases it is found to vary with direction and to behave as a tensor (§A.3).

An electron in an energy band behaves, perhaps not surprisingly, like a particle with a negative charge. A hole in an energy band, however, behaves like a particle with a positive charge. In an almost filled energy band it is simpler to consider the motion of the holes than to consider the motion of the electrons. A hole then acts as a positive charge carrier, with an appropriate effective mass, and can contribute to the current. Thus in the conduction band the current is carried by electrons and in the valence band the current is carried by holes.

The first experimental evidence for the existence of holes came from the occurrence of positive values of the Hall coefficient (§A.5). This implied that the current was being carried by positive charges. Cyclotron resonance experiments with circularly-polarized radiation (§A.3) established that holes and electrons rotate in opposite senses in a magnetic field, just as one would expect for charges of opposite sign.

The motion of electrons moving independently in a perfectly periodic potential is completely unimpeded, there being no mechanism by which electrons acted on by external forces can be returned to the equilibrium state. Electrical resistance arises because of deviations from a perfectly periodic structure caused by lattice vibrations and structural defects such as vacancies and impurity atoms. The scattering of electrons, or of holes, by such imperfections can in many cases be phenomenologically represented by means of a characteristic time constant known as the relaxation time (§A.2).

Another important time constant is the recombination or excess carrier lifetime (§A.6 and §A.7). An electron in the valence band can, by some appropriate means (e.g. illumination), be excited across the forbidden gap into the higher-lying conduction band, leaving a hole in the valence band and forming what is known as an electron–hole pair. The formation of these excess charge carriers is accompanied by a significant change in the conductivity of the crystal (§A.7). The electrons and holes can then recombine either by an electron losing energy and falling straight back into the valence band or through some intermediary states, usually lattice imperfections, known as recombination centres. The electron (hole) recombination lifetime may be thought of as being the mean time which elapses between the generation of an electron (hole) and its subsequent recombination.

The interpretation of the effects considered in this chapter in terms of the above-mentioned parameters is accomplished by means of the Boltzmann transport equation which is discussed in the next section.

A.2 THE BOLTZMANN TRANSPORT EQUATION

The macroscopic properties of a system containing a very large number of particles, such as a crystalline solid, can be calculated using a distribution function $f(\mathbf{p}, \mathbf{r}, t)$ which is defined in such a way that at time t the number of particles dN having position between $\mathbf{r}$ and $\mathbf{r} + d\mathbf{r}$ and momentum between $\mathbf{p}$ and $\mathbf{p} + d\mathbf{p}$ is given by

$$dN = f(\mathbf{p}, \mathbf{r}, t)\, d\mathbf{r}\, d\mathbf{p}. \tag{A.2.1}$$

The distribution function $f_0(\mathbf{p}, \mathbf{r}, t)$ for a system in thermal equilibrium can be obtained using the well-known methods of equilibrium statistical mechanics. When an external force is applied to the system, however, it is no longer in thermal equilibrium and the distribution function does not have its thermal equilibrium value. The evaluation of the distribution function when the system is in a nonequilibrium state is accomplished by means of the Boltzmann transport equation.

Consider an elementary volume $d\mathbf{r}\, d\mathbf{p}$ about the point $(\mathbf{r}, \mathbf{p})$ in the six-dimensional phase space x, y, z, p_x, p_y and p_z. The number of particles in this volume at time t is $f(\mathbf{p}, \mathbf{r}, t)\, d\mathbf{r}\, d\mathbf{p}$ and the time rate of change of this quantity, $(\partial f/\partial t)\, d\mathbf{r}\, d\mathbf{p}$, must be equal to the rate at which the number of particles at the point $(\mathbf{r}, \mathbf{p})$ increases with time. There are two contributions to this rate of increase:

(i) particles may move from one position in real space to another, or they may change from one momentum to another, on account of the external force applied to the system. This is the drift process. The changes in $f(\mathbf{p}, \mathbf{r}, t)$ caused by the drift process are smooth and continuous and, to a first order approximation, we can use Taylor's expansion and write

$$(\partial f/\partial t)_{\text{drift}} = -\nabla_{\mathbf{p}} f \cdot \frac{d\mathbf{p}}{dt} - \nabla_{\mathbf{r}} f \cdot \frac{d\mathbf{r}}{dt}, \tag{A.2.2}$$

or

$$(\partial f/\partial t)_{\text{drift}} = -\mathbf{F} \cdot \nabla_{\mathbf{p}} f - \frac{1}{m^*}\mathbf{p} \cdot \nabla_{\mathbf{r}} f, \tag{A.2.3}$$

where $\mathbf{F}$ is the applied force and m^* is an effective mass which, for simplicity, we have taken to be a constant scalar;

(ii) particles may change their state through collisions either with other particles of the distribution (e.g. electron–electron interactions) or with other types of particles (e.g. electron–phonon interactions, the phonon being the 'particle' associated with lattice vibrations). For the moment, we represent this contribution to the rate of increase by $(\partial f/\partial t)_{coll}\, d\mathbf{r}\, d\mathbf{p}$.

Combining these results, we now find that the distribution function $f(\mathbf{p}, \mathbf{r}, t)$ satisfies the equation

$$\frac{\partial f}{\partial t} + \mathbf{F} \cdot \nabla_{\mathbf{p}} f + \frac{1}{m^*} \mathbf{p} \cdot \nabla_{\mathbf{r}} f = (\partial f/\partial t)_{coll}. \tag{A.2.4}$$

This is the Boltzmann transport equation.

The quantum mechanical calculation of the collision term, $(\partial f/\partial t)_{coll}$, in the Boltzmann transport equation will not be discussed here. Specific scattering processes are described in detail in later chapters of this book (see §B.17d and §E.21). For our purposes, it is sufficient to say that in many cases, but not all, the scattering can be described by a relaxation time τ which is defined by the equation

$$(\partial f/\partial t)_{coll} = -(f - f_0)/\tau, \tag{A.2.5}$$

where $f_0(\mathbf{p}, \mathbf{r}, t)$ is the thermal equilibrium distribution function. τ may be thought of as being a measure of how quickly the system returns to thermal equilibrium after the removal of the external force. In most cases of interest τ is, in fact, energy-dependent, i.e. a function of the momentum $\mathbf{p}$, but here, in the spirit of the present chapter, we shall assume that τ is a constant and independent of energy[1].

Phenomenological equations, i.e. equations written in terms of macroscopically observable quantities, can be derived from the Boltzmann transport equation. The most important of these for our purposes is that for the current carried by electrons in a crystalline solid in the presence of electric and magnetic fields. In this case the applied force is given by the Lorentz force [see problems (B.13.10), (B.13.13) for a simple proof]

$$\mathbf{F} = -e\left[\mathscr{E} + \frac{1}{c}(\mathbf{v} \times \mathbf{B})\right], \tag{A.2.6}$$

where $\mathscr{E}$ is the electric field, $\mathbf{B}$ is the magnetic induction field, e is the magnitude of the electronic charge, c is the speed of light *in vacuo* and $\mathbf{v}$ is the velocity of an electron with momentum $\mathbf{p}$, i.e. $\mathbf{p} = m_e^* \mathbf{v}$, where we have now written m_e^* for the scalar effective mass of an electron. The phenomenological equation for the current is obtained by multiplying the Boltzmann transport equation by $\mathbf{v}$ and integrating over momentum.

[1] For more general treatments the reader is referred to the books listed at the end of this chapter. The author wishes to thank Dr. A. R. Beattie for his interest and advice during the writing of this chapter.

Using (A.2.5) and (A.2.6), this gives

$$\int \mathbf{v}\frac{\partial \mathrm{f}}{\partial t}\,d\mathbf{p} - e\int \mathbf{v}\left[\mathscr{E} + \frac{1}{c}(\mathbf{v}\times\mathbf{B})\right]\cdot\nabla_{\mathbf{p}}\mathrm{f}\,d\mathbf{p} + \frac{1}{m_e^*}\int \mathbf{v}\mathbf{p}\cdot\nabla_{\mathbf{r}}\mathrm{f}\,d\mathbf{p}$$

$$= -\int \mathbf{v}\frac{(\mathrm{f}-\mathrm{f}_0)}{\tau_e}\,d\mathbf{p}. \qquad \text{(A.2.7)}$$

The individual terms in this equation are:

(i)

$$\int \mathbf{v}\frac{\partial \mathrm{f}}{\partial t}\,d\mathbf{p} = \frac{\partial}{\partial t}(n\bar{\mathbf{v}}), \qquad \text{(A.2.8)}$$

where

$$n = \int \mathrm{f}\,d\mathbf{p} \qquad \text{(A.2.9)}$$

is the electron density, and

$$\bar{\mathbf{v}} = \frac{\int \mathbf{v}\mathrm{f}\,d\mathbf{p}}{\int \mathrm{f}\,d\mathbf{p}} \qquad \text{(A.2.10)}$$

is the mean electron velocity.

(ii)

$$-e\int \mathbf{v}\left[\mathscr{E} + \frac{1}{c}(\mathbf{v}\times\mathbf{B})\right]\cdot\nabla_{\mathbf{p}}\mathrm{f}\,d\mathbf{p} = \frac{ne}{m_e^*}\left[\mathscr{E} + \frac{1}{c}(\bar{\mathbf{v}}\times\mathbf{B})\right], \qquad \text{(A.2.11)}$$

where a partial integration has been performed and it is assumed that the distribution function tends to zero at infinite energies (momenta) faster than $1/|\mathbf{p}|$.

(iii)

$$\frac{1}{m_e^*}\int \mathbf{v}\mathbf{p}\cdot\nabla_{\mathbf{r}}\mathrm{f}\,d\mathbf{p} = \frac{2}{3m_e^*}\nabla_{\mathbf{r}}n\bar{E}_e, \qquad \text{(A.2.12)}$$

where $\bar{E}_e$ is the mean electron energy and it is assumed that the departure from equilibrium is small so that

$$\overline{p_x^2} = \overline{p_y^2} = \overline{p_z^2} = \overline{p^2}/3 = 2m_e^*\bar{E}_e/3, \qquad \text{(A.2.13)}$$

and

$$\overline{p_x p_y} = \overline{p_y p_z} = \overline{p_x p_z} = 0. \qquad \text{(A.2.14)}$$

(iv)

$$-\int \mathbf{v}\frac{(\mathrm{f}-\mathrm{f}_0)}{\tau_e}\,d\mathbf{p} = -n\bar{\mathbf{v}}/\tau_e, \tag{A.2.15}$$

where the electron relaxation time τ_e is taken to be constant and where the contribution to the integral from f_0 vanishes because f_0 is an even function of **v** or **p**.

Hence, when the electron density is independent of the time and the electron mean energy is independent of position,

$$\frac{\partial\bar{\mathbf{v}}}{\partial t} + \frac{\bar{\mathbf{v}}}{\tau_e} = -\frac{e}{m_e^*}\left[\mathscr{E} + \frac{1}{c}(\bar{\mathbf{v}} \times \mathbf{B})\right] - \frac{2\bar{E}_e}{3nm_e^*}\nabla_{\mathbf{r}}n. \tag{A.2.16}$$

The current density $\mathbf{J}_e$ due to the electrons is given by

$$\mathbf{J}_e = -ne\bar{\mathbf{v}}, \tag{A.2.17}$$

and, using equation (A.2.16), the phenomenological equation for the current is

$$\tau_e\frac{\partial\mathbf{J}_e}{\partial t} + \mathbf{J}_e = ne\mu_e\mathscr{E} - \frac{\mu_e}{c}(\mathbf{J}_e \times \mathbf{B}) + D_e\nabla_{\mathbf{r}}n, \tag{A.2.18}$$

where

$$\mu_e = e\tau_e/m_e^* \tag{A.2.19}$$

is the electron mobility and

$$D_e = 2\bar{E}_e\mu_e/3, \tag{A.2.20}$$

D_e/e being the diffusion constant for electrons.

The phenomenological equation for the current density $\mathbf{J}_h$ due to holes can be derived in a similar way with the result

$$\tau_h\frac{\partial\mathbf{J}_h}{\partial t} + \mathbf{J}_h = pe\mu_h\mathscr{E} + \frac{\mu_h}{c}(\mathbf{J}_h \times \mathbf{B}) - D_h\nabla_{\mathbf{r}}p, \tag{A.2.21}$$

where τ_h is the hole relaxation time, p is the hole density, μ_h is the hole mobility and D_h/e is the hole diffusion constant. Equations (A.2.18) and (A.2.21) will be used frequently in later sections of this chapter.

Before discussing the interpretation of some typical experiments, however, it is convenient here to say something about the effective mass m^* which appears in the Boltzmann transport equation. This effective mass was introduced through the equation

$$\mathbf{p} = m^*\mathbf{v} = \hbar\mathbf{k}, \tag{A.2.22}$$

where **k** is the free-particle wave vector and $\hbar = h/2\pi$, h being Planck's constant. For electrons in a crystalline solid it can be shown [see equation (B.14.6) and the discussion in §B.15] that

$$\mathbf{v} = \hbar^{-1}\nabla_{\mathbf{k}} E(\mathbf{k}), \tag{A.2.23}$$

where $E(\mathbf{k})$ is the electron energy. m^{*-1} is therefore related to the slope (first derivative with respect to **k**) of the $E(\mathbf{k})$ versus **k** curve.

If the energy surfaces are spherical, so that

$$E(\mathbf{k}) = \frac{\hbar^2 \mathbf{k}^2}{2m^{**}}, \tag{A.2.24}$$

then

$$\mathbf{v} = \hbar\mathbf{k}/m^{**} \tag{A.2.25}$$

and, as one would expect,

$$m^* = m^{**}. \tag{A.2.26}$$

If, however, the energy surfaces are ellipsoidal, so that

$$E(\mathbf{k}) = \frac{\hbar^2}{2}(k_x^2/m_x^{**} + k_y^2/m_y^{**} + k_z^2/m_z^{**}), \tag{A.2.27}$$

then

$$v_j = \hbar k_j/m_j^{**} \qquad (j = x, y, z) \tag{A.2.28}$$

and in the Boltzmann transport equation the reciprocal scalar effective mass m^{*-1} must be replaced by a diagonal reciprocal effective mass tensor $\overleftrightarrow{(1/m^{**})}$, defined by the equation

$$\bar{\mathbf{v}} = \overleftrightarrow{(1/m^{**})} \cdot \mathbf{p} = \overleftrightarrow{(1/m^{**})} \cdot (\hbar\mathbf{k}) = \hbar^{-1}\nabla_{\mathbf{k}} E(\mathbf{k}) \tag{A.2.29}$$

Such ellipsoidal energy surfaces will be discussed in more detail in the sections dealing with cyclotron resonance and the Hall effect.

It should be realized that the reciprocal effective mass tensor defined by equation (A.2.29) is not in general equal to the reciprocal effective mass tensor with components $(1/m^{***})_{ij}$ given by

$$(1/m^{***})_{ij} = \hbar^{-2}\frac{\partial^2 E(\mathbf{k})}{\partial k_i \partial k_j} \tag{A.2.30}$$

[cf. also equation (B.13.23)]. For the $E(\mathbf{k})$ versus **k** curves considered in this chapter, however, the two reciprocal effective mass tensors are identical.

A.3 CYCLOTRON RESONANCE

In the presence of a constant magnetic induction field **B** a free electron of mass m and charge $-e$ (e.s.u.) performs a circular motion with angular frequency $\omega_B = eB/mc$, where $B = |\mathbf{B}|$ and c is the speed of light *in vacuo*. If a radio frequency electric field, whose accompanying magnetic field may be ignored in comparison with **B**, is then applied in a plane perpendicular to the direction of **B**, resonant absorption of energy by the electron occurs when the frequency of the electric field equals $\omega_B/2\pi$. This is the cyclotron resonance frequency [cf. problem (B.13.7)].

If the electrons in a semiconductor have a single scalar effective mass m_e^*, resonant absorption occurs at a frequency $eB/2\pi m_e^* c$, i.e. the absorption versus frequency curve shows a single peak at the cyclotron resonance frequency. If B is known, a measurement of the frequency at which resonance takes place then yields the effective mass m_e^*. The cyclotron resonance experiment thus provides a direct determination of the effective mass.

In practice the situation is rather more complicated than that envisaged above. It is found that resonance generally occurs at a number of different frequencies and not just at one particular frequency. There are two basic reasons for this difference. Firstly, the complicated energy band structures encountered in practice lead to effective masses which are tensors and to sets of equivalent ellipsoidal energy surfaces in **k**-space. Secondly, since both electrons and holes exist in semiconductors, resonance peaks due to the absorption of energy by holes appear along with those due to electrons. The electrons and the holes rotate in opposite senses in a magnetic field and, as we shall see shortly, this means that one can distinguish between the two by using circularly-polarized radiation for the electric field. Before discussing ellipsoidal energy surfaces, however, it is worthwhile describing the semiclassical theory of cyclotron resonance for the case of spherical energy surfaces and a scalar effective mass.

Consider, therefore, electrons with a scalar effective mass m_e^* in the presence of a constant magnetic induction field $\mathbf{B} = B(0, 0, 1)$ and a radio frequency electric field $\mathscr{E} = \mathscr{E}(1, \alpha, 0)\,\mathrm{e}^{\mathrm{i}\omega t}$, where α is a constant. Using equation (A.2.18) and assuming a homogeneous distribution of electrons, so that $\nabla_{\mathbf{r}} n = 0$, the equation for the current density $\mathbf{J}_e$ due to these electrons becomes

$$\tau_e \frac{\partial \mathbf{J}_e}{\partial t} + \mathbf{J}_e = ne\mu_e \mathscr{E}(1, \alpha, 0)\,\mathrm{e}^{\mathrm{i}\omega t} - \frac{\mu_e}{c}(\mathbf{J}_e \times \mathbf{B}), \tag{A.3.1}$$

where τ_e is the electron relaxation time and $\mu_e \equiv e\tau_e/m_e^*$. Assuming that after a sufficiently long time $\mathbf{J}_e = \mathbf{J}_0\,\mathrm{e}^{\mathrm{i}\omega t}$, we find that

$$\mathbf{J}_0 = \frac{\sigma_e^0 \mathscr{E}}{[\omega_c^2 \tau_e^2 + (1 + \mathrm{i}\omega\tau_e)^2]}\{[1 + \mathrm{i}\omega\tau_e](1, \alpha, 0) - \omega_c \tau_e(\alpha, -1, 0)\}, \tag{A.3.2}$$

where

$$\sigma_e^0 = ne\mu_e \tag{A.3.3}$$

and $\omega_c = eB/m_e^*c$ is the angular cyclotron resonance frequency. Taking the current flow in the x-direction, the conductivity $\sigma_e = J_{01}/\mathscr{E}$ is then given by

$$\frac{\sigma_e}{\sigma_e^0} = \frac{1 + (\mathrm{i}\omega - \alpha\omega_c)\tau_e}{[\omega_c^2\tau_e^2 + (1 + \mathrm{i}\omega\tau_e)^2]}. \tag{A.3.4}$$

In this section the quantity of interest is the real part of the conductivity σ_{eR}, since it is this quantity which governs the absorption of energy by the electrons. Two cases will be considered: $\alpha = 0$, which corresponds to plane-polarized radiation, and $\alpha = \pm\mathrm{i}$, which correspond to circularly-polarized radiation.

When $\alpha = 0$ the real part of the conductivity, using equation (A.3.4), is given by

$$\frac{\sigma_{eR}}{\sigma_e^0} = \frac{1 + (\omega_c^2 + \omega^2)\tau_e^2}{\{[(\omega_c^2 - \omega^2)\tau_e^2 + 1]^2 + 4\omega^2\tau_e^2\}}. \tag{A.3.5}$$

Consider the case $\omega_c\tau_e \gg 1$, so that

$$\frac{\sigma_{eR}}{\sigma_e^0} \sim \frac{(\omega_c^2 + \omega^2)}{(\omega_c^2 - \omega^2)^2\tau_e^2 + 4\omega^2}. \tag{A.3.6}$$

Then for $\omega \ll \omega_c$ or $\omega \gg \omega_c$, $\sigma_{eR}/\sigma_e^0 \ll 1$ and the absorption is relatively small. For $\omega = \omega_c$, $\sigma_{eR}/\sigma_e^0 \sim \frac{1}{2}$ and the absorption is relatively much larger. Hence, when $\omega_c\tau_e \gg 1$, a fairly sharp peak appears in the absorption curve at $\omega = \omega_c$ and an experimental determination of the effective mass m_e^* is possible. As $\omega_c\tau_e$ gets smaller the sharpness of this peak decreases until eventually one is not able to determine from the absorption curve the precise frequency at which resonance takes place, i.e. the experimental determination of the effective mass becomes impossible. The relaxation time τ_e increases with decreasing temperature and thus the condition $\omega_c\tau_e \gg 1$, which must be satisfied for the cyclotron resonance experiment to be a success, is usually met in practice by using large magnetic fields at low temperatures.

When $\alpha = \pm\mathrm{i}$ the real part of the conductivity is given by

$$\frac{\sigma_{eR}}{\sigma_e^0} = \frac{1 + (\omega_c \mp \omega)^2\tau_e^2}{\{[(\omega_c^2 - \omega^2)\tau_e^2 + 1]^2 + 4\omega^2\tau_e^2\}}. \tag{A.3.7}$$

Assuming that the condition $\omega_c\tau_e \gg 1$ is satisfied,

$$\frac{\sigma_{eR}}{\sigma_e^0} \sim \frac{(\omega_c \mp \omega)^2}{(\omega_c^2 - \omega^2)^2\tau_e^2 + 4\omega^2}, \tag{A.3.8}$$

and at resonance, when $\omega = \omega_c$,

$$\sigma_{eR}/\sigma_e^0 \sim 0 \qquad \text{if } \alpha = +\mathrm{i}, \tag{A.3.9}$$

$$\sigma_{eR}/\sigma_e^0 \sim 1 \qquad \text{if } \alpha = -\mathrm{i}. \tag{A.3.10}$$

At the electron resonance frequency, therefore, a peak appears in the absorption curve when $\alpha = -\mathrm{i}$ but does not appear when $\alpha = +\mathrm{i}$. For holes, which rotate in the opposite sense to electrons in a magnetic field, the opposite is true. At the hole resonance frequency a peak appears in the absorption curve when $\alpha = +\mathrm{i}$ but does not appear when $\alpha = -\mathrm{i}$. Thus, by using circularly-polarized radiation, one can determine which of the peaks in the absorption curve are due to holes and which are due to electrons.

It is worthwhile mentioning here that plane-polarized radiation can be thought of as being right circularly-polarized radiation superimposed on left circularly-polarized radiation, i.e. the electric field may be written

$$\mathscr{E}(1, 0, 0)\,\mathrm{e}^{\mathrm{i}\omega t} = \tfrac{1}{2}\mathscr{E}(1, \mathrm{i}, 0)\,\mathrm{e}^{\mathrm{i}\omega t} + \tfrac{1}{2}\mathscr{E}(1, -\mathrm{i}, 0)\,\mathrm{e}^{\mathrm{i}\omega t} \tag{A.3.11}$$

Resonant absorption of energy by electrons takes place only if the sense of rotation of the circularly-polarized radiation is the same as that which the electrons have in the magnetic field. Hence for plane-polarized radiation one would expect the value of σ_{eR}/σ_e^0 at resonance to be one-half, as is the case.

Next let us consider ellipsoidal energy surfaces of the form

$$E = \frac{\hbar^2}{2}(k_x^2/m_t + k_y^2/m_t + k_z^2/m_l), \tag{A.3.12}$$

where m_t is the transverse effective mass and m_l is the longitudinal effective mass. Suppose that the magnetic induction field $\mathbf{B}$ makes an angle θ with the longitudinal axis of the ellipsoid,

$$\mathbf{B} = B(\sin\theta, 0, \cos\theta), \tag{A.3.13}$$

and that the electric field is in the y-direction, i.e.

$$\mathscr{E} = \mathscr{E}(0, 1, 0)\,\mathrm{e}^{\mathrm{i}\omega t}. \tag{A.3.14}$$

The equation for the current density $\mathbf{J}_e$ due to electrons with constant energy surfaces given by equation (A.3.12) and density n then becomes

$$\frac{\partial \mathbf{J}_e}{\partial t} + \frac{\mathbf{J}_e}{\tau_e} = \begin{pmatrix} m_t^{-1} & 0 & 0 \\ 0 & m_t^{-1} & 0 \\ 0 & 0 & m_l^{-1} \end{pmatrix} \cdot \left[ne^2\mathscr{E}(0, 1, 0)\,\mathrm{e}^{\mathrm{i}\omega t} - \frac{e}{c}(\mathbf{J}_e \times \mathbf{B}) \right]. \tag{A.3.15}$$

Assuming a solution of this equation of the form $\mathbf{J}_e = \mathbf{J}_0 \, e^{i\omega t}$, we find that

$$\mathbf{J}_0 = \frac{\sigma_e^0 \mathscr{E}}{[\omega_c^2 \tau_e^2 + (1 + i\omega\tau_e)^2]} \left[\frac{-eB\cos\theta}{cm_t}, (1 + i\omega\tau_e), \frac{eB\sin\theta}{cm_l} \right], \tag{A.3.16}$$

where

$$\sigma_e^0 = ne^2\tau_e/m_t \tag{A.3.17}$$

and

$$\omega_c^2 = \frac{e^2B^2}{c^2}\left(\frac{\sin^2\theta}{m_t m_l} + \frac{\cos^2\theta}{m_t^2}\right). \tag{A.3.18}$$

The conductivity $\sigma_e = J_{02}/\mathscr{E}$ is thus given by

$$\frac{\sigma_e}{\sigma_e^0} = \frac{1 + i\omega\tau_e}{[\omega_c^2\tau_e^2 + (1 + i\omega\tau_e)^2]}. \tag{A.3.19}$$

This is the same equation as equation (A.3.4) with $\alpha = 0$, and it follows that, for the ellipsoidal energy surfaces given in equation (A.3.12), the effective mass m_c^* as determined by the cyclotron resonance experiment is given by

$$\frac{1}{m_c^{*2}} = \frac{\sin^2\theta}{m_t m_l} + \frac{\cos^2\theta}{m_t^2} = \frac{m_t + (m_l - m_t)\cos^2\theta}{m_l m_t^2}, \tag{A.3.20}$$

i.e. the effective mass as determined by the cyclotron resonance experiment depends on the angle θ between the magnetic field and the longitudinal axis of the ellipsoid.

The six equivalent minima in the conduction band of silicon, for example, lie along the ⟨100⟩ axes in **k**-space. Two of the six equivalent constant energy ellipsoids lie along the [±1 0 0] axes and have the form

$$E = \frac{\hbar^2}{2}(k_x^2/m_l + k_y^2/m_t + k_z^2/m_t), \tag{A.3.21}$$

two others lie along the [0 ± 1 0] axes and have the form

$$E = \frac{\hbar^2}{2}(k_x^2/m_t + k_y^2/m_l + k_z^2/m_t), \tag{A.3.22}$$

and the remaining two lie along the [0 0 ± 1] axes and have the form

$$E = \frac{\hbar^2}{2}(k_x^2/m_t + k_y^2/m_t + k_z^2/m_l). \tag{A.3.23}$$

Here [ ] denotes an axis and ⟨ ⟩ a set of equivalent axes in **k**-space.

Suppose that in a cyclotron resonance experiment on silicon the magnetic induction field **B** lies in the (110) crystal plane (see §B.3) and makes an angle ϕ with the [001] axis, i.e.

$$\mathbf{B} = B\left(\frac{\sin\phi}{\sqrt{2}}, -\frac{\sin\phi}{\sqrt{2}}, \cos\phi\right). \tag{A.3.24}$$

The angle θ between the magnetic field and the longitudinal axis of an ellipsoid for the above ellipsoids is then given by

$$\cos\theta = \pm(1, 0, 0)\cdot\frac{\mathbf{B}}{B} = \pm\frac{\sin\phi}{\sqrt{2}} \tag{A.3.25}$$

for the first pair of ellipsoids,

$$\cos\theta = \pm(0, 1, 0)\cdot\frac{\mathbf{B}}{B} = \mp\frac{\sin\phi}{\sqrt{2}} \tag{A.3.26}$$

for the second pair of ellipsoids, and

$$\cos\theta = \pm(0, 0, 1)\cdot\frac{\mathbf{B}}{B} = \pm\cos\phi \tag{A.3.27}$$

for the third pair of ellipsoids. Hence for these six ellipsoids there are two distinct values of $\cos^2\theta$, i.e. $\cos^2\theta = \cos^2\phi$ and $\cos^2\theta = \frac{1}{2}\sin^2\phi$. This means that in a cyclotron resonance experiment on silicon, with the magnetic induction field in the (110) crystal plane, two peaks due to electrons appear, one peak corresponding to an effective mass m_{c1}^* given by

$$\frac{1}{m_{c1}^{*2}} = \frac{m_t + (m_l - m_t)\cos^2\phi}{m_l m_t^2}, \tag{A.3.28}$$

and the other peak corresponding to an effective mass m_{c2}^* given by

$$\frac{1}{m_{c2}^{*2}} = \frac{2m_t + (m_l - m_t)\sin^2\phi}{2m_l m_t^2}, \tag{A.3.29}$$

these peaks coalescing when $\phi = \phi_0 = \tan^{-1}\sqrt{2}$.

From such a cyclotron resonance experiment[2] it is found that for silicon $m_l = 0{\cdot}98m$ and $m_t = 0{\cdot}19m$, m being the free-electron mass.

The holes in the valence bands of silicon have warped spherical energy surfaces at $k = 0$ of the form

$$E = -\frac{\hbar^2}{2m}\{A_1k^2 \pm [A_2^2k^4 + A_3^2(k_x^2k_y^2 + k_x^2k_y^2 + k_y^2k_z^2)]^{\frac{1}{2}}\}, \tag{A.3.30}$$

[2] G. Dresselhaus, A. F. Kip and C. Kittel, *Phys. Rev.*, **98**, 368 (1955).

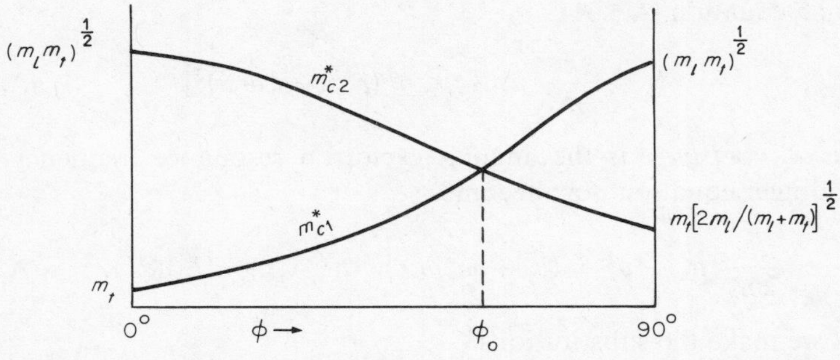

Fig. A.3.1. The cyclotron resonance effective masses m^*_{c1} and m^*_{c2} as functions of ϕ (schematic).

the positive sign being taken for the light mass holes and the negative sign for the heavy mass holes. From this expression the cyclotron resonance hole effective masses can be calculated in terms of the three constants A_1, A_2 and A_3. For silicon the cyclotron resonance experiment yields the values $A_1 = 4{\cdot}0$, $A_2 = 1{\cdot}1$ and $A_3 = 4{\cdot}1$.

From the above discussion it can be seen that the cyclotron resonance experiment not only provides a direct determination of effective masses but also yields information about the location in **k**-space of the minima of the conduction band and the maxima of the valence band, i.e. the electron and hole constant energy surfaces.

We end this section on cyclotron resonance with a brief discussion of the quantum mechanical behaviour of an electron in a crystalline solid in the presence of a uniform magnetic induction field $\mathbf{B} = B(0, 0, 1)$. A convenient vector potential for this field is given by $\mathbf{A} = B(0, x, 0)$, i.e.

$$\mathbf{B} = \text{curl}\,\mathbf{A} = B(0, 0, 1), \qquad \text{div}\,\mathbf{A} = 0 \tag{A.3.31}$$

Neglecting spin, the Hamiltonian operator for an electron moving in a periodic potential $V(\mathbf{r})$ and the magnetic induction field $\mathbf{B}$ is then given by

$$H = \frac{1}{2m}\left(\mathbf{p} + \frac{e}{c}\mathbf{A}\right)^2 + V(\mathbf{r}), \tag{A.3.32}$$

where m is the free-electron mass, e is the magnitude of the electronic charge and c is the speed of light in vacuo. In the effective mass approximation, which we now use, the effect of the periodic potential $V(\mathbf{r})$ is taken into account by replacing the free-electron mass m by an effective mass m^*_e which, for simplicity, we take to be a scalar. The Hamiltonian operator then becomes (cf. §B.13b)

$$H = \frac{1}{2m^*_e}\left(\mathbf{p} + \frac{e}{c}\mathbf{A}\right)^2 \tag{A.3.33}$$

or, using equation (A.3.31),

$$H = \frac{1}{2m_e^*}[p_x^2 + p_z^2 + (p_y + m_e^*\omega_c x)^2], \tag{A.3.34}$$

where $\omega_c = eB/m_e^*c$ is the angular cyclotron resonance frequency. The Schrödinger equation now becomes

$$\frac{1}{2m_e^*}[p_x^2 + p_z^2 + (p_y + m_e^*\omega_c x)^2]\psi(x, y, z) = E\psi(x, y, z), \tag{A.3.35}$$

and if we make the substitutions[3]

$$X = x + \hbar k_y/m_e^*\omega_c, \qquad \psi(x, y, z) = \phi(X)\,\mathrm{e}^{\mathrm{i}(k_y y + k_z z)} \tag{A.3.36}$$

we find that $\phi(X)$ satisfies the harmonic oscillator equation

$$\left(-\frac{\hbar^2}{2m_e^*}\frac{\mathrm{d}^2}{\mathrm{d}X^2} + \tfrac{1}{2}m_e^*\omega_c^2 X^2\right)\phi(X) = \left(E - \frac{\hbar^2 k_z^2}{2m_e^*}\right)\phi(X). \tag{A.3.37}$$

Hence the energy eigenvalues are given by

$$E_n(k_z) = \frac{\hbar^2 k_z^2}{2m_e^*} + (n + \tfrac{1}{2})\hbar\omega_c, \tag{A.3.38}$$

where n is a positive integer.

At a given k_z, therefore, there is an infinite number of equally spaced energy levels (the Landau levels). If a radio frequency electric field is now applied, absorption of energy takes place when the frequency of the electric field corresponds to the energy spacing $\hbar\omega_c$ between these levels, i.e. the electrons make transitions between the levels. This is the cyclotron resonance effect.

For these transitions to be observed, the uncertainty δE in the measurement of the energy must satisfy $\delta E \ll \hbar\omega_c$ and the time δt for the observation must be of the same order of magnitude as the relaxation time τ_e of the electrons. From the uncertainty principle

$$\delta E\,\delta t \sim \hbar, \tag{A.3.39}$$

and hence we again arrive at the condition

$$\omega_c\tau_e \gg 1. \tag{A.3.40}$$

Problems

(A.3.1) The equivalent minima in the conduction band of germanium lie along the principal diagonals in **k**-space (i.e. along the ⟨111⟩ axes), the ellipsoidal constant energy surfaces at these minima being such that $m_l = 20m_t$.

[3] J. Callaway, *Energy Band Theory*, Academic Press, New York, 1964.

In a cyclotron resonance experiment the magnetic field lies in the (110) crystal plane and makes an angle ϕ with the [001] crystal axis. Show that, in general, there will be three resonance peaks due to the electrons in germanium, and obtain expressions for the corresponding cyclotron resonance effective masses. When $\phi = 0$ the cyclotron resonance experiment yields an effective mass m_c given by $11m_c^2 = 2{\cdot}46m^2$, where m is the free-electron mass. Calculate the ratios m_l/m and m_t/m, and use the results to plot the cyclotron resonance effective masses versus ϕ curves for germanium.

(A.3.2) A cyclotron resonance experiment is performed on a hypothetical semiconductor whose conduction band has a single minimum at which the ellipsoidal constant energy surfaces are given by

$$E = \frac{\hbar^2}{2}(k_x^2/m_t + k_y^2/m_t + k_z^2/m_l).$$

The magnetic field makes an angle θ with the longitudinal axis of the above ellipsoid, i.e. $\mathbf{B} = B(\sin\theta, 0, \cos\theta)$, and the radio frequency field is elliptically polarized so that the electric field vector is given by

$$\mathscr{E} = (\mathscr{E}_1\,\mathrm{e}^{\mathrm{i}\phi}\cos\theta, \mathscr{E}_2, -\mathscr{E}_1\,\mathrm{e}^{\mathrm{i}\phi}\sin\theta)\,\mathrm{e}^{\mathrm{i}\omega t}.$$

Find a relationship between $\mathscr{E}_1$, $\mathscr{E}_2$ and ϕ to make the power absorption at resonance zero. Can you interpret your result?

(A.3.3) An electron with constant energy surfaces of the form

$$E = \frac{\hbar^2}{2}(k_x^2/m_t + k_y^2/m_t + k_z^2/m_l)$$

moves in the magnetic field $\mathbf{B} \equiv B(0, 0, 1)$. Calculate the cyclotron resonance effective mass:

(a) by using equation (A.3.20)

(b) by using the definition (B.13.21).

A.4 FARADAY ROTATION

If a plane-polarized electromagnetic wave is sent through a crystalline solid and a constant magnetic field is applied parallel to the direction of propagation the direction of polarization of the wave rotates through an angle θ whose magnitude depends on, amongst other things, the effective mass of the charge carriers in the solid. This rotation of the direction of polarization is known as the Faraday rotation and it provides a means for determining the effective mass of the charge carriers. In the cyclotron resonance experiment one looks at the effect of the radiation on the solid. In the Faraday rotation experiment one looks at the effect of the solid on the radiation.

Consider a plane electric wave $\mathscr{E} = \mathscr{E}_0\,\mathrm{e}^{\mathrm{i}(\omega t - kz)}$ propagating in the z-direction in a medium of dielectric constant ϵ, magnetic permeability κ and complex conductivity $\sigma_e = \sigma_{eR} + \mathrm{i}\sigma_{eI}$, the subscript e denoting that the charge carriers are electrons. From Maxwell's equations, $\mathscr{E}$ has to satisfy the equation

$$\nabla^2 \mathscr{E} = \frac{\epsilon\kappa}{c^2}\frac{\partial^2 \mathscr{E}}{\partial t^2} + \frac{4\pi\kappa\sigma_e}{c}\frac{\partial \mathscr{E}}{\partial t}, \tag{A.4.1}$$

and hence

$$\frac{c^2 k^2}{\omega^2} = \kappa\left(\epsilon - \frac{4\pi i c\sigma_e}{\omega}\right). \tag{A.4.2}$$

If we now define the refractive index n_1 and the extinction coefficient n_2 by the equation

$$\frac{ck}{\omega} = n_1 - in_2, \tag{A.4.3}$$

we find that

$$n_1^2 = \frac{\kappa}{2}(\epsilon + 4\pi c\sigma_{eI}/\omega) + \left[\frac{\kappa^2}{4}(\epsilon + 4\pi c\sigma_{eI}/\omega)^2 + 4\pi^2\kappa^2 c^2\sigma_{eR}^2/\omega^2\right]^{\frac{1}{2}}, \tag{A.4.4}$$

$$n_1 n_2 = 2\pi\kappa c\sigma_{eR}/\omega, \tag{A.4.5}$$

$$\mathscr{E} = \mathscr{E}_0\, e^{-\omega n_2 z/c}\, e^{i\omega(t - n_1 z/c)}, \tag{A.4.6}$$

where c/n_1 is the phase velocity of the wave.

Suppose that we have circularly-polarized radiation, so that

$$\mathscr{E}_0 = \mathscr{E}_0^{\pm} = \mathscr{E}_0(1, \pm i, 0),$$

and a constant magnetic induction field parallel to the direction of propagation of the wave, i.e. $\mathbf{B} = B(0, 0, 1)$. In this case, using equation (A.3.4) of the previous section,

$$\frac{\sigma_{eR}^{\pm}}{\sigma_e^0} = \frac{1 + (\omega_c \mp \omega)^2\tau_e^2}{\{[1 + (\omega_c^2 - \omega^2)\tau_e^2]^2 + 4\omega^2\tau_e^2\}}, \tag{A.4.7}$$

$$\frac{\sigma_{eI}^{\pm}}{\sigma_e^0} = \frac{(\omega \mp \omega_c)(\omega_c^2 - \omega^2)\tau_e^3 - (\omega \pm \omega_c)\tau_e}{\{[1 + (\omega_c^2 - \omega^2)\tau_e^2]^2 + 4\omega^2\tau_e^2\}}, \tag{A.4.8}$$

where $\omega_c = eB/m_e^* c$, $\sigma_e^0 = ne\mu_e = ne^2\tau_e/m_e^*$ and m_e^* is the scalar effective mass and τ_e the relaxation time of the electrons. It can now be seen, from equations (A.4.4), (A.4.7) and (A.4.8), that right and left circularly-polarized waves have different refractive indices, n_1^+ and n_1^- respectively, and hence different phase velocities.

A plane-polarized wave can be thought of as being a right circularly-polarized wave superimposed on a left circularly-polarized wave. The right circularly-polarized wave does not have the same phase velocity as the left circularly-polarized wave and hence the phase difference between these two waves increases as they move through the medium. It is this increase in the phase difference which causes a rotation of the direction of polarization of the plane-polarized wave.

To find an expression for the angle of rotation θ of the direction of polarization, let $\mathscr{E}^+(z, t)$ and $\mathscr{E}^-(z, t)$ denote the circularly-polarized waves which combine to form the plane-polarized wave $\mathscr{E}(z, t)$. Using equation (A.4.6) and assuming that the damping terms may be neglected, we find for the real parts of these waves

$$\operatorname{Re} \mathscr{E}^+(z, t) = \mathscr{E}_0(\cos \alpha_+, -\sin \alpha_+, 0), \tag{A.4.9}$$

$$\operatorname{Re} \mathscr{E}^-(z, t) = \mathscr{E}_0(\cos \alpha_-, \sin \alpha_-, 0), \tag{A.4.10}$$

$$\begin{aligned} \operatorname{Re} \mathscr{E}(z, t) &= \operatorname{Re}[\mathscr{E}^+(z, t) + \mathscr{E}^-(z, t)] \\ &= \mathscr{E}_0(\cos \alpha_+ + \cos \alpha_-, \sin \alpha_- - \sin \alpha_+, 0), \end{aligned} \tag{A.4.11}$$

where

$$\alpha_\pm = \omega(t - n_1^\pm z/c). \tag{A.4.12}$$

A unit vector in the direction of polarization is then given by

$$\hat{\mathscr{E}}(z, t) = \frac{1}{[2 + 2\cos(\alpha_+ + \alpha_-)]^{\frac{1}{2}}}(\cos \alpha_+ + \cos \alpha_-, \sin \alpha_- - \sin \alpha_+, 0). \tag{A.4.13}$$

Suppose that the medium (the specimen) extends from $z = 0$ to $z = d$. The angle θ through which the direction of polarization rotates in this distance is then given by

$$\cos \theta = \hat{\mathscr{E}}(0, t) \cdot \hat{\mathscr{E}}(d, t) = \cos \frac{\omega d}{2c}(n_1^+ - n_1^-), \tag{A.4.14}$$

i.e.

$$\theta = \frac{\omega d}{2c}(n_1^+ - n_1^-). \tag{A.4.15}$$

For the purpose of determining the effective mass m_e^*, the Faraday rotation experiment is performed at a high frequency ω which satisfies the conditions

$$\omega \tau_e \gg \omega_c \tau_e, 1. \tag{A.4.16}$$

Using equations (A.4.4), (A.4.7), (A.4.8) and (A.4.15), we find that at such a frequency $\sigma_{eI}^{\pm} \gg \sigma_{eR}^{\pm}$ and hence, neglecting the term involving σ_{eR},

$$n_1^{\pm 2} = \kappa(\epsilon + 4\pi c\sigma_{eI}^{\pm}/\omega), \tag{A.4.17}$$

$$n_1^{+2} - n_1^{-2} = \frac{4\pi c\kappa}{\omega}(\sigma_{eI}^{+} - \sigma_{eI}^{-}) = \frac{8\pi\kappa n e^3 B}{\omega^3 m_e^{*2}}, \tag{A.4.18}$$

$$\theta = \frac{2\pi\kappa d n e^3 B}{\bar{n}_1 \omega^2 c m_e^{*2}}, \tag{A.4.19}$$

where

$$\bar{n}_1 = \tfrac{1}{2}(n_1^{+} + n_1^{-}). \tag{A.4.20}$$

$\bar{n}_1$ can, in fact, be put equal to the refractive index in the absence of the magnetic field. Equation (A.4.19) is the required relationship between the angle of rotation θ and the effective mass m_e^*. This relationship is independent of $\omega_c\tau_e$, i.e. independent of the scattering mechanism, and hence the Faraday rotation can provide a determination of the effective mass when conditions are such that $\omega_c\tau_e \ll 1$ and the cyclotron resonance cannot be observed, e.g. at room temperature.

It should be mentioned that if there are two types of charge carrier present, with different scalar effective masses, the Faraday rotation experiment yields only an average effective mass [see problem (A.4.1) below].

Problem

(A.4.1) In a strongly extrinsic *p*-type semiconductor there are two types of holes. One type of hole has scalar effective mass m_1, relaxation time τ_1 and density p_1, and the other type of hole has scalar effective mass m_2, relaxation time τ_2 and density p_2. An electric field $\mathscr{E} = \mathscr{E}(1, \pm i, 0)\, e^{i\omega t}$ and a magnetic field $\mathbf{B} = B(0, 0, 1)$ are applied to a sample of this semiconductor. Obtain an expression for the complex conductivity of the sample.

Show, by making suitable approximations and by using the relation $p_1/p_2 = m_1^{\frac{3}{2}}/m_2^{\frac{3}{2}}$, that the angle of rotation θ in a Faraday rotation experiment on this semiconductor is given by

$$\theta = \frac{2d\pi\kappa e^3 Bp}{\omega^2 \bar{n}_1 c}\left[\frac{m_1^{-\frac{1}{2}} + m_2^{-\frac{1}{2}}}{m_1^{\frac{3}{2}} + m_2^{\frac{3}{2}}}\right],$$

where $p = p_1 + p_2$ and the other symbols are defined in the text [see equation (A.4.19)].

A.5 THE HALL EFFECT

When a magnetic field is applied to a semiconductor carrying a current in a direction at right angles to the current, a voltage is produced across the semiconductor in a direction perpendicular to the current and to the magnetic field. This is known as the Hall effect, and measurements of the Hall voltage yield the concentrations of the charge carriers in the semiconductor. If both Hall effect and conductivity measurements are made, the mobilities of the charge carriers in the semiconductor can be determined.

In the Hall effect the charge carriers are deflected by the magnetic field to one face of the semiconductor until a steady-state situation is reached in which the electric field due to these charges just balances the Lorentz force due to the magnetic field. Let $\mathbf{J}_e$ and $\mathbf{J}_h$ denote the current densities, due to electrons and holes respectively, in this steady-state situation. Assuming constant scalar effective masses, we then have

$$\frac{\partial \mathbf{J}_e}{\partial t} = 0, \qquad \mathbf{J}_e = ne\mu_e \mathscr{E} - \frac{\mu_e}{c}(\mathbf{J}_e \times \mathbf{B}), \tag{A.5.1}$$

$$\frac{\partial \mathbf{J}_h}{\partial t} = 0, \qquad \mathbf{J}_h = pe\mu_h \mathscr{E} + \frac{\mu_h}{c}(\mathbf{J}_h \times \mathbf{B}), \tag{A.5.2}$$

which can be solved to give

$$\mathbf{J}_e = \frac{1}{(1 + \mu_e^2 B^2/c^2)}\left[ne\mu_e\mathscr{E} - \frac{ne\mu_e^2}{c}(\mathscr{E} \times \mathbf{B}) + \frac{ne\mu_e^3}{c^2}(\mathscr{E} \cdot \mathbf{B})\mathbf{B}\right], \tag{A.5.3}$$

$$\mathbf{J}_h = \frac{1}{(1 + \mu_h^2 B^2/c^2)}\left[pe\mu_h\mathscr{E} + \frac{pe\mu_h^2}{c}(\mathscr{E} \times \mathbf{B}) + \frac{pe\mu_h^3}{c^2}(\mathscr{E} \cdot \mathbf{B})\mathbf{B}\right]. \tag{A.5.4}$$

If the magnetic induction field $\mathbf{B}$ is given by

$$\mathbf{B} = B(0, 0, 1) \tag{A.5.5}$$

and the electric field $\mathscr{E}$ is given by

$$\mathscr{E} = (\mathscr{E}_x, \mathscr{E}_y, 0), \tag{A.5.6}$$

the components of the total current density $\mathbf{J} = \mathbf{J}_e + \mathbf{J}_h$ become

$$J_x = \frac{1}{(1 + \mu_e^2 B^2/c^2)}\left[ne\mu_e\mathscr{E}_x - \frac{ne\mu_e^2}{c}\mathscr{E}_y B\right] + \frac{1}{(1 + \mu_h^2 B^2/c^2)}\left[pe\mu_h\mathscr{E}_x + \frac{pe\mu_h^2}{c}\mathscr{E}_y B\right], \tag{A.5.7}$$

$$J_y = \frac{1}{(1 + \mu_e^2 B^2/c^2)}\left[ne\mu_e\mathscr{E}_y + \frac{ne\mu_e^2}{c}\mathscr{E}_x B\right]$$
$$+ \frac{1}{(1 + \mu_h^2 B^2/c^2)}\left[pe\mu_h\mathscr{E}_y - \frac{pe\mu_h^2}{c}\mathscr{E}_x B\right], \qquad \text{(A.5.8)}$$

$$J_z = 0. \qquad \text{(A.5.9)}$$

Taking the current flow to be in the x-direction, the Hall field $\mathscr{E}_y$ is now obtained by putting $J_y = 0$. This gives

$$\mathscr{E}_y = \frac{[(p\mu_h^2 - n\mu_e^2)c^2 B + (p - n)\mu_e^2\mu_h^2 B^3]\mathscr{E}_x}{[(n\mu_e + p\mu_h)c^3 + (n\mu_h + p\mu_e)c\mu_e\mu_h B^2]}. \qquad \text{(A.5.10)}$$

Substituting this expression for $\mathscr{E}_y$ in equation (A.5.7), we now find that

$$J_x = \frac{e[(n\mu_e + p\mu_h)^2 c^2 + (n - p)^2\mu_e^2\mu_h^2 B^2]\mathscr{E}_x}{[(n\mu_e + p\mu_h)c^2 + (n\mu_h + p\mu_e)\mu_e\mu_h B^2]}. \qquad \text{(A.5.11)}$$

Hence the conductivity $\sigma_B = J_x/\mathscr{E}_x$ is given by

$$\sigma_B = \frac{e[(n\mu_e + p\mu_h)^2 c^2 + (n - p)^2\mu_e^2\mu_h^2 B^2]}{[(n\mu_e + p\mu_h)c^2 + (n\mu_h + p\mu_e)\mu_e\mu_h B^2]}. \qquad \text{(A.5.12)}$$

When $B = 0$ the conductivity is σ_0, where

$$\sigma_0 = ne\mu_e + pe\mu_h. \qquad \text{(A.5.13)}$$

If we now define the Hall coefficient R_B by the equation

$$R_B = \mathscr{E}_y/BJ_x, \qquad \text{(A.5.14)}$$

we find that

$$R_B = \frac{1}{ec}\frac{[(p - nb^2)c^2 + (p - n)b^2\mu_h^2 B^2]}{[(p + nb)^2 c^2 + (p - n)^2 b^2\mu_h^2 B^2]}, \qquad \text{(A.5.15)}$$

where

$$b = \mu_e/\mu_h. \qquad \text{(A.5.16)}$$

b is usually greater than unity. The small field ($B \to 0$) Hall coefficient R_0 is then given by

$$R_0 = \frac{1}{ec}\frac{(p - nb^2)}{(p + nb)^2}. \qquad \text{(A.5.17)}$$

If $nb \gg p$, the current is carried almost entirely by electrons and, using equations (A.5.13) and (A.5.17), we find that

$$\sigma_0 = ne\mu_e, \qquad \text{(A.5.18)}$$

$$R_0 = -1/nec, \tag{A.5.19}$$

$$R_0\sigma_0 = -\mu_e/c. \tag{A.5.20}$$

If $p \gg nb$, the current is carried almost entirely by holes and

$$\sigma_0 = pe\mu_h, \tag{A.5.21}$$

$$R_0 = 1/pec, \tag{A.5.22}$$

$$R_0\sigma_0 = \mu_h/c. \tag{A.5.23}$$

Equations (A.5.19), (A.5.20), (A.5.22) and (A.5.23) enable the density and the mobility of the dominant carriers in an *n*-type semiconductor or in a *p*-type semiconductor to be determined from Hall effect and conductivity measurements. It should be noted that the sign of the Hall coefficient tells us whether we have an *n*-type ($R_0 < 0$) or a *p*-type ($R_0 > 0$) semiconductor.

Values for n, p, μ_e and μ_h can be obtained in the above manner only at temperatures low enough for mixed conduction to be unimportant. At higher temperatures, when both electrons and holes contribute significantly to the current, the determination of these quantities is a much more difficult task. Approximate values may be obtained, however, by making simplifying assumptions about the mobility ratio b and about the difference $p - n$ between the hole density and the electron density.

In the case of a *p*-type semiconductor, for example, it is generally assumed that both b and $p - n$ are constant and temperature-independent. Let $p - n = N_A$. Then at low temperatures, when $p \gg n$,

$$p^L = N_A \tag{A.5.24}$$

$$\sigma_0^L = N_A e\mu_h^L \tag{A.5.25}$$

$$R_0^L = 1/N_A ec \tag{A.5.26}$$

$$R_0^L\sigma_0^L = \mu_h^L/c, \tag{A.5.27}$$

the superscript L signifying that these are low-temperature values. Hence N_A and μ_h^L may be determined from Hall effect and conductivity measurements at low temperatures. From equation (A.5.17), $R_0 = 0$ when $p = nb^2$. Suppose that this happens at a temperature T_0 when $p = p^0$ and $n = n^0$. Then

$$p^0 = n^0b^2, \tag{A.5.28}$$

$$p^0 - n^0 = N_A, \tag{A.5.29}$$

and hence

$$p^0 = N_A b^2/(b^2 - 1), \tag{A.5.30}$$

$$n^0 = N_A/(b^2 - 1). \tag{A.5.31}$$

The conductivity σ_0^0 at the temperature T_0 is then given by

$$\sigma_0^0 = e\mu_h^0(p^0 + bn^0) = bN_A e\mu_h^0/(b - 1). \tag{A.5.32}$$

For μ_h^0 one may use the extrapolated value at the temperature T_0 of the hole mobility as determined in the low temperature (extrinsic) region. Equation (A.5.32) then yields the value of b. When both N_A and b are known, equations (A.5.13), (A.5.16) and (A.5.17), together with the equation $p - n = N_A$, enable n, p, μ_e and μ_h to be determined at any given temperature.

For a p-type semiconductor it can be shown, using equation (A.5.17), that when both b and $p - n$ are constant the small field Hall coefficient versus temperature curve will exhibit a maximum at a temperature corresponding to a hole density of $N_A b/(b - 1)$, the value of R_0 at this maximum being given by

$$R_0 = R_0^M = (b - 1)^2/4N_A bec. \tag{A.5.33}$$

Using equation (A.5.26),

$$R_0^M/R_0^L = (b - 1)^2/4b \tag{A.5.34}$$

and hence b can be determined. This alternative method for finding the value of b has the advantage that it does not depend on conductivity measurements which may be very unreliable.

The variation of the small-field Hall coefficient with temperature can also be used to find a value for the forbidden energy gap of the semiconductor. At high temperatures (intrinsic region) $p \sim n$ and, using equation (A.5.17), the small field Hall coefficient is given by

$$R_0 = R_0^H = \frac{1}{p^H}\left(\frac{1 - b}{1 + b}\right), \tag{A.5.35}$$

the superscript H signifying that these are high-temperature values. If the semiconductor is assumed to be nondegenerate, it can be shown that[4]

$$p^H = AT^{\frac{3}{2}} \exp(-\Delta E/2kT), \tag{A.5.36}$$

where A is a constant which depends on the hole and electron effective

[4] R. A. Smith, *Semiconductors*, Cambridge University Press, 1961, p. 82.

masses, T is the absolute temperature, k is Boltzmann's constant and ΔE is the energy gap. Hence

$$R_0^H T^{\frac{3}{2}} = A^{-1}\left(\frac{1-b}{1+b}\right)\exp(\Delta E/2kT) \qquad \text{(A.5.37)}$$

and

$$\log_e(|R_0^H| T^{\frac{3}{2}}) = \Delta E/2kT + \text{a constant.} \qquad \text{(A.5.38)}$$

The (constant) high-temperature slope of the $\log_e(|R_0| T^{\frac{3}{2}})$ versus $1/T$ curve thus yields the forbidden energy gap ΔE. Note that if the energy gap varies linearly with temperature, so that $\Delta E = \Delta E_0 + \alpha T$, this method gives ΔE_0 and not ΔE.

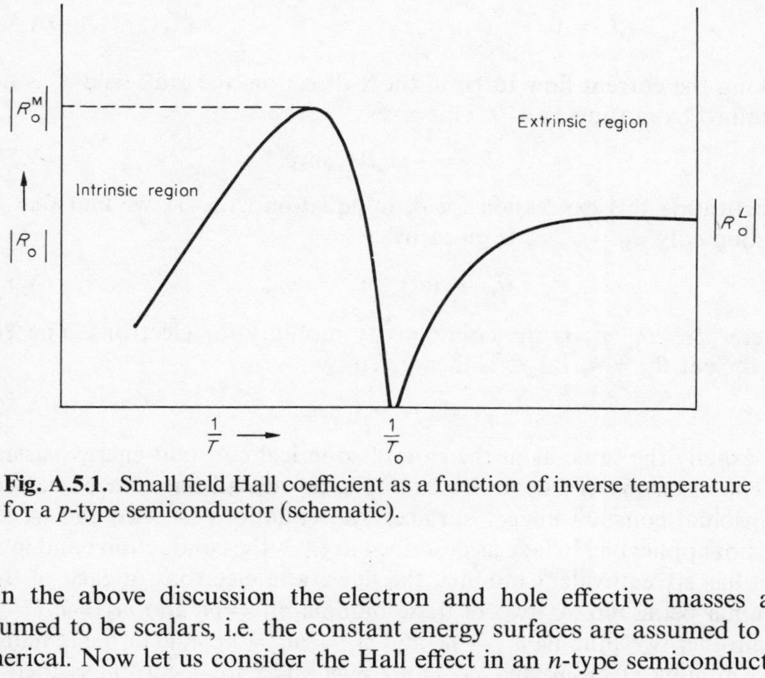

Fig. A.5.1. Small field Hall coefficient as a function of inverse temperature for a p-type semiconductor (schematic).

In the above discussion the electron and hole effective masses are assumed to be scalars, i.e. the constant energy surfaces are assumed to be spherical. Now let us consider the Hall effect in an n-type semiconductor whose electrons have the ellipsoidal constant energy surfaces

$$E = \frac{\hbar^2}{2}(k_x^2/m_1 + k_y^2/m_2 + k_z^2/m_3). \qquad \text{(A.5.39)}$$

Assuming that the temperature is so low that mixed conduction is unimportant, the equation for the steady-state current density $\mathbf{J}$ now becomes

$$\mathbf{J} = \begin{pmatrix} m_1^{-1} & 0 & 0 \\ 0 & m_2^{-1} & 0 \\ 0 & 0 & m_3^{-1} \end{pmatrix} \cdot \left\{ ne^2\tau_e(\mathscr{E}_x, \mathscr{E}_y, 0) - \frac{e\tau_e B}{c}[\mathbf{J} \times (0, 0, 1)] \right\}, \tag{A.5.40}$$

and this can be solved to give

$$J_x = \frac{nce^2\tau_e}{(m_1 m_2 c^2 + e^2\tau_e^2 B^2)}(m_2 c\mathscr{E}_x - e\tau_e B\mathscr{E}_y), \tag{A.5.41}$$

$$J_y = \frac{nce^2\tau_e}{(m_1 m_2 c^2 + e^2\tau_e^2 B^2)}(e\tau_e B\mathscr{E}_x + m_1 c\mathscr{E}_y), \tag{A.5.42}$$

$$J_z = 0. \tag{A.5.43}$$

Taking the current flow to be in the x-direction, the Hall field $\mathscr{E}_y$ is now obtained by putting $J_y = 0$. This gives

$$\mathscr{E}_y = -e\tau_e B\mathscr{E}_x/m_1 c. \tag{A.5.44}$$

Substituting this expression for $\mathscr{E}_y$ in equation (A.5.41), we find that the conductivity $\sigma_B = J_x/\mathscr{E}_x$ is given by

$$\sigma_B = ne^2\tau_e/m_1 = ne\mu_e^c, \tag{A.5.45}$$

where $\mu_e^c = e\tau_e/m_1$ is the conductivity mobility for electrons. The Hall coefficient $R_B = \mathscr{E}_y/B\sigma_B\mathscr{E}_x$ is then given by

$$R_B = -1/nec, \tag{A.5.46}$$

i.e. exactly the same as in the case of spherical constant energy surfaces.

The electrons in silicon, for example, do not all have the same set of ellipsoidal constant energy surfaces and equations (A.5.45) and (A.5.46) are not applicable. In fact, as described in §A.3, the conduction band in silicon has six equivalent minima, the density of electrons in each of these minima being $n/6$. At two of these minima $m_1 = m_l$ and $m_2 = m_3 = m_t$, at another two minima $m_2 = m_l$ and $m_1 = m_3 = m_t$, and at the remaining two minima $m_3 = m_l$ and $m_1 = m_2 = m_t$. The total current is then the sum of the contributions from the electrons in each of these minima and, using equations (A.5.41), (A.5.42) and (A.5.43), it follows that in this case

$$J_x = \frac{nce^2\tau_e}{3}\left\{\left[\frac{c(m_l + m_t)\mathscr{E}_x - 2e\tau_e B\mathscr{E}_y}{m_l m_t c^2 + e^2\tau_e^2 B^2}\right] + \left[\frac{cm_t\mathscr{E}_x - e\tau_e B\mathscr{E}_y}{m_t^2 c^2 + e^2\tau_e^2 B^2}\right]\right\}, \tag{A.5.47}$$

$$J_y = \frac{nce^2\tau_e}{3}\left\{\left[\frac{2e\tau_e B\mathscr{E}_x + c(m_l + m_t)\mathscr{E}_y}{m_l m_t c^2 + e^2\tau_e^2 B^2}\right] + \left[\frac{e\tau_e B\mathscr{E}_x + cm_t\mathscr{E}_y}{m_t^2 c^2 + e^2\tau_e^2 B^2}\right]\right\}, \tag{A.5.48}$$

$$J_z = 0. \tag{A.5.49}$$

Putting $J_y = 0$, we find that the Hall field $\mathscr{E}_y$ is now given by

$$\mathscr{E}_y = -\frac{e\tau_e B}{c}\left[\frac{c^2(2m_t^2 + m_l m_t) + 3e^2\tau_e^2 B^2}{c^2(2m_l m_t^2 + m_t^3) + (m_l + 2m_t)e^2\tau_e^2 B^2}\right]\mathscr{E}_x. \tag{A.5.50}$$

We could now substitute this expression for $\mathscr{E}_y$ in equation (A.5.47) and hence obtain the conductivity $\sigma_B = J_x/\mathscr{E}_x$ and the Hall coefficient $R_B = \mathscr{E}_y/B\sigma_B\mathscr{E}_x$. The resulting expressions for σ_B and R_B are, however, rather cumbersome and we do not give them here. Instead, we simply state the results obtained for the zero-field conductivity σ_0 and the small-field Hall coefficient R_0. These quantities are found to be given by

$$\sigma_0 = \frac{ne^2\tau_e}{3}(2/m_t + 1/m_l) \tag{A.5.51}$$

and

$$R_0 = -\frac{1}{nec}\frac{3K(K + 2)}{(2K + 1)^2}, \tag{A.5.52}$$

where

$$K = m_l/m_t. \tag{A.5.53}$$

If we define the conductivity mobility for electrons μ_e^c by the equation $\sigma_0 = ne\mu_e^c$, and the Hall mobility for electrons μ_e^H by the equation $-cR_0\sigma_0 = \mu_e^H$, we find that

$$\mu_e^H/\mu_e^c = 3K(K + 2)/(2K + 1)^2. \tag{A.5.54}$$

$\mu_e^H = \mu_e^c$ when the constant energy surfaces are spherical. It can be seen from equations (A.5.52) and (A.5.54) that n and μ_e^c can be determined from Hall effect and conductivity measurements only if K is known, e.g. from a cyclotron resonance experiment.

Problem

(A.5.1) The small-field Hall coefficient R_0 and the zero field conductivity σ_0 of a semiconductor sample vary with temperature as shown in the following table.

$10^3/T$ (°K^{-1})	eR_0 (cm^3)	σ_0/e (cm^{-1}V^{-1}s^{-1})
5·50	−0·50	52·00
6·00	−1·19	23·20
6·50	−3·24	
7·00	−6·80	3·30
7·20	−7·50	
7·50	−6·17	1·75
8·00	−2·24	1·40
8·25	−0·58	
8·40		1·38
8·45	0·00	
8·50	0·15	
8·75	0·56	
9·00	0·76	1·42
9·50	0·91	
10·00	0·98	1·55
10·50	1·00	
11·00	1·00	
11·50	1·00	
12·00	1·00	1·87
14·00		2·19
16·00		2·50
18·00		2·81
20·00		3·13

Using this table, plot $|eR_0|$ and σ_0/e against $10^3/T$. Deduce the value of the mobility ratio b (which may be assumed independent of temperature)

(i) by using both Hall and conductivity data
(ii) by using Hall data only.

Hence find the electron concentration, the hole concentration and the electron mobility at temperatures of 166·7°K and 125°K. Assuming a non-degenerate semiconductor, find a value for the energy gap at 0°K.

A.6 THE PHOTOMAGNETOELECTRIC EFFECT

When a semiconductor is illuminated on one surface with radiation of wavelength such that the photon energy exceeds the forbidden gap energy, electron–hole pairs are created near that surface owing to the high value of the absorption coefficient. These excess charge carriers then diffuse into the semiconductor and, if a magnetic field is applied in a direction parallel to the illuminated surface, the electrons and holes drift to opposite sides of the semiconductor and may then be observed either as a photo-voltage or as a short-circuit current. This effect is known as the photomagnetoelectric (PME) effect and was first observed in Cu_2O by I. K. Kikoin and M. M. Noskov[5]. From measurements of the PME short-circuit current one can obtain information about the recombination processes which take place in the bulk and on the surface of the semiconductor.

In this section we consider an idealized situation in which a strongly extrinsic p-type semiconductor is in the form of a semi-infinite block with $-\infty \leqslant x \leqslant \infty$, $-\infty \leqslant z \leqslant \infty$ and $0 \leqslant y \leqslant \infty$. The illuminated surface is given by $y = 0$, the magnetic induction field $\mathbf{B} = B(0, 0, 1)$ is in the z-direction and the short-circuit current is then measured in the x-direction. For this semi-infinite block the current densities and excess carrier concentrations vary only in the y-direction and, in the absence of an externally applied electric field, the only electric field is that due to the different mobilities and lifetimes of the electrons and holes[6]. The electric field $\mathscr{E} = \mathscr{E}(0, 1, 0)$ is thus in the y-direction. We assume scalar effective masses, constant relaxation times and that the excess carriers are created only at the surface, i.e. at $y = 0$. The analysis based on these assumptions is applicable when the sample dimensions are large enough for edge effects to be negligible, and when the carrier diffusion lengths are much smaller than the thickness of the sample and much larger than the absorption depth of the incident radiation.

Using equations (A.2.18) and (A.2.21), the steady-state current densities $\mathbf{J}_e$ and $\mathbf{J}_h$ now satisfy the equations

$$\mathbf{J}_e = ne\mu\mathscr{E}(0, 1, 0) - \frac{\mu B}{c}[\mathbf{J}_e \times (0, 0, 1)] + D(0, \frac{\mathrm{d}n}{\mathrm{d}y}, 0), \qquad \text{(A.6.1)}$$

$$\mathbf{J}_h = pe\mu_h\mathscr{E}(0, 1, 0) + \frac{\mu_h B}{c}[\mathbf{J}_h \times (0, 0, 1)] - D_h(0, \frac{\mathrm{d}p}{\mathrm{d}y}, 0), \qquad \text{(A.6.2)}$$

where we have now written μ for the electron mobility and D/e for the electron diffusion constant. Assuming the Einstein relations[7]

$$D/D_h = \mu/\mu_h = b \text{ (a constant)}, \qquad \text{(A.6.3)}$$

equation (A.6.2) may be written as

$$b\mathbf{J}_h = pe\mu\mathscr{E}(0, 1, 0) + \frac{\mu B}{c}[\mathbf{J}_h \times (0, 0, 1)] - D(0, \frac{\mathrm{d}p}{\mathrm{d}y}, 0). \qquad \text{(A.6.4)}$$

Since there is no continual accumulation of charge in the steady state

$$J_{ey} + J_{hy} = 0. \qquad \text{(A.6.5)}$$

[5] I. K. Kikoin and M. M. Noskov, *Phys. Z. Sowjet*, **5**, 586 (1934).
[6] H. Dember, *Z. Physik*, **32**, 554, 886 (1931); *Z. Physik*, **33**, 207 (1932).
[7] R. A. Smith, *Semiconductors*, Cambridge University Press, 1961, p. 235.

Using equations (A.6.1), (A.6.4) and (A.6.5), we now find that

$$J_{ey}\left(1 + \frac{\mu^2 B^2}{c^2}\right) = ne\mu\mathscr{E} + D\frac{\mathrm{d}n}{\mathrm{d}y}, \tag{A.6.6}$$

$$bJ_{ey}\left(1 + \frac{\mu^2 B^2}{b^2 c^2}\right) = -pe\mu\mathscr{E} + D\frac{\mathrm{d}p}{\mathrm{d}y}, \tag{A.6.7}$$

$$J_{ex} + J_{hx} = -\frac{\mu B}{c}\left(1 + \frac{1}{b}\right)J_{ey}, \tag{A.6.8}$$

and elimination of $\mathscr{E}$ yields

$$J_{ey} = \left[p\left(1 + \frac{\mu^2 B^2}{c^2}\right) + bn\left(1 + \frac{\mu^2 B^2}{b^2 c^2}\right)\right]^{-1} D\frac{\mathrm{d}}{\mathrm{d}y}(np). \tag{A.6.9}$$

The short-circuit current per unit sample width I_{sc} is given by

$$I_{sc} = \int_0^\infty (J_{ex} + J_{hx})\,\mathrm{d}y \tag{A.6.10}$$

$$= -\frac{\mu BD}{c}\left(1 + \frac{1}{b}\right)\int_0^\infty \left[p\left(1 + \frac{\mu^2 B^2}{c^2}\right) + bn\left(1 + \frac{\mu^2 B^2}{b^2 c^2}\right)\right]^{-1} \frac{\mathrm{d}}{\mathrm{d}y}(np)\,\mathrm{d}y, \tag{A.6.11}$$

where equations (A.6.8) and (A.6.9) have been used. If n_0 and p_0 are the thermal equilibrium densities and Δn and Δp are the excess carrier densities of the electrons and holes respectively,

$$p = p_0 + \Delta p, \tag{A.6.12}$$

$$n = n_0 + \Delta n, \tag{A.6.13}$$

where, for the *p*-type semiconductor under consideration,

$$p_0 \gg n_0, \Delta n, \Delta p. \tag{A.6.14}$$

The denominator of the integrand in equation (A.6.11) is then essentially constant, and the integral may be evaluated to give

$$I_{sc} = \frac{\mu BD}{c}\left(1 + \frac{1}{b}\right)\left(1 + \frac{\mu^2 B^2}{c^2}\right)^{-1}\left[1 + \frac{n_0(\Delta p)_0}{p_0(\Delta n)_0} + \frac{(\Delta p)_0}{p_0}\right](\Delta n)_0, \tag{A.6.15}$$

where $(\Delta n)_0$ and $(\Delta p)_0$ are the excess carrier densities at the surface $y = 0$. For a strongly extrinsic *p*-type semiconductor

$$\frac{n_0}{p_0}\frac{(\Delta p)_0}{(\Delta n)_0} + \frac{(\Delta p)_0}{p_0} \ll 1, \tag{A.6.16}$$

and if b^{-1} is small in comparison with unity, equation (A.6.15) may be simplified to

$$I_{sc} = \frac{\mu BD}{c}\left(1 + \frac{\mu^2 B^2}{c^2}\right)^{-1}(\Delta n)_0. \tag{A.6.17}$$

Equation (A.6.17) allows $(\Delta n)_0$ to be determined from a measurement of the short-circuit current I_{sc}.

If ϕ is the photon flux density and η is the quantum efficiency (the average number of electrons created per photon), the rate of generation of electrons per unit area of the illuminated surface is $\eta\phi$. The electrons may, however, recombine at the surface, and if we write the recombination rate per unit area of surface as $(\Delta n)_0 S_n$, where S_n is known as the surface recombination velocity, the resultant rate of generation of electrons per unit area of surface is $\eta\phi - (\Delta n)_0 S_n$. Hence the component J_{ey} of the electron current density vector has to satisfy the boundary conditions

$$J_{ey}(y = \infty) = 0, \tag{A.6.18}$$

$$J_{ey}(y = 0) = -e[\eta\phi - (\Delta n)_0 S_n]. \tag{A.6.19}$$

In general, S_n will be a function of $(\Delta n)_0$.

Consider next unit volume inside the semiconductor. Since we are assuming no bulk generation of excess carriers, the rate of increase of the number of electrons in this volume is equal to the rate at which electrons are supplied by the current $\mathbf{J}_e$ minus the rate at which electrons in this volume are lost through recombination. If we write the rate at which electrons are lost through recombination as $\Delta n/\tau_n$, where τ_n is the electron recombination lifetime, then

$$\frac{\partial n}{\partial t} = \frac{1}{e}\nabla \cdot \mathbf{J}_e - \frac{\Delta_n}{\tau_n}. \tag{A.6.20}$$

Similarly, if τ_p is the hole recombination lifetime,

$$\frac{\partial p}{\partial t} = -\frac{1}{e}\nabla \cdot \mathbf{J}_h - \frac{\Delta p}{\tau_p}. \tag{A.6.21}$$

In the steady state $\partial n/\partial t = \partial p/\partial t = 0$ and, using equation (A.6.5), the above continuity equations may be written as

$$\frac{\mathrm{d}J_{ey}}{\mathrm{d}y} = -\frac{\mathrm{d}J_{hy}}{\mathrm{d}y} = \frac{e\Delta n}{\tau_n} = \frac{e\Delta p}{\tau_p}. \tag{A.6.22}$$

In general, τ_n will be a function of Δn and τ_p a function of Δp.

Equations (A.6.18), (A.6.19) and (A.6.22) now enable us to express the short-circuit current I_{sc} in terms of τ_n and S_n. Let

$$r = \Delta p/\Delta n, \tag{A.6.23}$$

where, from equation (A.6.22), r may be regarded as a function of Δn. Equation (A.6.9) may then be written as

$$D\frac{\mathrm{d}\Delta n}{\mathrm{d}y} = J_{ey}\left[\frac{(p_0 + r\Delta n)(1 + \mu^2B^2/c^2) + b(n_0 + \Delta n)(1 + \mu^2B^2/b^2c^2)}{p_0 + r(n_0 + 2\Delta n) + r'\Delta n(n_0 + \Delta n)}\right], \tag{A.6.24}$$

where

$$r' = \frac{\mathrm{d}r}{\mathrm{d}\Delta n}. \tag{A.6.25}$$

Now

$$\frac{\mathrm{d}J_{ey}}{\mathrm{d}y} = \frac{e\Delta n}{\tau_n} = \frac{\mathrm{d}J_{ey}}{\mathrm{d}\Delta n}\frac{\mathrm{d}\Delta n}{\mathrm{d}y}, \tag{A.6.26}$$

and hence, using equation (A.6.24),

$$\begin{aligned} J_{ey}\frac{\mathrm{d}J_{ey}}{\mathrm{d}\Delta n} &= \frac{e\Delta n}{\tau_n}J_{ey}\frac{\mathrm{d}y}{\mathrm{d}\Delta n} \\ &= \frac{eD\Delta n}{\tau_n}\left[\frac{p_0 + r(n_0 + 2\Delta n) + r'\Delta n(n_0 + \Delta n)}{(p_0 + r\Delta n)\left(1 + \dfrac{\mu^2B^2}{c^2}\right) + b(n_0 + \Delta n)\left(1 + \dfrac{\mu^2B^2}{b^2c^2}\right)}\right]. \end{aligned} \tag{A.6.27}$$

Integration of equation (A.6.27) now yields

$$J_{ey} = -K(B)\mathrm{f}(\Delta n), \tag{A.6.28}$$

where, for $p_0 \gg n_0, \Delta n, \Delta p$,

$$K(B) = \left[2eD\left(1 + \frac{\mu^2B^2}{c^2}\right)^{-1}\right]^{\frac{1}{2}} \tag{A.6.29}$$

and

$$\mathrm{f}(\Delta n) = \left\{\int_0^{\Delta n}\left[1 + \frac{r}{p_0}(n_0 + 2\Delta n) + \frac{r'}{p_0}\Delta n(n_0 + \Delta n)\right]\frac{\Delta n}{\tau_n}\,\mathrm{d}\Delta n\right\}^{\frac{1}{2}}. \tag{A.6.30}$$

The boundary condition (A.6.19) thus becomes

$$K(B)\mathrm{f}((\Delta n)_0) = e\{\eta\phi - (\Delta n)_0 S_n((\Delta n)_0)\}. \tag{A.6.31}$$

If

$$p_0 \gg r_0[n_0 + 2(\Delta n)_0] + r_0'(\Delta n)_0[n_0 + (\Delta n)_0], \qquad \text{(A.6.32)}$$

where

$$r_0 = \frac{(\Delta p)_0}{(\Delta n)_0}, \qquad r_0' = \frac{\mathrm{d}r_0}{\mathrm{d}(\Delta n)_0}, \qquad \text{(A.6.33)}$$

equation (A.6.30) gives

$$\frac{\mathrm{d}}{\mathrm{d}(\Delta n)_0} \mathrm{f}^2((\Delta n)_0) = \frac{(\Delta n)_0}{\tau_n((\Delta n)_0)}, \qquad \text{(A.6.34)}$$

and the boundary condition (A.6.31) may be written as

$$\frac{K^2(B)(\Delta n)_0}{\tau_n((\Delta n)_0)} = e^2 \frac{\mathrm{d}}{\mathrm{d}(\Delta n)_0} \{\eta\phi - (\Delta n)_0 S_n((\Delta n)_0)\}^2. \qquad \text{(A.6.35)}$$

A simple method of analysis of the experimental results may be used when the surface recombination velocity S_n is negligible under all conditions. In this case, using equation (A.6.29), equation (A.6.35) becomes

$$\tau_n((\Delta n)_0) = \frac{D(\Delta n)_0}{e\eta^2(1 + \mu^2 B^2/c^2)\phi} \frac{\mathrm{d}(\Delta n)_0}{\mathrm{d}\phi}. \qquad \text{(A.6.36)}$$

From equations (A.6.17) and (A.6.36),

$$\frac{I_{sc}}{\phi} \times \frac{\mathrm{d}I_{sc}}{\mathrm{d}\phi} = \frac{\mu^2 B^2 e \eta^2 D}{c^2(1 + \mu^2 B^2/c^2)} \times \tau_n((\Delta n)_0), \qquad \text{(A.6.37)}$$

$$I_{sc} = \frac{\mu B D}{c(1 + \mu^2 B^2/c^2)} \times (\Delta n)_0. \qquad \text{(A.6.38)}$$

Thus, at constant magnetic field and varying photon flux, a plot of $(I_{sc}/\phi) \times (\mathrm{d}I_{sc}/\mathrm{d}\phi)$ against I_{sc} demonstrates, apart from constants of proportionality, the variation of $\tau_n((\Delta n)_0)$ with $(\Delta n)_0$.

If, in addition to S_n being negligible, τ_n is independent of excess carrier concentration and hence independent of photon flux and magnetic field, equation (A.6.37) can be integrated to give

$$I_{sc} = \frac{\mu\eta B e^{\frac{1}{2}} D^{\frac{1}{2}} \tau_n^{\frac{1}{2}} \phi}{c(1 + \mu^2 B^2/c^2)^{\frac{1}{2}}}. \qquad \text{(A.6.39)}$$

A plot of $(B\phi/I_{sc})^2$ against B^2 thus gives a straight line which is independent of photon flux, and the electron mobility μ may be determined by simply dividing the slope of this line by its intercept on the $(B\phi/I_{sc})^2$ axis. If

η and D are known, the constant electron recombination lifetime τ_n may then also be found.

When the surface recombination velocity S_n is not negligible, a different method of analysis of the experimental results is required. Suppose, for example, that μ, D and η are known and that experimental curves of I_{sc} against ϕ at several constant magnetic fields are available. A value for $(\Delta n)_0$ is first chosen and the corresponding short-circuit current calculated for each magnetic field according to equation (A.6.17). From the experimental I_{sc} against ϕ curves the corresponding photon fluxes are then found. Hence, for this particular $(\Delta n)_0$, ϕ is known as a function of magnetic field. But for constant $(\Delta n)_0$, ϕ is given as a function of magnetic field by equation (A.6.31). The unknown constants $f((\Delta n)_0)$ and $S_n((\Delta n)_0)$ in this equation may therefore be determined by an appropriate curve-fitting method. The above procedure is repeated for a number of different values of $(\Delta n)_0$, and this allows $f((\Delta n)_0)$ and $S_n((\Delta n)_0)$ to be plotted as functions of $(\Delta n)_0$. The variation of the electron recombination lifetime $\tau_n((\Delta n)_0)$ with $(\Delta n)_0$ is then found from the $f((\Delta n)_0)$ against $(\Delta n)_0$ curve with the help of equation (A.6.34), i.e. from

$$\tau_n((\Delta n)_0) = \frac{(\Delta n)_0}{2f((\Delta n)_0)} \frac{d(\Delta n)_0}{df((\Delta n)_0)} \tag{A.6.40}$$

From the photomagnetoelectric effect, therefore, one can obtain information about both the excess density and the recombination lifetime of the minority carriers in an extrinsic semiconductor. The excess majority-carrier density and the majority-carrier lifetime, however, cannot be determined by this method. A method for determining these latter quantities is described in the next section which deals with the photoconductive effect.

Problem

(A.6.1) A PME experiment is performed on a semiconductor whose properties are such that

(i) $\mu_e = \mu_h$,
(ii) $p_0/n_0 = \Delta p/\Delta n$ (= a constant),
(iii) $\Delta n = [\log_e(\tau_0/\tau_n)]^{\frac{1}{2}}$, where τ_0 is a constant,
(iv) the surface recombination velocity is zero.

Prove that, at constant magnetic field, a plot of I_{sc}^2 against $\log_e \phi$ will result in a straight line.

A.7 THE PHOTOCONDUCTIVE EFFECT

When a semiconductor is illuminated with electromagnetic radiation (light) of frequency such that the photon energy equals or exceeds the forbidden gap energy, electrons are excited across the forbidden gap and electron–hole pairs are created in the semiconductor. These excess electrons and holes contribute to the conductivity of the semiconductor, which is then increased under illumination, giving rise to what is known as the photoconductive effect. By combining the results of a photoconductive experiment with those of a preliminary photomagnetoelectric (§A.6) experiment, both experiments being performed with the same highly absorbed radiation, one can determine the excess density and recombination lifetime of the majority carriers in an extrinsic semiconductor.

The situation considered in this section is thus the same as that envisaged in §A.6, except that now there is no magnetic field and the electric field is given by $\mathscr{E} = (\mathscr{E}_x, \mathscr{E}_y, 0)$, where $\mathscr{E}_x$ is an external electric field which is applied in the x-direction to measure the increase in the conductivity. The steady-state current densities thus satisfy the equations

$$\mathbf{J}_e = ne\mu(\mathscr{E}_x, \mathscr{E}_y, 0) + D\left(0, \frac{\mathrm{d}n}{\mathrm{d}y}, 0\right), \tag{A.7.1}$$

$$b\mathbf{J}_h = pe\mu(\mathscr{E}_x, \mathscr{E}_y, 0) - D\left(0, \frac{\mathrm{d}p}{\mathrm{d}y}, 0\right), \tag{A.7.2}$$

$$J_{ey} + J_{hy} = 0, \tag{A.7.3}$$

$$J_{ey}(y = \infty) = 0, \tag{A.7.4}$$

$$J_{ey}(y = 0) = -e[\eta\phi - (\Delta n)_0 S_n], \tag{A.7.5}$$

$$\frac{\mathrm{d}J_{ey}}{\mathrm{d}y} = -\frac{\mathrm{d}J_{hy}}{\mathrm{d}y} = \frac{e\Delta n}{\tau_n} = \frac{e\Delta p}{\tau_p}, \tag{A.7.6}$$

the notation being the same as that used in §A.6.

From equations (A.7.1)–(A.7.3), we now find that

$$J_{ex} + J_{hx} = \frac{e\mu\mathscr{E}_x}{b}(p_0 + bn_0 + \Delta p + b\Delta n), \tag{A.7.7}$$

$$J_{ey} = D\left[\frac{p_0 + r(n_0 + 2\Delta n) + r'\Delta n(n_0 + \Delta n)}{p_0 + bn_0 + (r + b)\Delta n}\right]\frac{\mathrm{d}\Delta n}{\mathrm{d}y}, \tag{A.7.8}$$

where

$$r = \frac{\Delta p}{\Delta n} = \frac{\tau_p}{\tau_n}, \tag{A.7.9}$$

$$r' = \frac{\mathrm{d}r}{\mathrm{d}\Delta n}. \tag{A.7.10}$$

The photoconductive current per unit sample width I_{pc} (i.e. the increase in current over the dark current) is thus given by

$$I_{pc} = \frac{e\mu\mathscr{E}_x}{b}\int_0^\infty (\Delta p + b\Delta n)\,\mathrm{d}y \tag{A.7.11}$$

$$= \frac{-e\mu\mathscr{E}_x}{b}\int_0^{(\Delta n)_0} (r + b)\Delta n \frac{\mathrm{d}y}{\mathrm{d}\Delta n}\,\mathrm{d}\Delta n \tag{A.7.12}$$

$$= -e\mu\mathscr{E}_x D\int_0^{(\Delta n)_0}\left(1 + \frac{r}{b}\right)\left[1 + \frac{r}{p_0}(n_0 + 2\Delta n) + \frac{r'}{p_0}\Delta n(n_0 + \Delta n)\right] \times \frac{\Delta n}{J_{ey}}\,\mathrm{d}\Delta n, \tag{A.7.13}$$

where we have now used the fact that, for the p-type semiconductor under consideration, $p_0 \gg n_0, \Delta n, \Delta p$. Assuming that

$$p_0 \gg r_0(n_0 + 2(\Delta n)_0) + r'_0(\Delta n)_0(n_0 + (\Delta n)_0), \tag{A.7.14}$$

where

$$r_0 = \frac{(\Delta p)_0}{(\Delta n)_0}, \qquad r'_0 = \frac{\mathrm{d}r_0}{\mathrm{d}(\Delta n)_0}, \tag{A.7.15}$$

equation (A.7.13) now yields

$$J_{ey}(y = 0)\frac{\mathrm{d}I_{pc}}{\mathrm{d}(\Delta n)_0} = -e\mu\mathscr{E}_x D(1 + r_0/b)(\Delta n)_0. \tag{A.7.16}$$

Now, from equations (A.7.6) and (A.7.8),

$$J_{ey}\frac{\mathrm{d}J_{ey}}{\mathrm{d}\Delta n} = \frac{eD\Delta n}{\tau_n}\left[\frac{p_0 + r(n_0 + 2\Delta n) + r'\Delta n(n_0 + \Delta n)}{p_0 + bn_0 + (r + b)\Delta n}\right], \tag{A.7.17}$$

which can be integrated to give

$$J_{ey} = -[2eD]^{\frac{1}{2}}\mathrm{f}(\Delta n), \tag{A.7.18}$$

where

$$\mathrm{f}(\Delta n) = \left\{\int_0^{\Delta n}\left[1 + \frac{r}{p_0}(n_0 + 2\Delta n) + \frac{r'}{p_0}\Delta n(n_0 + \Delta n)\right]\frac{\Delta n}{\tau_n}\,\mathrm{d}\Delta n\right\}^{\frac{1}{2}}. \tag{A.7.19}$$

f(Δn) is the same function as that defined in equation (A.6.30). Equation (A.7.16) can thus be written as

$$\frac{\mathrm{d}I_{pc}}{\mathrm{d}(\Delta n)_0} = \mu\mathscr{E}_x(eD/2)^{\frac{1}{2}}(1 + r_0/b)\,\frac{(\Delta n)_0}{\mathrm{f}((\Delta n)_0)}, \tag{A.7.20}$$

and it is this equation which is used to determine r_0, and hence Δp and τ_p as functions of $(\Delta n)_0$, from measurements of the photoconductive current I_{pc}.

From the photomagnetoelectric experiment, which is performed with the same values of ϕ as are used in the photoconductive experiment, $f((\Delta n)_0)$ and $\tau_n((\Delta n)_0)$ are known (tabulated) functions of $(\Delta n)_0$. Also, from equation (A.6.17),

$$\left(\frac{I_{sc}}{B}\right)_{B=0} = \frac{\mu D}{c}(\Delta n)_0. \qquad \text{(A.7.21)}$$

Hence if, for each value of ϕ, one plots (I_{sc}/B) against B and extrapolates to $B = 0$, the ϕ–$(\Delta n)_0$ relationship for the photoconductive experiment can be found. In the photoconductive experiment, therefore, $(\Delta n)_0$, $f((\Delta n)_0)$ and $\tau_n((\Delta n)_0)$ are known for each value of ϕ. A plot of I_{pc} against ϕ can thus be converted into a plot of I_{pc} against $(\Delta n)_0$ and the slope $(dI_{pc}/d(\Delta n)_0)$ of the resulting curve measured. τ_p and $(\Delta p)_0$ can then be determined as functions of $(\Delta n)_0$ from equation (A.7.20) by using the fact that

$$r_0 = \frac{(\Delta p)_0}{(\Delta n)_0} = \frac{\tau_p((\Delta n)_0)}{\tau_n((\Delta n)_0)}. \qquad \text{(A.7.22)}$$

The analysis is, of course, much simpler if τ_n and τ_p are independent of excess carrier concentration and S_n is negligible. From equations (A.7.4)–(A.7.6) and (A.7.11)

$$\begin{aligned} I_{pc} &= \frac{e\mu\mathscr{E}_x}{b}\int_0^\infty (\tau_p/e + b\tau_n/e)\frac{dJ_{ey}}{dy}\,dy \\ &= -\frac{\mu\mathscr{E}_x}{b}(\tau_p + b\tau_n)J_{ey}(y = 0) \\ &= \frac{e\eta\mu\mathscr{E}_x}{b}(\tau_p + b\tau_n)\phi. \end{aligned} \qquad \text{(A.7.23)}$$

The quantity $(\tau_n + b^{-1}\tau_p)$ can thus be found from the slope of the straight line obtained by plotting I_{pc} against ϕ. τ_h is known from the photomagnetoelectric experiment and hence τ_p can be determined.

Further reading

Boltzmann Transport Equation and General Reading

Beer, A. C., 'Galvanomagnetic Effects in Semiconductors', *Solid State Phys., Suppl. 4*, (1963).

Kittel, C., '*Elementary Solid State Physics*', Wiley, New York, 1962.

Mckelvey, J. P., '*Solid State and Semiconductor Physics*', Harper and Row, New York, and Weatherhill, Tokyo, 1966.
Shyh Wang, '*Solid State Electronics*', International Series in Pure and Applied Physics, McGraw-Hill, New York, 1966.
Wilson, A. H., '*The Theory of Metals*', Cambridge University Press, 1953.

Cyclotron Resonance

Dexter, R. N., Zeiger, H. J. and Lax, B., *Phys. Rev.*, **104**, 637 (1956).
Lax, B. and Mavroids, J. G., '*Cyclotron Resonance*', in *Solid State Phys.*, **11**, 261 (1960).
Shockley, W., *Phys. Rev.*, **90**, 491 (1953).

Faraday Rotation

Mitchell, E. W. J., *Proc. Phys. Soc.* (*London*), **B68**, 973 (1955).
Rau, R. R. and Caspari, M. E., *Phys. Rev.*, **100**, 632 (1955).
Smith, S. D., Moss, T. S. and Taylor, K. W., *J. Phys. Chem. Solids*, **11**, 131 (1959).
Stephen, M. J. and Lidiard, A. B., *J. Phys. Chem. Solids*, **9**, 43 (1958).

Hall Effect

Brooks, H., *Advan. Electron. Electron Phys.*, **7**, 85 (1955).
Debye, P. P. and Conwell, E. M., *Phys. Rev.*, **93**, 693 (1954).
Howarth, D. J., Jones, R. H. and Putley, E. H., *Proc. Phys. Soc.* (*London*), **B70**, 124 (1957).
Willardson, R. K., Harman, T. C. and Beer, A. C., *Phys. Rev.*, **96**, 1512 (1954).

Photomagnetoelectric and Photoconductive Effects

Beattie, A. R. and Cunningham, R. W., *Phys. Rev.*, **125**, 533 (1962).
Beattie, A. R. and Cunningham, R. W., *J. Appl. Phys.*, **35**, 353 (1964).
Kurnick, S. W. and Zitter, R. N., *J. Appl. Phys.*, **27**, 278 (1956).
Moss, T. S., Pincherle, L. and Woodward, A. M., *Proc. Phys. Soc.*, **B66**, 743 (1953).
van Roosbroeck, W., *Phys. Rev.*, **119**, 636 (1960).

PART B

AN ELECTRON IN A PERFECT CRYSTAL

P. T. Landsberg

CHAPTER II
Waves in Lattices

B.1 THE OCCURRENCE OF PERIODIC POTENTIALS

In the theory of gases drastic simplifications have to be made to achieve a simple theory. For example, the potential energy $V(r)$ of two molecules a distance r apart is known to have an infinite range (fig. B.1.1). It is often convenient to treat such a potential as a rigid core potential (dotted line) or to ignore it altogether. When the simple theory has yielded some basic understanding, and its shortcomings are also clear, one drops the idealizations one by one so as to develop better theories.

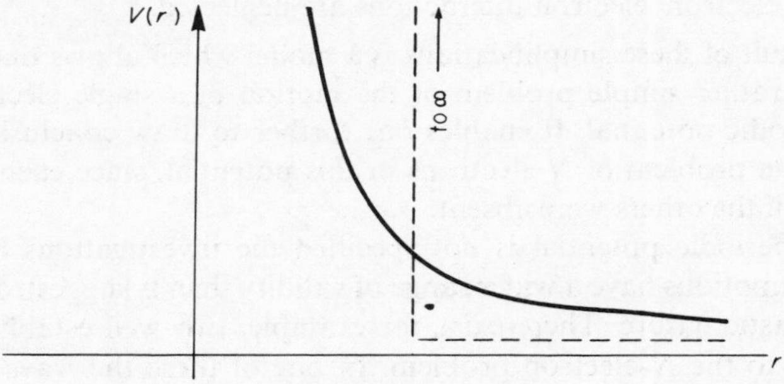

Fig. B.1.1.

An analogous procedure will serve for solids. In typical solids atoms are a few Angstroms (2–5 Å, say) apart. Each atom is therefore subject to forces exerted by its neighbours. In the case of the monovalent atoms (Na, K, etc.), this may be thought of as having the effect of delocalising the outer (valence) electron. The positive ions thus left behind give rise to a periodic potential. A number of physical effects will cause departures from such a potential, which must be ruled out by appropriate idealizations, to be brought back later when the simple theory has been studied. Thus one has to make assumptions about the ions:

(1) the ions are rigid;
(2) the ions are fixed in position.

By (1), the ions do not change their polarization if an electron approaches close to them. By (2), the thermal vibrations are neglected. To ensure that a rigid and undisturbed periodic potential exists in the crystal, it is assumed further that:

(3) no chemical impurities or physical lattice defects are present;
(4) the effect of the surface is neglected.

If N be the number of atoms present in volume V, and they are monovalent, then there are N electrons within this volume which are relatively free. For $N \geqslant 3$ the N-body problem presents great difficulties even in classical mechanics. These difficulties remain in quantum mechanics. Suppose now that there are N electrons, each electron being assumed to move in a medium in which the charge density is supposed uniform, and has the average value which would result from the ions and electrons actually present (this is zero in the simplest case). But apart from this restriction, one may suppose that each electron moves as if the others are all absent. Then the N-body problem is reduced to a one-body problem, and this will be assumed. Equivalently the assumption may be formulated:

(5) the electron–electron interactions are neglected.

The result of these simplifications is a model which allows one to consider the rather simple problem of the motion of a single electron in a rigid periodic potential. It enables one further to draw conclusions concerning the problem of N electrons in this potential, since each of them moves as if the others were absent.

If the periodic potential is not specified the investigations based on these assumptions have a wider range of validity than is suggested by their rather drastic nature. There exist, for example, two well established approaches to the N-electron problem. In one of these the wave function of the system is treated as a product of N one-electron functions, and the one-electron problem is solved by supposing the electron to move in the average potential due to all the other electrons. This is the Hartree scheme and clearly takes some account of the electron–electron interactions. It still leads to a one-electron Hamiltonian with a periodic potential. An improved method which also leads to this result is the Hartree–Fock scheme, in which antisymmetrized products of one-electron functions are used. Our considerations apply if the basis of discussion is improved by replacing assumption (5) in this way.

The difference between quantities calculated exactly and those calculated in the Hartree–Fock scheme are said to be due to *correlations.* The difference between quantities calculated in the Hartree–Fock scheme and those calculated in the Hartree scheme are said to be due to *exchange.* In this book we shall not be concerned with either. Our object is instead to give an introduction to the general principles of treating lattices (Chapter II), to the dynamics of electrons in lattices and to the formalism of band theory (Chapter III). The symmetry and group-theoretical aspects of this work are treated in Chapters IV and V.

B.2 DIRECT AND RECIPROCAL LATTICES

A lattice is a geometrical concept which may be discussed without introducing ideas of physics. Because a lattice is a regularly repeating pattern of sites it can be generated from basic blocks of sites. Such a block is called a *unit cell* and may in general be taken to be in the form of a parallelepiped (fig. B.2.1). This may be specified by the three angles ϵ_j and the lengths of the three vectors $\mathbf{a}_j$ along its edges. If there is one site at each of the eight corners and no others, the unit cell is called *simple* or *primitive*. Each lattice site belongs then to eight unit cells and the number of sites per unit cell is $8 \times \frac{1}{8} = 1$. A primitive unit cell is always the smallest unit cell capable of generating the lattice by translation.

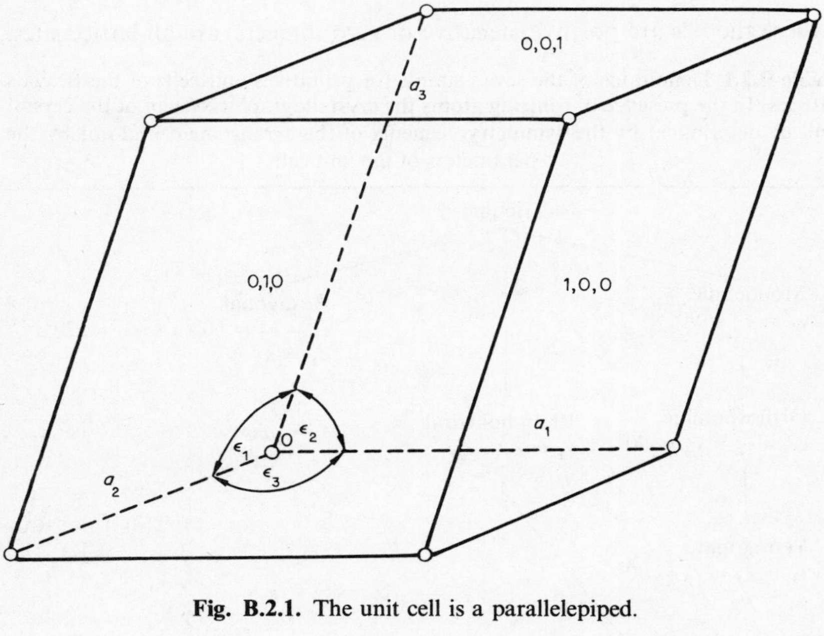

Fig. B.2.1. The unit cell is a parallelepiped.

A unit cell has three pairs of opposite faces and centring sites may be introduced in each pair, or there may be a site at the centre of the unit cell. This yields more complicated unit cells. A body-centred cell contains clearly two sites, and a face-centred cell contains $1 + \frac{6}{2} = 4$ sites. Much more complicated unit cells also occur. In order to make the number of sites per unit cell obvious, one may suppose the parallelepiped defining the unit cell to be shifted slightly so that it definitely includes the site at the position which one wants to regard as the origin. The two sites in a body-centred cell then have positions

$$(0, 0, 0) \quad \text{and} \quad (\tfrac{1}{2}, \tfrac{1}{2}, \tfrac{1}{2}) \tag{B.2.1}$$

or

$$\mathbf{r} = 0 \quad \text{and} \quad \mathbf{r} = \tfrac{1}{2} \sum_{j=1}^{3} \mathbf{a}_j.$$

Similarly the four sites for a face-centred cell can be taken to be

$$(0, 0, 0), \quad (\tfrac{1}{2}, \tfrac{1}{2}, 0), \quad (\tfrac{1}{2}, 0, \tfrac{1}{2}) \quad \text{and} \quad (0, \tfrac{1}{2}, \tfrac{1}{2}). \tag{B.2.2}$$

Some of the lattices which occur can be generated by specifying three primitive translations $\mathbf{a}_1, \mathbf{a}_2, \mathbf{a}_3$ such that the points

$$\mathbf{R_n} = \sum_{j=1}^{3} n_j \mathbf{a}_j. \tag{B.2.3}$$

(where the n_j's are positive, negative or zero integers) are all lattice sites.

Table B.2.1 Definitions of the seven simple (or primitive) unit cells of the Bravais lattices. In the presence of centring atoms the crystallographic system of the crystal will be determined by the symmetry elements of the arrangement and not by the parameters of the unit cell.

Triclinic

Monoclinic
$\epsilon_1 = \epsilon_3 = 90°$

Hexagonal
$\epsilon_1 = \epsilon_2 = 60°, \qquad \epsilon_3 = 120°,$
$a_1 = a_2.$

Orthorhombic
$\epsilon_1 = \epsilon_2 = \epsilon_3 = 90°$

Rhombohedral
$\epsilon_1 = \epsilon_2 = \epsilon_3$
$a_1 = a_2 = a_3$

Tetragonal
$\epsilon_1 = \epsilon_2 = \epsilon_3 = 90°$
$a_1 = a_2$

Cubic
$\epsilon_1 = \epsilon_2 = \epsilon_3 = 90°$

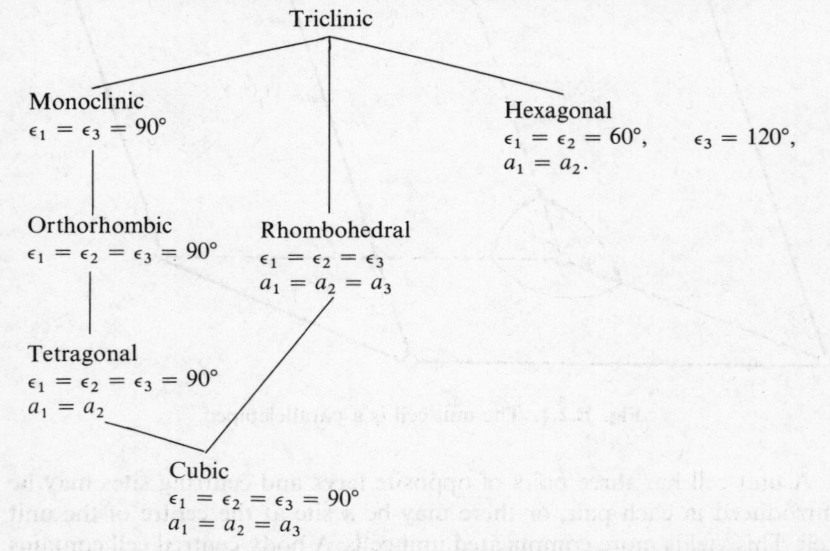

They are called the *Bravais lattices.* They may be classified by the symmetry operations under which they are invariant, and there are fourteen of them (see table B.2.2 and fig. B.2.2). The *simple* unit cells which can occur in this connexion are specified in table B.2.1. Their names are also associated with the seven crystallographic systems of the fourteen space lattices. Some caution is needed here since the addition of centring points may mean that the crystallographic system of the lattice is changed from one type to another type.

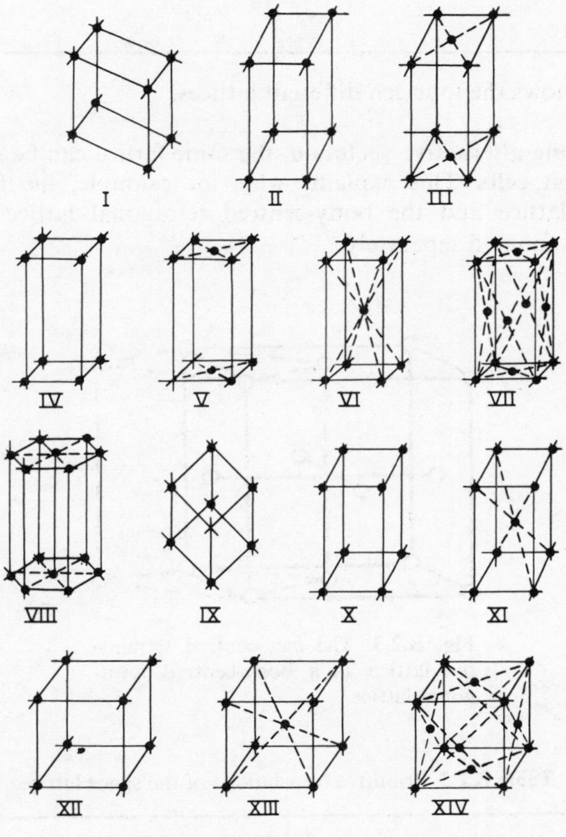

Fig. B.2.2. The 14 space lattices. I, triclinic; II, simple monoclinic; III, base-centred monoclinic; IV, simple orthorhombic; V, base-centred orthorhombic; VI, body-centred orthorhombic; VII, face-centred orthorhombic; VIII, hexagonal; IX, rhombohedral; X, simple tetragonal; XI, body-centred tetragonal; XII, simple cubic; XIII, body-centred cubic; XIV, face-centred cubic.

Table B.2.2 Enumeration of the fourteen space lattices (cf. Fig. B.1.3).

Symmetry	Triclinic	Monoclinic	Orthorhombic	Tetragonal	Cubic	Rhombohedral	Hexagonal
Simple	√	√	√	√	√	√	√
Side-centred or Base-centred		b.c.	s.c. and b.c.				
Face-centred			√		√		
Body-centred			√	√	√		

The table shows the fourteen different lattices.

By choosing alternative vectors $\mathbf{a}_j$ the same lattice can be specified by different unit cells. This explains why, for example, the face-centred tetragonal lattice and the body-centred tetragonal lattice (fig. B.2.3) need not be counted separately.

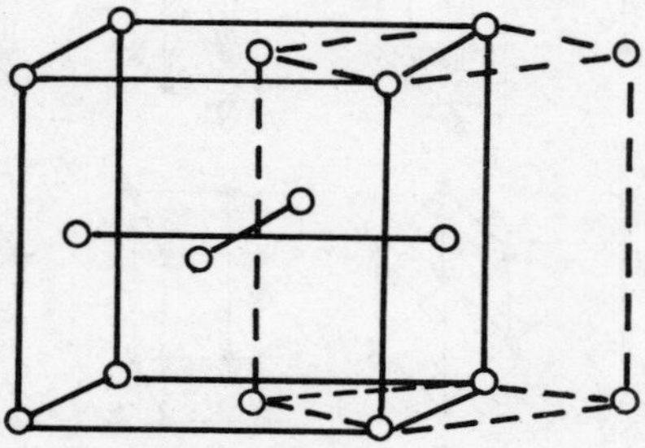

Fig. B.2.3. The face-centred tetragonal lattice as a body-centred tetragonal lattice.

Table B.2.3 Primitive translations of the space lattices.

	Primitive translations		
System	First	Second	Third
Simple	$2\tau_1$	$2\tau_2$	$2\tau_3$
Side-centred	$2\tau_1$	$\tau_2 + \tau_3$	$2\tau_3$
Base-centred	$2\tau_1$	$\tau_1 + \tau_2$	$2\tau_3$
Face-centred	$\tau_2 + \tau_3$	$\tau_3 + \tau_1$	$\tau_1 + \tau_2$
Body-centred	$2\tau_1$	$\tau_1 + \tau_2 + \tau_3$	$2\tau_3$

It is not obvious that equation (B.2.3) can, in fact, generate any but a simple lattice. One has, however, freedom in choosing the axes. Thus the face-centred and the body-centred cubic lattice can be shown to be equivalent to a *simple* rhombohedral lattice, which can certainly be described by equation (B.2.3) (see fig. B.2.4). However, this rhombohedral lattice has certain definite angles which impart to it more symmetries than are possessed by the ordinary simple rhombohedral lattice. Thus they have to be counted separately when the fourteen space lattices are enumerated.

To explain this situation more fully, it is convenient to distinguish the primitive translations $\mathbf{a}_j$ from the three vectors, $2\boldsymbol{\tau}_j$ say, which will represent the translations along the edge of the unit cell parallel to the three crystallographic axes. The primitive translations, which remain the vectors which join a site to its neighbouring sites, are then expressed quite simply in terms of the $\boldsymbol{\tau}_j$'s (see table B.2.3). If these expressions are substituted in equation (B.2.3) the appropriate lattice is generated.

If an additional lattice point is introduced non-centrally into a face of a unit cell, or if more complicated additions are made, the resulting lattice ceases in general to be a Bravais lattice, and equation (B.2.3) fails. One then has a *lattice with a basis.* While in a Bravais lattice, defined by (B.2.3), the neighbourhood of any site is the same as the neighbourhood of any other site, this may not be true in a lattice with a basis.

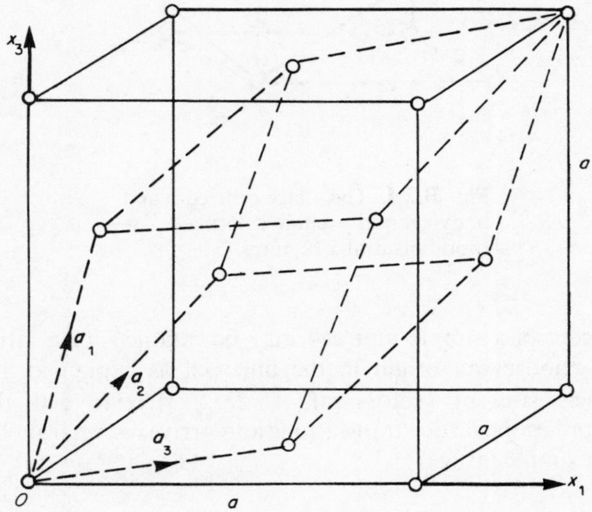

(a)

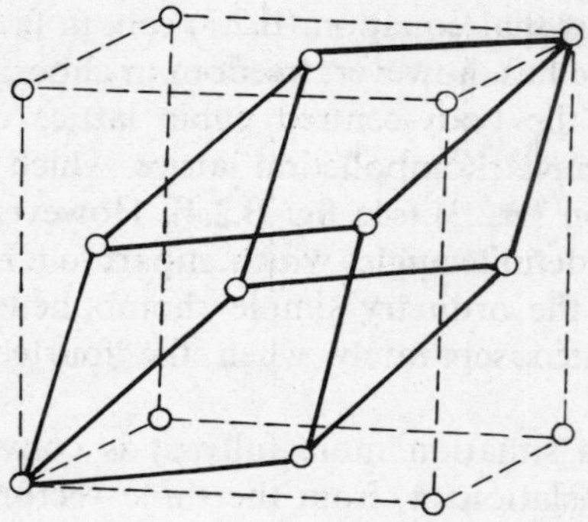

(b)

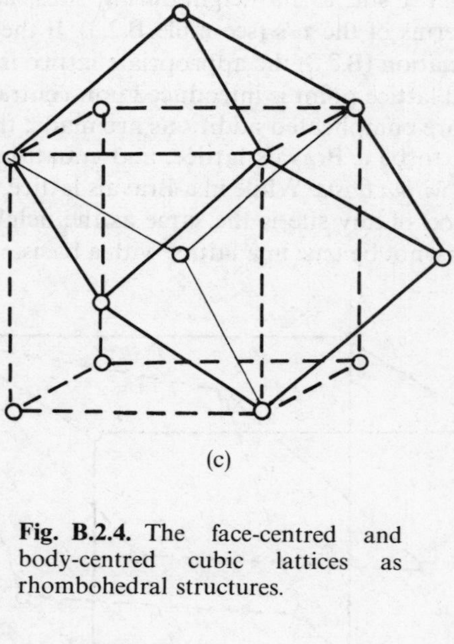

(c)

Fig. B.2.4. The face-centred and body-centred cubic lattices as rhombohedral structures.

The concept of a simple unit cell may be retained for a lattice with a basis. One chooses an origin in the unit cell as explained above, and specifies the s sites by vectors $\mathbf{r}_i (i = 1, 2, \ldots, s)$ relative to this origin. One now applies the lattice translations and arrives at s parallel and interpenetrating simple lattices:

$$\mathbf{R}_{i,\mathbf{n}} = \mathbf{r}_i + \sum_{j=1}^{3} n_j \mathbf{a}_j \qquad (i = 1, 2, \ldots, s). \tag{B.2.4}$$

The simple unit cell is still specified by the primitive translations $\mathbf{a}_j$, but with the origin of each cell one now has to associate s additional sites. The need to consider interpenetrating lattices arises, for example, from diatomic crystals. The atoms of type A may occupy the sites of one simple lattice and those of type B may occupy the sites of another simple lattice[1]. When associating atoms with sites, one makes the transitions from lattices to crystals, i.e. from geometry to physics.

Even if all the atoms in a lattice are the same, the description (B.2.4) may be convenient. From (B.2.1) and (B.2.2) this is so in the case of a body-centred cell ($s = 2$) and in the case of a face-centred cell ($s = 4$). In some such cases a simple unit cell may, however, be recovered by a change of axes (fig. B.2.4). Only when this is not possible is the lattice not a Bravais lattice.

We now recall that in a scalar product $\mathbf{u} \cdot \mathbf{v}$ it is understood that the first vector is regarded as a transpose of the type of vector occurring in the second factor as in

$$(u_1 \ \ u_2 \ \ . \ . \ .) \begin{pmatrix} v_1 \\ v_2 \\ \vdots \end{pmatrix} = u_1 v_1 + u_2 v_2 + \cdots$$

This procedure will not be emphasized explicitly in the sequel.

A *general* point in space, and therefore in a lattice, may be specified by a vector

$$\mathbf{r} = \sum_{j=1}^{3} r_j \mathbf{a}_j \qquad (r_j \text{ are numbers, not necessarily integers}). \quad \text{(B.2.5)}$$

Because the $\mathbf{a}_j$'s are not an orthonormal system of vectors, the important relation for scalar products of a vector $\mathbf{k} \equiv \Sigma k_j \mathbf{a}_j$ and $\mathbf{r}$

$$\mathbf{k} \cdot \mathbf{r} = \sum k_j r_j \qquad \text{(B.2.6)}$$

does not hold. This is inconvenient since for a plane wave $\exp(i\mathbf{k} \cdot \mathbf{r})$ of wave vector $\mathbf{k}$ scalar products with $\mathbf{r}$ are needed.

The relation (B.2.6) may be brought back for $\mathbf{k} \cdot \mathbf{r}$ products by writing

$$\mathbf{k} = \sum_{i=1}^{3} k_i \mathbf{b}_i \quad (k_i \text{ are numbers}) \text{ instead of } \mathbf{k} = \sum_j k'_j \hat{\mathbf{a}}_j, \qquad \text{(B.2.7)}$$

[1] For specific examples see § C.7.

where the $\hat{\mathbf{a}}_j$'s are unit vectors in the direction of $\mathbf{a}_j$, the k_j''s are reciprocal lengths, and where

$$\mathbf{b}_i \cdot \mathbf{a}_j = \epsilon\delta_{ij} \quad (\epsilon \text{ a number}). \tag{B.2.8}$$

It then follows that

$$\mathbf{k} \cdot \mathbf{r} = \epsilon \sum_j k_j r_j \tag{B.2.9}$$

in analogy with (B.2.6).

The reciprocal lattice is generated by an equation like (B.2.3):

$$\mathbf{K_m} = \sum_j m_j \mathbf{b}_j \tag{B.2.10}$$

where the m's are positive, negative or zero integers. By (B.2.9) this implies

$$\phi(\mathbf{K}, \mathbf{R}) \equiv \exp\left[\frac{2\pi i}{\epsilon} \mathbf{K} \cdot \mathbf{R}\right] = 1 \tag{B.2.11}$$

and

$$\phi(\mathbf{K}, \mathbf{r} + \mathbf{R}) = \phi(\mathbf{K}, \mathbf{r}), \tag{B.2.12}$$

where $\mathbf{K}$ is of the form (B.2.10) and $\mathbf{R}$ of the form (B.2.3). Thus $\phi(\mathbf{K}, \mathbf{r})$ is periodic in the direct lattice.

If Ω, Ω' are the unit cells of the direct and reciprocal lattices, we have

$$\Omega = \mathbf{a}_1 \cdot (\mathbf{a}_2 \times \mathbf{a}_3), \qquad \Omega' = \mathbf{b}_1 \cdot (\mathbf{b}_2 \times \mathbf{b}_3), \tag{B.2.13}$$

as may be seen from fig. B.2.5.

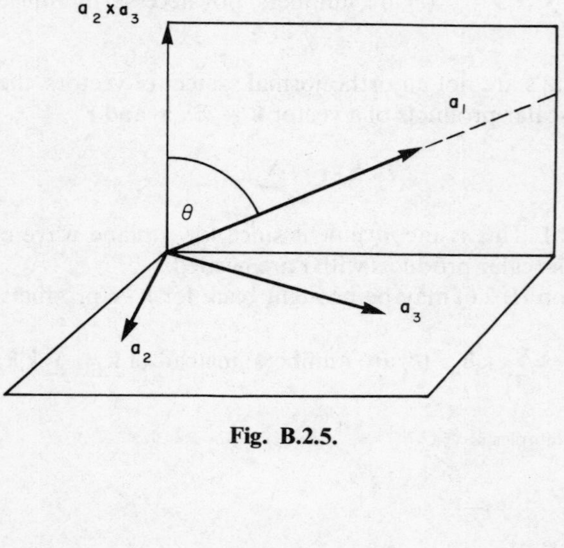

Fig. B.2.5.

We next obtain explicit expressions for the $\mathbf{b}_j$'s in terms of the $\mathbf{a}_j$'s. Since, by (B.2.8), $\mathbf{a}_1$ is perpendicular to both $\mathbf{b}_2$ and $\mathbf{b}_3$, a constant g_1 exists such that $\mathbf{a}_1 = g_1(\mathbf{b}_2 \times \mathbf{b}_3)$. More generally, taking suffices of modulus 3,

$$\mathbf{a}_j = g_j(\mathbf{b}_{j+1} \times \mathbf{b}_{j+2}).$$

Hence $\mathbf{b}_j\mathbf{a}_j = g_j\Omega' = \epsilon$. This is an equation for g_j, so that

$$\mathbf{a}_j = \frac{\epsilon}{\Omega'}(\mathbf{b}_{j+1} \times \mathbf{b}_{j+2}). \tag{B.2.14}$$

Similarly

$$\mathbf{b}_j = \frac{\epsilon}{\Omega}(\mathbf{a}_{j+1} \times \mathbf{a}_{j+2}). \tag{B.2.15}$$

By an expression theorem of vector algebra, one also has from (B.2.14), (B.2.15) and (B.2.8):

$$\begin{aligned}\mathbf{b}_1 \cdot \mathbf{a}_1 = \epsilon &= \frac{\epsilon^2}{\Omega\Omega'}(\mathbf{a}_2 \times \mathbf{a}_3)\cdot(\mathbf{b}_2 \times \mathbf{b}_3)\\ &= \frac{\epsilon^2}{\Omega\Omega'}[(\mathbf{b}_2 \cdot \mathbf{a}_2)(\mathbf{b}_3 \cdot \mathbf{a}_3) - (\mathbf{b}_3 \cdot \mathbf{a}_3)(\mathbf{b}_2 \cdot \mathbf{a}_3)].\end{aligned}$$

It follows that

$$\Omega\Omega' = \epsilon^3. \tag{B.2.16}$$

Strictly, the reciprocal lattice, particularly as used by crystallographers, is obtained from the above theory if ϵ is chosen as unity. For quantum theory a space is more convenient which is obtained from reciprocal space by simply multiplying each vector in that space by 2π. This is called **k**-space or wave vector space or, more loosely, reciprocal space. The above discussion covers this situation if one choses $\epsilon = 2\pi$.

For a lattice having mutually perpendicular vectors $\mathbf{a}_j$, $\Omega = a_1a_2a_3$ and equation (B.2.15) shows that the length b_j of $\mathbf{b}_j$ satisfies

$$b_j = \epsilon/a_j.$$

This relation suggests the origin of the term *reciprocal lattice.*

Problems

(B.2.1) Prove that the reciprocal to a reciprocal lattice is the original direct lattice.
(B.2.2) The three vectors $2\boldsymbol{\tau}_j$ represent the translations along the edges of a unit

cell of a lattice parallel to the three crystallographic axes (cf. table B.2.3). Let the m_j's be integers and let lattice sites be situated at

$$\mathbf{R_m} = \sum_{j=1}^{3} m_j \boldsymbol{\tau}_j.$$

What *restrictions* must be imposed on the m's in order that the lattice be (a) face-centred cubic, (b) body-centred cubic?

Show that these restrictions can be met by writing

$$\mathbf{R_n} = \sum_{j=1}^{3} n_j \mathbf{a}_j$$

where the n's are *arbitrary integers* and the $\mathbf{a}_j$'s are given by

$$\mathbf{a}_1 = \boldsymbol{\tau}_2 + \boldsymbol{\tau}_3, \qquad \mathbf{a}_2 = \boldsymbol{\tau}_3 + \boldsymbol{\tau}_1, \qquad \mathbf{a}_3 = \boldsymbol{\tau}_1 + \boldsymbol{\tau}_2 \text{ (face-centred).}$$

$$\mathbf{a}_1 = 2\boldsymbol{\tau}_1, \qquad \mathbf{a}_2 = \boldsymbol{\tau}_1 - \boldsymbol{\tau}_2 + \boldsymbol{\tau}_3, \qquad \mathbf{a}_3 = 2\boldsymbol{\tau}_3 \qquad \text{(body-centred).}$$

[Note that the argument is still valid even if the τ's are not all of equal lengths and are not perpendicular to each other, i.e. it is valid, for example, for an orthorhombic face-centred and body-centred lattice. For a cubic lattice appropriate conditions must be imposed on the τ_i's.]

(B.2.3) Obtain the reciprocal lattice of a face-centred cubic lattice by vector methods, using the results of problem (B.2.2). Illustrate your construction by a detailed diagram. Repeat for a body-centred cubic lattice.

(B.2.4) A Bravais lattice may be defined as a set of points L in three dimensional space fulfilling the following requirement: If $\mathbf{R}_1$, $\mathbf{R}_2$, $\mathbf{S}_1$ be position vectors of any three elements of L measured from any element of L then

$$\mathbf{S}_2 \equiv \mathbf{S}_1 + \mathbf{R}_2 - \mathbf{R}_1$$

is the position vector of another element of L.

Show that three vectors $\mathbf{a}_i (i = 1, 2, 3)$ exist such that any element of L has a position vector of the form $\mathbf{R_n} = \sum_{i=1}^{3} n_i \mathbf{a}_i$, where the n_i are integers.

(B.2.5) Explain the definition of problem (B.2.4) in simple language.

[The vista of lattice points seen from a lattice point $\mathbf{R}_1$ is identical with that seen from any other lattice point $\mathbf{S}_1$. This principle has been called the 'cosmological principle for lattices' (Ziman), because a similar principle is used in cosmology.]

B.3 PLANES IN THE DIRECT LATTICE AND LATTICE VECTORS IN THE RECIPROCAL LATTICE

To see the connexion suggested by the title of this section, consider

(i) the reciprocal lattice vector $\mathbf{K_m} = \Sigma m_j \mathbf{b}_j$ of equation (B.2.10);

(ii) the set of parallel planes which intercept, on the axes of the primitive translation vectors, the vectors

$$\mathbf{a}_j / M m_j \tag{B.3.1}$$

where each number M picks out one of the set of planes. One can then show (problem B.3.1) that $\mathbf{K_m}$ is normal to each of these planes. This is a key property much used in crystallography.

Since $\mathbf{K_m}$ is parallel to any multiple of it, it is desirable to single out one of these vectors. This is done by eliminating all common integral factors other than unity from the m's. This yields integers, h_j say, which are relatively prime. They are called Miller indices of the planes and can be positive, negative or zero, and are usually small. From the definition (B.2.10),

$$\mathbf{K}_{M\mathbf{h}} = M\mathbf{K_h}. \tag{B.3.2}$$

Examples are given in fig. B.3.1.

The problems (B.3.1) and (B.3.2) at the end of this section lead to the following result: let $\mathbf{K_h} = h_j\mathbf{b}_j$ be drawn from the origin of the direct lattice.

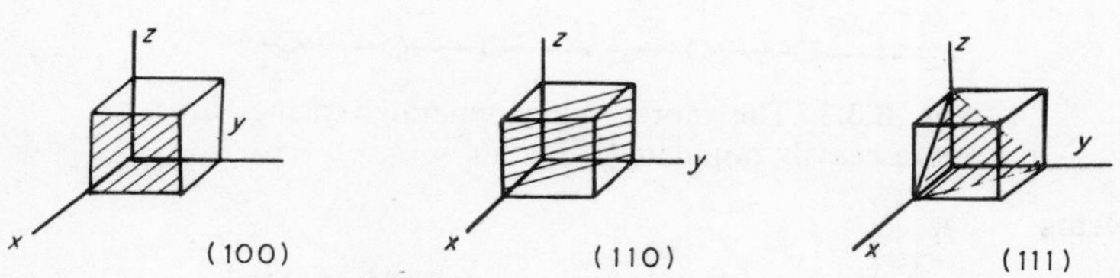

Fig. B.3.1.

It is perpendicular to the planes (h_1, h_2, h_3) of the direct lattice whose spacing is

$$d_\mathbf{h} = \epsilon/|\mathbf{K_h}|. \tag{B.3.3}$$

Consider the implications of choosing $M = 2$. $\mathbf{K_{2h}}$ is a longer vector in the reciprocal lattice than $\mathbf{K_h}$. But the vectors are parallel to each other by equation (B.3.2). The planes corresponding to $\mathbf{K_{2h}}$ are, however, twice as closely spaced as the planes corresponding to $\mathbf{K_h}$ (cf. fig. B.3.2). The result is that in general the set of planes corresponding to $\mathbf{K_{2h}}$ includes planes without any lattice points. Hence it is usual to put $M = 1$.

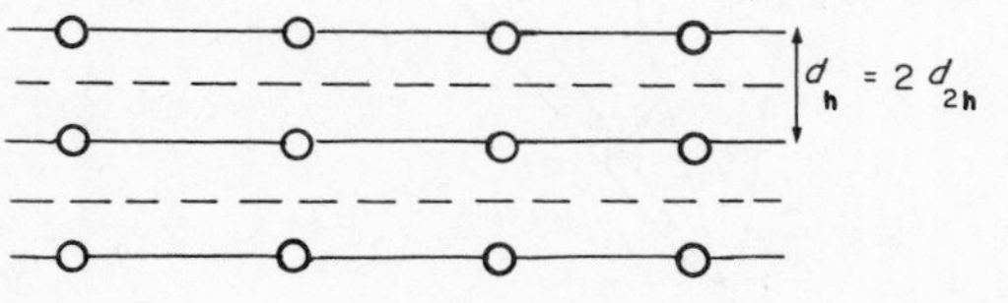

Fig. B.3.2.

Consider now changes of the Miller indices and hence of the direction of the $\mathbf{K_h}$'s. Suppose that change is such as to lengthen $\mathbf{K_h}$. The interplanar spacing decreases by (B.3.3). Now in a given part of a lattice the number of lattice points is fixed. It follows that the more closely spaced planes are less heavily populated by lattice points (fig. B.3.3).

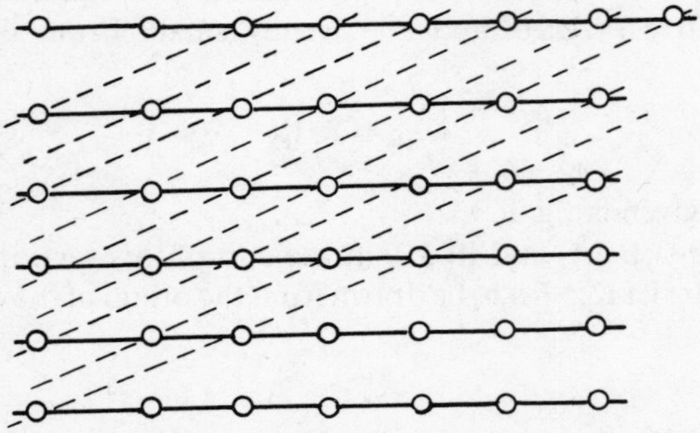

Fig. B.3.3. The more widely separated planes are more heavily populated.

Problems

(B.3.1) A plane drawn in a lattice makes intercepts $\mathbf{a}_j/Mm_j(j = 1, 2, 3)$ on the axes of the primitive translation vectors $\mathbf{a}_j$. Here m_j is a positive or negative integer and M is a positive real number. $K_{\mathbf{gm}} \equiv g\Sigma m_j \mathbf{b}_j$ is a vector in **k**-space where g is a positive real number. Assuming the usual condition $\mathbf{b}_j \cdot \mathbf{a}_l = \delta_{jl}$, prove that $K_{\mathbf{gm}}$ is normal to the plane for all M and g. (*Method* 1: Obtain an expression for a general vector in the plane first; see fig. B.3.4: $\mathbf{r} = r_l \mathbf{l} + r_m \mathbf{m}$. *Method* 2: Write the unit normal as $\mathbf{n} = \mathbf{l} \times \mathbf{m}/|\mathbf{l} \times \mathbf{m}|$.)

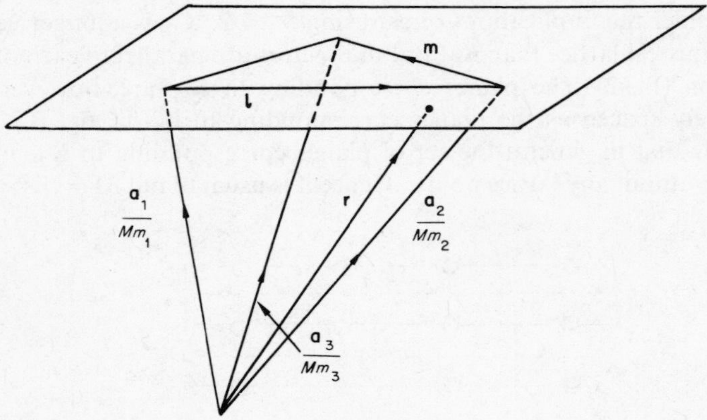

Fig. B.3.4.

(B.3.2) Show that the perpendicular distance of the plane in problem (B.3.1) from the origin of the direct lattice is

$$d_{M\mathbf{m}} = \epsilon/|\mathbf{K}_{M\mathbf{m}}| = \epsilon/M|\mathbf{K}_{\mathbf{m}}|.$$

Solution of problem (B.3.2).

Special case. Figure B.3.5 gives a specific example of a two-dimensional square lattice with lattice parameter *a*. Crystal planes having Miller indices (1,3) are drawn. Arrows will denote values which apply only for this figure.

$$\mathbf{b}_j \to \epsilon \mathbf{a}_j/a^2 \qquad (j = 1, 2).$$

$$\mathbf{K}_{M\mathbf{m}} = \sum Mm_j\mathbf{b}_j \to \tfrac{1}{3}\mathbf{b}_1 + \mathbf{b}_2.$$

$$|\mathbf{K}_{M\mathbf{h}}|^2 \to |\mathbf{K}_{\frac{1}{3}(1,3)}|^2 = \tfrac{1}{9}b_1^2 + b_2^2 = \frac{10\epsilon^2}{9a^2}.$$

$$\frac{\epsilon}{|\mathbf{K}_{M\mathbf{h}}|} \to \frac{\epsilon}{|\mathbf{K}_{\frac{1}{3}(1,3)}|} = \frac{3}{\sqrt{10}}a.$$

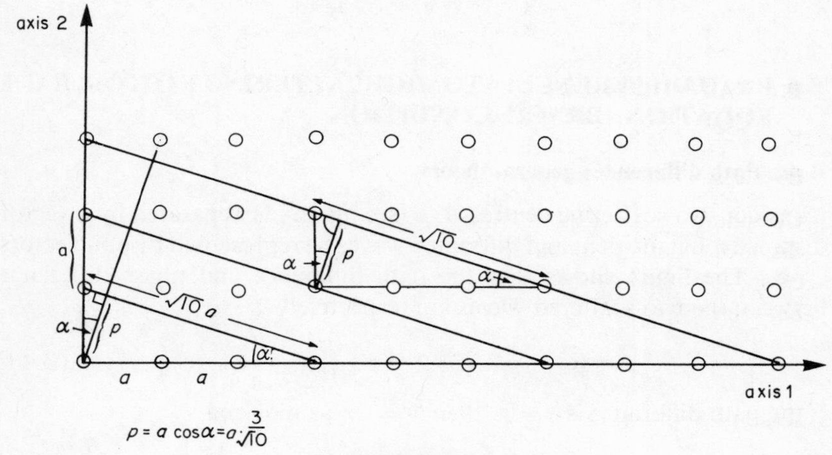

Fig. B.3.5.

The figure shows that the interplanar spacing is in fact

$$p = \frac{3}{\sqrt{10}}a = d_{\frac{1}{3}(1,3)}.$$

General case. The perpendicular distance of a plane with intercepts $\mathbf{a}_j/Mm_j$ from the origin is obtained by projecting any intercept on the unit normal $\mathbf{n}$, given by

$$\mathbf{n} = \mathbf{K}_{M\mathbf{m}}/|\mathbf{K}_{M\mathbf{m}}|.$$

Hence

$$d_{M\mathbf{m}} = \mathbf{n}\cdot\frac{\mathbf{a}_j}{Mm_j} = \frac{\epsilon Mm_j}{|\mathbf{K}_{M\mathbf{m}}|Mm_j} = \frac{\epsilon}{|\mathbf{K}_{M\mathbf{m}}|}.$$

(B.3.3) Obtain an expression for the perpendicular distance between two adjacent lattice planes having Miller indices (h_1, h_2, h_3).

(B.3.4) Prove that the reciprocal to a simple cubic lattice is a simple cubic lattice.

(B.3.5) Using problems (B.3.3) and (B.3.4), show that the perpendicular distance between neighbouring planes of type (h_1, h_2, h_3) in a simple cubic lattice of lattice parameter a is

$$\frac{a}{\sqrt{\left(\sum_{j=1}^{3} h_j^2\right)}}.$$

(B.3.6) Show that for any vector $\mathbf{v}$ in the direct lattice $\epsilon\mathbf{v} = \sum_{j=1}^{3} (\mathbf{b}_j \cdot \mathbf{v})\mathbf{a}_j$. Show also that for any vector $\mathbf{w}$ in the reciprocal lattice, $\epsilon\mathbf{w} = \sum_j (\mathbf{w} \cdot \mathbf{a}_j)\mathbf{b}_j$.

(B.3.7) Discuss the errors in the following argument: for any vector $\mathbf{v}$,

$$\mathbf{v} = \sum_j (\mathbf{v} \cdot \mathbf{b}_j)\mathbf{b}_j = \sum_j (\mathbf{v} \cdot \mathbf{a}_j)\mathbf{a}_j.$$

Hence

$$\mathbf{v} \cdot \mathbf{b}_j = \epsilon\mathbf{v} \cdot \mathbf{a}_j.$$

B.4 PATH DIFFERENCE: ATOMIC SCATTERING FACTOR, LAUE EQUATION, BRAGG CONDITION

B.4a Path difference: general theory

Consider two scattering centres A, B (see fig. B.4.1), separated by a vector distance $\mathbf{r}$. Let incident and diffracted waves be represented by unit vectors $\mathbf{c}_i$; $\mathbf{c}_d$. The figure shows that the path difference and phase difference between the two scattered waves are respectively

$$s = (\mathbf{c}_d - \mathbf{c}_i) \cdot \mathbf{r}, \qquad \Phi = \frac{2\pi}{\lambda}(\mathbf{c}_d - \mathbf{c}_i) \cdot \mathbf{r}. \tag{B.4.1}$$

If the path difference is $s = \lambda$, then $\Phi = 2\pi$ as required.

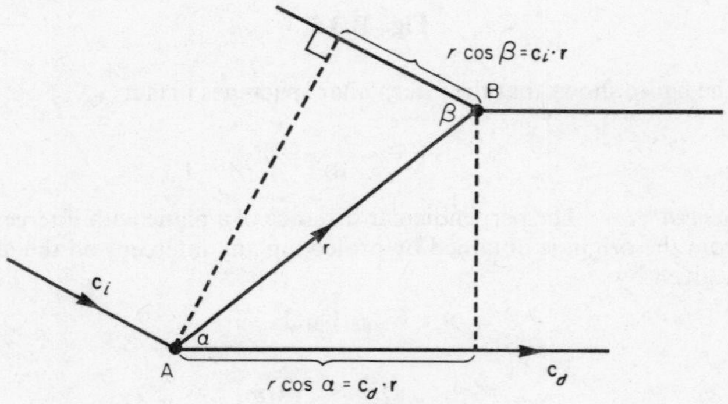

Fig. B.4.1.

The vector $\mathbf{c}_d - \mathbf{c}_i$ may be interpreted by drawing the plane XY which makes equal angles γ with $\mathbf{c}_i$ and $\mathbf{c}_d$ (fig. B.4.2). This is the plane which would reflect the incident into the scattered ray. This *reflecting plane* is seen to have $\mathbf{c}_d - \mathbf{c}_i$ as normal. Let θ be the angle between $\mathbf{c}_d - \mathbf{c}_i$ and $\mathbf{r}$. Then (B.4.1) becomes

$$|c_d - \mathbf{c}_i| = 2 \sin \gamma; \qquad \Phi = \frac{4\pi r}{\lambda} \sin \gamma \cos \theta. \qquad \text{(B.4.2)}$$

Thus

$$\Phi = \mu r \cos \theta, \qquad \text{where } \mu \equiv \frac{4\pi}{\lambda} \sin \gamma. \qquad \text{(B.4.3)}$$

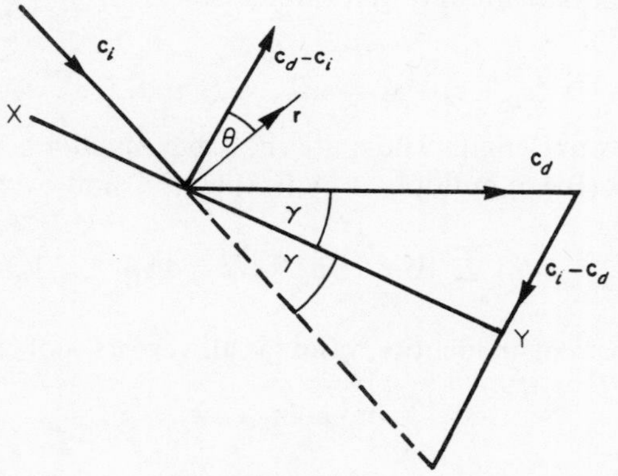

Fig. B.4.2.

The interpretation of Φ is that the space-dependent part of a wave scattered at B is $e^{i\Phi}$ times the space-dependent part of the same wave scattered by the same centre, but located at A instead of B.

B.4b First application of equation (B.4.2): atomic scattering factor

Suppose an atom is located with its centre at A, and that an electron has probability $P(\mathbf{r})\,\mathbf{dr}$ of being in a volume element $\mathbf{dr}$ at B, distance $\mathbf{r}$ from A. Let the space-dependent part of the wave scattered by Z electrons in the atom be f times the space-dependent part of the same wave scattered by one point electron at the atomic centre A. Then, by the meaning of Φ in (B.4.2),

$$f = Z \int P(\mathbf{r})\, e^{i\mu r \cos \theta}\, \mathbf{dr}, \qquad \text{(B.4.4)}$$

where the integral extends over the volume of the scattering material. This is the atomic scattering factor.

The expression (B.4.4) simplifies if the charge distribution is symmetrical (see problem B.4.1).

B.4c Second application of equations (B.4.1) and (B.4.2): Laue equations and Bragg's law

To obtain a diffraction maximum from identical atoms arranged in a Bravais lattice the scattering amplitudes for neighbouring lattice points must reinforce each other. Thus in (B.4.1) one needs merely to replace $\mathbf{r}$ by the three primitive translations $\mathbf{a}_j$. It is then seen that the condition for a diffraction maximum is that three integers m_j $(j = 1, 2, 3)$ exist such that

$$(\mathbf{c}_i - \mathbf{c}_d) \cdot \mathbf{a}_j = m_j\lambda \qquad (j = 1, 2, 3), \tag{B.4.5}$$

where λ is the wavelength. These are the *Laue equations.*

To interpret (B.4.5) multiply it by (ϵ/λ) $\mathbf{b}_j$ and sum over all j to obtain

$$(\epsilon/\lambda) \sum_{j=1}^{3} [(\mathbf{c}_i - \mathbf{c}_d) \cdot \mathbf{a}_j]\mathbf{b}_j = \epsilon\mathbf{K}_\mathbf{m}. \tag{B.4.6}$$

This has the form of an identity, valid for all vectors $\mathbf{w}$ of reciprocal space,

$$\sum_j (\mathbf{w} \cdot \mathbf{a}_j)\mathbf{b}_j = \epsilon\mathbf{w} \tag{B.4.7}$$

[see problem (B.3.6)]. Thus (B.4.6) implies that

$$\mathbf{c}_i - \mathbf{c}_d = (\lambda/\epsilon)\mathbf{K}_\mathbf{m}. \tag{B.4.8}$$

Removing any common integral factor M which may be present among the m_j's, the Laue equations are seen to be equivalent to

$$\mathbf{c}_i - \mathbf{c}_d = \frac{M\lambda}{\epsilon}\mathbf{K}_\mathbf{h}, \tag{B.4.9}$$

whence, taking absolute values and using $d_\mathbf{h} = \epsilon/|\mathbf{K}_\mathbf{h}|$ from problem (B.3.2),

$$d_\mathbf{h}|\mathbf{c}_d - \mathbf{c}_i| = M\lambda. \tag{B.4.10}$$

Here $d_\mathbf{h}$ is the interplanar spacing of the planes (h_1, h_2, h_3).

To interpret (B.4.9) and (B.4.10) now, we show in fig. B.4.3 the vector diagram for $\mathbf{c}_i$ and $\mathbf{c}_d$ and use the result of problem (B.3.1) that $\mathbf{K}_\mathbf{h}$ is perpendicular to the planes (h_1, h_2, h_3). The dotted line represents these planes. If $\gamma_{\lambda,\mathbf{h}}$ be the angle for strong reflexion of waves of wavelength λ

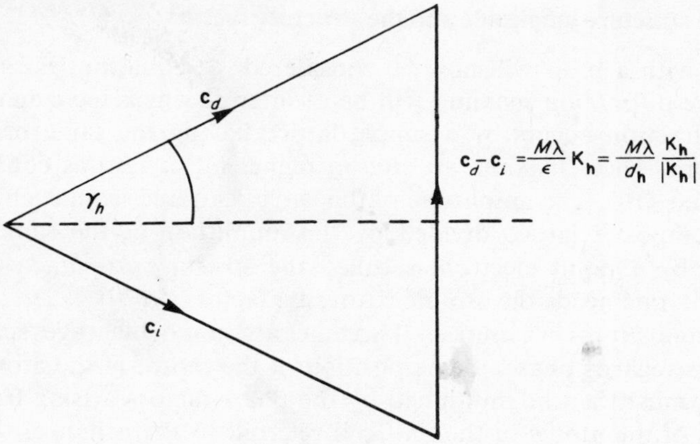

Fig. B.4.3.

from the planes (h_1, h_2, h_3), (B.4.10) is seen to imply

$$2d_{\mathbf{h}} \sin \gamma_{\lambda,\mathbf{h}} = M\lambda. \tag{B.4.11}$$

This is W. L. Bragg's law of 1912, which is easily obtainable directly (fig. B.4.4). It is implied by Laue's equations.

Given the beam direction and the orientation of the crystal an angle γ satisfying (B.4.11) is possible only for certain wavelengths λ. This leads to the so-called Laue patterns in X-ray crystallography.

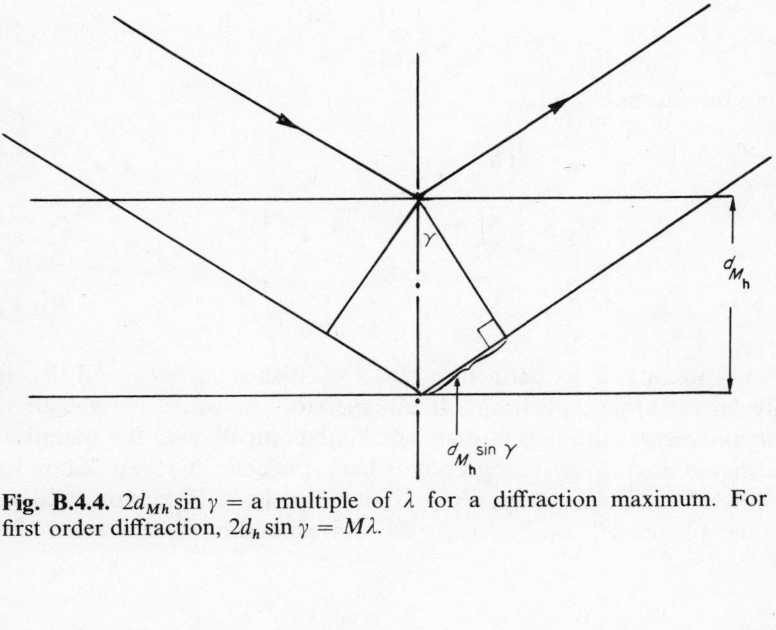

Fig. B.4.4. $2d_{Mh} \sin \gamma$ = a multiple of λ for a diffraction maximum. For first order diffraction, $2d_h \sin \gamma = M\lambda$.

B.4d The structure amplitude and the structure factor

A lattice with a basis will now be considered [cf. equation (B.2.4)]. But only those diffraction maxima will be of interest which have analogues when diffraction occurs by a simple lattice having the same primitive translations. These maxima are due to planes (h_1, h_2, h_3) as considered in §B.2 and §B.3. The amplitude of the wave scattered from such planes of the composite lattice divided by the amplitude of the same wave scattered by a point electron is called the *structure amplitude* $F_{\mathbf{h}}$. To evaluate it, one needs the atomic scattering factors f_j $(j = 1, 2, \ldots, s)$ for the s different atoms in a unit cell. This takes account of the waves scattered related, as regards phase, to the position of the centre of the atom. The factors f_j must then be multiplied by the phase factors arising from the positions of the atoms in the unit cell referred to the origin of the cell. Hence one finds

$$F_{\mathbf{h}} = \sum_{j=1}^{s} f_j \exp\left[\frac{2\pi i}{\lambda}(\mathbf{c}_d - \mathbf{c}_i) \cdot \mathbf{r}_j\right],$$

where

$$\mathbf{r}_j \equiv \sum_{l=1}^{3} u_{jl}\mathbf{a}_l$$

and the u_{jl} are not necessarily integers. For a first order (i.e. $M = 1$) diffraction maximum the Laue condition (B.4.9) must be fulfilled, so that

$$F_{\mathbf{h}} = \sum_{j=1}^{s} f_j \exp\left[-\frac{2\pi i}{\epsilon}\mathbf{K}_{\mathbf{h}} \cdot \mathbf{r}_j\right]. \tag{B.4.12}$$

For a monatomic solid $f_1 = f_2 = \cdots (= f \text{ say})$,

$$F_{\mathbf{h}} = sfS_h;$$

$$S_{\mathbf{h}} \equiv \frac{1}{s}\sum_{j=1}^{s} \exp\left[-\frac{2\pi i}{\epsilon}\mathbf{K}_{\mathbf{h}} \cdot \mathbf{r}_{\mathbf{j}}\right]$$

$$= \frac{1}{s}\sum_{\mathbf{j}=1}^{s} \exp\left[-2\pi i \sum_{l=1}^{3} h_l u_{jl}\right] \tag{B.4.13}$$

The quantity $S_{\mathbf{h}}$ as defined in (B.4.13) satisfies $|S_{\mathbf{h}}| \leqslant 1$, and depends only on the atomic positions. It can therefore be called the *geometrical structure factor*, although it does not quite coincide with the quantity of that name used in crystallography (where it is defined as $|F_{\mathbf{h}}|$). Suppose a primitive lattice of a certain type of unit cell yields a diffraction maximum for the planes (h_1, h_2, h_3); then the corresponding maximum may be absent when the lattice is converted to one which has a basis, by the provision of additional atoms. This can often be traced to the vanishing of the quantity $S_{\mathbf{h}}$. In crystallography such missing diffraction maxima give important clues regarding the crystal structure of the material being investigated [cf. problem (B.4.3)].

In a *non-periodic* situation, as in a liquid of s atoms, say, one can regard the whole system as just one very complicated unit cell. No question of reinforcements of waves scattered by different unit cells then arises and Laue's equations become irrelevant. One can, however, define a structure amplitude (cf. B.4.12):

$$F_{\mathbf{k}} = \sum_{j=1}^{s} f_j \exp\left[-\frac{2\pi i}{\epsilon}\mathbf{k}\cdot\mathbf{r}_j\right], \qquad \left(\mathbf{k} \equiv \frac{\epsilon}{\lambda}(\mathbf{c}_i - \mathbf{c}_d)\right), \qquad \text{(B.4.14)}$$

and a structure factor

$$S_{\mathbf{k}} = \frac{1}{s}\sum_{j=1}^{s} \exp\left[-\frac{2\pi i}{\epsilon}\mathbf{k}\cdot\mathbf{r}_j\right],$$

where $\mathbf{k}$ is replaced by $\mathbf{K}_{\mathbf{h}}$ in the case of diffraction by planes (h_1, h_2, h_3) of periodic structures. Thus (B.4.14) is more general than (B.4.13). It follows that

$$|S_{\mathbf{k}}|^2 = \frac{1}{s^2}\sum_{j,m=1}^{s} \exp\left[-\frac{2\pi i}{\epsilon}\mathbf{k}\cdot\mathbf{r}_{jm}\right] = \int P(\boldsymbol{\rho})\exp\left[-\frac{2\pi i}{\epsilon}\mathbf{k}\cdot\boldsymbol{\rho}\right]\mathrm{d}\boldsymbol{\rho}, \quad \text{(B.4.15)}$$

where

$$\mathbf{r}_{jm} = \mathbf{r}_j - \mathbf{r}_m.$$

In (B.4.15) summation has been converted to integration in terms of the probability $P(\boldsymbol{\rho})$ of finding two atoms a distance $\boldsymbol{\rho}$ apart in the system. This is called the pair correlation function and $|S_{\mathbf{k}}|^2$ is seen to be proportional to the $\mathbf{k}$th Fourier coefficient of $P(\boldsymbol{\rho})$. These results can be useful in developing a theory of the electrical properties of liquid metals[2].

Problems

(B.4.1) Let $p(\mathbf{r})\,\mathrm{d}\mathbf{r}$ be the probability of finding an electron in a Z-electron atom at a distance $\mathbf{r}$ from the centre of the atom. For a spherically symmetrical distribution the probability of an electron lying between radii r and $r + \mathrm{d}r$ is $4\pi r^2 p(r)\,\mathrm{d}r \equiv \mathrm{U}(r)\,\mathrm{d}r$. Show that the atomic scattering factor is in this case

$$f = Z\int_0^\infty \mathrm{U}(r)\frac{\sin \mu r}{\mu r}\,\mathrm{d}r$$

(where $\mu \equiv (4\pi/\lambda)\sin\gamma$, as in the main text).

[2] J. M. Ziman, *Phil. Mag.*, **6**, 1013–1034 (1961); N. H. March, *Liquid Metals*, Pergamon Press, 1968, p. 68.

(B.4.2) A body-centred and a face-centred lattice are represented in terms of the appropriate simple unit cells. Find expressions for the geometrical structure factors $S_{\mathbf{h}}$ and note that they are independent of the choice of ϵ.

(B.4.3) Show from problem (B.4.2) that for a body-centred lattice $S_{\mathbf{h}} = 0$ or 1 depending on whether Σh_j is odd or even.
Show that for a face-centred lattice $S_{100} = S_{110} = 0$ and $S_{111} = S_{200} = 1$.

B.5 BRILLOUIN ZONES

The condition for a diffraction maximum was given by the Laue equations, which in turn implied equation (B.4.8) for the unit vectors $\mathbf{c}_i$ and $\mathbf{c}_d$:

$$\frac{\epsilon}{\lambda}\mathbf{c}_d = \frac{\epsilon}{\lambda}\mathbf{c}_i - \mathbf{K}_{\mathbf{m}}. \tag{B.5.1}$$

Squaring, one obtains

$$2\mathbf{k} \cdot \mathbf{K}_{\mathbf{m}} = |\mathbf{K}_{\mathbf{m}}|^2 \tag{B.5.2}$$

where a propagation vector

$$\mathbf{k} \equiv (\epsilon/\lambda)\mathbf{c}_i \tag{B.5.3}$$

has been introduced. Equation (B.5.2) states that the orthogonal projection of $\mathbf{k}$ on the unit vector parallel to $\mathbf{K}_{\mathbf{m}}$ shall be equal to $\frac{1}{2}\mathbf{K}_{\mathbf{m}}$.

To interpret (B.5.2), let the perpendicular to a plane in $\mathbf{k}$-space be $\mathbf{p}$. The equation for that plane is then

$$\mathbf{k} \cdot \mathbf{p}/|\mathbf{p}| = |\mathbf{p}|, \qquad \text{i.e. } \mathbf{k} \cdot \mathbf{p} = |\mathbf{p}|^2. \tag{B.5.4}$$

Comparing (B.5.4) and (B.5.2), it is seen that the Laue equations imply that the propagation vectors $\mathbf{k}$ lie on a plane whose perpendicular distance from the origin in $\mathbf{k}$-space is given by

$$\mathbf{p} = \tfrac{1}{2}\mathbf{K}_{\mathbf{m}}. \tag{B.5.5}$$

The surfaces closest to the origin of $\mathbf{k}$-space formed by these planes enclose the so-called first Brillouin zone. The smallest volume enclosed by the next set of surfaces, but lying outside the first zone, is called the second zone, etc.

It is a simple matter to construct two-dimensional zones. Consider a rectangular lattice with primitive translations

$$\mathbf{a}_1 = \begin{pmatrix} a_1 \\ 0 \end{pmatrix}, \qquad \mathbf{a}_2 = \begin{pmatrix} 0 \\ a_2 \end{pmatrix},$$

so that the reciprocal or **k**-space lattice is given by

$$\mathbf{K_m} = \sum_{j=1}^{2} m_j \mathbf{b}_j = \epsilon(m_1/a_1, m_2/a_2), \tag{B.5.6}$$

since

$$\mathbf{b}_1 = \frac{\epsilon}{a_1}(1, 0), \qquad \mathbf{b}_2 = \frac{\epsilon}{a_2}(0, 1).$$

Writing

$$\mathbf{k} = (k_1, k_2), \tag{B.5.7}$$

(B.5.2) becomes, using (B.5.6) and (B.5.7),

$$\frac{2}{\epsilon}\left(\frac{m_1 k_1}{a_1} + \frac{m_2 k_2}{a_2}\right) = \left(\frac{m_1}{a_1}\right)^2 + \left(\frac{m_2}{a_2}\right)^2. \tag{B.5.8}$$

For given m_1 and m_2 this yields a linear relation between k_1 and k_2, and these represent lines in the **k**-plane (see figs. B.5.1–B.5.4). In three dimensions one has planes instead of lines. When a propagation vector lies on

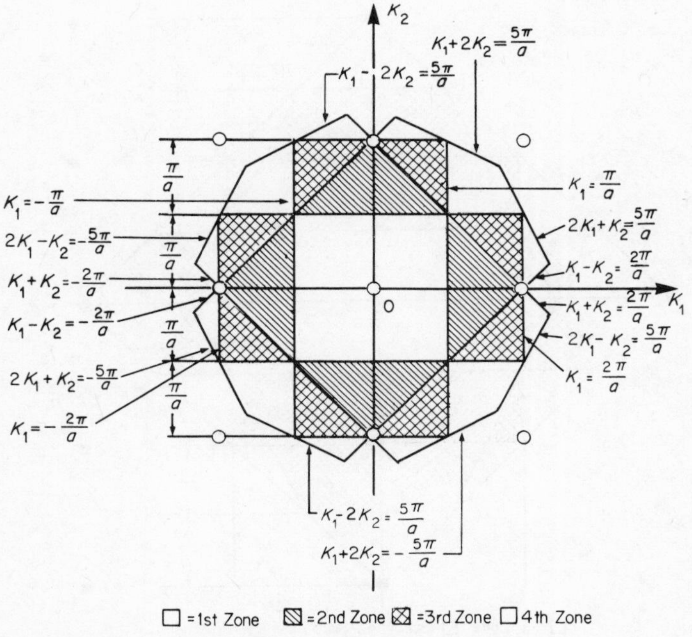

Fig. B.5.1. Brillouin zones of a square lattice.

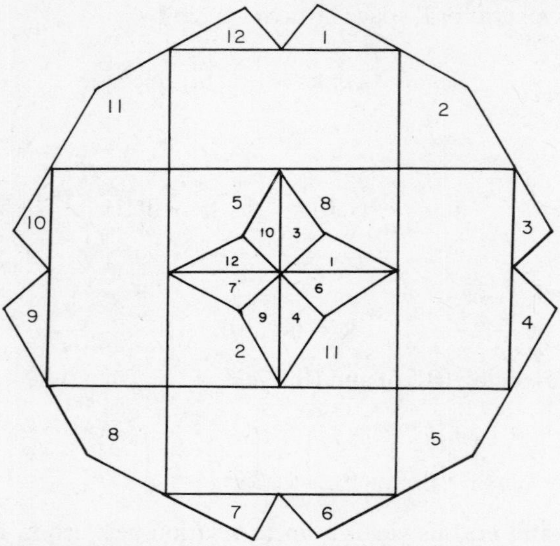

Fig. B.5.2. Projection of the fourth Brillouin zone of a square lattice on the first zone. Corresponding areas are numbered correspondingly.

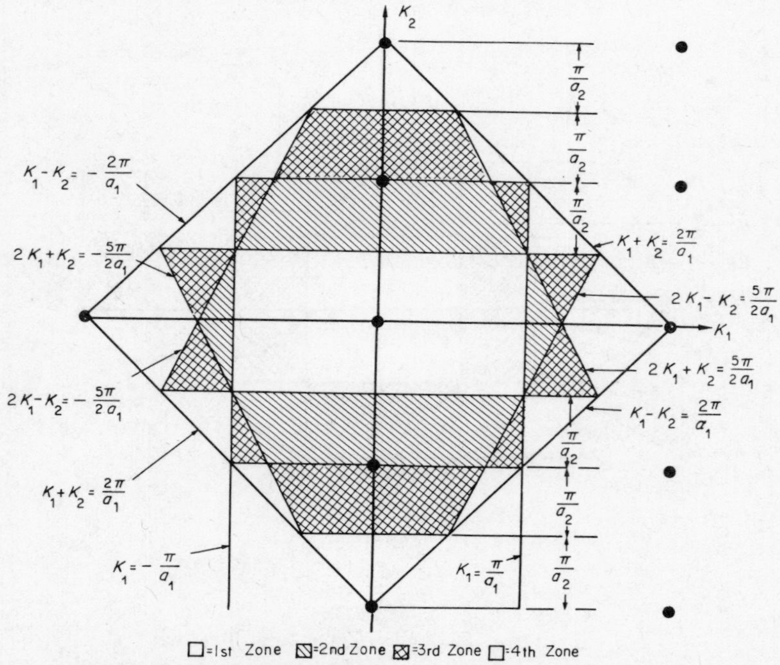

Fig. B.5.3. Brillouin zones of a rectangular lattice with $a_2/a_1 = 2$.

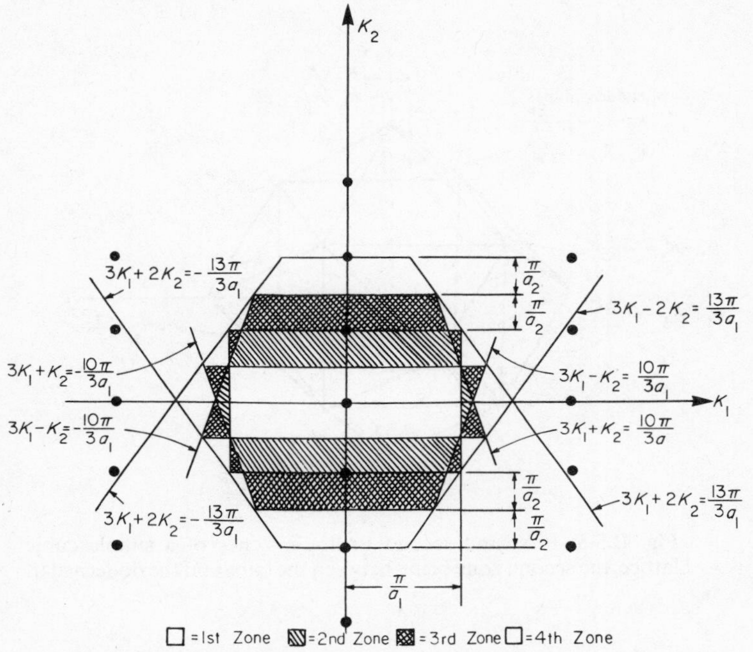

Fig. B.5.4. Brillouin zones of a rectangular lattice with $a_2/a_1 = 3$.

one of these lines or planes, the Laue condition for a diffraction maximum is satisfied.

Equation (B.5.5) gives a simple recipe for the construction of three-dimensional Brillouin zones in **k**-space: first indicate the lattice points and take one of them as origin. Join it to its neighbouring lattice points. The planes which are perpendicular bisectors of these vectors are then constructed for each **m**. The zones are then obtained from these planes. Illustrations are shown in figs. B.5.5–B.5.10.[3]

Problem

(B.5.1) A proximity cell of a lattice point is defined as the region of space whose points are closer to that lattice point than they are to any other.

Prove that such cells (a) are polyhedra; (b) fill all of space when stacked; (c) are identical in size, shape and orientation for all lattice points of a given Bravais lattice.

[In the direct lattice this is a Wigner–Seitz unit cell; in the reciprocal lattice it is the first Brillouin zone.]

[3] Reproduced from the author's report No. A.189 of the Research Laboratory of Associated Electrical Industries Ltd. issued in 1952.

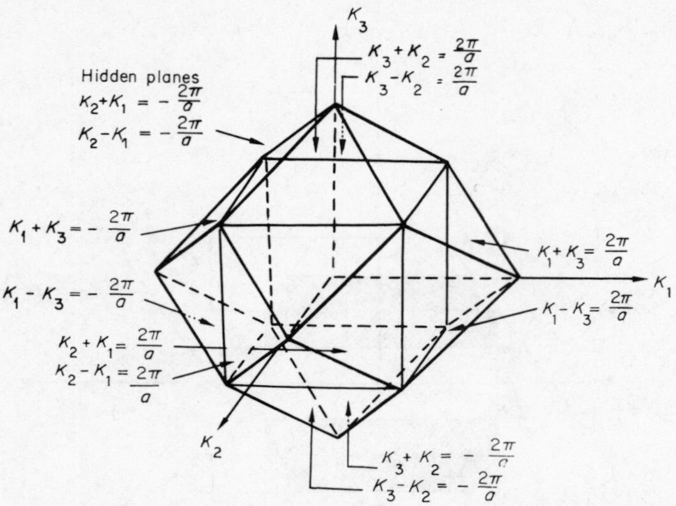

Fig. B.5.5. First and second Brillouin zones of a simple cubic lattice, the second zone being between the cube and the dodecaeder.

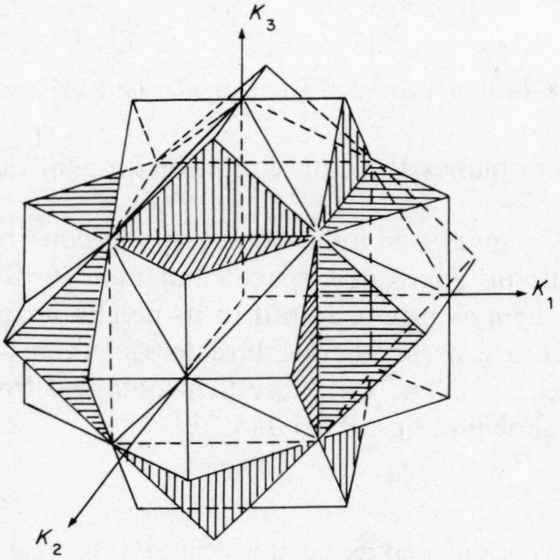

Fig. B.5.6. Outer surface of the third Brillouin zone of a simple cubic lattice. The twelve edges of the central cube are on the bounding surfaces of zones 1, 2 and 3. The shaded and blank planes are visible continuations of the planes bounding the first and second zones respectively. Hidden continuations of planes bounding the first zone are also left blank.

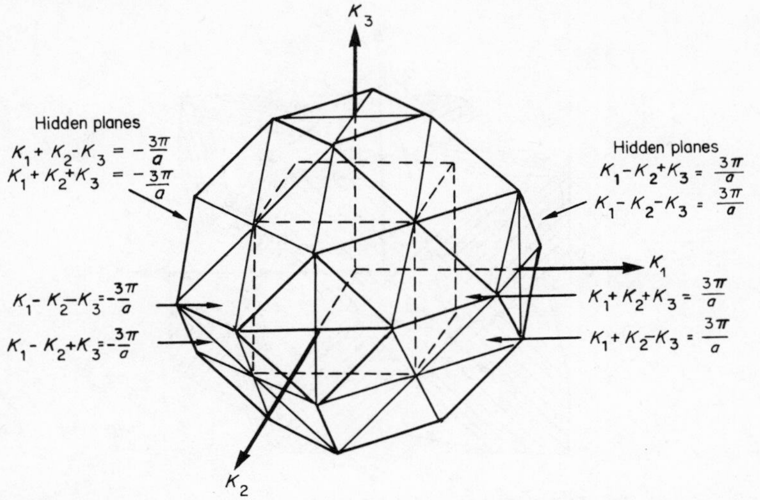

Fig. B.5.7. Outer surface of the fourth Brillouin zone of a simple cubic lattice. The thin lines are the bounding surfaces of both third and fourth zones. The eight corners of the central cube are on the bounding surfaces of zones 1, 2, 3 and 4. The six square faces have the equations $K_j = \pm 2\pi/a$ $(j = 1, 2, 3)$.

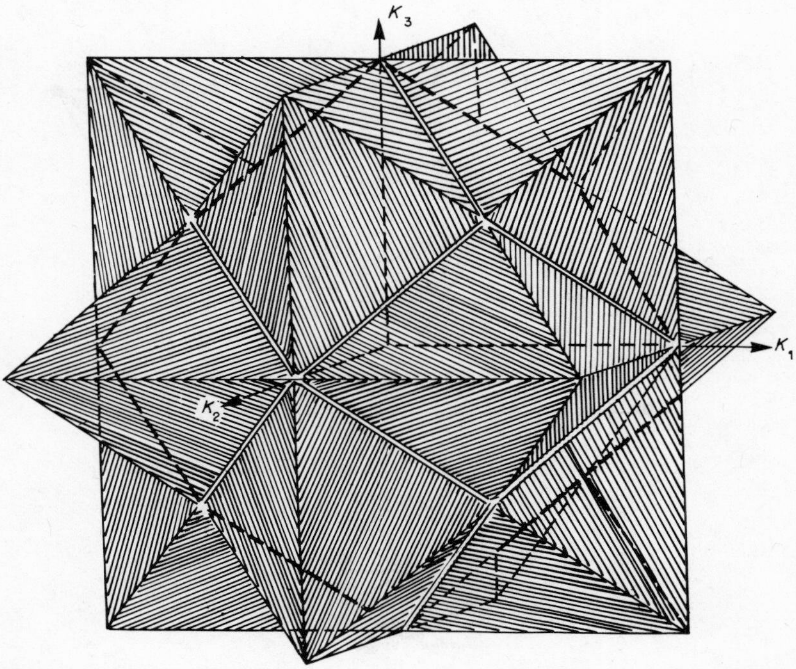

Fig. B.5.8. Second Brillouin zones of the body-centred cubic lattice. The thick lines indicate bounding the first Brillouin zone.

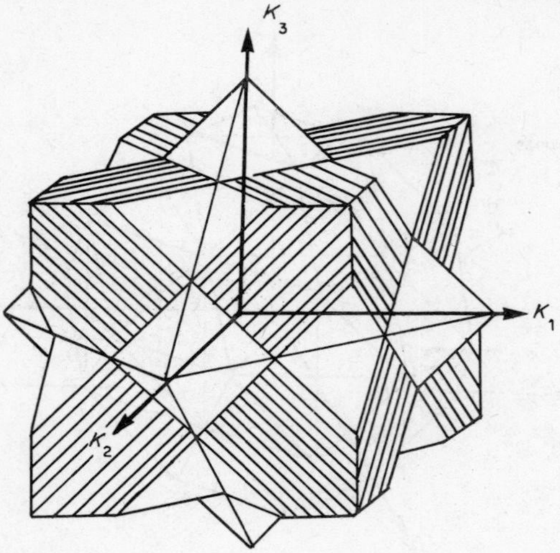

Fig. B.5.9. The second Brillouin zone of the face-centred cubic lattice. The shaded faces represent extensions of the square faces of the first zone.

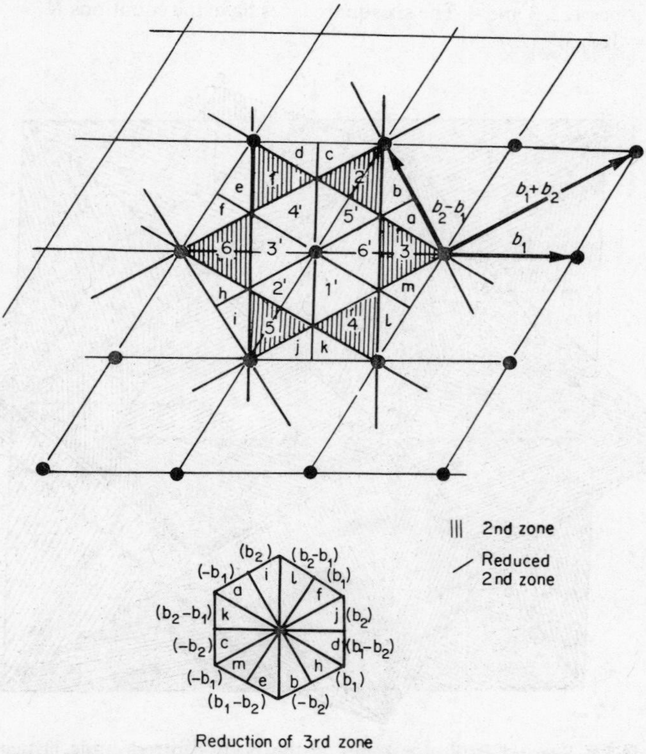

Fig. B.5.10. Zones in a general two-dimensional lattice.

B.6 THE PERIODIC BOUNDARY CONDITION

Suppose that the crystal under consideration is in the form of a parallelepiped of sides $N_j a_j$ ($j = 1, 2, 3$), where the N_j are integers and the $\mathbf{a}_j$ are again the primitive translations. The crystal contains then

$$N = N_1 N_2 N_3 \tag{B.6.1}$$

unit cells, and this region is referred to as the fundamental domain. The surface atoms are in markedly different environment from the atoms in the bulk of the crystal and have to be treated separately. Indeed, the finite extent of the crystal implies that pairs of atoms are only exceptionally in equivalent environments.

This brings up the assumption (4) which was made in §B.1. A free surface causes inconvenience in the theoretical treatment. On the other hand, the number of atoms at the surface is very small compared with those in the bulk material, and becomes increasingly so as the volume of crystal under consideration is increased. One may therefore hope that an incorrect boundary condition will introduce errors only for a small fraction of atoms. Assuming these to be in a layer of thickness Δa at the surface of a cube of side a, the required fraction is $6a^2\Delta a/a^3$ and will tend to zero as $a \to \infty$. Hence one would expect the distribution of electron energies, and the energies themselves, to be relatively insensitive to them. If this can be substantiated, one could use the simplest reasonable boundary condition in the hope that it should yield good results for a large crystal[4].

One therefore usually chooses a condition which implies that the finite crystal studied behaves exactly as if it were part of an infinite crystal, and as if it were delimited in the infinite crystal in a purely theoretical way. It is thus usual to suppose that all crystal properties satisfy

$$\mathrm{f}\left(\mathbf{r} + \sum_{j=1}^{3} N_j a_j\right) = \mathrm{f}(\mathbf{r}). \tag{B.6.2}$$

This is the periodic boundary condition. The infinite crystal is a special case of this, obtained by letting $N_j \to \infty$. For large N_j the assumption (B.6.2) is unlikely to affect the theory of the bulk properties of the crystal.

Applying (B.6.2) to a plane wave, one sees that

$$\exp\left[\frac{2\pi \mathrm{i}}{\epsilon}\mathbf{k}\cdot\left(\mathbf{r} + \sum_{j=1}^{3} N_j \mathbf{a}_j\right)\right] = \exp\left(\frac{2\pi \mathrm{i}}{\epsilon}\mathbf{k}\cdot\mathbf{r}\right). \tag{B.6.3}$$

Just as $\exp\left(\frac{2\pi \mathrm{i}}{\epsilon}\mathbf{K}\cdot\mathbf{r}\right)$ is periodic in the direct lattice, $\exp\left(\frac{2\pi \mathrm{i}}{\epsilon}\mathbf{k}\cdot\mathbf{r}\right)$ is

[4] See, for a simple direct calculation, G. Weinreich, *Solids: Elementary Theory for Advanced Students*, Wiley, New York, 1965, pp. 44–47.

periodic in the fundamental domain. It follows that integers t_j must exist such that

$$\mathbf{k} \equiv \sum_{j=1}^{3} k_j \mathbf{b}_j = \sum_{j=1}^{3} (t_j/N_j)\mathbf{b}_j \qquad (t_j = 0, \pm 1, \pm 2, \ldots). \qquad \text{(B.6.4)}$$

It is convenient to take as unit cell in **k**-space or in reciprocal space a parallelepiped extending from $-b_j/2$ to $b_j/2$ in the direction of $\mathbf{b}_j$ for each $j = 1, 2, 3$, and this can be achieved by supposing that N_j is odd:

$$-b_j/2 < k_j b_j \leqslant b_j/2, \qquad \text{i.e. } -N_j/2 < t_j \leqslant N_j/2. \qquad \text{(B.6.5)}$$

Thus the integers t_j assume the N_j values

$$t_j = -\frac{N_j - 1}{2}, -\frac{N_j - 3}{2}, \ldots, -1, 0, 1, \ldots, \frac{N_j - 1}{2} \qquad (j = 1, 2, 3) \qquad \text{(B.6.6)}$$

in the unit cell of the reciprocal space. The region may equivalently be specified by

$$-\frac{\epsilon}{2} < \mathbf{k} \cdot \mathbf{a}_j \leqslant \frac{\epsilon}{2} \quad \text{or} \quad -\tfrac{1}{2} < k_j \leqslant \tfrac{1}{2} \qquad (j = 1, 2, 3). \qquad \text{(B.6.7)}$$

The region thus specified is called the *reduced zone* or the first Brillouin zone. Its volume is

$$\mathbf{b}_1 \cdot (\mathbf{b}_2 \times \mathbf{b}_3) = \Omega'.$$

By (B.6.5) it contains $N_1 N_2 N_3 = N$ states if spin is neglected, and $2N$ states if spin is included. The number of wave vectors **k** per unit volume of **k**-space is

$$\nu(\mathbf{k}) = \frac{N}{\Omega'} = \frac{N\Omega}{\epsilon^3} = \frac{V}{\epsilon^3}, \qquad \text{(B.6.8)}$$

and are uniformly distributed.

A general wave vector $\mathbf{k}_1$ in **k**-space can always be represented in the first zone by adding to it a lattice vector $\mathbf{K}_\mathbf{m}$:

$$\mathbf{k} = \mathbf{k}_1 + \mathbf{K}_\mathbf{m}. \qquad \text{(B.6.9)}$$

The point of this procedure will be made clear later. For the moment it is sufficient to observe that by this method higher Brillouin zones can be mapped in to the first zone, as illustrated in fig. B.5.2.

Summarizing, the following vectors are of interest:

$$\mathbf{r} = \sum r_j \mathbf{a}_j \qquad (0 \leqslant r_j < N_j).$$

[General vector in the direct lattice.] (B.6.10)

$$\mathbf{R_n} = \sum n_j \mathbf{a}_j \qquad (n_j = 0, 1, \ldots N_j - 1).$$

[Lattice vector in the direct lattice.] (B.6.11)

$$\mathbf{k_t} = \sum \frac{t_j}{N_j} \mathbf{b}_j \qquad \left(t_j = -\frac{N_j - 1}{2}, -\frac{N_j - 3}{3}, \ldots, -1, 0, 1, \ldots \frac{N_j - 1}{2} \right).$$

[Reduced vector in the first zone.] (B.6.12)

$$\mathbf{K_m} = \sum m_j \mathbf{b}_j \qquad (m_j = 0, \pm 1, \pm 2, \ldots).$$

[Lattice vector in **k**-space.] (B.6.13)

Two sums

Consider now a reduced wave vector $\mathbf{k_t}$ and a direct lattice vector $\mathbf{R_n}$. Then

$$\frac{2\pi}{\epsilon} \mathbf{k_t} \cdot \mathbf{R_n} = \frac{2\pi}{\epsilon} \sum_{j=1}^{3} \left(\frac{t_j}{N_j} \mathbf{b}_j \right) \cdot \left(\sum_{l=1}^{3} n_l \mathbf{a}_l \right) = 2\pi \sum_{j=1}^{3} \frac{t_j n_j}{N_j}. \tag{B.6.14}$$

Writing

$$p_j \equiv \exp\left(\frac{2\pi i t_j}{N_j} \right), \qquad q_j \equiv \exp\left(\frac{2\pi i n_j}{N_j} \right), \tag{B.6.15}$$

one then has

$$\exp\left(\frac{2\pi}{\epsilon} i \mathbf{k_t} \cdot \mathbf{R_n} \right) = \prod_{j=1}^{3} p_j^{n_j} = \prod_{j=1}^{3} q_j^{t_j}. \tag{B.6.16}$$

If $p_j \neq 1$, $q_j \neq 1$, one can sum the following series:

$$\sum_{n_j=0}^{N_j - 1} p_j^{n_j} = \frac{1 - p_j^{N_j}}{1 - p_j} = 0; \tag{B.6.17}$$

$$\sum_{t_j = -\frac{1}{2}(N_j - 1)}^{\frac{1}{2}(N_j - 1)} q_j^{t_j} = q_j^{-\frac{1}{2}(N_j - 1)} (1 + q_j + \cdots + q_j^{N_j - 1}) = q_j^{-\frac{1}{2}(N_j - 1)} \frac{1 - q_j^{N_j}}{1 - q_j} = 0. \tag{B.6.18}$$

The first sum extends over all N vectors $\mathbf{R_n}$ in the fundamental domain V. The second sum extends over all N reduced wave vectors. Both sums are zero since, by (B.6.15), $p_j^{N_j} = q_j^{N_j} = 1$.

If t_j is a multiple of N_j then $p_j = 1$. But the sum (B.6.20), below, will still vanish unless t_j is a multiple of N_j for $j = 1, 2, 3$. Then $\mathbf{k_t}$ is a lattice vector in **k**-space and the sum (B.6.20) yields N. Similarly, if n_j is a multiple of N_j,

$q_j = 1$. Thus the sum (B.6.19) yields N if n_j is an integral multiple of N_j for $j = 1, 2, 3$. Summarizing,

$$\sum_{\mathbf{t}(\text{B.Z.})} \exp\left(\frac{2\pi}{\epsilon}\mathrm{i}\mathbf{k}_\mathbf{t} \cdot \mathbf{R}_\mathbf{n}\right) = \begin{cases} N \text{ (if } n_j = m_j N_j).^{(5)} \\ 0 \text{ (otherwise).} \end{cases} \tag{B.6.19}$$

$$\sum_{\mathbf{n}(V)} \exp\left(\frac{2\pi}{\epsilon}\mathrm{i}\mathbf{k}_\mathbf{t} \cdot \mathbf{R}_\mathbf{n}\right) = \begin{cases} N \text{ (if } \mathbf{k}_\mathbf{t} \text{ is a lattice vector in } \mathbf{k}\text{-space).} \\ 0 \text{ (otherwise).} \end{cases} \tag{B.6.20}$$

The range of the sums is illustrated by the symbols in brackets below the summation sign. (B.6.19) remains valid if $\mathbf{k}_\mathbf{t}$ be *any* **k**-space vector which is allowed to range over the N vectors of its zone. This follows from equation (B.2.11).

The most important results are covered in the following special cases:

$$\sum_{\mathbf{k}(\text{B.Z.})} \exp\left(\frac{2\pi\mathrm{i}}{\epsilon}\mathbf{k} \cdot \mathbf{R}\right) = N\delta_{\mathbf{R},0} \text{ (}\mathbf{R}\text{ in fundamental domain).} \tag{B.6.21}$$

$$\sum_{\mathbf{R}(V)} \exp\left(\frac{2\pi\mathrm{i}}{\epsilon}\mathbf{k} \cdot \mathbf{R}\right) = N\delta_{\mathbf{k},0} \text{ (}\mathbf{k}\text{ in first B.Z.).} \tag{B.6.22}$$

These sums prove to be very useful.

Two integrals

The following integrals are important. Writing

$$\mathbf{r} \equiv \mathbf{R}_\mathbf{n} + \boldsymbol{\rho},$$

and denoting the fundamental domain of N unit cells each of volume Ω by $V = N\Omega$,

$$I \equiv \int_V \exp\left(\frac{2\pi\mathrm{i}}{\epsilon}\mathbf{k} \cdot \mathbf{r}\right) \mathrm{d}\mathbf{r} = \sum_{\mathbf{n}(V)} \exp\left(\frac{2\pi\mathrm{i}}{\epsilon}\mathbf{k} \cdot \mathbf{R}_\mathbf{n}\right) \int_\Omega \exp\left(\frac{2\pi\mathrm{i}}{\epsilon}\mathbf{k} \cdot \boldsymbol{\rho}\right) \mathrm{d}\boldsymbol{\rho}.$$

This is zero by (B.6.20) if **k** is not a lattice vector in **k**-space. If $\mathbf{k} = \mathbf{K}_\mathbf{m}$ is a lattice vector in **k**-space, then

$$I = N \int_\Omega \exp\left(\frac{2\pi\mathrm{i}}{\epsilon}\mathbf{K}_\mathbf{m} \cdot \boldsymbol{\rho}\right) \mathrm{d}\boldsymbol{\rho}\, \delta_{\mathbf{k},\mathbf{K}_\mathbf{m}}. \tag{B.6.23}$$

We now deal with (B.6.23). Writing $\mathbf{K}_\mathbf{m} = \Sigma\, m_j \mathbf{b}_j$, $\boldsymbol{\rho} = \Sigma\, x_j \mathbf{a}_j$, I is clearly proportional to

$$\prod_{j=1}^{3}\left[\int_0^1 \exp(2\pi\mathrm{i}\, m_j x_j)\, \mathrm{d}x_j\right].$$

[5]This means that **R** must lie on lattice points in a direct lattice with lattice parameters $N_j a_j$.

(Let the constant of proportionality be B. Then B can be calculated at this stage, as explained below. But this is not needed.) The product is zero if at least one component of **m** is non-zero. This leaves as the only non-zero case the situation arising if $\mathbf{k} = 0$ in I. Hence, from (B.6.23) with $\mathbf{K_m} = 0$,

$$I = \int_V \exp\left(\frac{2\pi i}{\epsilon}\mathbf{k}\cdot\mathbf{r}\right) d\mathbf{r} = V\delta_{\mathbf{k},0} \qquad \text{(for all } \mathbf{k}\text{)}. \tag{B.6.24}$$

An incidental result, based on (B.6.23), is

$$\int_\Omega \exp\left(\frac{2\pi i}{\epsilon}\mathbf{K}\cdot\mathbf{r}\right) d\mathbf{r} = \Omega\delta_{\mathbf{K},0} \qquad (\mathbf{K} \text{ is a lattice vector in } \mathbf{k}\text{-space}). \tag{B.6.25}$$

The above argument shows that the constant of proportionality B is in fact $N\Omega = V$. This can also be shown directly by writing

$$\boldsymbol{\rho} = \sum_{j=1}^{3} x_j a_j = \sum_{j=1}^{3}\sum_{k=1}^{3} x_j a_{jk}\hat{\mathbf{X}}_k = \sum_{k=1}^{3} y_k \hat{\mathbf{X}}_k$$

where

$$y_k = \sum_{j=1}^{3} x_j a_{jk},$$

and where the $\hat{\mathbf{X}}_n$ are unit vectors along a *Cartesian* system of coordinates. In (B.6.23), $d\boldsymbol{\rho} = dy_1 dy_2 dy_3$. Now change the variables to the x_j. This involves the Jacobian

$$J = \begin{vmatrix} \partial y_1/\partial x_1 & \partial y_1/\partial x_2 & \partial y_1/\partial x_3 \\ \partial y_2/\partial x_1 & \partial y_2/\partial x_2 & \partial y_2/\partial x_3 \\ \partial y_3/\partial x_1 & \partial y_3/\partial x_2 & \partial y_3/\partial x_3 \end{vmatrix}$$

$$= \begin{vmatrix} a_{11} & a_{21} & a_{31} \\ a_{12} & a_{22} & a_{32} \\ a_{13} & a_{23} & a_{33} \end{vmatrix}$$

$$= \mathbf{a}_1 \cdot (\mathbf{a}_2 \times \mathbf{a}_3) = \Omega.$$

The integral (B.6.23) becomes

$$I = NJ \prod_{j=1}^{3} \int_0^1 \exp\left(\frac{2\pi i}{\epsilon} m_j x_j\right) dx_j,$$

showing that B is in fact V.

The potential box problem

It is informative to change the boundary conditions to an infinitely high potential at the walls of the fundamental domain. This domain then becomes a box with a perfectly reflecting wall. It is convenient to suppose the interior of the box to be field-free and to choose the energy zero so that it coincides with the constant potential energy inside the box. The time-independent Schrödinger equation is then

$$\frac{-\hbar^2}{2m}\nabla^2\psi(\mathbf{r}) = E\psi(\mathbf{r}). \tag{B.6.26}$$

For a rectangular box of sides $N_j a_j$ $(j = 1, 2, 3)$ the solution is

$$\psi_{\mathbf{k}}(\mathbf{r}) = C\prod_{j=1}^{3}\sin(k_j x_j + \alpha_j) \tag{B.6.27}$$

where the α_j's are constants. Substitution of (B.6.27) into (B.6.26) yields a result also appropriate to the free-electron wave function (B.6.3):

$$\frac{\hbar^2 k^2}{2m} = E. \tag{B.6.28}$$

If any one x_j lies at a wall, $\psi_{\mathbf{k}}$ must vanish. Thus

$$x_j = -\tfrac{1}{2}N_j a_j \Rightarrow -\frac{k_j N_j a_j}{2} + \alpha_j = n_j\pi \qquad (j = 1, 2, 3),$$

where the n_j's are integers. This yields an identification of the α_j's. The simplest choice is $n_1 = n_2 = n_3 = 0$, and this will be assumed. At the opposite walls $\psi_{\mathbf{k}}$ must also vanish, i.e.

$$\sin\left(\frac{k_j N_j a_j}{2} + \frac{k_j N_j a_j}{2}\right) = 0.$$

It follows that there must exist integers t_j such that

$$k_j = \frac{t_j\pi}{N_j a_j} \qquad (t_j = 1, 2, 3, \ldots). \tag{B.6.29}$$

If any one of the t_j is zero one finds $\psi_{\mathbf{k}} = 0$ by (B.6.27), and this is an inadmissible solution. On the other hand, if one uses periodic boundary conditions, the plane wave (B.6.3) must be used, and a value $t_j = 0$ is quite admissible.

There is also a difference for negative integers. In (B.6.29) they do not lead to linearly independent wave functions and hence they need not be considered when counting the number of independent one-electron states. In (B.6.4) or (B.6.6) they do lead to linearly independent wave functions.

This point is illustrated by requiring for all x_j and constants α, β

$$\alpha \sin\left[k_j\left(x_j + \frac{N_j a_j}{2}\right)\right] + \beta \sin\left[-k_j\left(x_j + \frac{N_j a_j}{2}\right)\right] = 0.$$

This is of the form $\alpha \sin\theta - \beta \sin\theta = 0$ and can be ensured by any choice $\alpha = \beta \neq 0$. Hence these two functions are dependent. On the other hand if one requires that for all $\mathbf{r}$

$$\alpha \exp(\mathrm{i}\mathbf{k}\cdot\mathbf{r}) + \beta \exp(-\mathrm{i}\mathbf{k}\cdot\mathbf{r}) = 0$$

it follows that $\alpha = \beta = 0$. Linear independence occurs only in this second case.

The formula (B.6.29) should be compared with the result obtained with periodic boundary conditions, equation (B.6.4). The lengths $N_j a_j$ of the sides of the box become the lengths of the sides of the fundamental parallelepiped of volume $V = N_1 N_2 N_3 a_1 a_2 a_3$. In addition, the sides must be taken to be mutually perpendicular in (B.6.4) in order to obtain a comparable geometry. One then finds with $\epsilon = 2\pi$ and $j = 1, 2, 3$:

$$\left.\begin{array}{ll} \text{Box:} & k_j = \dfrac{\pi}{N_j a_j}, \quad \dfrac{2\pi}{N_j a_j}, \quad \dfrac{3\pi}{N_j a_j}, \ldots, \quad \dfrac{(N_j - 1)\pi}{N_j a_j}, \quad \dfrac{\pi}{a_j}; \\ \text{Periodic boundary condition:} & k_j = 0, \quad \pm\dfrac{2\pi}{N_j a_j}, \quad \pm\dfrac{4\pi}{N_j a_j}, \ldots, \quad \pm\dfrac{(N_j - 1)\pi}{N_j a_j}. \end{array}\right\} \quad \text{(B.6.30)}$$

We have given only the k-values leading to linearly independent wavefunctions; they are equal in number for the two cases. The density of states per unit volume of $\mathbf{k}$-space is therefore the same in both cases: in the box problem, there are N_j linearly independent wavefunctions for $-\pi/a_j < k_j \leqslant \pi/a_j$, i.e. for a range of k_j equal to $2\pi/a_j$. The number of states per unit volume of $\mathbf{k}$-space is therefore

$$\prod_{j=1}^{3} \frac{N_j}{2\pi/a_j} = \frac{V}{(2\pi)^3},$$

as in (B.6.8).

Problems

(B.6.1) Let $\mathbf{k}$ be a vector in the first Brillouin zone; let $\mathbf{K}$ be a lattice vector in $\mathbf{k}$-space, and let $\mathbf{k}' = \mathbf{k} + \mathbf{K}$. Prove that:

(a) given $\mathbf{k}'$ is also in the first zone, $\mathbf{K} = 0$;

(b) given $\mathbf{k}'$ lies outside the first zone, $\mathbf{K} \neq 0$.

(B.6.2) The volume of an r-dimensional hyperellipsoid of semi-axes $a_1, a_2, \ldots, a_r$ is

$$\frac{\pi^{r/2} V}{(r/2)\Gamma(r/2)}$$

where $V \equiv a_1 a_2 \cdots a_r$, and $\Gamma(x)$ is the Gamma Function. With the aid of this formula obtain an expression for the density of states of a particle of mass m in a field-free r-dimensional box of sides $D_1, D_2, \ldots, D_r$. The potential at the walls of the box is infinite, and the r-dimensional Schrödinger equation is

$$\sum_{j=1}^{r} \frac{\partial^2 \psi}{\partial x_j^2} + \frac{8\pi^2 m}{h} E\psi = 0.$$

Solution

To satisfy the boundary conditions, $\psi = \prod_{j=1}^{r} \sin\left(\frac{\pi l_j x}{D_j}\right)$, where the l's are integers. In a Cartesian space with coordinates $l_1, l_2, \ldots, l_r$, each point represents a state. To count only linearly independent wave functions, the number of states N_0 with energies between 0 and E_0 is given by the number of points in a fraction 2^{-r} of a hyperellipsoid. Since

$$\sum \left(\frac{l_i}{D_j}\right)^2 = \frac{8mE}{h^2} \leqslant \frac{8mE_0}{h^2},$$

its semi-axes a_j satisfy

$$\left(\frac{a_j}{D_j}\right)^2 = \frac{8mE_0}{h^2}.$$

Hence

$$N_0 = \frac{V}{(r/2)\Gamma(r/2)} \left(\frac{2\pi m}{h^2}\right)^{r/2} E_0^{r/2}$$

and the density of states is

$$N_r(E) = \frac{V}{\Gamma(r/2)} \left(\frac{2\pi m}{h^2}\right)^{r/2} E^{r/2-1}.$$

If L and A are the one- and two-dimensional 'volumes,'

$$N_1(E) = L\sqrt{(2m)}/h\sqrt{E}$$

$$N_2(E) = 2\pi m\, A/h^2$$

$$N_3(E) = (2V/\sqrt{\pi})(2\pi m/h^2)^{3/2}\sqrt{E}.$$

B.7 FOURIER EXPANSIONS

It is now desirable to adopt definitely the convention $\epsilon = 2\pi$. It will be obvious that the general normalisation in terms of ϵ can be carried right through the theory, but that it leads to unnecessary complications.

It is convenient to state in summary form the main equations of this section:

$$\mathrm{f}(\mathbf{r}) = \mathrm{f}\left(\mathbf{r} + \sum_{j=1}^{3} N_j \mathbf{a}_j\right) \quad : \text{f } \textit{is periodic in the fundamental domain:}$$

$$\mathrm{f}(\mathbf{r}) = V^{-\frac{1}{2}} \sum_{\text{all } \mathbf{k}} \mathrm{f}_{\mathbf{k}}\, e^{\pm i\mathbf{k}\cdot\mathbf{r}}; \qquad \mathrm{f}_{\mathbf{k}} = V^{-\frac{1}{2}} \int_V \mathrm{f}(\mathbf{r})\, e^{\mp i\mathbf{k}\cdot\mathbf{r}}\, d\mathbf{r}. \qquad \text{(B.7.1a, b)}$$

$$\mathrm{f}(\mathbf{r}) = \mathrm{f}(\mathbf{r} + \mathbf{R}_{\mathbf{n}}) : \text{f } \textit{is periodic in the direct lattice:}$$

$$\mathrm{f}(\mathbf{r}) = \Omega^{-\frac{1}{2}} \sum_{\mathbf{K}} \mathrm{f}_{\mathbf{K}}\, e^{\pm i\mathbf{K}\cdot\mathbf{r}}; \qquad \mathrm{f}_{\mathbf{K}} = \Omega^{-\frac{1}{2}} \int_\Omega \mathrm{f}(\mathbf{r})\, e^{\mp i\mathbf{K}\cdot\mathbf{r}}\, d\mathbf{r}. \qquad \text{(B.7.2a, b)}$$

(**K** is here a lattice vector in the reciprocal lattice.)

f *is defined only at all lattice points R in the fundamental domain:*

$$\mathrm{f}(\mathbf{R}) = V^{-\frac{1}{2}} \sum_{\mathbf{k}(\text{B.Z.})} \mathrm{f}_{\mathbf{k}}\, e^{\pm i\mathbf{k}.\mathbf{R}}; \qquad \mathrm{f}_{\mathbf{k}} = (V^{\frac{1}{2}}/N) \sum_{\mathbf{R}(V)} \mathrm{f}(\mathbf{R})\, e^{\mp i\mathbf{k}.\mathbf{R}}. \qquad \text{(B.7.3a, b)}$$

f *is periodic in the* **k**-*space lattice:*

$$\mathrm{f}(\mathbf{k}) = (N)^{-\frac{1}{2}} \sum_{\mathbf{R}(V)} \mathrm{f}_{\mathbf{R}}\, e^{\mp i\mathbf{k}\cdot\mathbf{R}}; \qquad \mathrm{f}_{\mathbf{R}} = (N)^{-\frac{1}{2}} \sum_{k(\text{B.Z.})} \mathrm{f}(\mathbf{k})\, e^{\pm i\mathbf{k}\cdot\mathbf{R}}. \qquad \text{(B.7.4a, b)}$$

Note that (B.7.1a) is a Fourier expansion, the wave vectors forming a discrete set because of the periodic boundary condition. Either sign is allowed in the exponent, but the meaning of $\mathrm{f}_{\mathbf{k}}$ is changed by the change of sign. (B.7.1b) follows from (B.6.24). Lattice periodicity means that $\mathrm{f}_{\mathbf{k}} = 0$ unless **k** is a lattice vector. This follows from the first proof[6] of the theorem stated at the end of this section, and leads to (B.7.2a). (B.7.2b) follows by (B.6.25). To obtain (B.7.3a), use (B.7.1a) and put $\mathbf{k} = \mathbf{K} + \mathbf{k}_0$, where $\mathbf{k}_0$ is in the first Brillouin zone. Since $e^{i\mathbf{k}\cdot\mathbf{R}} = e^{i\mathbf{k}_0\cdot\mathbf{R}}$, one can collect together all the terms with the same $\mathbf{k}_0$. The result is that the remaining sum is over $\mathbf{k}_0$ and may be confined to the first Brillouin zone. One can use (B.6.22) to derive (B.7.3b). In (B.7.4a) one need not go outside the fundamental

[6] If the second proof were used, the argument would be circular.

domain, since were one to do so, one would have terms in the exponent of the type

$$\mathrm{i}\left[\mathbf{k}\cdot\left(\mathbf{R}+\sum s_jN_j\mathbf{a}_j\right)\right]=\mathrm{i}\left[\mathbf{k}\cdot\mathbf{R}+2\pi\sum_j\sum_l\frac{t_l}{N_l}\mathbf{b}_l\cdot s_jN_j\mathbf{a}_j\right]$$

$$=\mathrm{i}\left[\mathbf{k}\cdot\mathbf{R}+2\pi\sum s_jt_j\right],\quad \text{where the } s_j \text{ are integers.}$$

This shows that one can collect together in a single term all terms arising from various values of s_j in

$$\mathbf{R}=\mathbf{R}_1+\sum s_jN_j\mathbf{a}_j,$$

where $\mathbf{R}_i$ is in the fundamental domain. (B.7.4b) is obtained by using (B.6.21). Note that in the first of each pair of equations (B.7.1)–(B.7.4) the coefficients $V^{-\frac{1}{2}}$, $\Omega^{-\frac{1}{2}}$, $N^{-\frac{1}{2}}$ are part of the definition of the Fourier coefficients.

The similarity between (B.7.3) and (B.7.4) may be understood as follows. If f($\mathbf{R}$) is defined only at lattice points in V, the corresponding $f_{\mathbf{k}}$ need to be defined only for reduced wave vectors in Ω. One is therefore free to complete the definition of $f_{\mathbf{k}}$ so as to make it periodic in the $\mathbf{k}$-space lattice. If one does so one passes from the relation (B.7.3b) to the almost identical relation (B.7.4a). It will be seen later that in a band n the energy $E_n(\mathbf{k})$ is a function of $\mathbf{k}$ in the first zone only. This is therefore a function of the type (B.7.3b) and it may therefore be permissible to pass to (B.7.4a):

$$E_n(\mathbf{k})=E_n(\mathbf{k}+\mathbf{K}) \tag{B.7.5}$$

by extending the definition of the function $E_n(\mathbf{k})$.

Observe next that if the $\phi_n(\mathbf{r})$ form an orthonormal set, i.e. if

$$\int \phi_i^*(\mathbf{r})\phi_j(\mathbf{r})\,\mathrm{d}\mathbf{r}=\delta_{ij},$$

then one may write for functions in this space

$$f(\mathbf{r})=\sum_n c_n\phi_n(\mathbf{r}),\qquad c_m=\int f(\mathbf{r}')\phi_m^*(\mathbf{r}')\,\mathrm{d}\mathbf{r}'.$$

It follows that

$$f(\mathbf{r})=\int f(\mathbf{r}')\left[\sum_n \phi_n^*(\mathbf{r}')\phi_n(\mathbf{r})\right]\mathrm{d}\mathbf{r}'.$$

Hence, using the Dirac delta function,

$$\sum_n \phi_n^*(\mathbf{r}')\phi_n(\mathbf{r})=\delta(\mathbf{r}'-\mathbf{r}). \tag{B.7.6}$$

This is the completeness relation. It implies, in the case of plane waves,

$$\sum_{\text{all }\mathbf{k}} \mathrm{e}^{\mathrm{i}\mathbf{k}\cdot\mathbf{r}} = V\,\delta(\mathbf{r}). \tag{B.7.7}$$

Vector functions with the periodicity of the fundamental domain

For such functions the Fourier coefficients are themselves vectors:

$$\mathbf{V}(\mathbf{r}) = V^{-\frac{1}{2}} \sum_{\text{all }\mathbf{k}} \mathbf{V}_{\mathbf{k}}\,\mathrm{e}^{\mathrm{i}\mathbf{k}.\mathbf{r}}. \tag{B.7.8}$$

These coefficients can be decomposed into components parallel and perpendicular to $\mathbf{k}$. They are called longitudinal and transversal parts:

$$\mathbf{V}(\mathbf{r}) = \mathbf{V}_l(\mathbf{r}) + \mathbf{V}_t(\mathbf{r}) = V^{-\frac{1}{2}} \sum_{\text{all }\mathbf{k}} [\mathbf{V}_{l,\mathbf{k}} + \mathbf{V}_{t,\mathbf{k}}]\,\mathrm{e}^{\mathrm{i}\mathbf{k}.\mathbf{r}}. \tag{B.7.9}$$

It follows that if $\mathbf{V}_{t,\mathbf{k}} \perp \mathbf{k}$ and $\mathbf{V}_{l,\mathbf{k}} \parallel \mathbf{k}$ for all $\mathbf{k}$,

$$\operatorname{div} \mathbf{V}_t(\mathbf{r}) = \mathrm{i}V^{-\frac{1}{2}} \sum_{\text{all }\mathbf{k}} \mathbf{k}\cdot\mathbf{V}_{t,\mathbf{k}}\,\mathrm{e}^{\mathrm{i}\mathbf{k}.\mathbf{r}} = 0, \tag{B.7.10}$$

$$\operatorname{curl} \mathbf{V}_l(\mathbf{r}) = \mathrm{i}V^{-\frac{1}{2}} \sum_{\text{all }\mathbf{k}} \mathbf{k}\times\mathbf{V}_{l,\mathbf{k}}\,\mathrm{e}^{\mathrm{i}\mathbf{k}.\mathbf{r}} = 0. \tag{B.7.11}$$

Fourier integrals

If one goes to the limit $N_j \to \infty$, i.e. $V \to \infty$, the distribution of wave vectors becomes continuous and factors such as $\delta_{\mathbf{k},\mathbf{k}'}$ lose their meaning. To replace them by analogues, let $\gamma(\mathbf{k}')\Delta\mathbf{k}'$ be the number of such vectors in a small range $\Delta\mathbf{k}'$. The contribution of the value of some function $\mathrm{g}(\mathbf{k}')$ from this range may be written

$$\mathrm{g}(\mathbf{k}')\delta_{\mathbf{k},\mathbf{k}'} \to [\nu(\mathbf{k}')\Delta\mathbf{k}'][\mathrm{g}(\mathbf{k}')\delta_{\mathbf{k},\mathbf{k}'}],$$

where the number of wave vectors $\nu(\mathbf{k}')\Delta\mathbf{k}'$ in this small range is unity in the discrete case. The arrow indicates a preparation for the passage to the limit or an actual passage to the limit. It is more convenient to write this as

$$\mathrm{g}(\mathbf{k}')\delta_{\mathbf{k},\mathbf{k}'} \to \{\nu(\mathbf{k}')\delta_{\mathbf{k},\mathbf{k}'}\}\mathrm{g}(\mathbf{k}')\Delta\mathbf{k}'. \tag{B.7.12}$$

If $V \to \infty$, the expression in the brace diverges by (B.6.8) when $\mathbf{k} = \mathbf{k}'$. This suggests that one can put in the limit

$$\nu(\mathbf{k}')\delta_{\mathbf{k},\mathbf{k}'} \to A\delta(\mathbf{k} - \mathbf{k}'), \tag{B.7.13}$$

where A is to be determined and the Dirac delta function has been used. Hence (B.7.12) becomes

$$\mathrm{g}(\mathbf{k}')\delta_{\mathbf{k},\mathbf{k}'} \to A\delta(\mathbf{k} - \mathbf{k}')\mathrm{g}(\mathbf{k}')\,\mathrm{d}\mathbf{k}'. \tag{B.7.14}$$

To find A, sum the left hand side over all $\mathbf{k}'$, and integrate the right hand side over $\mathbf{k}'$ to find

$$g(\mathbf{k}) \to A g(\mathbf{k}).$$

Thus $A = 1$ in (B.7.13) and (B.7.14). Thus, by (B.6.8) and (B.7.14), (B.7.13) shows that in the limit $V \to \infty$ the replacement

$$\frac{V}{8\pi^3}\delta_{\mathbf{k},\mathbf{k}'} \to \delta(\mathbf{k} - \mathbf{k}') \tag{B.7.15}$$

is required. Note that both sides of (B.7.15) are dimensionally volumes.

Note also that the following replacements are needed:

$$V^{-1}\sum_{\mathbf{k}} g(\mathbf{k}) = V^{-1}\sum_{\Delta\mathbf{k}} v(\mathbf{k})g(\mathbf{k})\Delta\mathbf{k} \to V^{-1}\int v(\mathbf{k})g(\mathbf{k})\,d\mathbf{k}$$

$$= \frac{1}{8\pi^3}\int g(\mathbf{k})\,d\mathbf{k}. \tag{B.7.16}$$

In the case of (B.7.1a), for example, one has

$$V^{-1}\sum f_{\mathbf{k}}\,e^{i\mathbf{k}\cdot\mathbf{r}} \to \frac{1}{8\pi^3}\int f_{\mathbf{k}}\,e^{i\mathbf{k}\cdot\mathbf{r}}\,d\mathbf{k}. \tag{B.7.17}$$

The results valid for an infinite crystal may, of course, be established directly. The above procedure has, however, the following advantages: (a) it covers the important case of a finite crystal subject to periodic boundary conditions and (b) proofs of certain results such as the sums (B.6.19), (B.6.20) and the integral (B.6.24) are very simple.

A theorem

If $f(\mathbf{r})$ has the periodicity of the direct lattice, $\mathbf{k}$ is a general wave vector, and periodic boundary conditions have been imposed for a volume V of N unit cells of volume Ω,

$$I \equiv \int_V e^{i\mathbf{k}\cdot\mathbf{r}}\,f(\mathbf{r})\,d\mathbf{r} = N\sum_{\mathbf{m}}\left[\delta_{\mathbf{k},\mathbf{K_m}}\left(\int_\Omega e^{i\mathbf{K_m}\cdot\mathbf{r}}\,f(\mathbf{r})\,d\mathbf{r}\right)\right] \tag{B.7.18}$$

First proof (direct integration)

Let $\mathbf{r} = \mathbf{R} + \boldsymbol{\rho}$, where $\mathbf{R}$ goes to the origin of a unit cell, these origins being in standard positions for all unit cells in V. $\boldsymbol{\rho}$ ranges over the unit cell $\mathbf{R}$.

Then

$$I = \sum_{\substack{\mathbf{R} \\ \text{(all lattice} \\ \text{vectors in } V)}} \int_\Omega e^{i\mathbf{k}.(\mathbf{R}+\boldsymbol{\rho})} f(\boldsymbol{\rho})\, d\boldsymbol{\rho}$$

$$= \left[\sum_{\substack{\mathbf{R} \\ (V)}} e^{i\mathbf{k}.\mathbf{R}}\right]\left[\int_\Omega e^{i\mathbf{k}.\boldsymbol{\rho}} f(\boldsymbol{\rho})\, d\boldsymbol{\rho}\right].$$

The first sum vanishes by (B.6.20) unless **k** is a lattice vector in *k*-space. If it is such a vector the sum is N. So the sum is $N \sum_m \delta_{\mathbf{k},K_\mathbf{m}}$

Second proof, using the Fourier expansion of f(**r**)

By (B.7.2a),

$$I = \Omega^{-\frac{1}{2}} \sum_{\mathbf{K}} f_{\mathbf{K}} \int_V e^{i(\mathbf{k}+\mathbf{K}).\mathbf{r}}\, d\mathbf{r}$$

$$= \Omega^{-\frac{1}{2}} \sum_{\mathbf{K}} f_{\mathbf{K}} V\, \delta_{\mathbf{k},-\mathbf{K}}$$

by (B.6.24). Thus $I = 0$ if **k** is not a lattice vector in **k**-space. If **k** is such a vector, $\mathbf{k} = -\mathbf{K}$ say,

$$I = N\Omega^{\frac{1}{2}} f_{-\mathbf{K}} = N\Omega^{\frac{1}{2}} \cdot \frac{1}{\Omega^{\frac{1}{2}}} \int_\Omega f(\mathbf{r})\, e^{i\mathbf{K}.\mathbf{r}}\, d\mathbf{r}$$

by (B.7.2b). This is again (B.7.18).

If **k** is a reduced wave vector, i.e. in the first Brillouin zone, (B.7.18) becomes

$$\int_V e^{i\mathbf{k}\cdot\mathbf{r}} f(\mathbf{r})\, d\mathbf{r} = N\delta_{\mathbf{k},0}\left[\int_\Omega f(\mathbf{r})\, d\mathbf{r}\right]. \qquad \text{(B.7.19)}$$

If now $V \to \infty$, then, using (B.7.15),

$$\int_{\text{all space}} e^{i\mathbf{k}\cdot\mathbf{r}} f(\mathbf{r})\, d\mathbf{r} = \frac{8\pi^3}{\Omega}\, \delta(\mathbf{k})\left[\int_\Omega f(\mathbf{r})\, d\mathbf{r}\right]. \qquad \text{(B.7.20)}$$

B.8 SOME RESULTS FROM PERTURBATION THEORY

B.8a Simple theory[7]

Consider an unperturbed Hamiltonian with discrete eigenstates ψ_j ($j = 1, 2, \ldots$). If a small perturbation U is applied, the mth eigenvalue E_m

[7] The results of § B.8a are needed in § B.10.

is amended to second order in the perturbation to[8]

$$W_m = E_m + U_{mm} + \sum_{n(\neq m)} \frac{|U_{mn}|^2}{E_m - E_n}. \tag{B.8.1}$$

Similarly, the mth eigenstate becomes, to first order in the perturbation,

$$\phi_m = \psi_m + \sum_{n(\neq m)} \frac{U_{nm}}{E_m - E_n} \psi_n . \tag{B.8.2}$$

It is assumed here that at least ψ_m is a non-degenerate state.

B.8b Two-state approximation[9]

The above well-known theory requires amendment if two eigenstates, ψ_u and ψ_v say, can make a dominant contribution to the perturbed wave function ϕ. In that case one expresses ϕ as a superposition of the ψ_j's (assuming these to be a complete orthonormal set) and substitutes in the Schrödinger equation

$$H\phi \equiv (H_0 + U)\phi = E\phi, \tag{B.8.3}$$

$$\phi = \sum a_j \psi_j . \tag{B.8.4}$$

One finds

$$\sum_j (E_j - E) a_j \psi_j + \sum_j a_j U \psi_j = 0,$$

so that

$$(E_k - E) a_k + \sum_j a_j U_{kj} = 0. \tag{B.8.5}$$

If only a_u and a_v are considerable, one finds

$$(E_u - E + U_{uu}) a_u + U_{uv} a_v = 0. \tag{B.8.6}$$

$$U_{vu} a_u + (E_v - E + U_{vv}) a_v = 0. \tag{B.8.7}$$

For a non-zero solution the determinant of the coefficients must vanish. This yields a quadratic equation in E whose roots (E^+, E^-) satisfy the three equivalent relations

$$E^{\pm} = \tfrac{1}{2}(E_u + E_v + U_{uu} + U_{vv}) \pm \tfrac{1}{2}[(E_u - E_v + U_{uu} - U_{vv})^2 + 4|U_{uv}|^2]^{\frac{1}{2}}, \tag{B.8.8a}$$

[8] See any book on quantum mechanics.
[9] The results of § B.8b are needed in § B.9b.

$$E^{\pm} - E_v - U_{vv} = |U_{uv}| \frac{C \pm 1}{\sqrt{(1 - C^2)}}, \tag{B.8.8b}$$

$$E^{\pm} - E_u - U_{uu} = |U_{uv}| \frac{-C \pm 1}{\sqrt{(1 - C^2)}}, \tag{B.8.8c}$$

where

$$C \equiv \frac{E_u + U_{uu} - E_v - U_{vv}}{[(E_u + U_{uu} - E_v - U_{vv})^2 + 4|U_{uv}|^2]^{\frac{1}{2}}}. \tag{B.8.9}$$

One obtains the ratio a_v/a_u by putting (B.8.8) back into (B.8.6) or (B.8.7). A second equation for these coefficients is obtained by noting that the approximate expression

$$\phi = a_u\psi_u + a_v\psi_v \tag{B.8.10}$$

must be normalized:

$$|a_u|^2 + |a_v|^2 = 1. \tag{B.8.11}$$

The first step yields

$$\left(\frac{a_v}{a_u}\right)_{\pm} = \pm\frac{|U_{uv}|}{U_{uv}}\sqrt{\left(\frac{1 \mp C}{1 \pm C}\right)}; \tag{B.8.12}$$

the second step yields

$$|a_u|^2_{\pm} = \tfrac{1}{2}(1 \pm C); \qquad |a_v|^2_{\pm} = \tfrac{1}{2}(1 \mp C),$$

where $(a_u)_+$ and $(a_v)_+$ correspond to the eigenvalue E^+, and $(a_u)_-$ and $(a_v)_-$ correspond to the eigenvalue E^-.

Introducing arbitrary phase factors, these results show that the wave functions are given by

$$\sqrt{2}\phi^+ = e^{i\theta_u}(1 + C)^{\frac{1}{2}}\psi_u + e^{i\theta_v}(1 - C)^{\frac{1}{2}}\psi_v \to \sqrt{2}\, e^{i\theta_u}\psi_u, \tag{B.8.13}$$

$$\sqrt{2}\phi^- = e^{i\tau_u}(1 - C)^{\frac{1}{2}}\psi_u + e^{i\tau_v}(1 + C)^{\frac{1}{2}}\psi_v \to \sqrt{2}\, e^{i\tau_v}\psi_v. \tag{B.8.14}$$

The two limiting forms are obtained in the absence of a perturbation:

$$U \to 0, \quad \text{i.e.} \quad C \to 1.$$

If we make the convention $E_u > E_v$, the wavefunction for the upper state must converge to ψ_u and that for the lower state to ψ_v. Thus $\theta_u = \tau_v = 0$. Comparison of (B.8.13) and (B.8.14) with (B.8.12) now shows that

$$U_{uv}/|U_{uv}| = e^{-i\theta_v} = -e^{i\tau_u}. \tag{B.8.15}$$

Hence, if θ be an arbitrary phase factor, we have finally:

$$\sqrt{2}\phi^{+} = \sqrt{(1 + C)}\psi_u + e^{i\theta}\sqrt{(1 - C)}\psi_v, \tag{B.8.16}$$

$$\sqrt{2}\phi^{-} = -e^{-i\theta}\sqrt{(1 - C)}\psi_u + \sqrt{(1 + C)}\psi_v. \tag{B.8.17}$$

Equations (B.8.8), (B.8.16) and (B.8.17) represent the solution of this perturbation problem.

B.8c The relation between the perturbed and the unperturbed wave functions in the two-state theory[10]

Since ψ_u and ψ_v are assumed orthonormal, the scalar product $(\phi^+, \phi^-) = 0$. But the choice $\theta = 0$ is convenient. Making this choice, one can write (B.8.16) and (B.8.17) as a single matrix equation

$$\phi = A\psi, \qquad \text{where } \phi \equiv \begin{pmatrix} \phi^+ \\ \phi^- \end{pmatrix}, \quad \psi \equiv \begin{pmatrix} \psi_u \\ \psi_v \end{pmatrix} \tag{B.8.18}$$

and

$$A \equiv \frac{1}{\sqrt{2}} \begin{bmatrix} \sqrt{(1 + C)} & \sqrt{(1 - C)} \\ \sqrt{(1 - C)} & \sqrt{(1 + C)} \end{bmatrix}. \tag{B.8.19}$$

Matrices which, upon acting on an orthonormal set of vectors, yield another orthonormal set are called unitary. They may also be specified by the equation

$$A^{-1} = A^{+} \quad \text{or} \quad AA^{+} = I = \begin{pmatrix} 1 & 0 \\ 0 & 1 \end{pmatrix}, \tag{B.8.20}$$

where I is the identity operator. Introducing also the so-called Pauli spin operator

$$S_2 = \begin{pmatrix} 0 & -i \\ i & 0 \end{pmatrix} \tag{B.8.21}$$

it is seen that

$$A = [\cos\alpha]I + [i\sin\alpha]S_2 = \exp(iS_2\alpha), \tag{B.8.22}$$

[10] § B.8c and § B.8d may be omitted. They show how spin matrices can be introduced non-relativistically into the Hamiltonian. The corresponding relativistic theory is of current interest. For a recent publication with references to earlier work, see M. Suffczynski, *Phys. Letters*, **26A**, 325 (1968).

where

$$\cos \alpha \equiv \sqrt{(1 + C)/2}. \qquad \text{(B.8.23)}$$

The exponential function is defined by the exponential series, noting that $S_2{}^2 = I$.

One may think of the perturbation U as inducing a rotation through an angle α of ψ_u and ψ_v in their common plane, giving rise to the new eigenfunctions ϕ^+, ϕ^-. The following special cases are of interest:

(i) if $U = 0$, then $C = 1, \alpha = 0$; (B.8.24)

(ii) if $E_u + U_{uu} = E_v + U_{vv}$, then $C = 0$, $\alpha = \pi/4$. (B.8.25)

B.8d Alternative forms of the wave equation in the two-state theory

Equation (B.8.3) has the form $H\phi^{\pm} = E^{\pm}\phi^{\pm}$. These two equations imply

$$H\phi = (aI + bS_3)\phi, \qquad \text{(B.8.26)}$$

where

$$S_3 \equiv \begin{pmatrix} 1 & 0 \\ 0 & -1 \end{pmatrix},$$

$$a \equiv \tfrac{1}{2}(E_u + E_v + U_{uu} + U_{vv}),$$

$$b \equiv \tfrac{1}{2}[(E_u + U_{uu} - E_v - U_{vv})^2 + 4|U_{uv}|^2]^{\frac{1}{2}}. \qquad \text{(B.8.27)}$$

Multiplying (B.8.26) by $I - (b/a)S_3$ on the left, one finds

$$H'\phi = E'\phi, \qquad \text{(B.8.28)}$$

where

$$H' \equiv H - \sqrt{(1 - E'/a)}S_3 H, \qquad E' \equiv [1 - (b/a)^2]a. \qquad \text{(B.8.29)}$$

(B.8.28) is a two-component equation for the eigenfunctions ϕ and eigenvalue E', and involves a Hamiltonian H' which depends itself on E'. It is somewhat analogous to the equation for the two large components of the relativistic Dirac equation[11] which is given by (B.8.28) with

$$H' = \frac{1}{2m}\boldsymbol{\sigma}\cdot\mathbf{p}\left(1 + \frac{E' - V}{2mc^2}\right)^{-1}\boldsymbol{\sigma}\cdot\mathbf{p} + V.$$

[11] P. A. M. Dirac, *Proc. Roy. Soc.*, **117**, 610; *Proc. Roy. Soc.*, **118**, 341 (1928); E. Corinaldesi and F. Strocchi, *Relativistic Wave Mechanics*, North Holland, Amsterdam, 1963, p. 190.

Problems

(The following problems illustrate the important method of matrix diagonalisation by unitary transformations.

(B.8.1) Adding $S_1 \equiv \begin{pmatrix} 0 & 1 \\ 1 & 0 \end{pmatrix}$ and $S_3 \equiv \begin{pmatrix} 1 & 0 \\ 0 & -1 \end{pmatrix}$ to S_2 as defined, verify that $S_1^2 = S_3^2 = I$ and $S_1 S_2 = \mathrm{i}S_3 = -S_2 S_1$ together with relations obtained by cyclic permutation of the suffices.
[These are the **Pauli spin matrices**.]

(B.8.2) A Hermitian two-by-two matrix

$$H = \begin{pmatrix} a & be^{-\mathrm{i}\theta} \\ be^{\mathrm{i}\theta} & c \end{pmatrix} \qquad (a \neq c)$$

undergoes the unitary transformation

$$H' = A^{-1}HA,$$

where $A \equiv \exp(\mathrm{i}\alpha S_2)$.
Show that

$$H' = \tfrac{1}{2}(a + c)I + [\tfrac{1}{2}(a - c)\cos 2\alpha - b\cos\theta \sin 2\alpha]S_3 \\ +[b \sin\theta]S_2 + [b\cos\theta\cos 2\alpha + \tfrac{1}{2}(a - c)\sin 2\alpha]S_1 .$$

[Write $H = \frac{1}{2}(a + c)I + [b\cos\theta]S_1 + [b\sin\theta]S_2 + \frac{1}{2}(a - c)S_3$ and note that $A^{-1}S_1A = [\cos 2\alpha]S_1 - [\sin 2\alpha]S_3$

$$A^{-1}S_3A = [\cos 2\alpha]S_3 + [\sin 2\alpha]S_1 .]$$

(B.8.3) Show that the sum of the diagonal elements (the 'trace') of any matrix $M' \equiv A^{-1}MA$ is equal to the trace of M, where A is any matrix which possesses a reciprocal. Confirm this property for the matrix H' of problem (B.8.2).

(B.8.4) If the matrix H of problem (B.8.2) undergoes the unitary transformation $H' = B^{-1}HB$, where $B = \exp \mathrm{i}\beta S_3$, show that

$$H' = \tfrac{1}{2}(a + c)I + \tfrac{1}{2}(a - c)S_3 + bS_1$$

if β is chosen as $\theta/2$.

(B.8.5) If the unitary transformations B and A of the above problems are applied successively to the matrix H, show that the result of problem (B.8.2) may be used with $\theta = 0$ to yield

$$H' = \tfrac{1}{2}(a + c)I + [\tfrac{1}{2}(a - c)\cos 2\alpha - b\sin 2\alpha]S_3 \\ + [\tfrac{1}{2}(a - c)\sin 2\alpha + b\cos 2\alpha]S_1 .$$

Hence verify that H' is the diagonalised form of H if

$$\tan 2\alpha = -2b/(a - c)$$

and find the eigenvalues.

(B.8.6) Writing

$$a = E_u + U_{uu}, \quad b = |U_{uv}|, \quad c = E_v - U_{vv},$$

check that the solution of problem (B.8.5) is in agreement with the exposition given in the text, notably equation (B.8.23) in conjunction with equation (B.8.9).

[From (B.8.23), $\tan 2\alpha = \sqrt{(C^2 - 1)} = \pm 2b/(a - c)$.]

B.9 THE REASON FOR AND SIMPLE PROPERTIES OF BRILLOUIN ZONES OF ELECTRONS IN BRAVAIS LATTICES

B.9a Simple theory: the occurrence of Brillouin zones for electrons in Bravais lattices

Let

$$U(\mathbf{r}) = \sum_{\mathbf{K}} U_{\mathbf{K}} \, e^{i\mathbf{K}\cdot\mathbf{r}}, \qquad U_{\mathbf{K}} = V^{-1} \int_V U(\mathbf{r}) \, e^{-i\mathbf{K}\cdot\mathbf{r}} \, d\mathbf{r} \tag{B.9.1}$$

be a periodic potential. What effect does this have on the scattering between free electron states? Let

$$\psi_j \to \psi_{\mathbf{k}}(\mathbf{r}) = V^{-\frac{1}{2}} \, e^{i\mathbf{k}\cdot\mathbf{r}}, \qquad E_m \to E_{\mathbf{k}} = \hbar^2 k^2/2m. \tag{B.9.2}$$

Instead of a general scattering process from a state $\mathbf{k}$ to a state $\mathbf{k}'$, it is convenient to confine attention to elastic (i.e. constant energy) scattering $|\mathbf{k}| = |\mathbf{k}'|$. The relevant matrix elements are of the form

$$U_{\mathbf{k}\mathbf{k}'} = V^{-1} \sum_{\mathbf{K}} U_{\mathbf{K}} \int_V e^{i(\mathbf{K}+\mathbf{k}'-\mathbf{k})\cdot\mathbf{r}} \, d\mathbf{r} = \sum_{\mathbf{K}} U_{\mathbf{K}} \, \delta_{\mathbf{k}-\mathbf{k}',\mathbf{K}}. \tag{B.9.3}$$

The right-hand side vanishes unless there exists a lattice vector in the reciprocal lattice such that

$$\mathbf{k}' = \mathbf{k} - \mathbf{K}, \qquad \text{i.e. } 2\mathbf{k}\cdot\mathbf{K} = |\mathbf{K}|^2 + |\mathbf{k}|^2 - |\mathbf{k}'|^2 \tag{B.9.4}$$

and

$$\mathbf{k}^2 = \mathbf{k}'^2. \tag{B.9.5}$$

This implies that the state $\mathbf{k}$ lies near a plane of a Brillouin zone

$$2\mathbf{k}\cdot\mathbf{K} \sim |\mathbf{K}|^2, \tag{B.9.6}$$

or equivalently

$$E_{\mathbf{k}} \sim E_{\mathbf{k}-\mathbf{K}}. \tag{B.9.7}$$

Thus the free-electron approximation fails for a state $\mathbf{k}$ if elastic electron scattering is important between that state and some other state of the form $\mathbf{k}' = \mathbf{k} - \mathbf{K}$. If some states near a Brillouin zone boundary are occupied and others are empty, such scattering processes can in fact be important, and the above perturbation theory based on the free electron approximation for states near the boundary will certainly fail. Thus the concept of Brillouin zones is seen to be basic in the discussion of electronic states in crystals.

A process satisfying (B.9.6) is illustrated on the right-hand side of fig. (B.9.1). It is characterized by the fact that the wave-vector changes from one zone to another. On the left-hand side a scattering process is illustrated which violates (B.9.4). The two boldly drawn vectors represent the smallest non-zero lattice vectors in **k**-space.

The process on the right-hand side of the figure illustrates one of the so-called *Umklapp processes* (*umklappen* = to swing over). These processes are defined by $\mathbf{k} - \mathbf{k}' = \mathbf{K}$, where the initial **k** may lie anywhere in **k**-space. If, in addition, the scattering is elastic, the process will, in general, take a **k** from one face of a zone to another face of the same zone. The left-hand side of the figure illustrates a *normal process*, so described to distinguish it from an Umklapp process.

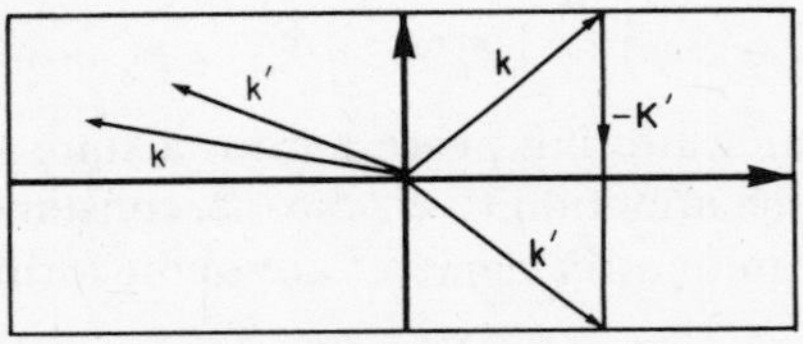

Fig. B.9.1.

We have not used the perturbation formula (B.8.1) because it assumes ψ_m (i.e. $\psi_{\mathbf{k}}$) to be a non-degenerate state. This condition is not satisfied because the energy depends only on $|\mathbf{k}|^2$. However, we now use the two-state perturbation theory of §B.8b to investigate the effect of the periodic potential when the state **k** lies near a Brillouin zone boundary.

B.9b Electron states near a Brillouin zone boundary in Bravais lattices

We now continue the theory of almost free electrons in the sense that the unperturbed functions remain free-electron functions, and the perturbation remains a periodic potential. However, it will be supposed that the main contributions to the perturbed states come from two unperturbed states whose wave-vectors satisfy (B.9.4):

$$\psi_u = \psi_{\mathbf{k}-\mathbf{K}}(\mathbf{r}), \qquad \psi_v = \psi_{\mathbf{k}}(\mathbf{r}). \tag{B.9.8}$$

It follows that

$$E_u \to E_{\mathbf{k}-\mathbf{K}} = V_0 + \frac{\hbar^2|\mathbf{k} - \mathbf{K}|^2}{2m}, E_v \to E_{\mathbf{k}} = V_0 + \frac{\hbar^2 k^2}{2m}, \tag{B.9.9}$$

where V_0 is the constant potential energy occurring in the free electron problem. The matrix elements needed are, by (B.8.9) and (B.9.3),

$$U_{uu} = U_{vv} \to U_0; \qquad |U_{uv}| = |U_{vu}| \to U_{\mathbf{K}}\left(\equiv \frac{\hbar^2 l^2}{2m} \neq 0\right). \quad \text{(B.9.10)}$$

Here l is a convenient variable (not an integer). In fact one merely needs to apply the theory of §B.8b with

$$C = \frac{K^2/2 - \mathbf{K}\cdot\mathbf{k}}{\sqrt{\{l^4 + [(K^2/2) - \mathbf{K}\cdot\mathbf{k}]^2\}}}$$

$$= \frac{\frac{1}{2}K^2 - k_\perp K}{\sqrt{[l^4 + (\frac{1}{2}K^2 - \mathrm{k}_\perp K)^2]}}, \quad \text{(B.9.11)}$$

where l is defined in (B.9.10), and we have put

$$\mathbf{k} = \mathbf{k}_{\parallel} + \mathbf{k}_{\perp},$$

giving

$$\mathbf{K}\cdot\mathbf{k}_{\parallel} = 0 \quad , \quad \mathbf{K}\cdot\mathbf{k}_{\perp} = Kk_{\perp}, \quad \text{(B.9.12)}$$

where $\mathbf{k}_{\parallel}$ is the component of $\mathbf{k}$ parallel to the reflecting planes of the direct lattice to which $\mathbf{K}$ is perpendicular, and $\mathbf{k}_{\perp}$ is the component perpendicular to these planes.

For small k and for $l^2 \ll K^2$, $C \sim 1$, and by (B.8.8b) and (B.8.8c)

$$E^+ = E_{\mathbf{k}-\mathbf{K}} + U_0, \qquad \mathrm{E}^- = E_{\mathbf{k}} + U_0 \quad \text{(B.9.13)}$$

so that the energy levels are merely shifted bodily by U_0. The condition for Bragg reflexion is, by (B.5.3) or (B.9.7),

$$2\mathbf{K}\cdot\mathbf{k} = |\mathbf{K}|^2, \qquad \text{i.e. } \mathbf{k}_{\perp} = \tfrac{1}{2}\mathbf{K} \quad \text{and} \quad C = 0. \quad \text{(B.9.14)}$$

In this case

$$E_{\mathbf{k}-\mathbf{K}} = E_{\mathbf{k}} \quad \text{and} \quad E^{\pm} = E_{\mathbf{k}} + U_0 \pm |U_{\mathbf{K}}|. \quad \text{(B.9.15)}$$

Near the zone boundary the upper unperturbed energy is further raised and the lower unperturbed energy is lowered by the perturbation. The energy gap at the boundary is $2|U_{\mathbf{K}}|$.

One may therefore think of the electron wave as fairly undisturbed by the crystal lattice unless its wave-vector fulfils approximately a condition for Bragg reflexion. As the condition is fulfilled more and more perfectly, the electron wave may be expected to suffer more and more internal reflexions between the appropriate crystal planes [those represented by **K** if the Bragg condition in question is (B.9.14)]. As a result the component of electron momentum normal to these planes must be expected to decrease. This is borne out quantitatively as follows: equation (B.8.8a) is

$$E^{\pm} = V_0 + U_0 + \frac{k^2}{4m}[K^2 - 2\mathbf{K}\cdot\mathbf{k} + 2k^2]$$
$$\pm\left\{|U_k|^2 + \left(\frac{\hbar^2}{4m}\right)^2[K^2 - 2\mathbf{K}\cdot\mathbf{k}]^2\right\}^{\frac{1}{2}}$$
$$= V_0 + U_0 + \frac{\hbar^2}{4m}[K^2 - 2Kk_\perp + 2k_\parallel^2 + 2k_\perp^2]$$
$$\pm\left\{|U_k|^2 + \left(\frac{\hbar^2}{4m}\right)^2[K^2 - 2k_\perp K]^2\right\}^{\frac{1}{2}}. \qquad \text{(B.9.16)}$$

It follows that

$$\frac{m}{\hbar}\frac{\partial E^{\pm}}{\partial k_\perp} = -\frac{\hbar}{2}(K - 2k_\perp)\left\{1 \pm \frac{K^2}{[4l^4 + (K^2 - 2k_\perp K)^2]^{\frac{1}{2}}}\right\}; \qquad \text{(B.9.17)}$$

$$\frac{m}{\hbar}\frac{\partial E^{\pm}}{\partial k_\parallel} = \hbar k_\parallel. \qquad \text{(B.9.18)}$$

If the condition for Bragg reflexion is fulfilled approximately,

$$\frac{m}{\hbar}\frac{\partial E^{\pm}}{\partial k_\perp} = -\frac{\hbar}{2}(K - 2k_\perp)[1 \pm K^2/2l^2] \qquad (l \neq 0). \qquad \text{(B.9.19)}$$

It will be seen later that the expressions on the left-hand sides of (B.9.17)–(B.9.19) do in fact represent the electron momentum perpendicular and parallel to the reflecting planes. (B.9.19) therefore establishes the required results.

It is often convenient to define effective masses for electrons in solids by

$$m_j = \hbar^{-2}\frac{\partial^2 E}{\partial k_j^2} \qquad (j = 1, 2, 3). \qquad \text{(B.9.20)}$$

If one regards these as a constants over a range of wave-vectors $\mathbf{k}$, one may say equivalently that in this range the connexion between electron kinetic energy $E(\mathbf{k})$ and $\mathbf{k}$ is assumed to be

$$E(\mathbf{k}) = E(0) + \mathbf{B}\cdot\mathbf{k} + \sum_{j=1}^{3}\frac{\hbar^2 k_j^2}{2m_j}, \qquad \text{(B.9.21)}$$

where $E(0)$ and $\mathbf{B}$ are constant. $\mathbf{B}$ can be neglected if one is near an extremum of an energy surface in $\mathbf{k}$-space. For this range of states the electron may then be thought of as having an energy–wave-vector relationship which is analogous to that for a free electron for each coordinate axis, except that the effect of the periodic potential is to yield a mass different from the free-electron mass. From (B.9.19) one finds

$$\frac{m}{m_\perp(\pm)} = 1 \pm K^2/2l^2, \qquad \frac{m}{m_{\parallel}(\pm)} = 1. \tag{B.9.22}$$

The upper sign applies to the electron state lying beyond the Brillouin zone. For the energetically lower state the effective mass $m_\perp$ is seen to be negative near the zone boundary.

The situation is illustrated in fig. B.9.2, drawn for a convenient direction in **k**-space. Points A and B have the perturbed energy E^-, while points C and D have the perturbed energy E^+. The points are equivalent in pairs. The dotted curve gives the unperturbed (free-electron) energies $E_{\mathbf{k}}$.

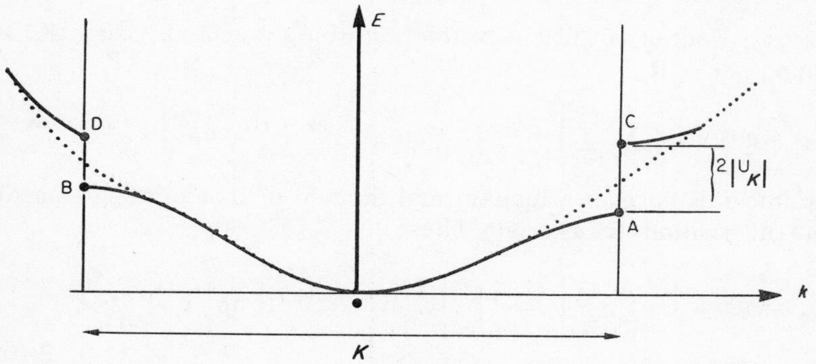

Fig. B.9.2.

B.10 ELECTRON SCATTERING IN MORE COMPLICATED CRYSTALS AND IN THE PRESENCE OF LATTICE WAVES[12]

B.10a General considerations

Consider the potential energy $U'(\mathbf{r})$ of an electron at position **r** due to atoms or ions at some set of points R_{jl} ($j = 1, 2, \ldots, s$; $l = 1, 2, \ldots, M_j$). The first suffix labels the s different types of atoms (the term atom includes ions in this section). The second suffix labels the different sites given the type of atom j. Contrary to the procedure in §B.9, a lattice is not now assumed. The potential energy U' is a superposition of contributions to the different atoms or ions:

$$U'(\mathbf{r}) = \sum_{j=1}^{s} \sum_{l=1}^{M_j} W_{jl}(\mathbf{r} - \mathbf{R}_{jl}).$$

[12] § B.10 may be omitted.

Subtracting the constant potential, u say, which gave rise to the free-electron wavefunction, yields the perturbation

$$U(r) = \sum_{j=1}^{s} \sum_{l=1}^{M_j} W_{jl}(\mathbf{r} - \mathbf{R}_{jl}) - u. \tag{B.10.1}$$

First-order perturbation theory, equation (B.8.2), shows that the scattering probability between unperturbed states

$$\psi(\mathbf{k}, \mathbf{r}) \equiv V^{-\frac{1}{2}} \mathrm{e}^{\mathrm{i}\mathbf{k}\cdot\mathbf{r}} \to \psi(\mathbf{k}', \mathbf{r}) \tag{B.10.2}$$

is proportional to the matrix element

$$U_{\mathbf{k},\mathbf{k}'} = V^{-1} \int_V U(\mathbf{r}) \exp[-\mathrm{i}(\mathbf{k} - \mathbf{k}') \cdot \mathbf{r}] \, \mathrm{d}\mathbf{r}. \tag{B.10.3}$$

This is a Fourier coefficient of the perturbing potential. Using (B.10.1) with $\boldsymbol{\rho}_{jl} \equiv \mathbf{r} - \mathbf{R}_{jl}$,

$$U_{\mathbf{k},\mathbf{k}'} = u\delta_{\mathbf{k},\mathbf{k}'} + \sum_{j=1}^{s} \sum_{l=1}^{M_j} \left[V^{-1} \int_V W_{jl}(\boldsymbol{\rho}_{jl}) \, \mathrm{e}^{-\mathrm{i}(\mathbf{k}-\mathbf{k}')\cdot\boldsymbol{\rho}_{jl}} \, \mathrm{d}\boldsymbol{\rho}_{jl} \right] \mathrm{e}^{-\mathrm{i}(\mathbf{k}-\mathbf{k}')\cdot\mathbf{R}_{jl}}.$$

The integrals go over a fundamental domain or its equivalent, starting from the positions $\mathbf{R}_{jl}$ as origin. Thus

$$U_{\mathbf{k},\mathbf{k}'} = u\delta_{\mathbf{k},\mathbf{k}'} + \sum_{j=1}^{s} \sum_{l=1}^{M_j} \left[V^{-1} \int_V W_{jl}(\boldsymbol{\rho}) \, \mathrm{e}^{-\mathrm{i}(\mathbf{k}-\mathbf{k}')\cdot\boldsymbol{\rho}} \, \mathrm{d}\boldsymbol{\rho} \right] \mathrm{e}^{-\mathrm{i}(\mathbf{k}-\mathbf{k}')\cdot\mathbf{R}_{jl}}. \tag{B.10.4}$$

B.10b First application of (B.10.4): The R_{jl} form s parallel and interpenetrating Bravais lattices

In this case the $\mathbf{R}_{jl}$ end on lattice sites, and one can put

$$\mathbf{R}_{jl} \to \mathbf{R}_{j\mathbf{n}} = \mathbf{r}_j + \mathbf{R}_{\mathbf{n}}^{(j)}$$
$$(j = 1, 2, \ldots, s;\ M_1 = M_2 = \cdots = M_j = N = N_1 N_2 N_3). \tag{B.10.5}$$

In (B.10.5), $N = N_1 N_2 N_3$ is the number of unit cells in the crystal. The unit cell which acts as origin has its s additional sites at the positions $\mathbf{r}_j$. Using (B.10.5) in (B.10.4),

$$U_{\mathbf{k},\mathbf{k}'} = u\delta_{\mathbf{k},\mathbf{k}'} + \sum_{j=1}^{s} \sum_{\mathbf{n}}$$
$$\times \left[\frac{1}{V} \int_V W_{j\mathbf{n}}(\boldsymbol{\rho}) \, \mathrm{e}^{-\mathrm{i}(\mathbf{k}-\mathbf{k}')\cdot\boldsymbol{\rho}} \, \mathrm{d}\boldsymbol{\rho} \right] \mathrm{e}^{-\mathrm{i}(\mathbf{k}-\mathbf{k}')\cdot\mathbf{r}_j} \, \mathrm{e}^{-\mathrm{i}(\mathbf{k}-\mathbf{k}')\cdot\mathbf{R}_n^{(j)}}. \tag{B.10.6}$$

To simplify this result one can assume either (a) that $W_{j\mathbf{n}}$ is independent of $\mathbf{n}$, or (b) that $W_{j\mathbf{n}}$ is independent of both j and $\mathbf{n}$. Because of the short range of the atomic potential it is reasonable to suppose that the potential energies $W_{j\mathbf{n}}$ depend only on the lattice j and not upon where on this lattice the origin of the integral is taken. This justifies (a). Assumption (b) is reasonable if all the atoms or ions give rise to very similar potentials as, for example, in a monatomic lattice. In either case sum (B.10.6) then involves a sum over $\mathbf{n}$ which is of type (B.6.20), and yields $N\delta_{\mathbf{k}-\mathbf{k}',\mathbf{K}^{(j)}}$, where $\mathbf{K}^{(j)}$ is *any* lattice vector in the $\mathbf{k}$-space for lattice j. Since, however, the lattices j are all similar the superscript may be omitted. Introduce the Fourier coefficient

$$W_j(\mathbf{k}) \equiv V^{-1} \int_V W_j(\boldsymbol{\rho})\, \mathrm{e}^{-i\mathbf{k}\cdot\boldsymbol{\rho}}\, \mathrm{d}\boldsymbol{\rho}. \tag{B.10.7}$$

Then (B.10.6) becomes either

$$U_{\mathbf{k},\mathbf{k}'} = u\delta_{\mathbf{k},\mathbf{k}'} + N\left[\sum_{j=1}^{s} W_j(\mathbf{k}-\mathbf{k}')\,\mathrm{e}^{-i(\mathbf{k}-\mathbf{k}')\cdot\mathbf{r}_j}\right]\delta_{\mathbf{k}-\mathbf{k}',\mathbf{K}} \tag{B.10.8}$$

[assumption (a)]

or

$$U_{\mathbf{k},\mathbf{k}'} = u\delta_{\mathbf{k},\mathbf{k}'} + sNW(\mathbf{k}-\mathbf{k}')S_{\mathbf{k}-\mathbf{k}'}\delta_{\mathbf{k}-\mathbf{k}',\mathbf{K}} \tag{B.10.9}$$

[assumption (b)].

The geometrical structure factor, defined in (B.4.13), has here been used. It is a characteristic of a unit cell and depends on the manner in which the lattices interpenetrate each other [see problem (B.4.3)]. If it vanishes, the scattering probability $k \to k'$ either vanishes or at least is small whenever (B.10.9) is a good approximation. (B.10.8) and (B.10.9) are clearly generalisations of the simple result (B.9.3) to lattices with a basis.

The physical significance of the result (B.10.9) is that although to each lattice vector $\mathbf{K}$ there corresponds a Brillouin zone plane given by (B.9.14), the energy gap, denoted by $2|U_{\mathbf{K}}|$ in (B.9.15), across the plane can vanish, or be small, if the structure factor $S_{\mathbf{K}}$ vanishes. $W(\mathbf{k}-\mathbf{k}')$ is the matrix element for the transition, assuming the only potential acting is that due to the free atom.

B.10c Second application of (B.10.4): The effect of lattice waves

One of the assumptions made in §B.1 to justify the use of a periodic potential was that the atoms (this term will include ions in this section) are at rest. In fact their oscillation disturbs the exact periodicity. It is therefore desirable to return to (B.10.4), while retaining the notation

(B.10.7). An assumption analogous to (a) in §B.10b will be made. The main contribution to the typical integral in (B.10.4) comes from the close neighbourhood of the position $\mathbf{R}_{jl}$ of the atom. An integral will therefore be relatively insensitive to the position l, given that it is on the lattice j. It will depend on which lattice is being considered, since different lattice sites may, for example, be occupied by different atoms. This approximation enables one to write

$$U_{\mathbf{k},\mathbf{k}'} = u\,\delta_{\mathbf{k},\mathbf{k}'} + s \sum_{j=1}^{s} W_j(\mathbf{k} - \mathbf{k}')S_{j,\mathbf{k}-\mathbf{k}'}; \tag{B.10.10}$$

$$S_{j,\mathbf{k}} \equiv \frac{1}{s}\sum_{l=1}^{N} \mathrm{e}^{-\mathrm{i}\mathbf{k}\,.\,\mathbf{R}_{jl}}. \tag{B.10.11}$$

Now for an atom of lattice j with an undisplaced position at $\mathbf{r}_j + \mathbf{R}_{\mathbf{n}}^{(j)}$, one can write the displacement as

$$\mathbf{u}_{\mathbf{n}}^{(j)}(\mathbf{R}_{\mathbf{n}}^{(j)}, t) = \sum_{\mathbf{q},\lambda} Q_{\lambda}^{(j)}(\mathbf{q}, t)\mathbf{l}_{\lambda}^{(j)}(\mathbf{q})\,\mathrm{e}^{\mathrm{i}\mathbf{q}\cdot\mathbf{R}_{\mathbf{n}}^{(j)}}. \tag{B.10.12}$$

The $\mathbf{l}_{\lambda}(\mathbf{q})$ are a triplet of orthonormal vectors for each $\mathbf{q}$, and give the amplitudes in the directions of polarization $\lambda = 1, 2, 3$. The Fourier components Q_{λ} and the $\mathbf{l}_{\lambda}$ are subject to conditions

$$[Q_{\lambda}^{(j)}(\mathbf{q}, t)\mathbf{l}_{\lambda}^{(j)}(\mathbf{q})]^* = Q_{\lambda}^{(j)}(-\mathbf{q}, t)\mathbf{l}_{\lambda}^{(j)}(-\mathbf{q}) \tag{B.10.13}$$

which ensure that the displacement is real. Thus (B.10.12) has the form

$$\mathbf{u}_{\mathbf{n}}^{(j)} = \frac{1}{2}\sum_{\mathbf{q},\lambda} [\mathbf{a}_{\mathbf{q}\lambda\mathbf{n}}^{(j)} + \mathbf{a}_{-\mathbf{q}\lambda\mathbf{n}}^{(j)}];$$

$$\mathbf{a}_{\mathbf{q}\lambda\mathbf{n}}^{(j)} \equiv Q_{\lambda}^{(j)}(\mathbf{q}, t)\mathbf{l}_{\lambda}^{(j)}(\mathbf{q})\,\mathrm{e}^{\mathrm{i}\mathbf{q}\cdot\mathbf{R}_{\mathbf{n}}^{(j)}}, \tag{B.10.14}$$

where each term in the square brackets is now real. Instead of (B.10.5) one now needs

$$\mathbf{R}_{jl} \to \mathbf{R}_{jn} = \mathbf{r}_j + \mathbf{R}_{\mathbf{n}}^{(j)} + \mathbf{u}_{\mathbf{n}}^{(j)} \tag{B.10.15}$$

In the exponent $-\mathrm{i}\mathbf{k}\cdot\mathbf{R}_{jl}$ of (B.10.11) the term arising from $\mathbf{u}_{\mathbf{n}}^{(j)}$ may be treated as small compared with the term arising from $\mathbf{r}_j + \mathbf{R}_{n}^{(j)}$. This enables one to use the fact that, with $b_{\mathbf{q}\lambda\mathbf{n}}^{(j)} \equiv \frac{1}{2}\mathbf{k}\cdot\mathbf{a}_{\mathbf{q}\lambda\mathbf{n}}^{(j)}$,

$$\begin{aligned}\mathrm{e}^{-\mathrm{i}(b_{\mathbf{q}}+b_{-\mathbf{q}})} &= 1 - \mathrm{i}(b_{\mathbf{q}} + b_{-\mathbf{q}}) - \tfrac{1}{2}(b_{\mathbf{q}}^2 + b_{-\mathbf{q}}^2) - b_{\mathbf{q}}b_{-\mathbf{q}} + \cdots \\ &\simeq 1 - |b_{\mathbf{q}}|^2 - \mathrm{i}(b_{\mathbf{q}} + b_{-\mathbf{q}}).\end{aligned} \tag{B.10.16}$$

Using (B.10.15) and (B.10.16) in (B.10.11), one finds

$$S_{j\mathbf{k}} = \frac{1}{s}\sum_{\mathbf{n}} \exp\left\{-\mathrm{i}\mathbf{k}\cdot[\mathbf{r}_j + \mathbf{R}_{\mathbf{n}}^{(j)}] - \mathrm{i}\sum_{\mathbf{q}\lambda}[b_{\mathbf{q}\lambda}^{(j)} + b_{-\mathbf{q}\lambda}^{(j)}]\right\}$$

$$\simeq \frac{1}{s}\mathrm{e}^{-\mathrm{i}\mathbf{k}\cdot\mathbf{r}_j}\sum_{\mathbf{n}}\mathrm{e}^{-\mathrm{i}\mathbf{k}\cdot\mathbf{R}_{\mathbf{n}}^{(j)}}\left\{1 - \sum_{\mathbf{q}\lambda}|b_{\mathbf{q}\lambda\mathbf{n}}^{(j)}|^2 - \mathrm{i}\sum_{\mathbf{q}\lambda}[b_{\mathbf{q}\lambda\mathbf{n}}^{(j)} + b_{-\mathbf{q}\lambda\mathbf{n}}^{(j)}]\right\}.$$

Using (B.6.20), and adopting the convention of real polarization vectors,

$$\mathbf{l}_\lambda^*(\mathbf{q}) = \mathbf{l}_\lambda(-\mathbf{q}) = \mathbf{l}_\lambda(\mathbf{q}); \tag{B.10.17}$$

$$S_{j\mathbf{k}}\,\mathrm{e}^{\mathrm{i}\mathbf{k}\cdot\mathbf{r}_j} = \left[1 - \frac{1}{4}\sum_{\mathbf{q}\lambda}|Q_\lambda^{(j)}(\mathbf{q},t)[\mathbf{k}\cdot\mathbf{l}_\lambda^{(j)}(\mathbf{q})]|^2\right]\delta_{\mathbf{k}\mathbf{K}}$$
$$-\frac{\mathrm{i}}{2}\sum_{\mathbf{q}\lambda}[\mathbf{k}\cdot\mathbf{l}_\lambda^{(j)}(\mathbf{q})][Q_\lambda^{(j)}(\mathbf{q},t)\delta_{\mathbf{k}-\mathbf{q},\mathbf{K}} + Q_\lambda^{(j)}(-\mathbf{q},t)\delta_{\mathbf{k}+\mathbf{q},\mathbf{K}}]. \tag{B.10.18}$$

The 1 in (B.10.18) gives, through (B.10.10), the result (B.10.8) for the matrix element in the absence of lattice waves. The first square bracket shows that one of the effects of the lattice vibrations is to add in (B.10.8) a factor which is approximately

$$\exp\left\{-\frac{1}{4}\sum_{\mathbf{q}\lambda}|Q_\lambda^{(j)}(\mathbf{q},t)[\mathbf{k}\cdot\mathbf{l}_\lambda^{(j)}(\mathbf{q})]|^2\right\}. \tag{B.10.19}$$

This shows that the matrix element (B.10.8) for electron scattering, i.e. for electron diffraction, is reduced by the square of (B.10.19). This leads to reduced X-ray diffraction intensities, the reduction increasing exponentially with the amplitude of the lattice vibrations, i.e. with temperature. Equation (B.10.19) is known in crystallography as the Debye–Waller factor. The last sum in (B.10.18) shows that the presence of lattice waves relaxes the condition

$$\mathbf{k} - \mathbf{k}' = \text{lattice vector in } \mathbf{k}\text{-space} \tag{B.10.20a}$$

to

$$\mathbf{k} - \mathbf{k}' \pm \mathbf{q} = \text{lattice vector in } \mathbf{k}\text{-space}. \tag{B.10.20b}$$

This is due to the fact that a phonon of wave-vector $\mathbf{q}$ may be emitted or absorbed during the scattering process. Higher order terms in the expansion (B.10.18) show that multi-phonon terms also contribute in principle to the matrix element $U_{\mathbf{k}\mathbf{k}'}$. In general they are, however, less important. Equations (B.10.20) are characteristic of momentum conservation conditions in the presence of a periodic lattice.

B.11 BLOCH'S THEOREM AND BLOCH FUNCTIONS

B.11a Bloch's theorem

Consider n quasi-free electrons and n ions in a perfect crystal. The Hamiltonian is a function of the electron momenta $\mathbf{p}_1, \ldots, \mathbf{p}_n$ and coordinates $\mathbf{r}_1, \ldots, \mathbf{r}_n$. The coordinates $\mathbf{R}_i$ of the ions enter as parameters and are not written down explicitly. Lattice periodicity is assumed, as explained in §B.1. Suppose now that each $\mathbf{r}_j$ is replaced by $\mathbf{r}_j + \mathbf{R}$, where $\mathbf{R}$ is a fixed lattice vector. For an infinite crystal, or for a finite crystal subject to periodic boundary conditions, this operation leaves the Hamiltonian unchanged since it merely moves each quasi-free electron from one unit cell into a corresponding position in a unit cell a distance $\mathbf{R}$ away.

This idea may be expressed mathematically as follows: let $\mathrm{f}(\mathbf{r}_1, \ldots, \mathbf{r}_n)$ be an arbitrary function of the coordinates of the n valence electrons in the crystal. Let $\mathbf{R}$ be a typical lattice vector in the direct lattice, and let $T(\mathbf{R})$ be the translation operator defined by

$$T(\mathbf{R})\mathrm{f}(\mathbf{r}_1, \mathbf{r}_2 \ldots) = \mathrm{f}(\mathbf{r}_1 + \mathbf{R}, \mathbf{r}_2 + \mathbf{R}, \ldots). \tag{B.11.1}$$

Since the operator T does not affect the electronic momenta, the exact n-electron Hamiltonian for a periodic potential satisfies

$$\begin{aligned} &T(\mathbf{R})H(\mathbf{p}_1, \mathbf{p}_2, \ldots, \mathbf{p}_n; \mathbf{r}_1, \mathbf{r}_2, \ldots, \mathbf{r}_n)\mathrm{f}(\mathbf{r}_1, \mathbf{r}_2, \ldots, \mathbf{r}_n) \\ &= H(\mathbf{p}_1, \ldots; \mathbf{r}_1 + \mathbf{R}, \ldots)\mathrm{f}(\mathbf{r}_1 + \mathbf{R}, \ldots) \\ &= H(\mathbf{p}_1, \ldots; \mathbf{r}_1, \ldots)T(\mathbf{R})\mathrm{f}(\mathbf{r}_1, \ldots). \end{aligned}$$

Since f is an arbitrary function, it follows that the operator equation

$$T(\mathbf{R})H = HT(\mathbf{R}) \qquad \text{(for all lattice vectors } \mathbf{R}) \tag{B.11.2}$$

holds. It expresses the invariance of the Hamiltonian under lattice translation.

By the definition (B.11.1) the nuclear coordinates are not changed by $T(\mathbf{R})$, so that internuclear distances which enter into the Hamiltonian are not affected by $T(\mathbf{R})$. However it does alter electron–nuclear distances, but in such a way that the change can be pictured as due to a translation of the lattice relative to the electrons through the lattice vector $-\mathbf{R}$. For an infinite crystal, or for a crystal model obtained by the use of periodic boundary conditions, this has no essential effect on the Hamiltonian.

Since all $T(\mathbf{R})$ and H commute by (B.11.2), they have common eigenfunctions. Let $\psi_l(\mathbf{r}_1, \ldots, \mathbf{r}_n)$ be one of them, where l stands for a set of quantum numbers. Then

$$T(\mathbf{R})\psi_l(\mathbf{r}_1, \ldots, \mathbf{r}_n) = \psi_l(\mathbf{r}_1 + \mathbf{R}, \ldots) = \lambda_l(\mathbf{R})\psi_l(\mathbf{r}_1, \ldots, \mathbf{r}_n). \tag{B.11.3}$$

As eigenvalues of $T(\mathbf{R})$, the $\lambda_l(\mathbf{R})$ are independent of the $\mathbf{r}_i$. By (B.11.3),

$$|\psi_l(\mathbf{r}_1 + \mathbf{R}, \ldots)|^2 = |\lambda_l(\mathbf{R})|^2 |\psi_l(\mathbf{r}_1, \ldots)|^2. \qquad \text{(B.11.4)}$$

Integrating over the n electron coordinates, and using the fact that the n-particle functions are normalized, i.e.

$$\int_V |\psi_l(\mathbf{r}_1, \ldots, \mathbf{r}_n)|^2 \, \mathrm{d}\mathbf{r}_1 \ldots \mathrm{d}\mathbf{r}_n = 1, \qquad \text{(B.11.5)}$$

one finds $|\lambda_l|^2 = 1$, i.e. there exist real functions θ_l such that

$$\lambda_l(\mathbf{R}) = \mathrm{e}^{\mathrm{i}\theta_l(\mathbf{R})}. \qquad \text{(B.11.6)}$$

To discuss θ, note that if $\mathbf{R}$ and $\mathbf{R}'$ are both lattice vectors,

$$T(\mathbf{R})T(\mathbf{R}')\psi_l(\mathbf{r}_1, \ldots) = \psi_l(\mathbf{r}_1 + \mathbf{R} + \mathbf{R}', \ldots) = \lambda_l(\mathbf{R} + \mathbf{R}')\psi_l(\mathbf{r}_1, \ldots).$$

But the left-hand side of this equation is $\lambda_l(\mathbf{R})\lambda_l(\mathbf{R}')\psi_l(\mathbf{r}_1, \ldots)$, so that

$$\theta_l(\mathbf{R}) + \theta_l(\mathbf{R}') = \theta_l(\mathbf{R} + \mathbf{R}'). \qquad \text{(B.11.7)}$$

Consider now a continuous function $\theta_l(\mathbf{r})$ agreeing with $\theta_l(\mathbf{R})$ at lattice points, and satisfying

$$\theta_l(\mathbf{r}) + \theta_l(\mathbf{r}') = \theta_l(\mathbf{r} + \mathbf{r}'). \qquad \text{(B.11.8)}$$

Then if $\alpha = 1, 2, 3$ denote components of $\mathbf{r}$,

$$\frac{\partial\theta_l(\mathbf{r})}{\partial r_\alpha} = \frac{\partial\theta_l(\mathbf{r}')}{\partial r'_\alpha} = \frac{\partial\theta_l(\mathbf{r} + \mathbf{r}')}{\partial(r_\alpha + r'_\alpha)} \qquad \text{(B.11.9)}$$

These differential coefficients are clearly independent of the vector argument, and may be put equal to a constant, k_α say. Hence, integrating,

$$\theta_l(\mathbf{r}) = \mathbf{k} \cdot \mathbf{r} + a, \qquad \text{so that } \theta_l(\mathbf{R}) = \mathbf{k} \cdot \mathbf{R} + a',$$

where a is a constant. Comparison with (B.11.7) shows that $a = 0$. Equation (B.11.3) may now be written in the form

$$\psi_{j,\mathbf{k}}(\mathbf{r}_1 + \mathbf{R}, \ldots, \mathbf{r}_n + \mathbf{R}) = \mathrm{e}^{\mathrm{i}\mathbf{k}\cdot\mathbf{R}}\psi_{j,\mathbf{k}}(\mathbf{r}_1, \ldots, \mathbf{r}_n). \qquad \text{(B.11.10)}$$

Thus $\mathbf{k}$ is a quantum number for the state $l \equiv (j, \mathbf{k})$, where j stands for the remaining quantum numbers (e.g. spin, degeneracy for given $\mathbf{k}$, etc.). Equation (B.11.10) is Bloch's theorem. It is a purely group-theoretical result due to the invariance of the problem under lattice translations. It will be used to determine the general form of a wavefunction which is determined by an eigenvalue equation involving a lattice-periodic Hamiltonian. Such a wavefunction must satisfy (B.11.10).

B.11b Bloch functions

Let Σ extend over a subset $[m]$ of m particles out of n. Then put

$$u_{j,\mathbf{k}[m]}(\mathbf{r}_1, \ldots) \equiv \psi_{j,\mathbf{k}}(\mathbf{r}_1, \ldots, \mathbf{r}_n) \exp\left[-\frac{\mathrm{i}}{m} \mathbf{k} \cdot \sum \mathbf{r}_i \right].$$

Hence, if $\mathbf{R}$ is again any vector in the direct lattice,

$$\begin{aligned} u_{j,\mathbf{k}[m]}(\mathbf{r}_1 + \mathbf{R}, \ldots) &= \psi_{j,\mathbf{k}}(\mathbf{r}_1 + \mathbf{R}, \ldots) \exp\left[-\mathrm{i}\mathbf{k} \cdot \mathbf{R} - \frac{\mathrm{i}}{m} \mathbf{k} \cdot \sum \mathbf{r}_i \right] \\ &= \psi_{j,\mathbf{k}}(\mathbf{r}_1, \ldots) \exp\left[-\frac{\mathrm{i}}{m} \mathbf{k} \cdot \sum \mathbf{r}_i \right] \\ &= u_{j,\mathbf{k}[m]}(\mathbf{r}_1, \ldots), \end{aligned}$$

where (B.11.10) has been used. Thus the n-electron wavefunction, determined by an n-electron lattice-periodic Hamiltonian, may be written in the form

$$\psi_{j,\mathbf{k}[m]}(\mathbf{r}_1, \ldots, \mathbf{r}_n) = u_{j,\mathbf{k}[m]}(\mathbf{r}_1, \ldots, \mathbf{r}_n) \exp\left[\frac{\mathrm{i}}{m} \mathbf{k} \cdot \sum_{[m]} \mathbf{r}_i \right] \quad \text{(B.11.11)}$$

where u is lattice periodic, and the sum extends over a selection $[m]$ of m particles. By summing (B.11.11) over all $n!/m!(n - m)!$ selections of m out of n particles, the wavefunction becomes symmetrical under interchange of particles. This sum degenerates to one term only in two important cases:

(a) a one-particle theory is considered ($n = 1$, so that $m = 1$); (B.11.12)
(b) an n-particle theory[13] is considered ($m = n$). (B.11.13)

For $m \neq 1, n$ (B.11.11) is a new result. The functional form (B.11.11) of the wavefunction is called a Bloch function.

For a group-theoretical approach to the one-particle results see §C.8.

B.11c Non-uniqueness of k

The quantum number $\mathbf{k}$ was introduced in (B.11.10) by virtue of the factor $e^{\mathrm{i}\mathbf{k}\cdot\mathbf{R}}$ relating $\psi_l(\mathbf{r}_1 + \mathbf{R}, \ldots)$ to $\psi_l(\mathbf{r}_1, \ldots)$. This factor is the same if $\mathbf{k}$ is replaced by $\mathbf{k} + \mathbf{K}$, where $\mathbf{K}$ is a lattice vector in $\mathbf{k}$-space, since $e^{\mathrm{i}\mathbf{K}\cdot\mathbf{R}} = 1$. It follows that instead of assigning the quantum number $\mathbf{k}$ to the eigenfunction ψ_l of the translation operator $T(\mathbf{R})$, one might equally

[13] H. Haken, *Z. Naturf.* **8a**, 228 (1954); H. Hasegawa, *Proc. Intern. Conf. Phys. Semiconductors*, Paris 1964, (Dunod, Paris, 1965) p. 23 and references quoted there.

well have assigned any of the quantum numbers $\mathbf{k} + \mathbf{K}$, where $\mathbf{K}$ is any lattice vector in $\mathbf{k}$-space. We therefore put

$$\psi_{j\mathbf{k}[m]}(\mathbf{r}_1, \ldots, \mathbf{r}_n) = \psi_{j\mathbf{k}+\mathbf{K}[m]}(\mathbf{r}_1, \ldots, \mathbf{r}_n). \tag{B.11.14}$$

By virtue of (B.11.2), $\psi_{j\mathbf{k}[m]}$ is an eigenfunction of the Hamiltonian, i.e.

$$H\psi_{j\mathbf{k}[m]}(\mathbf{r}_1, \ldots, \mathbf{r}_n) = E_{j[m]}(\mathbf{k})\psi_{j\mathbf{k}[m]}(\mathbf{r}_1, \ldots, \mathbf{r}_n). \tag{B.11.15}$$

From (B.11.14) and (B.11.15) it follows that

$$E_{j[m]}(\mathbf{k}) = E_{j[m]}(\mathbf{k} + \mathbf{K}). \tag{B.11.16}$$

These relations show that it is simplest to restrict the vector $\mathbf{k}$ in (B.11.14), (B.11.15) and (B.11.16) to the first zone. Then $E_j(\mathbf{k})$ is defined only in the first zone. (B.11.16) has then only the quite limited meaning that certain pairs of points on the zone boundary have equal energies, for example diagonally opposite points of a zone. $E_{j[m]}(\mathbf{k})$ may now be *defined* as a lattice-periodic function in $\mathbf{k}$-space, and (B.11.16) becomes valid throughout $\mathbf{k}$-space. A similar interpretation applies to equation (B.11.14) for the wavefunction. For a corresponding group-theoretical discussion see § C.8 and § C.9.

For the Bloch function (B.11.11), with $\mathbf{k}$ replaced by $\mathbf{k} + \mathbf{K}$, it follows from (B.11.14) that

$$u_{j\mathbf{k}[m]} \exp\left[\frac{\mathrm{i}}{m}\mathbf{k}\cdot\sum \mathbf{r}_i\right] = u_{j\mathbf{k}+\mathbf{K}[m]} \exp\left[\frac{\mathrm{i}}{m}\mathbf{K}\cdot\sum \mathbf{r}_i\right] \exp\left[\frac{\mathrm{i}}{m}\mathbf{k}\cdot\sum \mathbf{r}_i\right],$$

whence

$$u_{j\mathbf{k}+\mathbf{K}[m]}(\mathbf{r}_1, \ldots, \mathbf{r}_n) = u_{j\mathbf{k}[m]}(\mathbf{r}_1, \ldots \mathbf{r}_n) \exp\left[-\frac{\mathrm{i}}{m}\mathbf{K}\cdot\sum \mathbf{r}_i\right]. \tag{B.11.17}$$

For a given reduced wave vector there is still an infinite number of solutions of the Schrödinger equation, belonging in general to different energies. One can label these energies by a suffix n, so that $E_n(\mathbf{k})$ ($n = 1, 2, \ldots$) are these energies. The suffix n is commonly called the band number. Degeneracy of the wave function for given n and $\mathbf{k}$ will be denoted by the suffix j. The z-components of the spin angular momenta of the particles are further quantum numbers. Denoting by $\eta(\sigma)$ the corresponding spin function, the Bloch function in the single-particle scheme has the form

$$\psi_{nj\mathbf{k}}(\mathbf{r}, \sigma) = u_{nj\mathbf{k}}(\mathbf{r})\,\mathrm{e}^{\mathrm{i}\mathbf{k}\cdot\mathbf{r}}\eta(\sigma). \tag{B.11.18}$$

The energy satisfies (B.11.16) with $m = 1$. This implies $E_j(\frac{1}{2}\mathbf{K}) = E_j(-\frac{1}{2}\mathbf{K})$. However, the stronger result

$$E_j(\mathbf{k}) = E_j(-\mathbf{k})$$

is always valid, as shown in §C.10 for spinless particles and more generally in the Appendix.

Problems

(B.11.1) Let $\mathbf{p} \equiv (\hbar/\mathrm{i})\nabla$ be the operator representing the electron momentum. Show that the general matrix element of $\mathbf{p}$ with respect to single-particle Bloch functions is

$$\langle n'\mathbf{k}'|\mathbf{p}|n\mathbf{k}\rangle = \left[\hbar\mathbf{k}\delta_{nn'} - \mathrm{i}\hbar N\int_{\Omega} u^*_{n'\mathbf{k}}\nabla u_{n\mathbf{k}}\,\mathrm{d}\mathbf{r}\right]\delta_{\mathbf{k}\mathbf{k}'}.$$

[The results (B.7.19) and (B.11.17) have to be used.]

(B.11.2) Since $u_{n\mathbf{k}}$ is lattice periodic, its Fourier expansion can be taken in the form

$$u_{n\mathbf{k}}(\mathbf{r}) = V^{-\frac{1}{2}}\sum_{\mathbf{K}} v_n(\mathbf{k}, \mathbf{K})\,\mathrm{e}^{\mathrm{i}\mathbf{k}\cdot\mathbf{r}}.$$

If the u's are normalized so that $\int_V |u_{n\mathbf{k}}|\,\mathrm{d}\mathbf{r} = 1$, show that

$$\sum_{\mathbf{K}} |v_n(\mathbf{k}, \mathbf{K})|^2 = 1.$$

Show also from problem (B.11.1) that

$$\langle n\mathbf{k}|\mathbf{p}|n\mathbf{k}\rangle = \sum_{\mathbf{K}} \hbar(\mathbf{k} + \mathbf{K})P(\mathbf{K}) = \hbar\left[\mathbf{k} + \sum_{\mathbf{K}} \mathbf{k}P(\mathbf{k})\right],$$

where

$$P(\mathbf{K}) \equiv |v_n(\mathbf{k}, \mathbf{K})|^2.$$

[$P(\mathbf{K})$ is a probability. The interpretation is as follows: an electron in state $(n, \mathbf{k})$ has probability $P(\mathbf{K})$ of having momentum $\hbar(\mathbf{k} + \mathbf{K})$. For this reason one refers to the v's as momentum eigenfunctions. If an electron with momentum $\hbar\mathbf{k}$ incident on a lattice acquires momentum $\hbar(\mathbf{k} + \mathbf{K})$, then by momentum conservation the lattice acquires momentum $-\hbar\mathbf{K}$.]

(B.11.3) Show that the momentum eigenfunction $v_n(\mathbf{k}, \mathbf{K})$ depends only on $\mathbf{k} + \mathbf{K}$, i.e. $v_n(\mathbf{k}, \mathbf{K}) = v_n(\mathbf{k} + \mathbf{K}), 0)$. [Use equation (B.11.15).]

(B.11.4) From the Schrödinger equation for an electron in a periodic potential $U(\mathbf{r}) = \sum_{\mathbf{K}} U(\mathbf{K})\,\mathrm{e}^{\mathrm{i}\mathbf{K}\cdot\mathbf{r}}$, show that the momentum eigenfunctions must satisfy the infinite set of equations

$$\left[\frac{\hbar^2}{2m}|\mathbf{k} + \mathbf{K}|^2 - E_n(\mathbf{k})\right]v_n(\mathbf{k} + \mathbf{K}) + \sum_{\mathbf{K}'} U(\mathbf{K}')v_n(\mathbf{k} + \mathbf{K} - \mathbf{K}') = 0$$

for all n and $\mathbf{K}$.

[Substitute in the Schrödinger equation, multiply by $\mathrm{e}^{-\mathrm{i}(\mathbf{k}+\mathbf{K})\cdot\mathbf{r}}$ and integrate over $\mathbf{r}$. The system of equations is Schrödinger's equation in $\mathbf{k}$-space.]

(B.11.5) Verify that for some fixed $\mathbf{k}_0$ the functions

$$\chi_{n\mathbf{k}}(\mathbf{r}) \equiv u_{n\mathbf{k}_0}\,\mathrm{e}^{\mathrm{i}\mathbf{k}\cdot\mathbf{r}}$$

form an orthonormal set for all n and $\mathbf{k}$ if the corresponding Bloch functions do.

[These are the Kohn–Luttinger functions[14]. The set of functions $\{u_{n\mathbf{k}_0}\}$ ($\mathbf{k}_0$ fixed, n variable) is in fact complete with respect to functions which are periodic in the direct lattice.]

(B.11.6) Verify that for some fixed $\mathbf{k}_0$ the functions

$$\chi_{n\mathbf{k},n'\mathbf{k}'}(\mathbf{r},\mathbf{r}') \equiv \chi_{n\mathbf{k}}(\mathbf{r})\chi_{n'\mathbf{k}'}(\mathbf{r}')$$

form also an orthonormal set if n, n', $\mathbf{k}$, $\mathbf{k}'$ are variable.

[Such functions are useful for two-particle complexes such as excitons.]

(B.11.7) In the next section a transition will be made from a crystal subject to periodic boundary conditions to an infinite crystal. This can be effected by letting the volume V of the fundamental domain tend to infinity. In preparation for this, consider the following simple question.

The Dirac delta function has the property that for a function $\mathrm{f}(x)$

$$\int_{-\infty}^{\infty} \mathrm{f}(x)\,\delta(x-x')\,\mathrm{d}x = \mathrm{f}(x').$$

Show that

$$\int_{-\infty}^{\infty} \delta(x-x')\,\mathrm{d}x = 1 \quad \text{and} \quad \delta(ax) = \frac{1}{a}\,\delta(x).$$

What is the dimension associated with $\delta(x)$?

If $\mathbf{r} = x_1\mathbf{i} + x_2 j + x_3\mathbf{k}$, $\delta(\mathbf{r}-\mathbf{r}')$ is defined by

$$\delta(\mathbf{r}-\mathbf{r}') = \delta(x_1-x_1')\,\delta(x_2-x_2')\,\delta(x_3-x_3').$$

What are the dimensions of (i) $\delta(\mathbf{r})$; (ii) $\delta(\mathbf{k})$; (iii) the Krönecker delta $\delta_{\mathbf{k}\mathbf{k}'}$?

(B.11.8) The figure shows the first Brillouin zone of a plane rectangular lattice. Prove that the energies on the boundary satisfy

$$E(P) = E(Q); \qquad E(R) = E(S); \qquad E(A) = E(B) = E(C) = E(D).$$

[Use equation (B.11.16)].

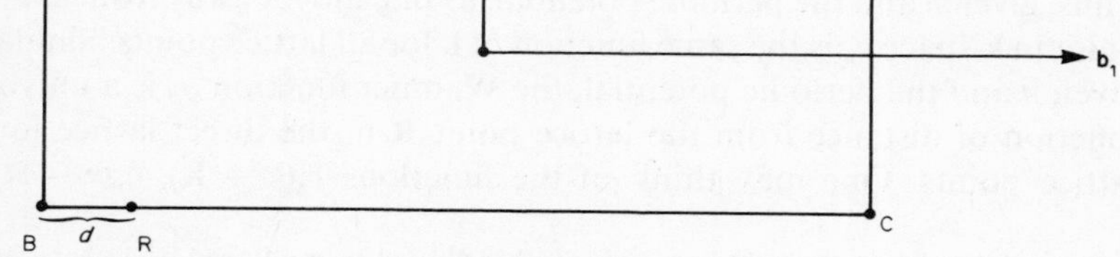

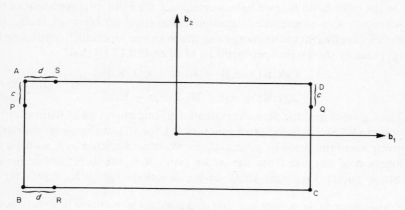

Fig. B.11.1.

[14] W. Kohn and J. M. Luttinger, *Phys. Rev.*, **97**, 869 (1955).

B.12 SYSTEMATIC FOURIER EXPANSIONS OF THE MAIN VARIABLES

For the present purpose the main variables to be considered are four in number and are the periodic potential $U(r)$, the modulating part $u_{n\mathbf{k}}(\mathbf{r})$ of a single-particle Bloch function, the energy $E_n(\mathbf{k})$ of a Bloch electron of wave-vector $\mathbf{k}$ in band n and the Bloch function of such an electron. The first two quantities can be Fourier expanded by equation (B.7.2). The last two quantities can be treated by (B.7.4) since they are defined only for wave-vectors $\mathbf{k}$ lying in the first zone. The additional suffix which distinguishes digenerate functions has been suppressed in $u_{n\mathbf{k}}$ and $\psi_{n\mathbf{k}}$. Thus one can write, with A, B, C, D arbitrary constants depending on the normalization adopted,

$$U(\mathbf{r}) = A\sum_{\mathbf{K}} U(\mathbf{K})\,\mathrm{e}^{\mathrm{i}\mathbf{K}\cdot\mathbf{r}}, \qquad U(\mathbf{K}) = \frac{1}{A\Omega}\int_{\Omega} U(\mathbf{r})\,\mathrm{e}^{-\mathrm{i}\mathbf{K}\cdot\mathbf{r}}\,\mathrm{d}\mathbf{r}; \qquad \text{(B.12.1a,b)}$$

$$u_{n\mathbf{k}}(\mathbf{r}) = B\sum_{\mathbf{K}} v_n(\mathbf{k},\mathbf{K})\,\mathrm{e}^{\mathrm{i}\mathbf{K}\cdot\mathbf{r}}, \qquad v_n(\mathbf{k},\mathbf{K}) = \frac{1}{B\Omega}\int_{\Omega} u_{n\mathbf{k}}(\mathbf{r})\,\mathrm{e}^{-\mathrm{i}\mathbf{K}\cdot\mathbf{r}}\,\mathrm{d}\mathbf{r}; \qquad \text{(B.12.2a,b)}$$

$$\epsilon_n(\mathbf{k}) = C\sum_{\mathbf{R}(V)} \epsilon_n(\mathbf{R})\,\mathrm{e}^{\mathrm{i}\mathbf{k}\cdot\mathbf{R}}, \qquad \epsilon_n(\mathbf{R}) = \frac{1}{CN}\sum_{\mathbf{k}(\mathrm{B.Z.})} \epsilon_n(\mathbf{k})\,\mathrm{e}^{\mathrm{i}\mathbf{k}\cdot\mathbf{R}}; \qquad \text{(B.12.3a, b)}$$

$$\psi_{n\mathbf{k}}(\mathbf{r}) = D\sum_{\mathbf{R}(V)} a_n(\mathbf{r},\mathbf{R})\,\mathrm{e}^{\mathrm{i}\mathbf{k}\cdot\mathbf{R}}, \qquad a_n(\mathbf{r},\mathbf{R}) = \frac{1}{DN}\sum_{\mathbf{k}(\mathrm{B.Z.})} \psi_{n\mathbf{k}}(\mathbf{r})\,\mathrm{e}^{-\mathrm{i}\mathbf{k}\cdot\mathbf{R}}. \qquad \text{(B.12.4a, b)[15]}$$

On the right-hand side we have respectively the Fourier component of the potential, the momentum eigenfunctions [(see problem (B.11.2)], the Fourier coefficient of the energy and the Wannier function[16] centred on $\mathbf{R}$. It is readily shown [see problems (B.11.3) and (B.12.1)] that

$$\begin{aligned} v_n(\mathbf{k},\mathbf{K}) &= v_n(\mathbf{k}+\mathbf{K}) \equiv v_n(\mathbf{k}+\mathbf{K},0); \\ a_n(\mathbf{r},\mathbf{R}) &= a_n(\mathbf{r}-\mathbf{R}) \equiv a_n(\mathbf{r}-\mathbf{R},0). \end{aligned} \qquad \text{(B.12.5a, b)}$$

Thus, given n and the periodic potential, as one moves away from a lattice point in $\mathbf{k}$-space v_n is the same function of $\mathbf{k}$ for all lattice points. Similarly, given n and the periodic potential, the Wannier function a_n is a universal function of distance from the lattice point $\mathbf{R}$ in the direct lattice for all lattice points. One may think of the functions $v_n(\mathbf{k}+\mathbf{K})$, $a_n(\mathbf{r}-\mathbf{R})$ as

[15] Since each term on the right-hand side of (B.12.4b) can be multiplied by a phase factor, there is a lack of uniqueness in the definition of the Wannier function. These problems are still under study; see, for instance, J. des Cloizeaux, *Phys. Rev.*, **129**, 554 (1963).
[16] G. H. Wannier, *Phys. Rev.*, **52**, 191 (1937).

located at the lattice points **K** and **R** of reciprocal and direct space respectively. Indeed the two functions are Fourier transforms of each other, as shown in problem (B.12.2). It is clear from the expressions given in this problem that the further one is away from the lattice point under consideration the more rapid are the oscillations due to the exponential term. Thus both functions have their maximum values at the lattice point at which they are located, and decrease in magnitude as one moves away from it, otherwise they would not be normalizable. The Wannier functions in fact decrease exponentially at large distances from the lattice point on which they are centred[17].

It is readily seen that the Wannier functions $a_n(\mathbf{r} - \mathbf{R})$ are orthogonal for different sites **R**, and that they do not satisfy the Schrödinger equation [see problems (B.12.3) and (B.12.6)]. The Wannier functions form a complete set, as do the Bloch functions. Hence, when normalized, the completeness relation (B.7.6) must be satisfied:

$$\sum_{n,\mathbf{R}} a_n^*(\mathbf{r}' - \mathbf{R})a_n(\mathbf{r} - \mathbf{R}) = \delta(\mathbf{r}' - \mathbf{r}). \qquad \text{(B.12.6)}$$

In the last relation $\mathbf{r}'$ and $\mathbf{r}$ must lie in the same unit cell.

Keeping the lattice site at **R** fixed, one can pass to the case of an isolated atom by letting the lattice parameters go to infinity. In this case the site at **R** is the only one not at infinity and the result of problem (B.12.6) yields

$$\int_V a_n^*(\mathbf{r} - \mathbf{R})Ha_n{}'(\mathbf{r} - \mathbf{R}')\,\mathrm{d}\mathbf{r} = \epsilon_n(\mathbf{R}' - \mathbf{R})\,\delta_{nn'} \to \epsilon_n(0)\,\delta_{\mathbf{R}\mathbf{R}'}\,\delta_{nn'}, \qquad \text{(B.12.7)}$$

since there is in this case no overlap between neighbouring lattice sites. Thus, in this limit, the Hamiltonian is diagonal in the Wannier functions, which must therefore represent in this limit the electron wave functions $\phi_n(\mathbf{r} - \mathbf{R})$ for the isolated atom. Thus, in the limit considered in (B.12.7),

$$a_n(\mathbf{r} - \mathbf{R}) \to \phi_n(\mathbf{r} - \mathbf{R}). \qquad \text{(B.12.8)}$$

The replacement (B.12.8) is possible, even without going to the limit of infinite lattice parameters. It is then an *approximation* which is justified if there is little interaction between electrons on different lattice sites. In this approximation a typical eigenfunction of an electron in a periodic lattice is, by (B.12.4a) and (B.12.8),

$$\chi_{n\mathbf{k}}(\mathbf{r}) = D\sum_{\mathbf{R}(V)} \phi_n(\mathbf{r} - \mathbf{R})\,\mathrm{e}^{\mathrm{i}\mathbf{k}\cdot\mathbf{R}}. \qquad \text{(B.12.9)}$$

[17] E. I. Blount, in *Solid-State Physics*, (eds. Seitz and Turnbull), **13**, 305 (1962).

This is the basis of the *tight-binding approximation*[18]. It has the advantage that the ϕ_n's are known at least approximately for many atoms, and the disadvantage that $\phi_n(\mathbf{r} - \mathbf{R})$ is not orthogonal to $\phi_n(\mathbf{r} - \mathbf{R}')$ when $\mathbf{R} \neq \mathbf{R}'$ are two lattice sites. It was in order to obtain functions with this orthogonality property, but otherwise similar to the ϕ's, that the Wannier functions were introduced. It will be assumed that the functions ϕ_n are normalized over the volume V.

The tight-binding approximation

If an isolated nucleus is at $\mathbf{R}$ and there is a single outer electron with position vector $\mathbf{r}$, then its Hamiltonian is

$$H_{\mathbf{R}} \equiv -\frac{\hbar^2}{2m}\nabla^2 + U(\mathbf{r} - \mathbf{R}), \tag{B.12.10}$$

$$H_{\mathbf{R}}\phi_n(\mathbf{r} - \mathbf{R}) = \mathrm{E}_n\phi_n(\mathbf{r} - \mathbf{R}) \qquad (n = 1, 2, \ldots). \tag{B.12.11}$$

Here U is the potential energy of the outer electron in the field of the ion. Suppose now that N identical ions are brought together to form a lattice with the ion at $\mathbf{R}$ on one lattice site, and another ion at the origin. Suppose also that the different states j are non-degenerate in the atom (apart from spin degeneracy) and do not overlap in the solid. The meaning of n is then well-defined on the right-hand sides of (B.12.9) and (B.12.11). Each atomic orbital, specified by a given n, then leads to a band of N atomic states of type n. One may thus think of the label n as specifying a band in the lattice (in agreement with our earlier notation). The wavefunction of an electron in the solid is now approximately given by (B.12.9). This function satisfies the Bloch theorem [see problem (B.12.10)] and is approximately normalised if $D = 1/\sqrt{N}$.

The wavefunction (B.12.9) is a superposition of atomic functions, all of the same type n, and centred on the sites of a Bravais lattice. We will therefore discuss an electron in a *monatomic* crystal which is based on a Bravais lattice. (B.12.9) is clearly not an exact eigenfunction of the appropriate Hamiltonian H. It is, however, meaningful to work out its matrix elements with respect to functions such as (B.12.9). This will be done without approximation. One finds

$$H_{jt}(\mathbf{k}) = \frac{1}{N}\sum_{\mathbf{R},\mathbf{R}'} \mathrm{e}^{\mathrm{i}\mathbf{k}\cdot(\mathbf{R}'-\mathbf{R})} \int_V \phi_j^*(\mathbf{r} - \mathbf{R}) H \phi_t(\mathbf{r} - \mathbf{R}')\,\mathrm{d}\mathbf{r}.$$

[18] Due to F. Bloch, *Z. Physik*, **52**, 555 (1928).

If $V(\mathbf{r}) = V(\mathbf{r} - \mathbf{R}')$ (where $\mathbf{R}'$ is any lattice site) is the periodic potential, then for any given $\mathbf{R}$ and $\mathbf{R}'$, express H in the form

$$H = -\frac{\hbar^2\nabla^2}{2m} + V(\mathbf{r} - \mathbf{R}') = H_{\mathbf{R}} - [U(\mathbf{r} - \mathbf{R}) - V(\mathbf{r} - \mathbf{R}')]. \tag{B.12.12}$$ [19]

The term in square brackets can be regarded as the perturbation. Then

$$H_{jt}(\mathbf{k}) = E_j \int \chi_{j\mathbf{k}}^*(\mathbf{r})\chi_{t\mathbf{k}}(\mathbf{r})\,\mathrm{d}\mathbf{r}$$
$$- \frac{1}{N}\sum_{\boldsymbol{\rho}} \mathrm{e}^{-\mathrm{i}\mathbf{k}\cdot\boldsymbol{\rho}} \sum_{\mathbf{R}'} \int_V \phi_j^*(\mathbf{r} - \mathbf{R}' - \boldsymbol{\rho})$$
$$\times [U(\mathbf{r} - \mathbf{R}' - \boldsymbol{\rho}) - V(\mathbf{r} - \mathbf{R}')]\phi_t(\mathbf{r} - \mathbf{R}')\,\mathrm{d}\mathbf{r}.$$

In the first term we have used (B.12.11); in the second we have replaced $\mathbf{R}$ by $\boldsymbol{\rho} \equiv \mathbf{R} - \mathbf{R}'$. Changing the variable of integration to $\mathbf{r}' = \mathbf{r} - \mathbf{R}'$,

$$H_{jt}(\mathbf{k}) = E_j I_{jt}(\mathbf{k}) - \sum_{\boldsymbol{\rho}} \mathrm{e}^{-\mathrm{i}\mathbf{k}\cdot\boldsymbol{\rho}} W_{jt}(\boldsymbol{\rho}), \tag{B.12.13a}$$

where

$$I_{jt}(\mathbf{k}) \equiv \int \chi_{j\mathbf{k}}^*(\mathbf{r})\chi_{t\mathbf{k}}(\mathbf{r})\,\mathrm{d}\mathbf{r} = \delta_{jt} + \sum_{\boldsymbol{\rho}\neq 0} \mathrm{e}^{-\mathbf{k}\cdot\boldsymbol{\rho}} \int_V \phi_j^*(\mathbf{r}' - \boldsymbol{\rho})\phi_t(\mathbf{r}')\,\mathrm{d}\mathbf{r}' \tag{B.12.13b}$$

and

$$W_{jt}(\boldsymbol{\rho}) \equiv \int \phi_j^*(\mathbf{r} - \boldsymbol{\rho})[U(\mathbf{r} - \boldsymbol{\rho}) - V(\mathbf{r})]\phi_t(\mathbf{r})\,\mathrm{d}\mathbf{r}. \tag{B.12.13c}$$

The expectation value of the energy in state $\chi_{j\mathbf{k}}$ of (B.12.9) is, for all j,

$$\epsilon_j(\mathbf{k}) \equiv \frac{H_{jj}(\mathbf{k})}{I_{jj}(\mathbf{k})} = E_j - \frac{\sum_{\boldsymbol{\rho}} \mathrm{e}^{-\mathbf{k}\cdot\boldsymbol{\rho}} W_{jj}(\boldsymbol{\rho})}{I_{jj}(\mathbf{k})}. \quad (I_{jj}(\mathbf{k}) \text{ is real.}) \tag{B.12.13d}$$

This has been denoted by $\epsilon_j(\mathbf{k})$ to distinguish it from an energy eigenvalue[20]. The distinction is needed since the $\chi_{j\mathbf{k}}$ of (B.12.9) are not eigenfunctions of the Hamilton for an electron in a lattice.

A relation exists between the two expressions (B.12.13b and c) because of the hermiticity of H. This implies $H_{jt}(\mathbf{k}) - H_{tj}^*(\mathbf{k}) = 0$, i.e.

$$(E_j - E_t)I_{jt}(\mathbf{k}) = \sum_{\boldsymbol{\rho}} \mathrm{e}^{-\mathrm{i}\mathbf{k}\cdot\boldsymbol{\rho}}[W_{jt}(\boldsymbol{\rho}) - W_{tj}^*(-\boldsymbol{\rho})]. \tag{B.12.14a}$$

[19] Since the potential is periodic, $V(\mathbf{r} - \mathbf{R}') = V(\mathbf{r} - \mathbf{R})$.
[20] Confusion with $\epsilon_n(\mathbf{R})$ of equation (B.12.3) is unlikely since one quantity depends on $\mathbf{R}$, the other on $\mathbf{k}$.

In particular, if $j = t$, the right-hand side vanishes for all $\mathbf{k}$, whence

$$W_{jj}(\boldsymbol{\rho}) = W_{jj}^*(-\boldsymbol{\rho}). \tag{B.12.14b}$$

The relations (B.12.13) and (B.12.14) are exact. Using the fact that a monatomic crystal based on a Bravais lattice is considered, one can also use (fig. B.12.1)

$$\phi_t(\mathbf{r} + \boldsymbol{\rho}) = \phi_t(\mathbf{r} - \boldsymbol{\rho}) \tag{B.12.15a}$$

for all functions ϕ_t and lattice vectors $\boldsymbol{\rho}$. This enables one to give simple relations for matrix elements under reversal of $\mathbf{k}$, since it implies

$$I_{jt}(\mathbf{k}) - I_{jt}(-\mathbf{k}) = \sum_{\boldsymbol{\rho}} e^{-i\mathbf{k}\cdot\boldsymbol{\rho}} \int_V \phi_j^*(\mathbf{r})[\phi_t(\mathbf{r} + \boldsymbol{\rho}) - \phi_t(\mathbf{r} - \boldsymbol{\rho})]\, d\mathbf{r} = 0,$$

$$W_{jt}(\boldsymbol{\rho}) - W_{jt}(-\boldsymbol{\rho}) = \int_V \phi_j^*(\mathbf{r})[U(\mathbf{r}) - V(\mathbf{r})][\phi_t(\mathbf{r}+\boldsymbol{\rho}) - \phi_t(\mathbf{r}-\boldsymbol{\rho})]\, d\mathbf{r} = 0.$$

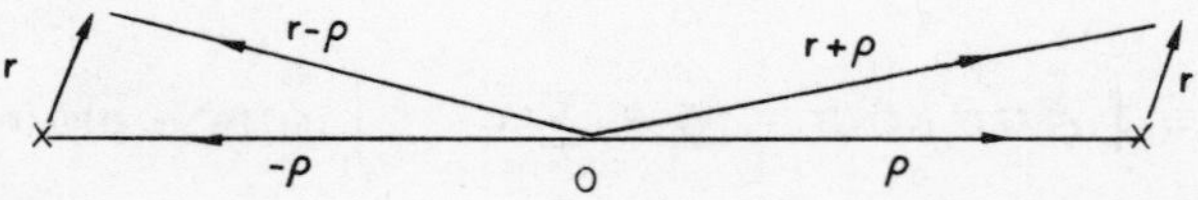

Fig. B.12.1.

It follows that

$$H_{jt}(\mathbf{k}) = H_{jt}(-\mathbf{k}), \qquad \epsilon_j(\mathbf{k}) = \epsilon_j(-\mathbf{k}), \tag{B.12.15b}$$

and, by (B.12.14b), $W_{jj}(\boldsymbol{\rho}) = W_{jj}(-\boldsymbol{\rho})$ is real. Thus one can re-express the energy expectation value as

$$\epsilon_j(\mathbf{k}) = E_j - [2/I_{jj}(\mathbf{k})] \sum_{\boldsymbol{\rho}}{}' W_{jj}(\boldsymbol{\rho}) \cos \mathbf{k}\cdot\boldsymbol{\rho}, \tag{B.12.15c}$$

where $\sum'$ denotes a sum over half the volume V only. All expressions occurring in (B.12.15c) are real. The essence of the approximation is to interpret the quantity (B.12.15c) as an energy eigenvalue $E_j(\mathbf{k})$ of an electron in the lattice.

If the bands overlap, degenerate perturbation theory is needed. The wavefunction to be used is then a linear combination of functions such as (B.12.9), and becomes a linear combination of different types of atomic orbitals. This is the L.C.A.O. method. The allowed energy values are in all cases obtainable from a secular equation

$$\begin{vmatrix} H_{11}-E & H_{12} & \cdots & H_{1n} \\ H_{21} & H_{22}-E & \cdots & H_{2n} \\ \vdots & \vdots & \vdots & \vdots \\ H_{n1} & H_{n2} & \cdots & H_{nn}-E \end{vmatrix} = 0.$$

Note that

$$U(\mathbf{r}) - V(\mathbf{r}) \geqslant 0 \qquad \text{(B.12.16)}$$

since the potential energy of an electron in a lattice is lowered by the presence of all the other ions; the atomic potential energy $U(\mathbf{r})$ takes account only of the ion at $\mathbf{r} = 0$. Thus $W_{jj}(\boldsymbol{\rho})$ is the integral of a positive quantity.

The terms $W_{jj}(\boldsymbol{\rho})$ are largest for $\boldsymbol{\rho} = 0$ and decrease rapidly as $\boldsymbol{\rho}$ increases. As an approximation, the sum in (B.12.15c) will be confined to the a nearest neighbours, assumed to be equidistant: $\boldsymbol{\rho} = \boldsymbol{\rho}_1, \boldsymbol{\rho}_2, \ldots \boldsymbol{\rho}_n$; $|\boldsymbol{\rho}_1| = |\boldsymbol{\rho}_2| = \cdots$. For a spherically symmetrical state such as an s-state one then finds that $W_{ss}(\boldsymbol{\rho}_j)$ has the same value for $j = 1, 2, \ldots n$, so that

$$\epsilon_s(\mathbf{k}) = E_s - W_{ss}(0) - W_{ss}(\boldsymbol{\rho}_1) \sum_{i=1}^{n} \cos(\mathbf{k} \cdot \boldsymbol{\rho}_i). \qquad \text{(B.12.17)}$$

The components k_i of $\mathbf{k}$ may, for an orthorhombic lattice, be defined by

$$\mathbf{k} \equiv \sum_{i=1}^{3} k_i' \mathbf{b}_i = \sum_{i=1}^{3} \frac{2\pi k_i'}{a_i} \frac{\mathbf{b}_i}{b_i} \equiv \sum_{i=1}^{3} k_i \frac{\mathbf{b}_i}{b_i}$$

so that

$$\mathbf{k} \cdot \mathbf{a}_i = k_i \frac{2\pi}{b_i} = k_i a_i.$$

Consider now a simple cubic lattice (6 nearest neighbours), so that

$$a = 6, \qquad \boldsymbol{\rho}_i = \pm \mathbf{a}_i, \quad H_{ss} = E_s - W_{ss}(0) - 2W_{ss}(\boldsymbol{\rho}_1) \sum_{i=1}^{3} \cos k_i a. \qquad \text{(B.12.18)}$$

Introduce effective masses m_b, m_t for the bottom and the top of the band by

$$\epsilon_s(\mathbf{k}) = E_s - W_{ss}(0) \begin{cases} -6W_{ss}(a) + \dfrac{\hbar^2 k^2}{2m_b} & (0 \leqslant k_i a \ll 1). \\ +6W_{ss}(a) + \dfrac{\hbar^2 k'^2}{2m_t} & (0 \leqslant \pi - k_i a = k_i' a \ll 1). \end{cases} \qquad \text{(B.12.19)}$$

Here $\mathbf{k}'$ is the wave-vector of a hole near the top of the band. Noting that, for small x, $\cos x \simeq 1 - x^2/2 + \cdots$, one finds from (B.12.18) and (B.12.19) that

$$\frac{1}{m_b} = \frac{2a^2 W_{ss}(a)}{\hbar^2} = -\frac{1}{m_t}. \tag{B.12.20}$$

Thus the effective mass is negative near the top of the band. The width of the band of allowed energies goes up with the overlap integral:

$$E_w = 12 W_{ss}(a), \tag{B.12.21}$$

while the effective mass decreases numerically as the overlap integral goes up. Analogous results hold for other cubic lattices (see table B.12.1).

The limit $V \to \infty$

This limit has already been studied in (B.7.17) and (B.7.20). Some additional results will now be added to these, using the same method. The completeness relation (B.7.7) for plane waves

$$\frac{1}{V}\sum_{\mathbf{k}} e^{i\mathbf{k}\cdot\mathbf{r}} = \delta(\mathbf{r}) \quad \text{becomes} \quad \frac{1}{8\pi^3}\int e^{i\mathbf{k}\cdot\mathbf{r}}\, d\mathbf{k} = \delta(\mathbf{r}). \tag{B.12.22}$$

The equation (B.12.4b) for the Wannier function

$$a_n(\mathbf{r}, \mathbf{R}) = \frac{1}{DN} \sum_{\mathbf{k}(\text{B.Z.})} \psi_{n\mathbf{k}}(\mathbf{r})\, e^{-i\mathbf{k}\cdot\mathbf{R}}$$

becomes

$$\frac{\Omega}{8\pi^3 D}\int_{\text{all space}} \psi_{n\mathbf{k}}\, e^{-i\mathbf{k}\cdot\mathbf{R}}\, d\mathbf{k}. \tag{B.12.23}$$

Other analogues of (B.12.1)–(B.12.4) are readily derived.

Problems

(B.12.1) Show that the Wannier function $a_n(\mathbf{r}, \mathbf{R})$ depends only on $\mathbf{r} - \mathbf{R}$, i.e. $a_n(\mathbf{r}, \mathbf{R}) = a_n(\mathbf{r} - \mathbf{R}, 0)$.

(B.12.2) Show that $a_n(\mathbf{r})$ and $v_n(\mathbf{k})$ are Fourier transforms of each other in the sense

$$a_n(\mathbf{r}) = \frac{B}{DN}\sum_{\text{all }\mathbf{k}} v_n(\mathbf{k})\, e^{i\mathbf{k}\cdot\mathbf{r}}, \qquad v_n(\mathbf{k}) = \frac{D}{B\Omega}\int_V a_n(\mathbf{r})\, e^{-i\mathbf{k}\cdot\mathbf{r}}\, d\mathbf{r}.$$

(B.12.3) Suppose the modulating parts of the Bloch functions have been chosen so that

$$\int_\Omega u^*_{n\mathbf{k}}(\mathbf{r}) u_{n'\mathbf{k}}(\mathbf{r})\, d\mathbf{r} \equiv I_{nn'} \text{ (independent of } \mathbf{k}\text{)}.$$

Table B.12.1 Further use of equation (B.12.17)

	Body-centred cubic lattice	Face-centred cubic lattice
Number of nearest neighbours	8	12
Coordinates of neighbours nearest to origin in units of the length a of the side of the cube	$(\pm\frac{1}{2}, \pm\frac{1}{2}, \pm\frac{1}{2})$; $(\pm\frac{1}{2}, \pm\frac{1}{2}, \mp\frac{1}{2})$; $(\pm\frac{1}{2}, \mp\frac{1}{2}, \pm\frac{1}{2})$; $(\mp\frac{1}{2}, \pm\frac{1}{2}, \pm\frac{1}{2})$;	$(\pm\frac{1}{2}, \pm\frac{1}{2}, 0)$; $(\pm\frac{1}{2}, 0, \pm\frac{1}{2})$; $(0, \pm\frac{1}{2}, \pm\frac{1}{2})$; $(\pm\frac{1}{2}, \mp\frac{1}{2}, 0)$; $(\pm\frac{1}{2}, 0, \mp\frac{1}{2})$; $(0, \pm\frac{1}{2}, \mp\frac{1}{2})$;
$\epsilon_s(k)$	$E_s - W_{ss}(0) - 8W_{ss}\left(\frac{\sqrt{3}a}{2}\right)\cos\frac{ak_1}{2}\cos\frac{ak_2}{2}\cos\frac{ak_3}{2}$	$E_s - W_{ss}(0) - 4W_{ss}(a)\left(\cos\frac{ak_2}{2}\cos\frac{ak_3}{2} + \cos\frac{ak_3}{2}\cos\frac{ak_1}{2} + \cos\frac{ak_1}{2}\cos\frac{ak_2}{2}\right)$
Width of allowed band	$16W_{ss}(a)$	$24W_{ss}(a)$
$\epsilon_{ss}(\mathbf{k})$ for $ak \ll 1$	$E_s - W_{ss}(0) - 8W_{ss}\left(\frac{\sqrt{3}a}{2}\right) + W_{ss}\left(\frac{\sqrt{3}a}{2}\right)a^2k^2$	$E_s - W_{ss}(0) - 12W_{ss}\left(\frac{a}{\sqrt{2}}\right) + W_{ss}\left(\frac{a}{\sqrt{2}}\right)a^2k^2$

Show that the following orthonormality expressions are obtained:

$$\int_{N\Omega} \psi^*_{n\mathbf{k}}(\mathbf{r})\psi_{n'\mathbf{k}'}(\mathbf{r})\,\mathrm{d}\mathbf{r} = NI_{nn'}\,\delta_{\mathbf{k}\mathbf{k}'}$$

$$\int_{N\Omega} a^*_n(\mathbf{r}-\mathbf{R})a_{n'}(\mathbf{r}-\mathbf{R}')\,\mathrm{d}\mathbf{r} = |D|^{-2}I_{nn'}\,\delta_{\mathbf{R}\mathbf{R}'}$$

$$\sum_{\mathbf{K}} v^*_n(\mathbf{k}+\mathbf{K})v_{n'}(\mathbf{k}+\mathbf{K}) = |B|^{-2}\Omega^{-1}I_{nn'}.$$

If $I_{nn'} = N^{-1}\delta_{nn'}$, verify that the above expressions can, by a suitable choice of B and D, be given the orthonormality properties

$$\int_{N\Omega} \psi^*_{n\mathbf{k}}\psi_{n'\mathbf{k}'}\,\mathrm{d}\mathbf{r} = \delta_{\mathbf{k}\mathbf{k}'}\,\delta_{nn'};$$

$$\int_{N\Omega} a^*_n(\mathbf{r}-\mathbf{R})a_{n'}(\mathbf{r}-\mathbf{R}')\,\mathrm{d}\mathbf{r} = \delta_{nn'}\,\delta_{\mathbf{R}\mathbf{R}'} \qquad (D = N^{-\frac{1}{2}});$$

$$\sum_{\mathbf{K}} v^*_n(\mathbf{k}+\mathbf{K})v_{n'}(\mathbf{k}+\mathbf{K}) = \delta_{nn'} \qquad [B = (N\Omega)^{-\frac{1}{2}} = V^{\frac{1}{2}}].$$

(B.12.4) Show that in the limit of an infinite crystal,

$$\epsilon_n(\mathbf{R}) = \frac{1}{C\Omega'}\int_{\Omega} E_n(\mathbf{k})\,\mathrm{e}^{-\mathrm{i}\mathbf{k}\cdot\mathbf{R}}\,\mathrm{d}\mathbf{k};$$

$$a_n(\mathbf{r}-\mathbf{R}) = \frac{1}{D\Omega'}\int_{\Omega'} u_{n\mathbf{k}}(\mathbf{r})\,\mathrm{e}^{\mathrm{i}\mathbf{k}\cdot(\mathbf{r}-\mathbf{R})}\,\mathrm{d}\mathbf{k};$$

$$a_n(\mathbf{r}) = \frac{B}{D\Omega'}\int_{\text{all }\mathbf{k}\text{-space}} v_n(\mathbf{k})\,\mathrm{e}^{\mathrm{i}\mathbf{k}\cdot\mathbf{r}}\,\mathrm{d}\mathbf{k}.$$

(B.12.5) The transformation matrix $U_{\mathbf{R}\mathbf{k}}$, which yields the Wannier function $a_n(\mathbf{r}-\mathbf{R})$ from the Bloch function $\psi_{n\mathbf{k}}(\mathbf{r})$, and the matrix $V_{\mathbf{k}\mathbf{R}}$ effecting the opposite transformation are both obtainable from equation (B.12.4). Verify that $V = U^{-1}$, and that U is a unitary matrix provided the normalization constant D is chosen as $N^{-\frac{1}{2}}$, as suggested in problem (B.12.3).

(B.12.6) Show that, if the Bloch functions satisfy the Schrödinger equation

$$H\psi_{n\mathbf{k}}(\mathbf{r}) = E_n(\mathbf{k})\psi_{n\mathbf{k}}(\mathbf{r}),$$

then the Wannier functions satisfy

$$\int_V a^*_n(\mathbf{r}-\mathbf{R})Ha_{n'}(\mathbf{r}-\mathbf{R}')\,\mathrm{d}\mathbf{r} = \epsilon_n(\mathbf{R}'-\mathbf{R})\,\delta_{nn'}$$

where H is the same Hamiltonian in both cases. The normalization suggested in problem (B.12.3), together with $C = 1$, has been adopted.

[The Hamiltonian can have off-diagonal terms with respect to Wannier functions centred on different sites. These functions therefore do not satisfy the Schrödinger equation.]

(B.12.7) Show, with the normalization suggested in problem (B.12.3), that

$$\sum_{\mathbf{k}(\text{B.Z.})} |\psi_{n\mathbf{k}}(\mathbf{r})|^2 = \sum_{\mathbf{R}(V)} |a_n(\mathbf{r}-\mathbf{R})|^2.$$

[This equation gives two ways of expressing the probability density of finding an electron in band n and at position **r**.]

(B.12.8) Let x_1, x_2, x_3 be the coordinates along the cubic axes of an electron position vector **r** when the electron is in a cubic lattice of lattice parameter a. By using a plane wave for $\psi_{n\mathbf{k}}$ in equation (B.12.4b), and summing geometrical progressions, show that the Wannier function of a free electron in this lattice satisfies

$$a(x_1, x_2, x_3) = F \prod_{j=1}^{3} \frac{\sin \pi x_j/a}{\sin \pi x_j/aN_j}$$

where F is a normalizing constant and $N_j a$ denote the lengths of the edges of the fundamental domain.

(B.12.9) By integrating instead of summing over a Brillouin zone show that one finds in problem (B.12.8)

$$a(x_1, x_2, x_3) = \frac{\sqrt{\Omega}}{\pi^3} \prod_{j=1}^{3} \frac{\sin \pi x_j/a}{x_j}$$

where the constant has been chosen so that

$$\int_V |a(\mathbf{r})|^2 \, d\mathbf{r} = 1.$$

Verify that this agrees with problem (B.12.8) in the limit $N_j \to \infty$.

(B.12.10) Show that if the overlap integral between atomic orbitals on different lattice points can be neglected then the function given in equation (B.12.9) is approximately normalized if $D = 1/\sqrt{N}$. Show that this wavefunction satisfies Bloch's theorem exactly.

(B.12.11) What is the normal derivative $\partial \epsilon_s/\partial k_n$ near a Brillouin zone plane of the quantity given by equation (B.12.17)?

(B.12.12) In the tight-binding approximation put, in generalization of equation (B.12.9),

$$\chi_{n\mathbf{k}}(\mathbf{r}) = A_{n\mathbf{k}} \sum_{\mathbf{R}(V)} \phi_n(\mathbf{r} - \mathbf{R}) \, e^{i\mathbf{k}\cdot\mathbf{R}}$$

where the atomic functions are normalized over the volume V of the crystal. Treating the overlap integral

$$\alpha(\mathbf{R}) \equiv \int_V \phi_n^*(\mathbf{r})\phi_n(\mathbf{r} - \mathbf{R}) \, d\mathbf{r}$$

as small if $\mathbf{R} \neq 0$, show that

$$|A_{n\mathbf{k}}| \simeq N^{-\frac{1}{2}} \left[1 - \frac{1}{2} \sum_{\mathbf{R}\neq 0} \alpha(\mathbf{R}) \, e^{i\mathbf{k}\cdot\mathbf{R}} \right].$$

Show that in this approximation the Wannier function is a superposition

$$a_n(\mathbf{r} - \mathbf{R}) = \sum_{\mathbf{R}'} \beta(\mathbf{R}')\phi_n(\mathbf{r} - \mathbf{R} + \mathbf{R}')$$

of atomic functions, where

$$D\beta(0) = \frac{1}{\sqrt{N}}, \qquad D\beta(\mathbf{R}') = -\frac{\alpha(\mathbf{R}')}{2\sqrt{N}} \qquad (\mathbf{R}' \neq 0).$$

(B.12.13) Let $\phi_n(\mathbf{r} - \mathbf{R})$ be the nth of an orthonormal set of atomic functions for an electron at $\mathbf{r}$ in the field of an isolated ion which is located at a lattice site with position vector $\mathbf{R}$ from the lattice site at the origin. Justify in broad terms the use of the wavefunction

$$\chi_{n\mathbf{k}}(\mathbf{r}) = (1/\sqrt{N}) \sum_{\mathbf{R}} \phi_n(\mathbf{r} - \mathbf{R})\, e^{i\mathbf{k}\cdot\mathbf{R}}$$

for an electron in a crystal based on a Bravais lattice, the sum being extended over all N lattice sites of a fundamental domain of volume V subjected to periodic boundary conditions.

Show that the wavefunction satisfies Bloch's theorem and that it is approximately normalized.

If $U(\mathbf{r})$ be the potential energy of an electron in the field of an isolated ion at the origin, and if $V(\mathbf{r})$ be the periodic potential, let

$$I_{jt}(\mathbf{k}) \equiv \delta_{jt} + \sum_{\mathbf{R}(\neq 0)} e^{-i\mathbf{k}\cdot\mathbf{R}} \int_V \phi_j^*(\mathbf{r}' - \mathbf{R})\phi_t(\mathbf{r}')\, d\mathbf{r}',$$

$$W_{jt}(\mathbf{R}) \equiv \int_V \phi_j^*(\mathbf{r})[U(\mathbf{r}) - V(\mathbf{r})]\phi_t(\mathbf{r} + \mathbf{R})\, d\mathbf{r}.$$

Prove that:

(a) $I_{jt}(\mathbf{k}) = [I_{tj}(\mathbf{k})]^*$,

(b) $W_{jt}(\mathbf{R}) - [W_{tj}(-\mathbf{R})]^* = \int_V \phi_j^*(\mathbf{r} - \mathbf{R})[U(\mathbf{r} - \mathbf{R}) - U(\mathbf{r})]\phi_t(\mathbf{r})\, d\mathbf{r}$,

(c) $W_{jj}(\mathbf{R}) = [W_{jj}(-\mathbf{R})]^*$,

(d) $\sum_{\mathbf{R}} e^{-i\mathbf{k}\cdot\mathbf{R}}\{W_{jt}(\mathbf{R}) - [W_{tj}(-\mathbf{R})]^*\} = (E_j - E_t)I_{jt}(\mathbf{k})$,

where E_n is the eigenvalue belonging to the state $\phi_n(\mathbf{r} - \mathbf{R})$.

[This confirms (B.12.14a) and (B.12.14b) of the text by direct calculation.]

CHAPTER III

Dynamics of an Electron in a Perfect Crystal

B.13 THE DYNAMICS OF ELECTRONS IN LATTICES (BASED ON WANNIER'S THEOREM)

B.13a Wannier's theorem

The fields which act on electrons may be internal or external. So far the (internal) field acting on an electron as a result of the periodic potential in a crystal has been discussed. The effect of externally applied fields may be understood in an approximate manner with the aid of Wannier's theorem. This is based on the following observations:

(i) Recall from equations (B.7.1a) and (B.7.4a) that if f is periodic in the **k**-space lattice and g is periodic in the direct lattice, and both functions are unique and continuous,

$$f(\mathbf{k}) = \sum_{\mathbf{R}(V)} f_{\mathbf{R}}\, e^{i\mathbf{k}\cdot\mathbf{R}}, \qquad g(\mathbf{r}) = \sum_{\mathbf{K}} g_{\mathbf{K}}\, e^{i\mathbf{K}\cdot\mathbf{r}}, \tag{B.13.1}$$

where **k** is any vector in **k**-space.

(ii) If $u(\mathbf{r})$ be any differentiable function and **s** an ordinary vector,

$$\begin{aligned} e^{\mathbf{s}\cdot\nabla}u(\mathbf{r}) &\equiv [1 + \mathbf{s}\cdot\nabla + \tfrac{1}{2}(\mathbf{s}\cdot\nabla)^2 + \cdots]u(\mathbf{r}) \\ &= u(\mathbf{r}) + \mathbf{s}\cdot\nabla u(\mathbf{r}) + \frac{1}{2}\sum_{i,j=1}^{3} \mathbf{s}_i\mathbf{s}_j \frac{\partial}{\partial x_i}\frac{\partial}{\partial x_j}u(\mathbf{r}) + \cdots. \end{aligned}$$

This is a Taylor expansion so that

$$e^{\mathbf{s}\cdot\nabla}u(\mathbf{r}) = u(\mathbf{r} + \mathbf{s}). \tag{B.13.2}$$

(iii) If H be a Hamiltonian for an electron in a periodic lattice, and $\psi_{n\mathbf{k}}$ be a Bloch function for eigenvalue $E_n(\mathbf{k})$, then by (B.11.10) and (B.11.17)

$$\psi_{n\mathbf{k}}(\mathbf{r} + \mathbf{R}) = e^{i\mathbf{k}\cdot\mathbf{R}}\psi_{n\mathbf{k}}(\mathbf{r}); \tag{B.13.3}$$

$$H\psi_{n\mathbf{k}}(\mathbf{r}) = E_n(\mathbf{k})\psi_{n\mathbf{k}}(\mathbf{r}). \tag{B.13.4}$$

Suppose now that $f(\mathbf{k})$ be a unique and continuous function of **k** with lattice periodicity in **k**-space. Then using in turn (i), (ii), (iii), and again (i),

$$f(-\mathrm{i}\nabla)\psi_{n\mathbf{k}}(\mathbf{r}) = \sum_{\mathbf{R}(V)} f_{\mathbf{R}}\, e^{\mathbf{R}\cdot\nabla}\psi_{n\mathbf{k}}(\mathbf{r})$$
$$= \sum_{\mathbf{R}(V)} f_{\mathbf{R}}\psi_{n\mathbf{k}}(\mathbf{r} + \mathbf{R})$$
$$= \sum_{\mathbf{R}(V)} f_{\mathbf{R}}\, e^{\mathrm{i}\mathbf{k}\cdot\mathbf{R}}\psi_{n\mathbf{k}}(\mathbf{r}),$$

so that

$$f(-\mathrm{i}\nabla)\psi_{n\mathbf{k}}(\mathbf{r}) = f(\mathbf{k})\psi_{n\mathbf{k}}(\mathbf{r}). \tag{B.13.5}$$

This is the required theorem.

In applying this theorem we wish to interpret $f(\mathbf{k})$ as the single-particle Bloch energy $E(\mathbf{k})$. $f(\mathbf{k})$ must, however, have a unique value for each $\mathbf{k}$, so that the theorem can be applied only for a given band n, assuming it not to overlap with other bands.

B.13b The effective mass theorem

Suppose an internal or external field leads to an electronic potential energy $F(\mathbf{r})$ in addition to the crystal potential $U(\mathbf{r})$, so that the total single-electron Hamiltonian is now

$$H_{\mathrm{T}} \equiv H + F \equiv H_0 + U + F, \qquad H_0 \equiv -\frac{\hbar^2\nabla^2}{2m}. \tag{B.13.6}$$

The corresponding wave function ψ must be a superposition of Bloch functions since these form a complete set:

$$\psi(\mathbf{r}) = \sum_{n}\sum_{\mathbf{k}(\mathrm{B.Z.})} c_{n\mathbf{k}}\psi_{n\mathbf{k}}(\mathbf{r}). \tag{B.13.7}$$

where the c's are expansion parameters. Using (B.13.4) it follows that

$$H_{\mathrm{T}}\psi(\mathbf{r}) = \sum_{n}\sum_{\mathbf{k}} c_{n\mathbf{k}}[E_n(\mathbf{k}) + F]\psi_{n\mathbf{k}}(\mathbf{r}).$$

By (B.11.16) the theorem (B.13.5) can be applied for given n, so that

$$H_{\mathrm{T}}\psi(\mathbf{r}) = \sum_{n}[E_n(-\mathrm{i}\nabla) + F]\left[\sum_{\mathbf{k}(\mathrm{B.Z.})} c_{n\mathbf{k}}\psi_{n\mathbf{k}}(\mathbf{r})\right]. \tag{B.13.8}$$

If it is an adequate approximation to use in (B.13.7) Bloch functions from one band only, the nth band say, then (B.13.8) shows that in its action on such an approximate wavefunction $\psi(\mathbf{r})$,

$$H_{\mathrm{T}} = H_0 + U + F \text{ can be replaced by } E_n(-\mathrm{i}\nabla) + F. \tag{B.13.9}$$

The periodic potential has disappeared. The equivalent Hamiltonian is more easily handled if $E_n(\mathbf{k})$ is an approximately known function, as will now be shown.

It is often a very good approximation to assume ellipsoidal surfaces of constant energy in **k**-space. These may conveniently be written

$$E_n(\mathbf{k}) = E_{n0} + \sum_{j=1}^{3} \frac{\hbar^2 k_j^2}{2m_j}, \tag{B.13.10}$$

where the m_j are three effective masses. The Schrödinger equation $H_T\psi(\mathbf{r}) = E\psi(\mathbf{r})$ becomes

$$\left[\sum_{j=1}^{3}\left(-\frac{\hbar^2}{2m_j}\frac{\partial^2}{\partial x_j^2}\right) + F\right]\psi(\mathbf{r}) = (E - E_{n0})\psi(\mathbf{r}), \tag{B.13.11}$$

subject to the approximation that only one band contributes to the wave-function $\psi(\mathbf{r})$. If the energy surfaces in **k**-space are spheres, (B.13.11) is just the Schrödinger equation of a free electron in a potential energy field $F(\mathbf{r})$. In both cases the effect of the periodic potential energy $W(\mathbf{r})$ is absorbed in the effective mass m_j. Important consequences of this theorem are illustrated in §B.13f.

The replacement

$$H_0 + U + F = \frac{\mathbf{p}^2}{2m} + U + F \quad \to \quad E_n(-\mathrm{i}\nabla) + F = E_n\left(\frac{\mathbf{p}}{\hbar}\right) + F \tag{B.13.12}$$

suggests that in the presence of a magnetic field, when **p** is replaced by $\mathbf{p} + e\mathbf{A}/c$, where **A** is the vector potential,

$$\frac{1}{2m}\left(\mathbf{p} + \frac{e}{c}\mathbf{A}\right)^2 + U + F \quad \to \quad E_n\left(\frac{\mathbf{p} + e\mathbf{A}/c}{\hbar}\right) + F. \tag{B.13.13}$$

The equivalent Hamiltonian (B.13.13) can be used to discuss the dia-magnetism of an electron gas in a periodic field subject to the effective mass approximation [see, for example, equation (A.3.33)].

Note that (B.13.12) suggests that if only one band (the nth, say) is important, and if the electron is in a state of reduced wave vector **k** and treated as a free electron, then its momentum can, to this approximation, be written as

$$\mathbf{p}_{n\mathbf{k}} = \hbar\mathbf{k}. \tag{B.13.14}$$

More rigorously, one should discuss the expectation value of the momentum operator in a Bloch state $(n, \mathbf{k})$; this contains contributions from all states of the type $\mathbf{k} + \mathbf{K}$ [see problem (B.11.2)].

The uncertainty relation

$$\Delta p \Delta x \geqslant \hbar, \quad \text{i.e.} \quad \Delta x \geqslant \frac{\hbar}{\Delta p}$$

implies that the additional potential energy F must vary by a small percentage in a distance Δx, since the theorem can be accurate only for distances in excess of Δx. Its magnitude may be estimated as follows: as for free particles, put $\mathbf{p} \sim \hbar\mathbf{k}$. By (B.6.5), a component of $\mathbf{k}$ lies in the range $\pm\frac{1}{2}b_j = \pm\pi/a_j$. Hence

$$\Delta x \gtrsim \frac{\hbar}{\Delta p} > \frac{\hbar}{\hbar(2\pi/a_j)} = \frac{a_j}{2\pi}. \tag{B.13.15}$$

Thus F must vary by a small fraction in a distance which exceeds, or has the order of magnitude of, the lattice parameters.

B.13c The equations of motion

Assuming the potential energy F to depend on the electron coordinate $\mathbf{r}$, but not its momentum $\mathbf{p}$, then if (B.13.12) is regarded as a classical Hamiltonian $H(\mathbf{p}, \mathbf{r})$, the Hamiltonian equations of motion are

$$\dot{\mathbf{r}} = \nabla_{\mathbf{p}}H, \qquad \dot{\mathbf{p}} = -\nabla_{\mathbf{r}}H. \tag{B.13.16}$$

Writing $v_{n\mathbf{k}}$ for the velocity of an electron in state $(n, \mathbf{k})$, and using (B.13.12) and (B.13.14), one finds

$$\mathbf{v}_{n\mathbf{k}} = \frac{1}{\hbar}\nabla_{\mathbf{k}}E_n(\mathbf{k}), \tag{B.13.17}$$

$$\hbar\dot{\mathbf{k}} = -\nabla_{\mathbf{r}}F(\mathbf{r}) = \mathbf{f}(\mathbf{r}), \tag{B.13.18}$$

where $\mathbf{f}(\mathbf{r})$ is the applied force. The velocity is seen to be normal to the surface of constant energy at the appropriate $\mathbf{k}$, and is therefore parallel to $\mathbf{k}$ only in special cases.

Strictly speaking, equations obtained by the use of classical analogies can be expected to hold in quantum mechanics only in the sense that they apply to the centre of a wave packet. This is the case also in (B.13.17) and (B.13.18). One would therefore expect a simple wave packet theory to exist which can also lead to these equations. This will be given in §B.14. All results of this section can, however, be obtained also by more precise considerations (see §B.15). The velocity $\mathbf{v}_{n\mathbf{k}}$ will be seen to be strictly the quantum mechanical expectation value of a velocity operator in a Bloch state $(n, \mathbf{k})$.

In the presence of an applied electric field $\mathbf{E}$ and applied magnetic field $\mathbf{H}$, the expression for the Lorentz force on a particle of charge q may be assumed:

$$\mathbf{f} = q\left(\mathbf{E} + \frac{1}{c}\mathbf{v}_{n\mathbf{k}} \times \mathbf{H}\right) \quad (q < 0 \text{ for electrons}). \tag{B.13.19}$$

This is an important example of a velocity-dependent force. Its derivation from a Lagrangian is given in problem (B.13.10). The derivation of the magnetic field term from the law $\mathbf{f} = q\mathbf{E}$ in the absence of a magnetic field is given in problem (B.13.13).

If only a magnetic field is acting, (B.13.18) and (B.13.19) yield

$$\dot{k} = \frac{q}{c\hbar} v_\perp H \tag{B.13.20}$$

where $v_\perp$ is the component of $\mathbf{v}_{n\mathbf{k}}$ perpendicular to the field $\mathbf{H}$. Assuming the electron is not scattered, it completes an orbit about the field direction in a time $t = 2\pi/\omega$ given by [see also problem (A.3.3)]

$$m_{cycl} \equiv \frac{qH}{\omega c} = \frac{\hbar}{2\pi} \oint \frac{\mathrm{d}k}{v_\perp}. \tag{B.13.21}$$

For a *free* electron it is easy to show [problem (B.13.7)] that $v_\perp$ is constant in a circuit, and that $\mathbf{v} = \hbar\mathbf{k}/m$. In this case $\mathrm{d}k = (m/\hbar)\,\mathrm{d}v = (m/\hbar)v_\perp\,\mathrm{d}\theta$ and

$$m_{cycl} = \frac{\hbar}{2\pi}\frac{m}{\hbar}\int_0^{2\pi} \frac{v_\perp\,\mathrm{d}\theta}{v_\perp} = m.$$

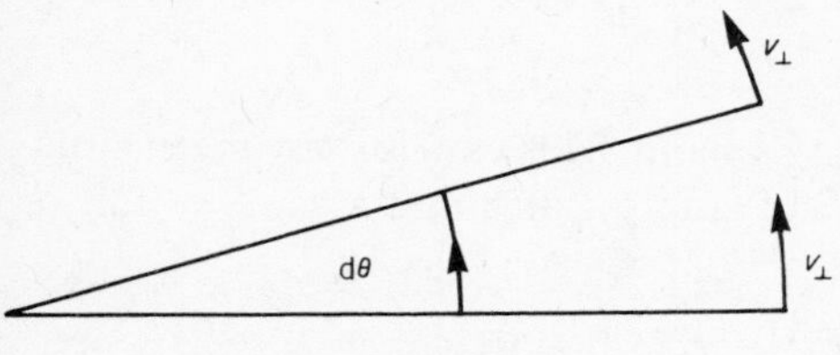

Fig. B.13.1.

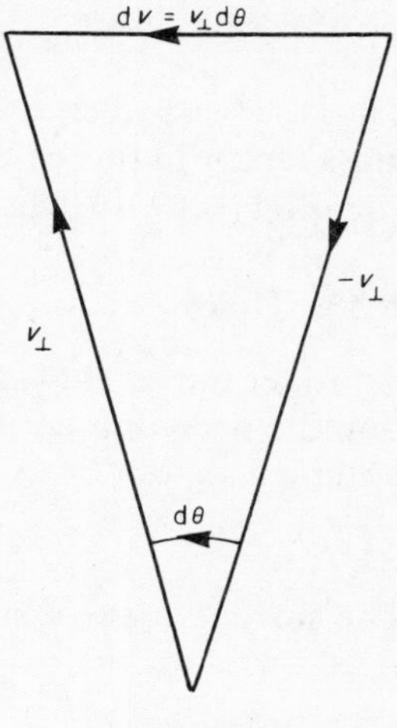

Fig. B.13.2.

This suggests that for a charge q in a *periodic lattice* a cyclotron effective mass, which is an effective mass averaged over an orbit, can be defined by the expression (B.13.21). The method of cyclotron resonances has been used to determine effective masses with great accuracy (see problem A.3.3). Equation (B.13.21) can be applied to yield information about the shape of Fermi surfaces in metals[21].

[21] J. M. Ziman, *Principles of the Theory of Solids*, Cambridge University Press, 1964, p. 250.

B.13d The effective mass tensor

We now use the equations of motion to obtain the acceleration **a** of an electron in state $(n, \mathbf{k})$. In component form $(i = 1, 2, 3)$, and using (B.13.17) and (B.13.18),

$$(\mathbf{a}_{n\mathbf{k}})_i \doteq \frac{\mathrm{d}}{\mathrm{d}t}(\mathbf{v}_{n\mathbf{k}})_i = \frac{\mathrm{d}}{\mathrm{d}t}\left[\hbar^{-1}\frac{\partial E_n(\mathbf{k})}{\partial k_i}\right]$$

$$= \sum_{j=1}^{3}\hbar^{-1}\frac{\partial^2 E_n(\mathbf{k})}{\partial k_i \partial k_j}\dot{k}_j$$

$$= \hbar^{-2}\sum_{j=1}^{3}\frac{\partial^2 E_n(\mathbf{k})}{\partial k_i \partial k_j} f_j.$$

One can compare this with a free particle of mass m when

$$a_i = m^{-1} f_i. \tag{B.13.22}$$

This suggests the rewriting of the result for an electron in a crystal:

$$(\mathbf{a}_{n\mathbf{k}})_i = \sum_{j=1}^{3}\left(\frac{1}{m}\right)_{ij}^{n\mathbf{k}} f_j, \qquad \left(\frac{1}{m}\right)_{ij}^{n\mathbf{k}} \equiv \hbar^{-2}\frac{\partial^2 E_n(\mathbf{k})}{\partial k_i \partial k_j}. \tag{B.13.23}$$

Thus the scalar reciprocal mass of a free particle is replaced by a reciprocal mass tensor in this case. It generalizes the effective mass as introduced in equation (B.13.10) (where it was treated as a diagonal tensor).

B.13e Holes

Consider the set B of states forming a band. This can, for any given situation, be decomposed into a set E of empty states and a set F of full or occupied states

$$B = E \cup F, \qquad E \cap F = 0.$$

The current due to a small number of particles in a band n is given by

$$I = q \sum_{\mathbf{k}(F)} \mathbf{v}_{n\mathbf{k}} \qquad (q < 0 \text{ for electrons}).$$

Now (a) by (B.13.17) and (B.13.23), $\mathbf{v}_{n\mathbf{k}}$ and $\mathbf{a}_{n\mathbf{k}}$ do not depend on whether the state $(n, \mathbf{k})$ is occupied or not. Also (b) a full band cannot carry a current if other bands are sufficiently removed in energy, there being no available empty states into which an electron can be excited by gaining energy from the applied electric field. These two facts enable one to put in general

$$q\left[\sum_{\mathbf{k}(F)} \mathbf{v}_{n\mathbf{k}} + \sum_{\mathbf{k}(E)} \mathbf{v}_{n\mathbf{k}}\right] = 0.$$

Hence if one uses each form of (B.13.24) only if the set of states involved is small,

$$\mathbf{I} = q \sum_{\mathbf{k}(F)} \mathbf{v}_{n\mathbf{k}} = -q \sum_{\mathbf{k}(E)} \mathbf{v}_{n\mathbf{k}}. \tag{B.13.24}$$

Considering the ith component and differentiating with respect to time,

$$\frac{\mathrm{d}I_i}{\mathrm{d}t} = q \sum_{\mathbf{k}(F)} \sum_{j=1}^{3} \left(\frac{1}{m}\right)_{ij}^{n\mathbf{k}} f_j = -q \sum_{\mathbf{k}(E)} \sum_{j=1}^{3} \left(\frac{1}{m}\right)_{ij}^{n\mathbf{k}} f_j \tag{B.13.25}$$

where the applied force is, in general, given by (B.13.19). Using suffices e and h for occupied states (electrons) and vacant states ('holes') respectively, (B.13.25) can be written

$$q_e \sum_{\mathbf{k}(F)} \sum_{j=1}^{3} \left(\frac{1}{m_e}\right)_{ij}^{n\mathbf{k}} f_j = -q_h \sum_{\mathbf{k}(E)} \sum_{j=1}^{3} \left(\frac{1}{m_h}\right)_{ij}^{n\mathbf{k}} f_j,$$

where

$$q = q_e = -q_h, \qquad \left(\frac{1}{m}\right)_{ij}^{n\mathbf{k}} = \left(\frac{1}{m_e}\right)_{ij}^{n\mathbf{k}} = -\left(\frac{1}{m_h}\right)_{ij}^{n\mathbf{k}}. \tag{B.13.26}$$

The relation for the charges may be written equivalently if $-|e|$ is put for q_e,

$$q_e = -|e|, \qquad q_h = |e|.$$

The conventions (B.13.26) can be explained by considering an almost full band. The important states are then those near the top of the energy band. Suppose that the maximum permitted energy occurs at $\mathbf{k} = \mathbf{k}_0$; then on moving away from $\mathbf{k}_0$ the electron energy must decrease. A possible set of energy surfaces near $\mathbf{k}_0$ is therefore given by

$$E_n(\mathbf{k}) = E_n(\mathbf{k}_0) - \sum_{i=1}^{3} \frac{\hbar^2 (k_i - k_{0i})^2}{2\mu_i}, \tag{B.13.27}$$

where the μ_i are positive constants. The effective mass is given by

$$\left(\frac{1}{m}\right)_{ij}^{n\mathbf{k}_0} = \hbar^{-2} \frac{\partial^2 E_n(\mathbf{k})}{\partial k_i \partial k_j} = -\frac{\delta_{ij}}{\mu_i}, \tag{B.13.28}$$

and is seen to have negative non-zero components. This has the inconvenient result that, by (B.13.23), applied forces induce accelerations in a direction opposite to that of the applied force. It is therefore more convenient to describe an almost full band in terms of the empty states, since there are fewer of them and their reciprocal effective mass components are positive. By (B.13.26) and (B.13.27)

$$\left(\frac{1}{m_h}\right)_{ij}^{n\mathbf{k}_0} = \frac{\delta_{ij}}{\mu_i}.$$

To summarize: the empty states in an almost full band represent possible states of (negatively charged) electrons whose mass components tend to be negative, or, alternatively, of positively charged particles called holes, whose mass components tend to be positive. The words 'tend to be' have been used because complicated situations can arise in which some mass components are positive and others are negative [see problem (B.13.6)].

B.13f Hydrogen-like donors[22] in a semiconductor

As an application of the preceding theory, consider the following problem:

An electron in a perfect lattice is acted upon by an ion of positive charge q. Assuming a scalar effective mass m^*, apply the effective mass theorem to discuss the bound energy levels and the Bohr radii of the orbits.

Solution of the problem

The effect of the charge q is to give rise to a potential energy $-eq/\epsilon r$ for an electron of charge $-|e|$ at a distance r, where the intervening medium has been treated as a dielectric continuum. This is justifiable for large 'orbits'. For a scalar effective mass m^* the Schrödinger equation

$$\left(-\frac{\hbar^2}{2m}\nabla^2 + U - \frac{eq}{\epsilon r}\right)\psi(\mathbf{r}) = E\psi(\mathbf{r})$$

becomes

$$\left(-\frac{\hbar^2}{2m^*}\nabla^2 - \frac{eq}{\epsilon r}\right)\psi(\mathbf{r}) = E\psi(\mathbf{r}), \tag{B.13.29}$$

the periodic potential energy U having disappeared. Now

$$\nabla^2 \equiv \frac{1}{r^2}\frac{\partial}{\partial r}\left(r^2\frac{\partial}{\partial r}\right) + \frac{1}{r^2 \sin\theta}\frac{\partial}{\partial\theta}\left(\sin\theta\frac{\partial}{\partial\theta}\right) + \frac{1}{r^2\sin^2\theta}\frac{\partial}{\partial\phi^2}$$

in spherical polar coordinates. The solution

$$\psi(r, \theta, \phi) = R(r)Y(\theta, \phi)$$

in (B.13.29) yields

$$\frac{1}{r^2}\frac{\mathrm{d}}{\mathrm{d}r}\left(r^2\frac{\mathrm{d}R}{\mathrm{d}r}\right) + \left[\frac{2m^*}{\hbar^2}\left(E - \frac{eq}{\epsilon r}\right) - \frac{\lambda}{r^2}\right]R = 0, \tag{B.13.30}$$

$$\frac{1}{\sin\theta}\frac{\partial}{\partial\theta}\left(\sin\theta\frac{\partial Y}{\partial\theta}\right) + \frac{1}{\sin^2\theta}\frac{\partial^2 Y}{\partial\phi^2} + \lambda Y = 0. \tag{B.13.31}$$

[22] For an introductory discussion of donors and acceptors, see §D.2.

The solutions of (B.13.31) are the well-known spherical harmonics

$$Y(\theta, \phi) = Y_{l,\pm m}(\theta, \phi), \qquad l \geqslant |m|, \qquad \lambda = l(l+1),$$

where l and m are integers. (B.13.30) contains the potential and yields

$$\frac{d^2R}{dr^2} + \frac{2}{r}\frac{dR}{dr} + \left\{\frac{2m^*}{\hbar^2}\left[E - \frac{eq}{\epsilon r}\right] + \frac{l(l+1)}{r^2}\right\}R = 0. \qquad \text{(B.13.32)}$$

For negative E we define n and x by

$$E \equiv -\frac{m^*e^2q^2}{2n^2\hbar^2\epsilon^2}, \qquad r \equiv \frac{n\hbar^2\epsilon}{2m^*eq}x, \qquad \text{(B.13.33)}$$

whence (B.13.32) is

$$\frac{d^2R}{dx^2} + \frac{2}{x}\frac{dR}{dx} + \left[-\frac{1}{4} + \frac{n}{x} - \frac{l(l+1)}{x^2}\right]R = 0.$$

Looking for solutions of the form

$$R = u(x)x^l\,e^{-x/2}, \qquad \text{(B.13.34)}$$

the equation for u is found to be

$$x\frac{d^2u}{dx^2} + (2l + 2 - x)\frac{du}{dx} + (n - l - 1)u = 0. \qquad \text{(B.13.35)}$$

Satisfactory solutions are found only if $n - l - 1$ is a non-negative integer. Since l can have values 0, 1, 2, . . . , the allowed values of n are

$$n = 1, 2, 3, \ldots \qquad \text{with } n \geqslant l + 1.$$

Hence the bound energies are given by

$$E_n = -\frac{m^*e^2q^2}{2n^2\hbar^2\epsilon^2} \qquad (n = 1, 2, \ldots) \qquad \text{(B.13.36)}$$

For a potential varying as r^s the virial theorem states that the expectation value of the kinetic energy and that of the potential energy in eigenstate n of the Hamiltonian are related by

$$2T_{nn} = sV_{nn}.$$

But $T_{nn} + V_{nn} = E_n$, whence $(s/2 + 1)V_{nn} = E_n$, so that from (B.13.36) and putting $s = -1$,

$$V_{nn} = -\frac{m^*e^2q^2}{n^2\hbar^2\epsilon^2}.$$

We now define the 'radius' a_n of the nth 'orbit' by

$$-\frac{eq}{\epsilon a_n} \equiv V_{nn}, \qquad \text{whence } a_n = \frac{n^2\hbar^2\epsilon}{m^*eq}. \tag{B.13.37}$$

One can also obtain a_n from the Bohr theory by equating the electrostatic attraction between the ion and the electron to the centrifugal force

$$\frac{qe}{\epsilon a_n^2} = m^*v^2/a_n.$$

Applying the Bohr quantum condition to the angular momentum one obtains

$$m^*va_n = n\hbar.$$

Eliminating v between these two equations,

$$\frac{qe}{\epsilon a_n^2} = \frac{n^2\hbar^2}{m^*a_n^3},$$

so that

$$a_n = \frac{n^2\hbar^2\epsilon}{m^*qe} \qquad \text{as before.}$$

A third method of obtaining a_n is to evaluate the expectation value of the distance r between electron and nucleus. This is a kind of equivalent radius of the orbit, requiring merely the radial part of the wavefunction of an electron in a Coulomb field. This is

$$R_{nl}(r) = \sqrt{\left\{\frac{(n-l-1)!}{2n[(n+l)!]^3}\left(\frac{2}{na_1}\right)\right\}^3}\,(2\rho)^l\,\mathrm{e}^{-\rho}L_{n+l}^{2l+1}(2\rho)$$

where the L's are associated Laguerre functions, $\rho \equiv r/na_1$, and a_1 is given by (B.13.37). The normalization is here such that

$$\int_0^\infty R_{nl}^2 r^2\,\mathrm{d}r = 1.$$

Note that n is the principal quantum number and $l = 0, 1, \ldots, n-1$ is the angular momentum quantum number. Now

$$\begin{aligned}(r^k)_{n,l} &= \int_0^\infty r^{k+2}R_{nl}^2\,\mathrm{d}r \\ &= 2^{-k}(na_1)^k\frac{(n-l-1)!}{2n[(n+l)!]^3}\int_0^\infty x^{k+2+2l}\,\mathrm{e}^{-x}[L_{n+l}^{2l+1}(x)]^2\,\mathrm{d}x.\end{aligned}$$

The following formula is given in discussions of Laguerre functions:

$$\int_0^\infty x^p\, e^{-x}[L_k^n(x)]^2\, dx = p!(k!)^2 \sum_{s=0}^{k-n} (-1)^{2k-s} \binom{p-n}{k-n-s}^2 \binom{-p-1}{s}.$$

Hence

$$(r^k)_{nl} = \left(\frac{na_1}{2}\right)^k \frac{(n-l-1)!}{2n[(n+l)!]}(k+2l+2)!$$

$$\times \sum_{s=0}^{n-l-1} (-1)^{2(n+l)+s} \binom{k+1}{n-l-1-s}^2 \binom{-k-2l-3}{s}.$$

The following cases are noteworthy. If $l = n-1$,

$$(r^k)_{nn-1} = \left(\frac{na_1}{2}\right)^k (2n+k)(2n+k-1)\cdots(2n+1);$$

$$(r)_{nn-1} = \frac{na_1}{2}(2n+1) = a_1 n^2\left(1 + \frac{1}{2n}\right) = a_n\left(1 + \frac{1}{2n}\right).$$

Thus for large n, $r_{nn-1} \to a_n$.

The above is essentially based on the standard theory of the hydrogen atom. The bound energy levels E_n of principal quantum number n are related to the bottom E_c of the conduction band by

$$E_c - E_n = \frac{m^*}{\epsilon^2 m}\left(\frac{q}{e}\right)^2 \frac{1}{n^2} \times 13{\cdot}6\,(\text{eV}) \qquad (n = 1, 2, \ldots), \quad (\text{B.13.38})$$

where e and m are free electron charge and mass, and $I_0 \equiv e^4 m/2\hbar^2$ has been replaced by 13·6 eV. The spread of the wave function in such a localized state is given by the Bohr radius

$$a_n = \frac{\epsilon m}{m^*}\frac{e}{q} n^2 \times 0{\cdot}53\ \text{Å} \qquad (\text{B.13.39})$$

where $a_0 \equiv \hbar^2/me^2$ has been replaced by 0·53 Å. For group V donors in Ge ($m^* \sim 0{\cdot}2\,m$, $\epsilon = 16$, $|q| = e$)

$$E_c - E_n \sim \frac{0{\cdot}011}{n^2}\,\text{eV}, \qquad a_n \sim 43n^2\ \text{Å}.$$

For Si ($m^* \sim 0{\cdot}4m$, $\epsilon = 12$, $|q| = e$)

$$E_c - E_n \sim \frac{0{\cdot}038}{n^2}\,\text{eV}, \qquad a_n \sim 16n^2\ \text{Å}.$$

The larger the radius the better the approximation used here, which supposes that an electron is acted upon by a positive point charge. Strictly, the more complicated interaction between an electron and a polarizable ion should be used.

Problems

Energy surfaces and effective mass theory

(B.13.1) Show that near point $\mathbf{k}_0$ of the first Brillouin zone the electron energy for band n can be written as

$$E_n(\mathbf{k}) = E_n(\mathbf{k}_0) + \hbar(\mathbf{k} - \mathbf{k}_0) \cdot \mathbf{v}_{n\mathbf{k}_0}$$
$$+ \tfrac{1}{2}\hbar^2 \sum_{ij=1}^{3} \left(\frac{1}{m}\right)_{ij}^{n\mathbf{k}} (\mathbf{k} - \mathbf{k}_0)_i(\mathbf{k} - \mathbf{k}_0)_j + \cdots.$$

(B.13.2) Two particles of masses m_1 and m_2 whose coordinates are denoted by $\mathbf{x}_1$ and $\mathbf{x}_2$ have an interaction energy $V(\mathbf{x}_1 - \mathbf{x}_2)$. Show that in coordinate representation

$$\frac{p_1^2}{2m_1} + \frac{p_2^2}{2m_2} = \frac{P^2}{2(m_1 + m_2)} + \frac{p^2}{2\mu}, \quad \frac{1}{\mu} \equiv \frac{1}{m_1} + \frac{1}{m_2},$$

where $\mathbf{p}$ is the momentum conjugate to $\mathbf{x} \equiv \mathbf{x}_1 - \mathbf{x}_2$ and P is conjugate to

$$\mathbf{X} = \frac{m_1\mathbf{x}_1 + m_2\mathbf{x}_2}{m_1 + m_2}.[23]$$

(B.13.3) A weakly bound exciton is an electron–hole complex in a periodic lattice whose wavefunction extends over many lattice atoms and having, in the notation of problem (B.13.2), $V = -e^2/\epsilon x$. Assuming that the effect of the periodic potential is to give the two particles scalar effective masses, show that for a given wave-vector $\mathbf{k}$ of the exciton there exist hydrogen-like bound states at energies

$$E_n(\mathbf{k}) = E_G - \frac{\mu e^4}{2\hbar^2\epsilon^2 n^2} + \frac{\hbar^2 k^2}{2(m_1 + m_2)},$$

where E_G is the bandgap[24].

(B.13.4) Let J be the current density in the x-direction and v be the velocity of the carriers in the absence of an applied magnetic field; let H be the applied magnetic field in the z-direction and E_y be the electric field induced by the Lorentz force in the y-direction. Give diagrams to indicate the situations for (i) negatively charged, (ii) positively charged current carriers, both of positive mass, and check that the induced potential difference in the direction between two points differing only in the y-coordinate is opposite in the two cases.

[What used to be called the 'anomalous' Hall effect in semiconductors was found to be due to the fact that conduction by holes dominated conduction by electrons; see §A.5.]

[23] L. I. Schiff, *Quantum Mechanics*, McGraw-Hill, New York, 1949, § 16.
[24] R. S. Knox, *Theory of Excitons*, Academic Press, New York, 1963, p. 38.

(B.13.5) An 'ellipsoidal' surface of constant energy has
(i) $m_1 \gg m_2, m_3 > 0$;
(ii) $m_1, m_2 > 0, m_3 < 0$.
Discuss the resulting shapes in the two cases.

[(i) The ellipsoids are elongated in the k_1-direction. (ii) This case arises near the zone boundary in Cu, Ag and Au. The surfaces are hyperbolic and form 'necks'.]

(B.13.6) A family of surfaces of constant energy has the form

$$E_n(\mathbf{k}) = \frac{\hbar^2}{2m}[\alpha k^2 - \beta(k_1^2k_2^2 + k_2^2k_3^2 + k_3^2k_1^2)k^{-2}]$$

where $\alpha, \beta > 0$. Show that

$$m\left(\frac{1}{m}\right)_{11}^{nk} \equiv \frac{m}{\hbar^2}\frac{\partial^2 E}{\partial k_1^2} = \alpha - \beta(k_2^4 + k_2^2k_3^2 + k_3^4)(-3k_1^2 + k_2^2 + k_3^2)k^{-6};$$

$$m\left(\frac{1}{m}\right)_{12}^{nk} = -\frac{2\beta k_1 k_2}{k^6}(2k_1^2k_2^2 + k_1^2k_3^2 + k_2^2k_3^2 - k_3^4).$$

Hence verify that there are points in **k**-space for which the components of the reciprocal effective mass tensor can have opposite signs.

[Such energy surfaces cannot be reduced to the ellipsoidal form discussed in the text. The formulae are approximate representations near the six (100) axes of **k**-space for Ge, where $\alpha = 4{\cdot}8$, $\beta = 9{\cdot}4$.[25]]

The Lorentz force

(B.13.7) An electron is acted upon only by a homogeneous magnetic field H in the y-direction, subject to the initial conditions $x = z = \dot{x} = 0$, $\dot{z} = u$ at $t = 0$. Assuming that no electric field is present use the Lorentz force to show the following:
(i) $\dot{x} = \omega z$, $\dot{z} = \omega x + u$ where $\omega \equiv |e|H/mc$.
(ii) By considering $\phi = z + ix - iu/\omega$, show that

$$x = \frac{u}{\omega}(1 - \cos \omega t); \qquad z = \frac{u}{\omega}\sin \omega t.$$

(iii) $$\left(x - \frac{u}{\omega}\right)^2 + z^2 = \left(\frac{u}{\omega}\right)^2,$$

so that the particle which moved along the z-axis is deflected into part of a circle of radius $r_0 = u/\omega$ whose centre lies on the x-axis at r_0, Fig. B.13.3.
(iv) If the field acts over a small enough distance so that on emerging from it $x = x_f \ll z = z_f$, show that the deflexion is

$$x_f = z_f^2/2r_0.$$

(v) $\dot{x}^2 + \dot{z}^2 = u^2 =$ constant at all times $t > 0$.

[This is the principle of the magnetic lens. The result (v) is referred to in connexion with equation (B.13.21).]

[25] H. Krömer, *Progress in Semiconductors*, Heywood, London, 1960, vol. 4, p. 1.

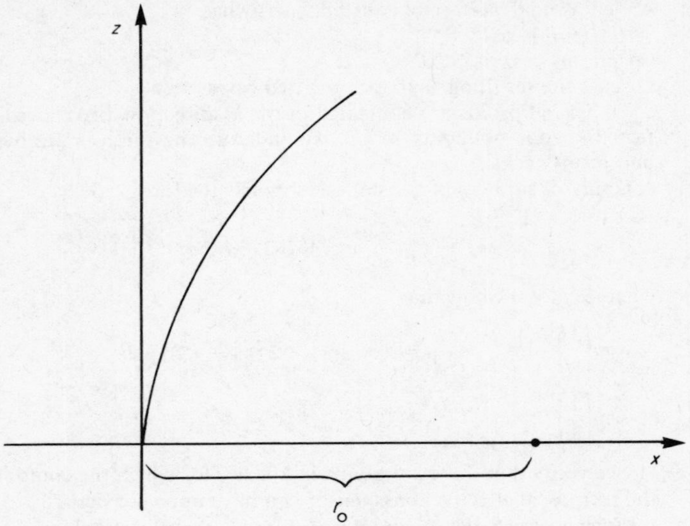

Fig. B.13.3.

(B.13.8) From (B.13.17) and (B.13.20) show that

$$v_\perp = \hbar^{-1} \mathrm{d}E_n(\mathbf{k})/\mathrm{d}k_\perp,$$

where $\mathrm{d}k_\perp$ is an increment of $\mathbf{k}$ normal to the surface of constant energy and in the plane of the orbit induced by the magnetic field.

Hence show from (B.13.2) that the cyclotron effective mass satisfies

$$m_{cycl} = \frac{\hbar^2}{2\pi} \frac{\partial A}{\partial E_n(\mathbf{k})},$$

where A is the area enclosed by the orbit.

$$\left[\frac{\hbar}{2\pi}\int \frac{\mathrm{d}k}{v_\perp} = \frac{\hbar^2}{2\pi}\int \frac{\mathrm{d}k_\perp}{\mathrm{d}E_n(\mathbf{k})}\mathrm{d}k.\right]$$

(B.13.9) Consider a set $t_1, t_2, \ldots, t_v$ of independent variables, and a set $q_i(t_1, \ldots, t_v)$ $(i = 1, 2, \ldots a)$ of argument functions. Let

$$L \equiv L\left(t_1, \ldots t_v; q_1, \ldots, q_a; \quad \frac{\partial q_1}{\partial t_1}, \frac{\partial q_1}{\partial t_2}, \ldots \frac{\partial q_a}{\partial t_v}\right),$$

and suppose that

$$\int_{t_{1a}}^{t_{1b}} \cdots \int_{t_{va}}^{t_{vb}} L \, \mathrm{d}t, \ldots \mathrm{d}t_v$$

has an extremum with respect to arbitrary small variations

$$q_i(t_1, \ldots, t_v) \to q_i(t_1, \ldots, t_v) + \alpha\eta(t_1, \ldots, t_v)$$

of the argument functions. Here the t_{ia} and t_{jb} are fixed and given values and it is also supposed that

$$\eta(t_{1a}, \ldots, t_{va}) = \eta(t_{1b}, \ldots, t_{vb}) = 0.$$

Prove that Lagrange's equations follow ($r = 1, 2, \ldots a$; $s = 1, 2, \ldots v$):

$$\frac{\partial L}{\partial q_r} - \sum_{s=1}^{s=v} \frac{\partial}{\partial t_s}\left(\frac{\partial L}{\partial q_{rs}}\right) = 0, \qquad q_{rs} \equiv \partial q_r/\partial t_s.$$

[This is a generalization of Hamilton's principle and can be written $\delta\int\ldots\int L\mathrm{d}t_1\ldots\mathrm{d}t_v = 0$. It states that the integral is an extremum when its value, taken along a path for which Lagrange's equation are satisfied, is compared with its value for neighbouring paths having the same terminal points.]

(B.13.10) A Lagrangian with one independent variable (time t) and three argument functions [components of the position vector of a particle, $x_r(t)$] is given by

$$L = m_0c^2\left[1 - \sqrt{\left(1 - \frac{v^2}{c^2}\right)}\right] - V(x_1, x_2, x_3) + \frac{e}{c}\sum_{r=1}^{3} A_r\dot{x}_r - e\phi.$$

Here ϕ is a scalar potential and A_r are Cartesian components of a vector potential both depending on the position (x_1, x_2, x_3) of a particle of rest mass m_0, charge e, and velocity

$$v \equiv \sqrt{\left[\sum_{r=1}^{3} (\mathrm{d}x_r/\mathrm{d}t)^2\right]} \equiv \sqrt{\left[\sum_{r=1}^{3} (\dot{x}_r)^2\right]}.$$

Prove the following:

(i) $p_s \equiv \dfrac{\partial L}{\partial \dot{x}_s} = m\dot{x}_s + \dfrac{e}{c}A_s \qquad [m \equiv m_0/\sqrt{(1 - v^2/c^2)}].$

(ii) $(m^2 - m_0^2)c^2 = \left(\mathbf{p} - \dfrac{e}{c}\mathbf{A}\right)^2.$

(iii) The Hamiltonian H is

$$\sum p_r\dot{x}_r - L = c\sqrt{\left[\left(\mathbf{p} - \frac{e}{c}\mathbf{A}\right)^2 + m_0^2c^2\right]} - m_0c^2 + e\phi + V.$$

(iv) Lagrange's equations can be expressed as the rate of change of mechanical momentum

$$\frac{\mathrm{d}}{\mathrm{d}t}(m\dot{\mathbf{x}}) = e\left(\mathbf{E} + \frac{1}{c}\mathbf{v} \times \mathbf{H}\right) - \operatorname{grad} V,$$

where $\mathbf{E} = -\operatorname{grad}\phi$, $\mathbf{H} = \operatorname{curl}\mathbf{A}$.

[The mechanical equations of classical and of relativistic mechanics are obtained if $\phi = A = 0$. The Lorentz force equation for a free particle is also a special case.]

(B.13.11) If, in the preceding problem, $v/c \to 0$, show that

(i) $(m - m_0)c^2 = \frac{1}{2}m_0v^2.$

(ii) $H = \dfrac{1}{2m_0}\left(p - \dfrac{e}{c}A\right)^2 + e\phi + V = \frac{1}{2}m_0v^2 + e\phi + V.$

(B.13.12) Two inertial frames S, S' are in standard configuration if (i) the coordinate plane $y = 0$ of S coincides permanently with the coordinate plane $y' = 0$ of S' (similarly, the planes $z = 0$ and $z' = 0$ coincide permanently); (ii) the times in S and S' have a common zero $t = t' = 0$, and at this instant the two origins O and O' coincide; (iii) the x-axis of S and the x-axis of S' coincide and have the same sense; (iv) S moves with constant velocity v along the positive x-axis of S.

With these conventions, a particle has velocity u parallel to the x-axis of S where it is acted upon by a force f. Prove from the Lorentz transformation that in S'

$$\mathbf{f}' = \left[f_x \ , \ \frac{f_y}{\beta(1 - uv/c^2)} \ , \ \frac{f_z}{\beta(1 - uv/c^2)}\right]$$

where $\beta \equiv (1 - v^2/c^2)^{-\frac{1}{2}}$.

If electric and magnetic fields $\mathbf{E}$, $\mathbf{H}$ act in S, prove that

$$\mathbf{E}' = \left[E_x \ , \ \beta\left(E_y - \frac{v}{c}H_z\right) \ , \ \beta\left(E_z + \frac{v}{c}H_y\right)\right];$$

$$\mathbf{H}' = \left[H_x \ , \ \beta\left(H_y + \frac{v}{c}E_z\right) \ , \ \beta\left(H_z - \frac{v}{c}E_y\right)\right].$$

(B.13.13) Deduce the expression for the Lorentz force with non-zero magnetic field from the expression with zero magnetic field.

[Use the preceding problem. If S' is the instantaneous rest frame, $u = v$ and $\mathbf{f}' = q\mathbf{E}'$, where q is the charge on the particle. Substituting for $\mathbf{f}'$ and $\mathbf{E}'$,

$$(f_x, \beta f_y, \beta f_z) = \left[qE_x \ , \ \beta\left(E_y - \frac{u}{c}H_z\right) \ , \ \beta\left(E_z + \frac{u}{c}H_y\right)\right],$$

whence, in S,

$$f = q\left(\mathbf{E} + \frac{1}{c}\mathbf{u} \times \mathbf{H}\right).\Big]$$

The hydrogen atom problem

(B.13.14) N hydrogen-like donors are randomly distributed in a semiconductor of dielectric constant ϵ and effective mass m^* for the conduction band. If a_n be the radius of the nth Bohr orbit and r be the mean distance between donors, and if N is expressed in cm^{-3}, show that

$$\frac{a_n}{r} = 0{\cdot}852 \times 10^{-8} \frac{\epsilon n^2}{m^*/m} N^{\frac{1}{3}}.$$

[Use equations (B.13.38) and (B.13.39).]

(B.13.15) Suppose equation (B.13.38) for the activation energy $I_n \equiv E_c - E_n$ of an impurity electron out of the nth Bohr orbit is not in good agreement with experiment. Show that a second estimate of a_n is then obtainable from the empirical value of I_n, and is given by

$$a_n = \frac{n}{\sqrt{(m^*/m)}} a_0 \sqrt{\left(\frac{I_0}{I_n}\right)}.$$

Show that, with r defined as in problem (B.13.14), this leads to

$$\frac{a_n}{r} = 3{\cdot}14 \times 10^{-8} \frac{n}{(m^*/m)I_n} N^{\frac{1}{3}},$$

where I_n is expressed in eV and N in cm^{-3}.

(B.13.16) For a certain semiconductor $\epsilon = 11$, $m^* = \frac{1}{3}m$, $I_1 = 0{\cdot}1$ eV, $N = 10^{17}$ cm^{-3} (this applies roughly to silicon in GaP). Show that $a_1/r = 0{\cdot}13$ or $0{\cdot}08$ depending on whether the result of problem (B.13.14) or (B.13.15) is used respectively.

[At low temperatures the 'hopping' of electrons from donor to donor is found to start at values of $a_1/r \sim 0{\cdot}10$ and to become more pronounced for $a_1/r > 0{\cdot}10$].

(B.13.17) Given the dielectric constant ϵ and the empirical activation energy $I_1 \equiv E_c - E_1$, show that a *donor electron effective mass* may be defined by

$$\frac{m^*}{m} = \frac{\epsilon^2 n^2 I_n}{I_0}.$$

B.14 THE DYNAMICS OF ELECTRONS IN LATTICES (SIMPLE THEORY OF THE EQUATIONS OF MOTION)

B.14a The velocity of a Bloch wave packet

The time-dependent Bloch function is of the form

$$\psi_{n\mathbf{k}}(\mathbf{r}, t) = \psi_{n\mathbf{k}}(\mathbf{r})\,\mathrm{e}^{-\mathrm{i}E_n(\mathbf{k})t/\hbar} = u_{n\mathbf{k}}(\mathbf{r})\,\mathrm{e}^{\mathrm{i}\mathbf{k}\cdot\mathbf{r}}\,\mathrm{e}^{-\mathrm{i}E_n(\mathbf{k})t/\hbar}. \quad \text{(B.14.1)}$$

A wave packet based on Bloch states from band n and therefore offering a representation of the wavefunction (B.13.7) is

$$\psi(\mathbf{r}, t) = \int_{\Omega'} a(\mathbf{k})\psi_{n\mathbf{k}}(\mathbf{r}, t)\,\mathrm{d}\mathbf{k}, \quad \text{(B.14.2)}$$

where the integral extends over a Brillouin zone and $a(\mathbf{k})$ is an amplitude function. (B.14.2) is the analogue of equation (B.13.7). The wave packet will, in general, have a strong maximum at a wave vector $\mathbf{k}_0$, so that it is convenient to expand

$$E_n(\mathbf{k}) = E_n(\mathbf{k}_0) + [\nabla_{\mathbf{k}}E_n(\mathbf{k})]_{\mathbf{k}=\mathbf{k}_0}\cdot\boldsymbol{\eta} + \cdots (\boldsymbol{\eta} \equiv \mathbf{k} - \mathbf{k}_0). \quad \text{(B.14.3)}$$

Suppose $u_{n\mathbf{k}}$ varies sufficiently slowly with $\mathbf{k}$ to be taken out of the integral (B.14.2) and that $\boldsymbol{\eta}$ is sufficiently small for all important $\mathbf{k}$. Then

$$\begin{aligned}\psi(\mathbf{r}, t) &= u_{n\mathbf{k}_0}(\mathbf{r})\exp\{\mathrm{i}[\mathbf{k}_0\cdot\mathbf{r} - E_n(\mathbf{k}_0)t/\hbar]\}\\ &\quad\times \int_{\Omega'} a(\mathbf{k})\exp(\mathrm{i}\{\mathbf{r} - [\nabla_{\mathbf{k}}E_n(\mathbf{k}_0)]_{\mathbf{k}=\mathbf{k}_0}t/\hbar\}\cdot\boldsymbol{\eta})\,\mathrm{d}\mathbf{k}\\ &\equiv \psi_{n\mathbf{k}_0}(\mathbf{r}, t)A, \end{aligned} \quad \text{(B.14.4)}$$

where A is an amplitude which determines the group behaviour. It varies more slowly than $\psi_{n\mathbf{k}_0}$ and depends on the second term in braces. Thus the wave packet has a constant amplitude for points r satisfying

$$\mathbf{r} - [\nabla_{\mathbf{k}} E_n(\mathbf{k})]_{\mathbf{k}=\mathbf{k}_0} t/\hbar = \text{constant}.$$

It has constant phase for points satisfying

$$\mathbf{k}_0 \cdot \mathbf{r} - E_n(\mathbf{k}_0)t/\hbar = \text{constant}.$$

Differentiation yields group and phase velocity respectively:

$$\mathbf{v}_g = \dot{\mathbf{r}} = \frac{1}{\hbar}[\nabla_{\mathbf{k}} E_n(\mathbf{k})]_{\mathbf{k}=\mathbf{k}_0}. \tag{B.14.5}$$

Thus for a sharp wave packet centred on $\mathbf{k}$ the group velocity is

$$v_{n\mathbf{k}} = \frac{1}{\hbar}\nabla_{\mathbf{k}} E_n(\mathbf{k}). \tag{B.14.6}$$

This phase velocity satisfies

$$\mathbf{v}_p \cdot \mathbf{k}_0 = \frac{1}{\hbar}E_n(\mathbf{k}_0). \tag{B.14.7}$$

The first of these results confirms (B.13.17) as expected.

We have worked from first principles. One might have argued more quickly, but less instructively, by observing that

$$\hbar\omega = h\nu = E_n(\mathbf{k}). \tag{B.14.8}$$

The group velocity formula

$$\mathbf{v}_g = \nabla_{\mathbf{k}}\omega(\mathbf{k}) \tag{B.14.9}$$

again leads to (B.14.5).

B.14b The $\mathbf{k}\dot{\mathbf{k}}$-formula from the velocity

Suppose a force $\mathbf{f}(\mathbf{r})$ acts on an electron which is represented by a wave packet formed of Bloch states of band n contributed to mainly by states whose wave-vectors lie near $\mathbf{k}$. The work done by the force in time δt is

$$\delta W = \mathbf{f} \cdot \mathbf{v}_{n\mathbf{k}}\, \delta t.$$

Suppose that as a result of this force the wave-vector of the electron changes by $\delta\mathbf{k}$ in this period. Then the change in the energy of the electron is

$$\nabla_{\mathbf{k}} E_n(\mathbf{k}) \cdot \delta\mathbf{k} = \hbar\mathbf{v}_{n\mathbf{k}} \cdot \delta\mathbf{k}.$$

Since the two expressions must be equal,

$$\mathbf{f} \cdot \mathbf{v}_{n\mathbf{k}}\, \delta t = \hbar \mathbf{v}_{n\mathbf{k}} \cdot \delta \mathbf{k}.$$

This must hold for forms **f** of wide variety, so that

$$\hbar \dot{\mathbf{k}} = \mathbf{f}.$$

This is equation (B.13.8).

B.15 THE DYNAMICS OF ELECTRONS IN LATTICES (RIGOROUS INTRODUCTION OF A VELOCITY OPERATOR AND DERIVATION OF THE EFFECTIVE MASS SUM RULE)

A more abstract, but more exact, theory of some basic equations for electrons in lattices will now be given. A third approach to the formula (B.13.17) or (B.14.6) for the velocity of an electron will thus be provided. This approach will also enable one to derive an exact equation relating the effective mass tensor [see equation (B.13.23)] and matrix elements of the velocity operator. These derivations will be based on general properties of Hamiltonians which depend on parameters, and we shall turn to these first.

B.15a General theory[26]

Consider a Hermitian operator $H_{\boldsymbol{\alpha}}$ with eigenvalues and eigenfunctions specified by the equation

$$H_{\boldsymbol{\alpha}} u_{nl}(\boldsymbol{\alpha}) = E_n(\boldsymbol{\alpha}) u_{nl}(\boldsymbol{\alpha}). \tag{B.15.1}$$

The suffix l numbers degenerate eigenfunctions, and the components $\alpha_1, \alpha_2, \ldots$ of $\boldsymbol{\alpha}$ represent parameters entering into H. This may be written

$$A_n(\boldsymbol{\alpha}) u_{nl}(\boldsymbol{\alpha}) = 0, \qquad A_n(\boldsymbol{\alpha}) \equiv H_{\boldsymbol{\alpha}} - E_n(\boldsymbol{\alpha}). \tag{B.15.2}$$

Differentiating with respect to α_i and taking matrix elements

$$\left[A_n(\boldsymbol{\alpha}) \frac{\partial}{\partial \alpha_i} + \frac{\partial A_n}{\partial \alpha_i} \right] u_{nl} = 0, \tag{B.15.3}$$

$$\begin{aligned} \left\langle n'l' \left| \frac{\partial A_n}{\partial \alpha_i} \right| nl \right\rangle &= - \left\langle n'l' \left| A_n(\boldsymbol{\alpha}) \frac{\partial}{\partial \alpha_i} \right| nl \right\rangle \\ &= [E_n(\boldsymbol{\alpha}) - E_{n'}(\boldsymbol{\alpha})] \left\langle n'l' \left| \frac{\partial}{\partial \alpha_{i}} \right| nl \right\rangle . \end{aligned} \tag{B.15.4}$$

[26] D. J. Morgan and P. T. Landsberg, *Proc. Phys. Soc.*, **86**, 261 (1965).

Differentiating (B.15.3) again,

$$\left(\frac{\partial A_n}{\partial \alpha_j}\frac{\partial}{\partial \alpha_i} + A_n \frac{\partial^2}{\partial \alpha_i\, \partial \alpha_j} + \frac{\partial^2 A_n}{\partial \alpha_i\, \partial \alpha_j} + \frac{\partial A_n}{\partial \alpha_i}\frac{\partial}{\partial \alpha_j}\right) u_{nl} = 0. \qquad \text{(B.15.5)}$$

Hence

$$-\left\langle nl' \left| \frac{\partial^2 A_n}{\partial \alpha_i\, \partial \alpha_j} \right| nl \right\rangle = \boxed{2} \sum_{\substack{n'',l \\ (n'' \neq n)}} \left\langle nl' \left| \frac{\partial A_n}{\partial \alpha_i} \right| n''l'' \right\rangle \left\langle n''l'' \left| \frac{\partial}{\partial \alpha_j} \right| nl \right\rangle, \qquad \text{(B.15.6)}$$

noting that the second term in (B.15.5) does not contribute by (B.15.2), and that the diagonal term in (B.15.6) can be omitted by (B.15.4). The $\boxed{2}$ means that to each term which follows there should be added another term, not written down, obtained by interchanging i and j. Eliminating the $\partial/\partial\alpha_j$ term by (B.15.4),

$$-\left\langle nl' \left| \frac{\partial^2 A_n}{\partial \alpha_i\, \partial \alpha_j} \right| nl \right\rangle = \boxed{2} \sum_{n''(\neq n), l''} \frac{\langle nl' | (\partial A_n/\partial \alpha_i) | n''l'' \rangle \langle n''l'' | (\partial A_n/\partial \alpha_j) | nl \rangle}{E_n(\boldsymbol{\alpha}) - E_{n''}(\boldsymbol{\alpha})}. \qquad \text{(B.15.7)}$$

Equations (B.15.4) and (B.15.7) are the main results needed here.

B.15b Application to Bloch functions

For a given reduced wave-vector $\mathbf{k}$ and Bloch function

$$|n\mathbf{k}) \equiv \psi_{n\mathbf{k}}(\mathbf{r}) \equiv u_{n\mathbf{k}}(\mathbf{r})\, e^{i\mathbf{k}\cdot\mathbf{r}} \qquad \text{(B.15.8)}$$

the Schrödinger equation for $n = 1, 2, \ldots$ is

$$H\psi_{n\mathbf{k}}(\mathbf{r}) = E_n(\mathbf{k})\psi_{n\mathbf{k}}(\mathbf{r}). \qquad \text{(B.15.9)}$$

We now introduce $\mathbf{k}$ as a parameter vector of three components into the Hamiltonian by the transformation[27]

$$H_{\mathbf{k}} \equiv e^{-i\mathbf{k}\cdot\mathbf{r}} H\, e^{i\mathbf{k}\cdot\mathbf{r}}. \qquad \text{(B.15.10)}$$

The Schrödinger equation is then

$$H_{\mathbf{k}} u_{n\mathbf{k}}(\mathbf{r}) = E_n(\mathbf{k}) u_{n\mathbf{k}}(\mathbf{r}). \qquad \text{(B.15.11)}$$

Thus the theory based on (B.15.1) applies, with

$$A_n(\boldsymbol{\alpha}) \to A_n(\mathbf{k}) \equiv H_{\mathbf{k}} - E_n(\mathbf{k}). \qquad \text{(B.15.12)}$$

The kets of the general theory are defined by

$$|n\mathbf{k}\rangle \equiv u_{n\mathbf{k}}(\mathbf{r}), \qquad \text{(B.15.13)}$$

and are distinguished notationally from the kets in (B.15.8). Equation

[27] E. I Blount, *Lectures in Theoretical Physics, Vol. 5*, Interscience, New York, 1963.

(B.15.11) suggests that, for given **k**, any periodic functions can be expressed in the form

$$\sum_n c_n u_{n\mathbf{k}}(\mathbf{r})$$

In fact the completeness relation (B.7.6)

$$\sum_n u^*_{n\mathbf{k}}(\mathbf{r}')u_{n\mathbf{k}}(\mathbf{r}) = \delta(\mathbf{r}' - \mathbf{r})$$

is valid.

We now note two properties of the transformation (B.15.10). Let $B(\mathbf{p}, \mathbf{r})$ be an operator which depends on the electron momentum and coordinate, and can be represented as a power series in $\mathbf{p}$. Let $g(\mathbf{k}, \mathbf{r})$ be a well-behaved function. Then

(i) $$B_{\mathbf{k}}(\mathbf{p}, \mathbf{r}) = B(\mathbf{p} + \hbar\mathbf{k}, \mathbf{r}). \qquad \text{(B.15.14)}$$

Using the definition (B.15.10)

(ii) $$\mathrm{e}^{-\mathrm{i}\mathbf{k}\cdot\mathbf{r}}\nabla_{\mathbf{p}}B(\mathbf{p}, \mathbf{r})\,\mathrm{e}^{\mathrm{i}\mathbf{k}\cdot\mathbf{r}}g(\mathbf{k}, \mathbf{r}) = \hbar^{-1}[\nabla_{\mathbf{k}}B_{\mathbf{k}}(\mathbf{p}, \mathbf{r})]g(\mathbf{k}, \mathbf{r}), \qquad \text{(B.15.15a)}$$

where the square bracket means that $\nabla_{\mathbf{k}}$ does not also act on $g(\mathbf{k}, \mathbf{r})$.

The result (i) is established by observing that

$$\mathbf{p}\,\mathrm{e}^{\mathrm{i}\mathbf{k}\cdot\mathbf{r}} = \frac{\hbar}{\mathrm{i}}\nabla\,\mathrm{e}^{\mathrm{i}\mathbf{k}\cdot\mathbf{r}} = \frac{\hbar}{\mathrm{i}}(\mathrm{i}\mathbf{k}\,\mathrm{e}^{\mathrm{i}\mathbf{k}\cdot\mathbf{r}} + \mathrm{e}^{\mathrm{i}\mathbf{k}\cdot\mathbf{r}}\nabla) = \mathrm{e}^{\mathrm{i}\mathbf{k}\cdot\mathbf{r}}(\mathbf{p} + \hbar\mathbf{k});$$

$$(\mathbf{p})^2\,\mathrm{e}^{\mathrm{i}\mathbf{k}\cdot\mathbf{r}} = \mathbf{p}\,\mathrm{e}^{\mathrm{i}\mathbf{k}\cdot\mathbf{r}}\cdot(\mathbf{p} + \hbar\mathbf{k}) = \mathrm{e}^{\mathrm{i}\mathbf{k}\cdot\mathbf{r}}(\mathbf{p} + \hbar\mathbf{k})^2,$$

and (i) follows by continuing this process. Alternatively this can be proved by induction, assuming

$$(\mathbf{p}^n)_{\mathbf{k}} = (\mathbf{p} + \hbar\mathbf{k})^n$$

for some n. For it follows that

$$\begin{aligned}(\mathbf{p}^{n+1})_{\mathbf{k}} &= \mathrm{e}^{-\mathrm{i}\mathbf{k}\cdot\mathbf{r}}(\mathbf{p})^n\,\mathrm{e}^{\mathrm{i}\mathbf{k}\cdot\mathbf{r}}\,\mathrm{e}^{-\mathrm{i}\mathbf{k}\cdot\mathbf{r}}\mathbf{p}\,\mathrm{e}^{\mathrm{i}\mathbf{k}\cdot\mathbf{r}}\\ &= (\mathbf{p} + \hbar\mathbf{k})^n\cdot(\mathbf{p} + \hbar\mathbf{k}) = (\mathbf{p} + \hbar\mathbf{k})^{n+1}.\end{aligned}$$

To establish (ii), note that the transformation (B.15.10) does not affect the operator $\nabla_{\mathbf{p}}$ since $\mathrm{e}^{\mathrm{i}\mathbf{k}\cdot\mathbf{r}}$ is independent of $\mathbf{p}$. Hence the left-hand side of (ii) is

$$\nabla_{\mathbf{p}}B_{\mathbf{k}}(\mathbf{p}, \mathbf{r}) = \nabla_{\mathbf{p}}B(\mathbf{p} + \hbar\mathbf{k}, \mathbf{r}) = \hbar^{-1}\nabla_{\mathbf{k}}B_{\mathbf{k}}(\mathbf{p}, \mathbf{r}), \qquad \text{(B.15.15b)}$$

which establishes the result

Consider now the implications of the diagonal forms of (B.15.4) and (B.15.7). For this purpose ignore the quantum number l, which needs to be used only if one wishes to consider the effect of degeneracy due to spin or band overlap. The quantum number n which specifies the state of the electron becomes now $(n, \mathbf{k})$. Then (B.15.4) states, for $i = 1, 2, 3$,

$$\left\langle n\mathbf{k}\left|\frac{\partial H_{\mathbf{k}}}{\partial k_i}\right|n\mathbf{k}\right\rangle = \frac{\partial E_n(\mathbf{k})}{\partial k_i}. \tag{B.15.16}$$

The result (B.15.7) becomes, for $i, j = 1, 2, 3$,

$$\left\langle n\mathbf{k}\left|\frac{\partial^2 H_{\mathbf{k}}}{\partial k_i \partial k_j}\right|n\mathbf{k}\right\rangle = \frac{\partial^2 E_n(\mathbf{k})}{\partial k_i \partial k_j} + \boxed{2} \sum_{n'(\neq n)} \frac{\left\langle n\mathbf{k}\left|\frac{\partial H_{\mathbf{k}}}{\partial k_i}\right|n'\mathbf{k}\right\rangle\left\langle n'\mathbf{k}\left|\frac{\partial H_{\mathbf{k}}}{\partial k_j}\right|n\mathbf{k}\right\rangle}{E_{n'}(\mathbf{k}) - E_n(\mathbf{k})} \tag{B.15.17}$$

In the last term on the right of (B.15.17) we have used the fact that since only non-diagonal elements occur, there is, by the orthogonality of the Bloch functions, no contribution due to the terms $E_n(\mathbf{k})$ in (B.15.11).

B15c The velocity operator

With these auxiliary results available, the velocity operator is easily introduced and discussed. We merely assume that the Hamiltonian is periodic in the direct lattice. It can otherwise be arbitrary.

The velocity operator for a particle is, by standard quantum mechanics,

$$\mathbf{v} = \frac{\mathrm{i}}{\hbar}[H(\mathbf{p}, \mathbf{r}), \mathbf{r}] = \nabla_{\mathbf{p}} H(\mathbf{p}, \mathbf{r}). \tag{B.15.18}$$

Applying the transformation (B.15.10), and using (B.15.15),

$$\mathbf{v}_{\mathbf{k}} = \hbar^{-1}\nabla_{\mathbf{k}} H_{\mathbf{k}}(\mathbf{p}, \mathbf{r}), \quad \text{or} \quad v_{\mathbf{k}i} = \hbar^{-1}[\partial H_{\mathbf{k}}(\mathbf{p}, \mathbf{r})/\partial k_i], \tag{B.15.19}$$

where the component form has also been given ($i = x, y, z$). We need the expectation value of the velocity to establish (B.13.17). Using (B.15.8), (B.15.13), (B.15.16) and (B.15.19),

$$(n\mathbf{k}|v_i|n\mathbf{k}) = \langle n\mathbf{k}|v_{\mathbf{k}i}|n\mathbf{k}\rangle = \hbar^{-1}\left\langle n\mathbf{k}\left|\frac{\partial H_{\mathbf{k}}}{\partial k_i}\right|n\mathbf{k}\right\rangle = \hbar^{-1}\frac{\partial E_n(\mathbf{k})}{\partial k_i}. \tag{B.15.20}$$

Thus one finally has the *exact* result

$$(n\mathbf{k}|\mathbf{v}|n\mathbf{k}) = \hbar^{-1}\nabla_{\mathbf{k}} E_n(\mathbf{k}). \tag{B.15.21}$$

B.15d The effective mass sum rule

The result (B.15.7) implies a statement about the reciprocal effective mass tensor. This may be seen by noting from (B.13.23) that the first term on the right of (B.15.17) is $\hbar^2(1/m)^{n\mathbf{k}}_{ij}$. In addition, the operator on the left-hand side of (B.15.17) is

$$\frac{\partial^2 H_{\mathbf{k}}}{\partial k_i \partial k_j} = \frac{\partial^2 H(\mathbf{p} + \hbar\mathbf{k}, \mathbf{r})}{\partial k_i\, \partial k_j} = \frac{\hbar^2\, \delta_{ij}}{m}, \tag{B.15.22}$$

provided only that the Hamiltonian has the general form

$$H(\mathbf{p}, \mathbf{r}) = \frac{p^2}{2m} + \text{term linear in } \mathbf{p} + \text{term independent of } \mathbf{p}. \tag{B.15.23}$$

Here m is the free-electron mass. This leaves the terms in the last sum of (B.15.17) for discussion. By (B.15.19) they involve the operator

$$\frac{\partial H_{\mathbf{k}}}{\partial k_i} = \hbar v_{\mathbf{k}i}.$$

Thus, if (B.15.17) is multiplied by $m/\hbar^2$, one finds

$$m\left(\frac{1}{m}\right)^{n\mathbf{k}}_{ij} + \sum_{n'(\neq n)} f^{n'n}_{\mathbf{k}ij} = \delta_{ij}, \tag{B.15.24}$$

where

$$f^{n'n}_{\mathbf{k}ij} \equiv \boxed{2}\; m\frac{(n\mathbf{k}|v_i|n'\mathbf{k})(n'\mathbf{k}|v_j|n\mathbf{k})}{E_{n'}(\mathbf{k}) - E_n(\mathbf{k})} = -f^{nn'}_{\mathbf{k}ij} \tag{B.15.25}$$

is a generalized oscillator strength.

The origin of the oscillator strengths is as follows. Suppose an atom is regarded as a rigid ion with a single outer electron. Then the behaviour of this electron in the presence of light can often be simulated by a number of oscillators each of charge e and mass $m/f_{n'n}$. There is one oscillator for each admissible frequency $(E_{n'} - E_n)/h$. Under these conditions the quantity

$$f_{ii} \equiv 1 - \sum_{n'(\neq n)} f^{n'n}_{ii} \tag{B.15.26}$$

vanishes by a sum rule due to Thomas, Reiche and Kuhn. For an electron in a periodic field

$$f_{ii} = \frac{m}{\hbar^2}\frac{\partial^2 E_n(\mathbf{k})}{\partial k_i^2}, \tag{B.15.27}$$

in generalization of the atomic result.

We may sum over n in (B.15.24) and find[28]

$$\sum_{n=1}^{\infty}\left[\left(\frac{1}{m}\right)^{n\mathbf{k}}_{ij} - \frac{\delta_{ij}}{m}\right] = 0.$$

[28] A. R. Beattie and G. Smith, *Phys. Stat. Solids*, **19**, 577 (1967).

If b is the number of bands to be considered, one has approximately

$$\sum_{n=1}^{b}\left(\frac{1}{m}\right)_{ij}^{n\mathbf{k}} \simeq \frac{b}{m}\,\delta_{ij}. \tag{B.15.24$'$}$$

For scalar effective masses at $(n, \mathbf{k})$ and $(n', \mathbf{k})$ and $b = 2$ one finds the approximate relation

$$\frac{2}{m} \simeq \left(\frac{1}{m}\right)^{n\mathbf{k}} + \left(\frac{1}{m}\right)^{n'\mathbf{k}}. \tag{B.15.24$''$}$$

As already pointed out, strictly speaking additional superfixes are needed to allow for degeneracy[29]. This would lead to an additional sum and/or additional superfixes in the formulae following (B.15.15b).

The relation (B.15.24) admits of a simple discussion. At infinite band separations $f_{\mathbf{k}ij}^{n'n} \to 0$ for all n', $\mathbf{k}$, i and j when n is given. Thus

$$\left(\frac{1}{m}\right)_{ij}^{n\mathbf{k}} \to \frac{\delta_{ij}}{m}.$$

The mass tensor becomes the scalar free-electron mass. Suppose now that two bands are brought together; then one (t) is on top on the energy scale, the other (b) is below it. (B.15.24) now yields two equations:

$$m\left(\frac{1}{m}\right)_{ij}^{t\mathbf{k}} + f_{\mathbf{k}ij}^{bt} = \delta_{ij}; \tag{B.15.28}$$

$$m\left(\frac{1}{m}\right)_{ij}^{b\mathbf{k}} + f_{\mathbf{k}ij}^{tb} = \delta_{ij}. \tag{B.15.29}$$

Adding these and noting (B.15.25),

$$\frac{m}{2}\left[\left(\frac{1}{m}\right)_{ij}^{t\mathbf{k}} + \left(\frac{1}{m}\right)_{ij}^{b\mathbf{k}}\right] = \delta_{ij}. \tag{B.15.30}$$

Now, in the simplest case [see problem (B.15.2)], $\mathbf{v} = \mathbf{p}/m$, so that

$$f_{\mathbf{k}ii}^{tb} = \frac{2}{m}\,\frac{|(b\mathbf{k}|p_i|t\mathbf{k})|^2}{E_t(\mathbf{k}) - E_b(\mathbf{k})} \tag{B.15.31}$$

and therefore

$$f_{\mathbf{k}ii}^{tb} = -f_{\mathbf{k}ii}^{bt} \geqslant 0. \tag{B.15.32}$$

As the bands are pushed together both f's increase numerically, and

[29] This has been allowed for and group theoretical considerations have been included in D. J. Morgan and J. A. Galloway, *Phys. Stat. Solidi*, **23**, 97 (1967). (B.15.24$''$) has already been noticed in E. Antončik and P. T. Landsberg, *Proc. Phys. Soc.*, **82**, 337 (1963).

(B.15.29) and (B.15.32) show that for the lower band the diagonal components of the effective mass tensor become eventually negative. There is no reason to expect this phenomenon for the top band. This situation can be pictured as a repulsion between bands in **k**-space, with the result that they eventually curve in opposite directions. This was explicitly recognized already in equations such as (B.13.27).

Problems

(B.15.1) For a Dirac electron with a lattice periodic potential energy $U(\mathbf{r})$,

$$H(\mathbf{p}, \mathbf{r}) = c\boldsymbol{\alpha} \cdot \mathbf{p} + U(\mathbf{r}),$$

where c is the velocity of light and the three components of $\boldsymbol{\alpha}$ are the four-by-four Dirac matrices. Show that

$$H_{\mathbf{k}}(\mathbf{p} \cdot \mathbf{r}) = H(\mathbf{p}, \mathbf{r}) + c\hbar\mathbf{k} \cdot \boldsymbol{\alpha}$$

and

$$\mathbf{v} = c\boldsymbol{\alpha} = \mathbf{v}_{\mathbf{k}}.$$

(B.15.2) The Hamiltonian for a non-relativistic electron with a lattice periodic potential energy $U(\mathbf{r})$ is

$$H(\mathbf{p}, \mathbf{r}) = \frac{p^2}{2m} + a(\boldsymbol{\sigma} \times \nabla U) \cdot \mathbf{p} + U(\mathbf{r}),$$

where $a = \hbar/4m^2c^2$ and the second term represents the spin–orbit interaction. Show that

$$\mathbf{v}_{\mathbf{k}} = \mathbf{v} + \hbar\mathbf{k}/m,$$

where

$$\mathbf{v} = \frac{\mathbf{p}}{m} + a\boldsymbol{\sigma} \times \nabla U.$$

(B.15.3) An operator has the form $\sum_{j=0}^{m} f_j(\mathbf{r})(\mathbf{p})^j$ where the f_j are lattice periodic functions. Show that the operator is **k**-diagonal with respect to Bloch functions. Hence confirm that the operators $\mathrm{H}(\mathbf{p}, \mathbf{r})$, $\mathbf{v}$, $\mathbf{v}_{\mathbf{k}}$ of problem (B.15.2) are **k**-diagonal.

[*cf.* problem (B.11.1).]

(B.15.4) Show, by a partial integration and using the periodic boundary conditions, that for any wave functions ϕ and ψ,

$$\int_V \phi^* p_x \psi \, d\mathbf{r} = \frac{\hbar}{\mathrm{i}} \int_V \phi^* \frac{\partial \psi}{\partial x} \, d\mathbf{r} = -\frac{\hbar}{\mathrm{i}} \int_V \psi \frac{\partial \phi^*}{\partial x} \, d\mathbf{r}.$$

Hence show that

$$\int_V \phi^* \mathbf{p} \psi \, d\mathbf{r} = \frac{\hbar}{2\mathrm{i}} \int_V [\phi^* \nabla \psi - \psi \nabla \phi^*] \, d\mathbf{r}.$$

Note that the formula

$$\int_V \phi^* \mathbf{v}\psi \, d\mathbf{r} = \frac{\hbar}{2im}\int_V [\phi^*\nabla\psi - \psi\nabla\phi^*]\, d\mathbf{r}$$

holds only in the absence of spin–orbit interaction.

(B.15.5) Show that

$$\frac{i}{\hbar}[v_{\mathbf{k}i}, r_j] = \hbar^{-1}\frac{\partial v_{\mathbf{k}i}}{\partial k_j} = \hbar^{-2}\frac{\partial^2 H_{\mathbf{k}}}{\partial k_i \partial k_j}$$

where i and j denote Cartesian components.

[Use equations (B.15.15a, b), (B.15.19).]

(B.15.6) Let

$$I\begin{pmatrix} n'\mathbf{k}' \\ n\mathbf{k} \end{pmatrix}_l \equiv \int_V u^*_{n'\mathbf{k}'}(\mathbf{r})\, e^{i\mathbf{l}\cdot\mathbf{r}} u_{n\mathbf{k}}(\mathbf{r})\, d\mathbf{r}.$$

Show that, with the notation of equation (B.15.8),

$$\hbar q \cdot (n'\mathbf{k}'|\, e^{i\mathbf{q}\cdot\mathbf{r}}\mathbf{v}|n\mathbf{k}) = \left[E_{n'}(\mathbf{k}') - E_n(\mathbf{k}) - \frac{\hbar^2 q^2}{2m}\right] I\begin{pmatrix} n'\mathbf{k}' \\ n\mathbf{k} \end{pmatrix}_{\mathbf{q}+\mathbf{k}-\mathbf{k}'}$$

where $\mathbf{q}$ is any wave vector[30].

[Since the right-hand side vanishes if $\mathbf{q} + \mathbf{k} - \mathbf{k}'$ is not a lattice vector in $\mathbf{k}$-space, equation (B.7.18), the result proved here agrees with the $\mathbf{k}$-diagonality of $\mathbf{v}$ as established in problem (B.15.3).]

(B.15.7) Prove that, with the notation of equation (B.15.8), and for $q \ll k_0$,

$$(n\mathbf{k}_0 + \mathbf{q}|v_i|n\mathbf{k}_0 + \mathbf{q}) = (n\mathbf{k}_0|v_i|n\mathbf{k}_0) + \sum_{j=1}^{3} \hbar q_j \left(\frac{1}{m}\right)_{ij}^{n\mathbf{k}_0}.$$

[See equation (B.17.13).]

(B.15.8) Prove that for positive integral n,

$$\frac{i}{\hbar}[(\mathbf{p})^{2n}, x_1] = 2np_1(\mathbf{p})^{2n-2} = \frac{\partial}{\partial p_1}(\mathbf{p})^{2n};$$

$$\frac{i}{\hbar}[(\mathbf{p})^{2n+1}, x_1] = (\mathbf{p})^{2n-2}((\mathbf{p})^2 + 2np_1^2, \quad 2np_1p_2, \quad 2np_1p_3)$$

$$= \frac{\partial}{\partial p_1}(\mathbf{p})^{2n+1}.$$

B.16 EFFECTIVE MASS THEORY BASED ON BLOCH FUNCTIONS[31]

B.16a General theory

A more precise theory will now be attempted of the effect of a perturbing potential $F(\mathbf{r})$ which acts on an electron in a perfect lattice [*cf.* equation

[30] P. T. Landsberg, *Lectures in Theoretical Physics*, Boulder Colorado (University of Colorado Press, 1965); W. Brauer, *Phys. Stat. Solidi*, **5**, 139 (1964).

[31] I am indebted to Mr. M. J. Adams for discussion of this section.

(B.13.11) which was based on Wannier's theorem]. To do so suppose that there exists a wave vector $\mathbf{k}_0$ in the Brillouin zone such that it is a good approximation to regard the surfaces of constant energy as ellipsoidal in its neighbourhood. The effective masses will be denoted by m_{nj} where n labels the band under consideration and $j = 1, 2, 3$. We shall denote by m_n some average effective mass. It is then convenient to pass from **k**-space to what may be called **κ**-space by the simple change of variable

$$\boldsymbol{\kappa} \equiv \left[\sqrt{\left(\frac{m_n}{m_{n1}}\right)}(k_1 - k_{01}), \sqrt{\left(\frac{m_n}{m_{n2}}\right)}(k_2 - k_{02}), \sqrt{\left(\frac{m_n}{m_{n3}}\right)}(k_3 - k_{03})\right]. \quad \text{(B.16.1)}$$

Functions which have previously been regarded as functions of **k** will now be regarded as functions of **κ** without introducing new symbols to indicate the change in the functional form. Thus the equation

$$E(\mathbf{k}) - E(\mathbf{k}_0) = \sum_{j=1}^{3} \frac{\hbar^2 (k_j - k_{0j})^2}{2m_{nj}} \quad \text{(B.16.2)}$$

becomes simply

$$E(\boldsymbol{\kappa}) - E(0) = \frac{\hbar^2}{2m_n}|\boldsymbol{\kappa}|^2. \quad \text{(B.16.3)}$$

Let $H = (p^2/2m) + U(\mathbf{r})$ be the unperturbed Hamiltonian, U being the periodic potential. The Bloch functions $\psi_{n\mathbf{k}}(\mathbf{r})$ are eigenfunctions of H. This fact can be turned to advantage by expanding the unknown wave function ψ in terms of Bloch functions, since these form a complete set of functions. The Schrödinger equation to be solved is

$$H_{\mathrm{T}}\psi(\mathbf{r}) = E\psi(\mathbf{r}), \qquad H_{\mathrm{T}} \equiv H + F(\mathbf{r}), \quad \text{(B.16.4)}$$

where H_{T} is again the perturbed Hamiltonian (B.13.6). If the $A_n(\boldsymbol{\kappa})$ are expansion parameters defined only when **k** lies in the first Brillouin zone of **k**-space, we have

$$\psi(\mathbf{r}) = \sum_{n'}\sum_{\boldsymbol{\kappa}'} A_{n'}(\boldsymbol{\kappa}')\psi_{n'\boldsymbol{\kappa}'}(\mathbf{r}), \quad \text{(B.16.5)}$$

$$H\psi_{n\boldsymbol{\kappa}}(\mathbf{r}) = E_n(\boldsymbol{\kappa})\psi_{n\boldsymbol{\kappa}}(\mathbf{r}). \quad \text{(B.16.6)}$$

Substituting (B.16.5) into (B.16.4),

$$\sum_{n'}\sum_{\boldsymbol{\kappa}'} [E_{n'}(\boldsymbol{\kappa}') + F]A_{n'}(\boldsymbol{\kappa}')\psi_{n'\boldsymbol{\kappa}'} = E\psi(\mathbf{r}). \quad \text{(B.16.7)}$$

Since the relabelling (B.16.1) does not affect the orthonormality of the ψ's, one finds on multiplying by $\psi_{n\boldsymbol{\kappa}}^*$ and integrating over the fundamental domain V,

$$[E_n(\boldsymbol{\kappa}) - E_n(0)]A_n(\boldsymbol{\kappa}) + \sum_{n'\boldsymbol{\kappa}'} \langle n\boldsymbol{\kappa}|F|n'\boldsymbol{\kappa}'\rangle A_{n'}(\boldsymbol{\kappa}') = [E - E_n(0)]A_n(\boldsymbol{\kappa}). \quad \text{(B.16.8)}$$

Here

$$\langle n\boldsymbol{\kappa}|F|n'\boldsymbol{\kappa}'\rangle \equiv \int_V \psi^*_{n\boldsymbol{\kappa}}(\mathbf{r})F(\mathbf{r})\psi_{n'\boldsymbol{\kappa}'}(\mathbf{r})\,\mathrm{d}\mathbf{r}. \quad \text{(B.16.9)}$$

Now multiply (B.16.8) by $e^{i\boldsymbol{\kappa}\cdot\mathbf{r}}$ and sum over the first Brillouin zone, so that

$$\sum_{\boldsymbol{\kappa}} [E_n(\boldsymbol{\kappa}) - E_n(0)]A_n(\boldsymbol{\kappa})\,e^{i\boldsymbol{\kappa}\cdot\mathbf{r}} + \sum_{n'\boldsymbol{\kappa}\boldsymbol{\kappa}'} \langle n\boldsymbol{\kappa}|F|n'\boldsymbol{\kappa}'\rangle\,e^{i\boldsymbol{\kappa}\cdot\mathbf{r}}A_{n'}(\boldsymbol{\kappa}')$$
$$= [E - E_n(0)]\sum_{\boldsymbol{\kappa}} A_n(\boldsymbol{\kappa})\,e^{i\boldsymbol{\kappa}\cdot\mathbf{r}}. \quad \text{(B.16.10)}$$

This is the Schrödinger equation in $\boldsymbol{\kappa}$-space.

Introduce now the function

$$G_n(\mathbf{r}) \equiv V^{-\frac{1}{2}} \sum_{\boldsymbol{\kappa}} A_n(\boldsymbol{\kappa})\,e^{i\boldsymbol{\kappa}\cdot\mathbf{r}}. \quad \text{(B.16.11)}$$

This summation is for all $\mathbf{k}$ in the first Brillouin zones. One finds, using (B.16.3),

$$\frac{-\hbar^2}{2m_n}\nabla^2 G_n(\mathbf{r}) = V^{-\frac{1}{2}} \sum_{\boldsymbol{\kappa}} [E_n(\boldsymbol{\kappa}) - E_n(0)]A_n(\boldsymbol{\kappa})\,e^{i\boldsymbol{\kappa}\cdot\mathbf{r}}. \quad \text{(B.16.12)}$$

Introducing (B.16.12) into (B.16.10),

$$-\frac{\hbar^2}{2m_n}\nabla^2 G_n(\mathbf{r}) + C_n(\mathbf{r}) = [E - E_n(0)]G_n(\mathbf{r}), \quad \text{(B.16.13)}$$

where

$$C_n(\mathbf{r}) \equiv V^{-\frac{1}{2}} \sum_{n'\boldsymbol{\kappa}'} A_{n'}(\boldsymbol{\kappa}')\,e^{i\boldsymbol{\kappa}'\cdot\mathbf{r}} \sum_{\boldsymbol{\kappa}} \langle n\boldsymbol{\kappa}|F|n'\boldsymbol{\kappa}'\rangle\,e^{i(\boldsymbol{\kappa}-\boldsymbol{\kappa}')\cdot\mathbf{r}}.$$
$$\text{(B.16.14)}$$

The rather general theory here set up has to be simplified by appropriate approximations, which reduce $C_n(\mathbf{r})$ to a product of $G_n(\mathbf{r})$ and a simple function of $\mathbf{r}$ which acts like a potential in (B.16.13), viewed as a Schrödinger equation. If this decomposition is simple enough (B.16.13) can be solved for $G_n(\mathbf{r})$. The expansion parameters $A_n(\boldsymbol{\kappa})$ of (B.16.5) can then be identified in virtue of (B.16.11). For example, if

$$\sum_{\boldsymbol{\kappa}} \langle n\boldsymbol{\kappa}|F|n'\boldsymbol{\kappa}'\rangle\,e^{i(\boldsymbol{\kappa}-\boldsymbol{\kappa}')\cdot\mathbf{r}} \equiv W_n(\mathbf{r}) \quad \text{(B.16.15)}$$

and is independent of n' and $\boldsymbol{\kappa}'$, then

$$C_n(\mathbf{r}) = W_n(\mathbf{r})G_n(\mathbf{r}). \qquad \text{(B.16.16)}$$

B.16b Approximations

We now seek reasonable approximations which can lead to the decomposition (B.16.16).

Write

$$\psi_{n\boldsymbol{\kappa}} = V^{-\frac{1}{2}} u_{n\boldsymbol{\kappa}}\, \mathrm{e}^{\mathrm{i}\mathbf{k}\cdot\mathbf{r}}, \qquad \text{(B.16.17)}$$

which is consistent with a pure change of notation (B.16.1). Then a product of the u's, being lattice periodic, can be expanded by (B.7.2a):

$$u^*_{n\boldsymbol{\kappa}}(\mathbf{r})u_{n'\boldsymbol{\kappa}'}(\mathbf{r}) = \sum_{\mathbf{K}} I\begin{pmatrix} n\boldsymbol{\kappa} \\ n'\boldsymbol{\kappa}' \end{pmatrix}_{\mathbf{K}} \mathrm{e}^{\mathrm{i}\mathbf{K}\cdot\mathbf{r}},$$

where

$$I\begin{pmatrix} n\boldsymbol{\kappa} \\ n'\boldsymbol{\kappa}' \end{pmatrix}_{\mathbf{K}} \equiv V^{-1}\int_V u^*_{n\boldsymbol{\kappa}}(\mathbf{r})u_{n'\boldsymbol{\kappa}'}(\mathbf{r})\,\mathrm{e}^{-\mathrm{i}\mathbf{K}\cdot\mathbf{r}}\,\mathrm{d}\mathbf{r} \qquad \text{(B.16.18)}$$

is a Fourier coefficient. It follows that

$$\langle n\boldsymbol{\kappa}|F|n'\boldsymbol{\kappa}'\rangle = V^{-1}\sum_{\mathbf{K}} I\begin{pmatrix} n\boldsymbol{\kappa} \\ n'\boldsymbol{\kappa}' \end{pmatrix}_{\mathbf{K}} \int_V F(\mathbf{r}')\,\mathrm{e}^{\mathrm{i}(\mathbf{K}+\mathbf{k}'-\mathbf{k})\cdot\mathbf{r}'}\,\mathrm{d}\mathbf{r}'.$$

The $\boldsymbol{\kappa}$-sum in $C_n(\mathbf{r})$ is accordingly

$$V^{-1}\sum_{\boldsymbol{\kappa}} \mathrm{e}^{\mathrm{i}(\boldsymbol{\kappa}-\boldsymbol{\kappa}')\cdot\mathbf{r}} \sum_{\mathbf{K}} I\begin{pmatrix} n\boldsymbol{\kappa} \\ n'\boldsymbol{\kappa}' \end{pmatrix}_{\mathbf{K}} \int_V F(\mathbf{r}')\,\mathrm{e}^{\mathrm{i}(\mathbf{K}+\mathbf{k}'-\mathbf{k})\cdot\mathbf{r}'}\,\mathrm{d}\mathbf{r}'$$

$$= \sum_{\boldsymbol{\kappa}} \mathrm{e}^{\mathrm{i}(\boldsymbol{\kappa}-\boldsymbol{\kappa}')\cdot\mathbf{r}} \sum_{\mathbf{K}} I\begin{pmatrix} n\boldsymbol{\kappa} \\ n'\boldsymbol{\kappa}' \end{pmatrix}_{\mathbf{K}} F_{\mathbf{K}+\mathbf{k}'-\mathbf{k}}, \qquad \text{(B.16.19)}$$

where we have introduced a notation for the Fourier coefficients of $F(\mathbf{r})$.

By (B.16.17) and (B.16.18)

$$I\begin{pmatrix} n\boldsymbol{\kappa} \\ n'\boldsymbol{\kappa}' \end{pmatrix}_0 = V^{-1}\int_V u^*_{n\boldsymbol{\kappa}}(\mathbf{r})u_{n'\boldsymbol{\kappa}}(\mathbf{r})\,\mathrm{d}\mathbf{r} = \int_V \psi^*_{n\boldsymbol{\kappa}}(\mathbf{r})\psi_{n'\boldsymbol{\kappa}}(\mathbf{r})\,\mathrm{d}\mathbf{r} = \delta_{n'n}. \qquad \text{(B.16.20)}$$

It will now be assumed that for $\mathbf{k} - \mathbf{k}'$ or $\boldsymbol{\kappa} - \boldsymbol{\kappa}'$ small enough we can put, by (B.16.20),

$$I\begin{pmatrix} n\boldsymbol{\kappa} \\ n'\boldsymbol{\kappa}' \end{pmatrix}_{\mathbf{K}} \simeq \delta_{n'n}\,\delta_{\mathbf{K},0} \quad \text{(Approximation I)}, \qquad \text{(B.16.21)}$$

whence the $\boldsymbol{\kappa}$-sum in $C_n(\mathbf{r})$, (B.16.19), becomes

$$\delta_{n'n} \sum_{\boldsymbol{\kappa}} \mathrm{e}^{\mathrm{i}(\boldsymbol{\kappa}-\boldsymbol{\kappa}')\cdot\mathbf{r}} F_{\mathbf{k}'-\mathbf{k}}$$
$$= \delta_{n'n} \sum_{\mathbf{k}} \mathrm{e}^{\mathrm{i}(\mathbf{k}-\mathbf{k}')\cdot\boldsymbol{\rho}} F_{\mathbf{k}'-\mathbf{k}}, \tag{B.16.22}$$

where $\boldsymbol{\rho}$ is defined by

$$\boldsymbol{\rho} \equiv \left[\sqrt{\left(\frac{m_n}{m_{n1}}\right)} r_1, \sqrt{\left(\frac{m_n}{m_{n2}}\right)} r_2, \sqrt{\left(\frac{m_n}{m_{n3}}\right)} r_3\right]$$

and the r_j are Cartesian components of the vector $\mathbf{r}$. Hence (B.16.21) yields

$$C_n(\mathbf{r}) = V^{-\frac{1}{2}} \sum_{\boldsymbol{\kappa}'} A_n(\boldsymbol{\kappa}')\, \mathrm{e}^{\mathrm{i}\boldsymbol{\kappa}'\cdot\mathbf{r}} \sum_{\mathbf{k}} F_{\mathbf{k}'-\mathbf{k}}\, \mathrm{e}^{-\mathrm{i}(\mathbf{k}'-\mathbf{k})\cdot\boldsymbol{\rho}}. \tag{B.16.23}$$

If the $\mathbf{k}$-sum in (B.16.23) covered the whole of $\mathbf{k}$-space, that sum would simply be the Fourier expansion of $F(\boldsymbol{\rho})$. Our assumption II is that the $\mathbf{k}$-sum, although it goes over the first zone only, contains the main contribution to $F(\boldsymbol{\rho})$:

$$\sum_{\mathbf{k}} F_{\mathbf{k}'-\mathbf{k}}\, \mathrm{e}^{-\mathrm{i}(\mathbf{k}'-\mathbf{k})\cdot\boldsymbol{\rho}} \simeq F(\boldsymbol{\rho}) \qquad \text{(Approximation II).} \tag{B.16.24}$$

With these assumptions, $C_n(\mathbf{r}) = F(\boldsymbol{\rho})G_n(\mathbf{r})$.

Equation (B.16.13) becomes a Schrödinger equation with the perturbing potential $F(\mathbf{r})$ modified to $F(\boldsymbol{\rho})$:

$$\left[-\frac{\hbar^2}{2m_n}\nabla^2 + F(\boldsymbol{\rho})\right] G_n(\mathbf{r}) = [E - E_n(\mathbf{k}_0)]G_n(\mathbf{r}). \tag{B.16.25}$$

The periodic potential has been removed. The cost is the replacement of the electron mass by the effective mass.

B.16c Interpretation of the results

We have, by (B.16.5) and (B.16.17), found the wavefunction

$$\psi(\mathbf{r}) = V^{-\frac{1}{2}} \sum_{n\boldsymbol{\kappa}} A_n(\boldsymbol{\kappa}) u_{n\boldsymbol{\kappa}}(\mathbf{r})\, \mathrm{e}^{\mathrm{i}\mathbf{k}\cdot\mathbf{r}}. \tag{B.16.26}$$

The expansion coefficients A are, by (B.16.11), seen to be approximate Fourier coefficients of a function $G_n(\mathbf{r})$. This function is, by (B.16.25), a solution of a Schrödinger equation which has as potential a modification, $F(\boldsymbol{\rho})$, of the original perturbing potential $F(\mathbf{r})$. This modification arises from the nature of the band shape near the minimum $\mathbf{k} = \mathbf{k}_0$ in which one is interested.

Assume now that only one band makes appreciable contributions, and that the u corresponding to the dominant minimum, i.e. $\boldsymbol{\kappa} = 0$ or $\mathbf{k} = \mathbf{k}_0$, can be taken outside the sum by retaining only the first term in a Taylor expansion about $\mathbf{k} = \mathbf{k}_0$. One then finds, from (B.16.11) and (B.16.26),

$$\psi(\mathbf{r}) = [u_n(\mathbf{r})]_{\mathbf{k}=\mathbf{k}_0} V^{-\frac{1}{2}} \sum_{n\boldsymbol{\kappa}} A_n(\boldsymbol{\kappa})\, \mathrm{e}^{\mathrm{i}\mathbf{k}\cdot\mathbf{r}}. \tag{B.16.27}$$

To obtain a simple interpretation of the sum one wants the exponents to be of the form $\boldsymbol{\kappa}\cdot\mathbf{r}$. This is achieved by defining

$$\mathbf{r}' \equiv \left[\sqrt{\left(\frac{m_{n1}}{m_n}\right)} r_1, \sqrt{\left(\frac{m_{n2}}{m_n}\right)} r_2, \sqrt{\left(\frac{m_{n3}}{m_n}\right)} r_3\right].$$

Then

$$\begin{aligned}\psi(\mathbf{r}) &= [u_n(\mathbf{r})]_{\mathbf{k}=\mathbf{k}_0} V^{-\frac{1}{2}} \mathrm{e}^{\mathrm{i}\mathbf{k}_0\cdot\mathbf{r}} \sum_{n\mathbf{k}} A_n(\boldsymbol{\kappa})\, \mathrm{e}^{\mathrm{i}\boldsymbol{\kappa}\cdot\mathbf{r}'} \\ &= [u_n(\mathbf{r})]_{\mathbf{k}=\mathbf{k}_0} \mathrm{e}^{\mathrm{i}\mathbf{k}_0\cdot\mathbf{r}}\, G_n(\mathbf{r}').\end{aligned}$$

B.16d Discussion

In arguing from Wannier's theorem (equation B.13.11) it appeared that the perturbed wave function $\psi(\mathbf{r})$ itself can be determined from a Schrödinger equation with an effective mass. The present argument suggests that such an equation must indeed be solved, but that it yields only the auxiliary function $G_n(\mathbf{r})$. Its Fourier coefficients, $A_n(\boldsymbol{\kappa})$, are then the expansion parameters in the expression (B.16.5) for the perturbed wave-function.

Approximation II can be justified by the circumstance that the higher Fourier coefficients are in fact often less important than the lower ones for smoothly varying potentials. But Approximation I shows that a restriction to one band is still involved to some extent in the argument.

If the impurity binding energy is small, the assumptions made are reasonable. The potential for the electron is slowly varying and the other bands may be expected to cause only small errors. If the binding energy is large both assumptions can break down. Equation (B.16.13) is still valid, but assumptions leading to (B.16.16) cannot then be justified.

More sophisticated theories exist in which the interband coupling terms are reduced to some desired accuracy by applying canonical transformations[32], but they again yield equation (B.16.25). The effect of

[32] J. M. Luttinger and W. Kohn, *Phys. Rev.*, **97**, 869 (1955). Reviews of the theory are given by W. Kohn, *Solid State Physics*, **5**, 257 (1957) and by T. P. McLean in *Proc. Intern. School of Physics*, '*E. Fermi*', Academic Press, New York, 1963, course 22, p. 479.

magnetic fields on the effective mass theory leads to additional complications, which are also under active study[33].

B.16e Coulomb potentials

At large distances from an impurity ion, the perturbing potential due to this ion may be taken as an attractive Coulomb potential

$$F(\mathbf{r}) = \frac{-e^2}{\epsilon r},$$

where ϵ is the static dielectric constant. Thus equation (B.16.25) becomes

$$\left(\frac{-\hbar^2}{2m_n}\nabla r^2 - \frac{e^2}{\epsilon|\boldsymbol{\rho}|}\right)G_n(\mathbf{r}) = [E - E_n(\mathbf{k}_0)]G_n(\mathbf{r}), \tag{B.16.28}$$

and

$$|\boldsymbol{\rho}| = \sqrt{\left(\frac{m_n}{m_{n1}}r_1^2 + \frac{m_n}{m_{n2}}r_2^2 + \frac{m_n}{m_{n3}}r_3^2\right)}.$$

If we now change coordinates by putting

$$\rho_1 = \sqrt{\left(\frac{m_n}{m_{n1}}\right)}r_1, \qquad \rho_2 = \sqrt{\left(\frac{m_n}{m_{n2}}\right)}r_2, \qquad \rho_3 = \sqrt{\left(\frac{m_n}{m_{n3}}\right)}r_3,$$

then equation (B.16.28) assumes the more usual form:

$$\left[-\frac{\hbar^2}{2m_{n1}}\frac{\partial^2}{\partial\rho_1^2} - \frac{\hbar^2}{2m_{n2}}\frac{\partial^2}{\partial\rho_2^2} - \frac{\hbar^2}{2m_{n_3}}\frac{\partial^2}{\partial\rho_3^2} - \frac{e^2}{\epsilon\rho}\right]G_n(\mathbf{r}) = [E - E_n(\boldsymbol{\kappa}_0)]G_n(\mathbf{r}). \tag{B.16.29}$$

The operator on the left can be written as

$$E_n\left(\frac{1}{i}\nabla_{\boldsymbol{\rho}}\right), \quad \text{where} \quad E_n(\mathbf{k}) \equiv \sum_{j=1}^{3}\frac{\hbar^2 k_j^2}{2m_{nj}},$$

so that yet another way of writing the general G-equation is

$$\left[E_n\left(\frac{1}{i}\nabla_{\boldsymbol{\rho}}\right) + F(\boldsymbol{\rho})\right]G_n(\mathbf{r}) = [E - E_n(\mathbf{k}_0)]G_n(\mathbf{r}). \tag{B.16.30}$$

Approximate solutions of this equation to determine the energy levels have been obtained[34] using either perturbation theory or variational

[33] See, for instance, J. Zak and W. Zawadzki, *Phys. Rev.*, **145**, 536 (1966), and references given there.

[34] C. Kittel and A. H. Mitchell, *Phys. Rev.*, **96**, 1488 (1954); M. Lampert, *Phys. Rev.*, **97**, 352 (1955); W. Kohn and J. M. Luttinger, *Phys. Rev.*, **97**, 1721 (1955); *ibid.* **98**, 915 (1955); E. M. Conwell and B. W. Levinger, in *Proc. Intern. Conf. Phys. of Semiconductors, Exeter, 1962*, (London: Inst. of Phys. and the Phys. Soc., 1962, p. 227); P. J. Roberts, *J. Phys. Chem. Solids*, **28**, 1353 (1967).

methods, with the additional assumption, applicable to Si and Ge, that $m_{n1} = m_{n2}$.

If all three effective masses are equal, equation (B.16.28) reduces to the Schrödinger equation for the hydrogen atom. Then $G_n(\mathbf{r})$ is a hydrogen-like function whose Fourier coefficients are known, and are most conveniently represented in terms of Gegenbauer polynomials[35]. The required impurity wavefunction is then given by equation (B.16.26).

B.16f Donors and acceptors

In the preceding theory we may take all effective masses as positive. The theory then applies to shallow donors. If all effective masses are interpreted as negative quantities, the theory applies to shallow acceptors. In that case both sides of equations (B.16.2) and (B.16.3) are negative, since $E_n(0)$ is the energy at a valence band maximum. The theory goes through for acceptors on this basis.

Consider, for example, the equation (B.16.28) for $G_n(\mathbf{r})$. The energy $E - E_n(\mathbf{k}_0)$ must be negative for donors in order that there be a bound state (in analogy with the hydrogen atom). Now while an ionized donor is positively charged and becomes neutral when it has captured an electron, an ionized acceptor is negatively charged and becomes neutral on capturing a hole. Thus on going from donors to acceptors the sign of the Coulomb potential changes. Hence all terms in (B.16.28) change sign [except $G_n(\mathbf{r})$], and one finds simply the normal hydrogen-type equation multiplied throughout by -1. This yields the hydrogen-type bound states for acceptors as a mirror image (except for the difference in the effective masses) of the excited states for the donors if the energies of these states are plotted.

B.17 SIMPLE PERTURBATION METHODS IN BAND THEORY

B.17a Generalities

Suppose the point $\mathbf{k}$ in $\mathbf{k}$-space is a symmetry point, or in any case a point at which are known the Bloch energies $E_n(\mathbf{k})$ (assumed to be all different as the band number is varied) and the periodic parts $u_{n\mathbf{k}}(\mathbf{r})$ of the Bloch functions. One would then expect to be able to infer approximate information concerning the neighbouring states $\mathbf{k} + \mathbf{q}$ by non-degenerate perturbation theory [see equations (B.8.1), (B.8.2)]. The expressions obtained would be expected to depend on the first and second wave-vector derivatives of the functions $E_n(\mathbf{k})$. It will now be shown how these ideas can be given precise form.

[35] M. E. Cohen and P. T. Landsberg, *Phys. Rev.*, **154**, 683 (1967).

Treating the $E_n(\mathbf{k})$ and the $u_{n\mathbf{k}}$ as unperturbed functions and $E_n(\mathbf{k}+\mathbf{q})$, $u_{n\mathbf{k}+\mathbf{q}}$ as the perturbed functions, stationary or time independent perturbation theory yields at once to first and second order in the perturbation respectively:

$$u_{n\mathbf{k}+\mathbf{q}} = u_{n\mathbf{k}} + \sum_{n'(\neq n)} \frac{H'_{n'n}}{E_n(\mathbf{k}) - E_{n'}(\mathbf{k})} u_{n'\mathbf{k}}, \tag{B.17.1}$$

$$E_n(\mathbf{k} + \mathbf{q}) = E_n(\mathbf{k}) + H'_{nn} + \sum_{n'(\neq n)} \frac{H'_{nn'}H'_{n'n}}{E_n(\mathbf{k}) - E_{n'}(\mathbf{k})}. \tag{B.17.2}$$

From (B.15.11) one has the Schrödinger equations

$$H_{\mathbf{k}}u_{n\mathbf{k}}(\mathbf{r}) = E_n(\mathbf{k})u_{n\mathbf{k}}(\mathbf{r}),$$

$$H_{\mathbf{k}+\mathbf{q}}u_{n\mathbf{k}+\mathbf{q}}(\mathbf{r}) = E_n(\mathbf{k} + \mathbf{q})u_{n\mathbf{k}+\mathbf{q}}(\mathbf{r}).$$

Hence the perturbation is

$$H' \equiv H_{\mathbf{k}+\mathbf{q}} - H_{\mathbf{k}}. \tag{B.17.3}$$

B.17b Energy differences

In order to define the problem more completely, one needs to go beyond the restriction (B.15.23) so far imposed on the Hamiltonian $H(\mathbf{p}, \mathbf{r})$. If $\boldsymbol{\lambda}$ be a vector independent of momentum, suppose that

$$H(\mathbf{p}, \mathbf{r}) \equiv \frac{p^2}{2m} + \boldsymbol{\lambda} \cdot \mathbf{p} + \mu. \tag{B.17.4}$$

Then

$$H_{\mathbf{k}} = H(\mathbf{p} + \hbar\mathbf{k}, \mathbf{r}) = H_0 + \frac{\hbar^2 k^2}{2m} + \hbar\mathbf{k} \cdot \boldsymbol{\lambda} + \frac{\hbar\mathbf{k} \cdot \mathbf{p}}{m}. \tag{B.17.5}$$

From (B.15.19), the velocity operator $\mathbf{v}_{\mathbf{k}}$ satisfies

$$\hbar\mathbf{v}_{\mathbf{k}} = \frac{\hbar^2}{m}\mathbf{k} + \hbar\boldsymbol{\lambda} + \frac{\hbar\mathbf{p}}{m}. \tag{B.17.6}$$

The vector $\boldsymbol{\lambda}$ is non-zero if spin-orbit effects are taken into account. The scalar μ depends on the energy origin. The perturbation is

$$\begin{aligned} H' &= \frac{\hbar^2 q^2}{2m} + \frac{\hbar^2}{m}\mathbf{k} \cdot \mathbf{q} + \hbar\mathbf{q} \cdot \boldsymbol{\lambda} + \frac{\hbar\mathbf{q} \cdot \mathbf{p}}{m} \\ &= \frac{\hbar^2 q^2}{2m} + \hbar\mathbf{q} \cdot \mathbf{v}_{\mathbf{k}}. \end{aligned} \tag{B.17.7}$$

Taking matrix elements with respect to the $u_{n\mathbf{k}}(\mathbf{r})$, and using the notation (B.15.13) and the result (B.15.20),

$$\langle n'\mathbf{k}|H'|n\mathbf{k}\rangle = \frac{\hbar^2 q^2}{2m}\delta_{n'n} + (1 - \delta_{n'n})\hbar\mathbf{q}\cdot\langle n'\mathbf{k}|\mathbf{v}_{\mathbf{k}}|n\mathbf{k}\rangle + \mathbf{q}\cdot\nabla_{\mathbf{k}}E_n(\mathbf{k})\,\delta_{n'n}. \tag{B.17.8}$$

Substituting in (B.17.2) one finds finally a perturbation expression for the change in energy corresponding to the displacement $\mathbf{q}$ in $\mathbf{k}$-space:

$$E_n(\mathbf{k}+\mathbf{q}) - E_n(\mathbf{k}) = \frac{\hbar^2 q^2}{2m} + \mathbf{q}\cdot\nabla_{\mathbf{k}}E_n(\mathbf{k}) + \sum_{n'(\neq n)}\sum_{i,j=1}^{3}\frac{\hbar^2 q_i q_j \langle n\mathbf{k}|\mathbf{v}_{\mathbf{k},i}|n'k\rangle\langle n'\mathbf{k}|v_{\mathbf{k},j}|n\mathbf{k}\rangle}{E_n(\mathbf{k}) - E_{n'}(\mathbf{k})}. \tag{B.17.9}$$

The suffices i, j distinguish Cartesian components. Often the state $\mathbf{k}$ will be at an energy extremum so that the second term on the right vanishes. Note that the additive constant μ in (B.17.4) has disappeared, and that the vector $\boldsymbol{\lambda}$ (which vanishes in the simpler cases) has been absorbed in the velocity operator.

One can use the oscillator strength (B.15.25) to write (B.17.9) in the form

$$E_n(\mathbf{k}+\mathbf{q}) - E_n(\mathbf{k}) = \frac{\hbar^2 q^2}{2m} + \mathbf{q}\cdot\nabla_{\mathbf{k}}E_n(\mathbf{k}) + \frac{\hbar^2}{2m}\sum_{n'(\neq n)}\sum_{i,j} q_i q_j f_{\mathbf{k}ij}^{nn'}. \tag{B.17.10}$$

This is valid for arbitrary $\mathbf{q}$, if they are small enough.

A Taylor expansion in $\mathbf{q}$ on the left-hand side yields therefore

$$\left(\frac{1}{m}\right)_{ij}^{n\mathbf{k}} \equiv \hbar^{-2}\frac{\partial^2 E_n(\mathbf{k})}{\partial k_i\,\partial k_u} = \frac{\delta_{ij}}{m} - \frac{1}{m}\sum_{n'(\neq n)} f_{\mathbf{k}ij}^{n'n}. \tag{B.17.11}$$

This is again the sum rule (B.15.24), but this time it appears *as an approximate perturbation theory result*, whereas we saw in §B.15 that it is exact.

It is useful to put (B.17.11) into (B.17.10) to find

$$E_n(\mathbf{k}+\mathbf{q}) - E_n(\mathbf{k}) = \mathbf{q}\cdot\nabla_{\mathbf{k}}E_n(\mathbf{k}) + \frac{\hbar^2}{2}\sum_{i,j} q_i q_j\left(\frac{1}{m}\right)_{ij}^{n\mathbf{k}}. \tag{B.17.12}$$

Applying the operator $\hbar^{-1}\partial/\partial q_i$ to (B.17.10), and noting that

$$\hbar^{-1}\frac{\partial E_n(\mathbf{k}+\mathbf{q})}{\partial q_i} = (n\mathbf{k}+\mathbf{q}|v_i|n\mathbf{k}+\mathbf{q}),$$

one finds for a symmetrical effective mass tensor

$$(n\mathbf{k} + \mathbf{q}|v_i|n\mathbf{k} + \mathbf{q}) = (n\mathbf{k}|v_i|n\mathbf{k}) + \hbar \sum_j q_j \left(\frac{1}{m}\right)_{ij}^{n\mathbf{k}}. \qquad \text{(B.17.13)}$$

This result will prove helpful later.

B.17c Perturbed-wave functions and overlap integrals

We now turn to the other perturbation result, (B.17.1), in order to give it a more specific form. For this purpose use (B.17.8), noting that only terms which are non-diagonal in the band numbers are needed. Hence let

$$C_{n'n}^{\mathbf{k}i} \equiv \frac{\hbar(n'\mathbf{k}|v_i|n\mathbf{k})}{E_n(\mathbf{k}) - E_{n'}(\mathbf{k})}. \qquad \text{(B.17.14)}$$

Then

$$u_{n\mathbf{k}+\mathbf{q}} = u_{n\mathbf{k}} + \sum_{n'(\neq n)} \sum_i C_{n'n}^{\mathbf{k}i} q_i u_{n'\mathbf{k}} + \mathrm{O}(q^2). \qquad \text{(B.17.15)}$$

This procedure contains the essence of what is called the $\mathbf{k}\cdot\mathbf{p}$-approximation. We shall neglect numbers involving q^2 compared with terms of order unity. Then (B.17.15) is still normalized in the same sense as $u_{n\mathbf{k}}$.

As an application of these formulae, we shall establish that the integrals

$$I_{n\mathbf{k}}^{n'\mathbf{k}+\mathbf{q}} \equiv \int_V \overset{*}{u}_{n'\mathbf{k}+\mathbf{q}}(\mathbf{r}) u_{n\mathbf{k}}(\mathbf{r})\, d\mathbf{r} \qquad \text{(B.17.16)}$$

satisfy the sum rule

$$\sum_{n'(\neq n)} [E_{n'}(\mathbf{k}) - E_n(\mathbf{k})]|I_{n\mathbf{k}}^{n'\mathbf{k}+\mathbf{q}}|^2 = \frac{\hbar^2}{2m}\left[q^2 - m\sum_{ij} q_i q_j \left(\frac{1}{m}\right)_{ij}^{n\mathbf{k}}\right]. \qquad \text{(B.17.17)}$$

Substitution of (B.17.15) into (B.17.16) shows that

$$I_{n\mathbf{k}}^{n\mathbf{k}+\mathbf{q}} = 1 + \mathrm{O}(q^2) \sim 1. \qquad \text{(B.17.18)}$$

This leaves the integrals (B.17.16) with $n \neq n'$ for investigation. Here one finds

$$I_{n\mathbf{k}}^{n'\mathbf{k}+\mathbf{q}} = \sum_{i=1}^{3} q_i C_{nn'}^{\mathbf{k}i}. \qquad \text{(B.17.19)}$$

A little calculation yields next

$$[E_{n'}(\mathbf{k}) - E_n(\mathbf{k})]|I_{n\mathbf{k}}^{n'\mathbf{k}+\mathbf{q}}|^2 = \frac{\hbar^2}{2m}\sum_{ij} q_i q_j f_{\mathbf{k}ij}^{n'n} \quad (n \neq n') \qquad \text{(B.17.20)}$$

Summing over all $n'(\neq n)$ and using the sum rule (B.17.11) yields the desired result.

These integrals occur in band theory investigations[36] and in the discussion of electron collision processes in semiconductors. The sum rule (B.17.20) can give assistance in estimating the values of the overlap integrals[37].

We shall now make some remarks about the behaviour of these overlap integrals as the two reduced wave-vectors which enter into it range over the zone. This can be done if the periodic potential is assumed to be small. Let

$$\mathbf{K} = \mathbf{L} + \mathbf{k} \tag{B.17.21}$$

be a wave-vector in the extended zone scheme and let $\mathbf{k}$ be the corresponding wave-vector in the reduced zone scheme. Then the Bloch wavefunction for the state $\mathbf{K}$ may be written in the two schemes

$$\psi(\mathbf{K}, \mathbf{r}) = U(\mathbf{K}, \mathbf{r})\, \mathrm{e}^{\mathrm{i}\mathbf{K}\cdot\mathbf{r}}, \tag{B.17.22}$$

$$\psi_{\mathbf{L}}(\mathbf{k}, \mathbf{r}) = u_{\mathbf{L},\mathbf{k}}(\mathbf{r})\, \mathrm{e}^{\mathrm{i}\mathbf{k}\cdot\mathbf{r}}, \tag{B.17.23}$$

where instead of a band number the reciprocal lattice vector $\mathbf{L}$ has been used as an auxiliary label. In the limit of a small periodic potential $U(\mathbf{K}, \mathbf{r}) \to 1$ because (B.17.22) becomes a plane wave. Since (B.17.22) and (B.17.23) are strictly equal,

$$u_{\mathbf{L},\mathbf{k}}(\mathbf{r}) = U(\mathbf{k} + \mathbf{L}, \mathbf{r})\, \mathrm{e}^{\mathrm{i}\mathbf{L}\cdot\mathbf{r}} \to \mathrm{e}^{\mathrm{i}\mathbf{L}\cdot\mathbf{r}}.$$

In any one zone different vectors $\mathbf{L}$ will be appropriate to different parts which may be called *subzones.* It then follows that in the limit of a vanishing periodic potential (the so-called empty lattice case)

$$\frac{1}{V}\int_V u^*_{\mathbf{L}',\mathbf{k}'}(\mathbf{r})u_{\mathbf{L},\mathbf{k}}(\mathbf{r})\, \mathrm{d}\mathbf{r} = \frac{1}{V}\int_V \mathrm{e}^{\mathrm{i}(\mathbf{L}-\mathbf{L}')\cdot\mathbf{r}}\, \mathrm{d}\mathbf{r} = \delta_{\mathbf{L},\mathbf{L}'}.$$

Thus, in this limit, one would expect a step function behaviour of the integral in the sense that it is unity if the states are in the same subzone of a given zone and zero otherwise[38].

This situation is illustrated in the figures. For a one-dimensional crystal the reciprocal lattice points are indicated by large dots in fig. B.17.2, and the reciprocal lattice vectors are $\mathbf{L}_0 = 0$ or $n\mathbf{L}_1$ ($n = \pm 1, \pm 2, \ldots$) where $\mathbf{L}_1$ is shown in fig. B.17.1. Regions of $\mathbf{k}$-space labelled $2a$, $2b$ refer to two distinct subzones of the second zone. For each subzone

[36] See, for example, M. Cordona and F. H. Pollak, *Phys. Rev.*, **142**, 530 (1966).
[37] For more details see, for example, P. T. Landsberg in *Lectures in Theoretical Physics, Vol. 8A*, Univ. of Colorado Press, 1966, §6.
[38] C. J. Hearn, Ph.D. Thesis, University College, Cardiff, Wales, 1964.

the appropriate vector **L** from equation B.17.21 is given in brackets in figs. B.17.2 and B.17.4. Fig. B.17.3 shows the extended (full curves) and the reduced (dotted curves) wave-vector scheme. Sections of the graph corresponding to states between which the overlap integral is unity in the

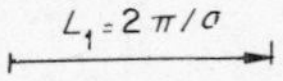

Fig. B.17.1.

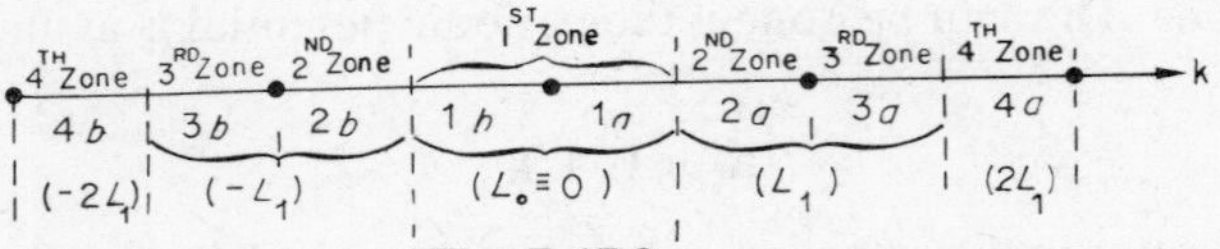

Fig. B.17.2.

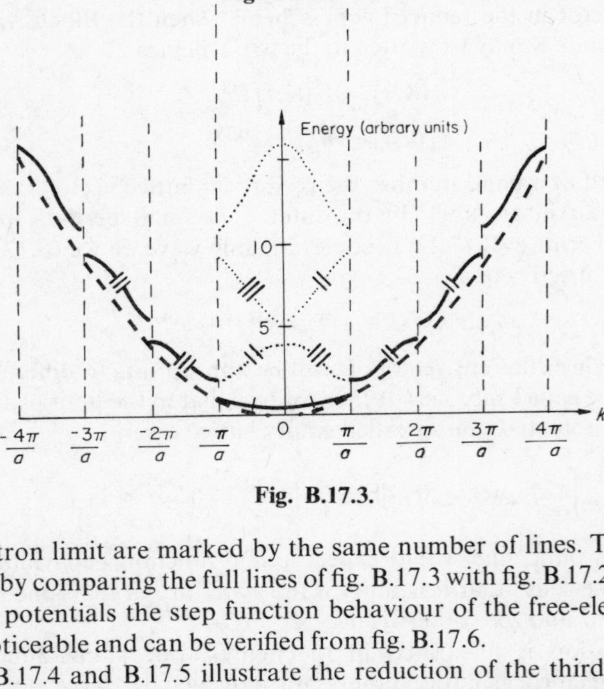

Fig. B.17.3.

free-electron limit are marked by the same number of lines. This may be checked by comparing the full lines of fig. B.17.3 with fig. B.17.2. For small periodic potentials the step function behaviour of the free-electron limit is still noticeable and can be verified from fig. B.17.6.

Figs. B.17.4 and B.17.5 illustrate the reduction of the third zone for a two-dimensional lattice. One would expect the overlap integrals between any two states in the region which is made up of subzones *a*, *b*, *j* to be unity in the free-electron limit. The same applies to the region made up of subzones *f*, *g* and *e*, to the region made up of subzones *c* and *d* and to the region made up of subzones *i* and *h*. This splits the third zone into ten subzones or four regions. States in different regions have zero overlap in the free-electron limit.

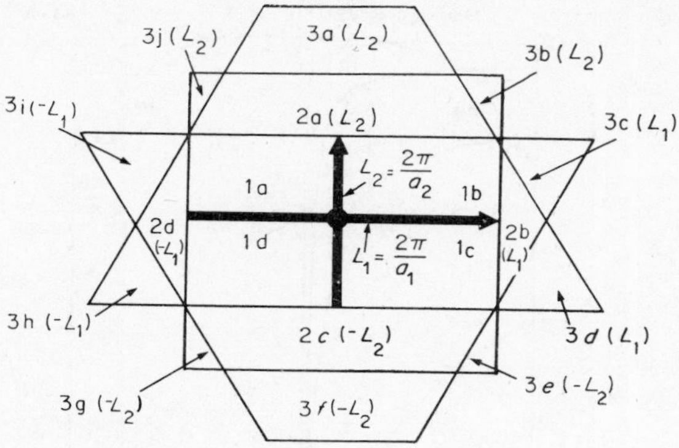

Fig. B.17.4.

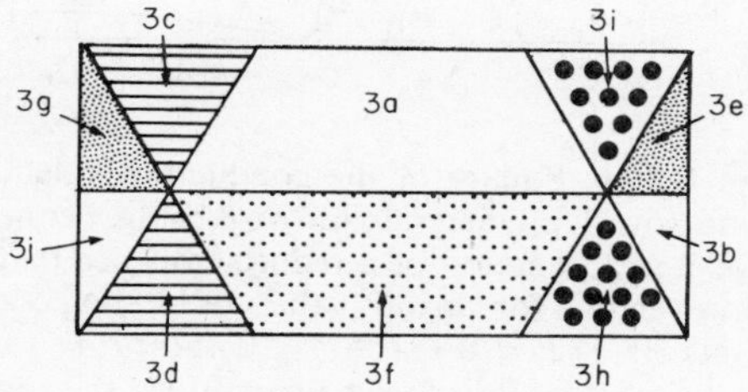

Fig. B.17.5.

B.17d Radiative transitions in a localized centre

The matrix element of the momentum operator $\mathbf{p}$ is well-known to govern radiative transition probabilities. Using (B.15.18) and (B.17.4), it is closely related to the velocity operator

$$\mathbf{v} = \nabla_{\mathbf{p}} H(\mathbf{p}, \mathbf{r}) = \frac{1}{m}\mathbf{p} + \boldsymbol{\lambda}, \qquad \text{(B.17.24)}$$

particularly if spin–orbit interaction is neglected. We therefore consider the matrix element of $\mathbf{v}$ (a) between two hydrogen-like states and (b) between two states given by an effective mass approximation in a crystal. The object is to discover in what sense, if any, the calculation (b) can be replaced by the simpler calculation (a).

As the localized states in volume V we take

$$F^{(\alpha)}(\mathbf{r}) = f \sum_{\mathbf{k}} A^{(\alpha)}(\mathbf{k})\, e^{i\mathbf{k}\cdot\mathbf{r}} \qquad (\alpha = 1, 2), \qquad \text{(B.17.25)}$$

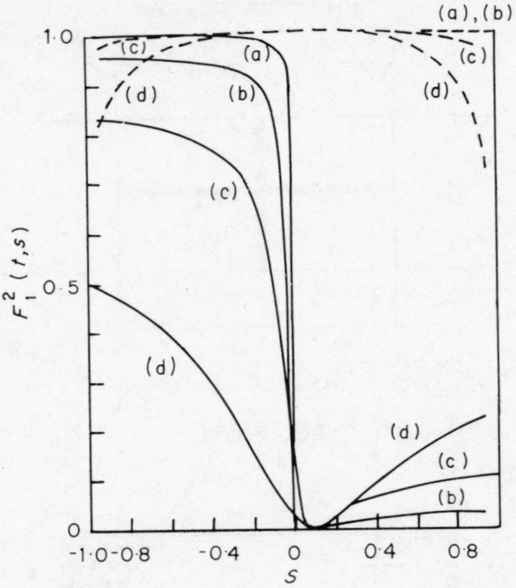

Fig. B.17.6. Square of the conduction–valence band overlap integral in a Kronig–Penney model as a function of a reduced valence band wave vector **s** for various values of binding constants P; (a) $P = 0{\cdot}1$; (b) $P = 0{\cdot}4$; (c) $P = 1{\cdot}0$; (d) $P = 5{\cdot}0$. **s** attains the value unity at the zone boundary. These curves hold for a reduced conduction band wave vector $\mathbf{t} = 0{\cdot}1$. The broken curve gives $F_1^2(\mathbf{t},\mathbf{s}) + F_2^2(\mathbf{t},\mathbf{s})$, where $F_2(\mathbf{t},\mathbf{s})$ is the corresponding overlap integral when both wave vectors are in the conduction band. (From A. R. Beattie and P. T. Landsberg, *Proc. Roy. Soc.*, **A258**, 486, 1960).

where f is a normalizing constant ($V|f|^2 = 1$), and the Fourier expansion is terminated at **k**-values corresponding to the Brillouin zone boundaries arising in calculation (b). The matrix element for the localized states is

$$L_j \equiv \int_V F^{(1)}(\mathbf{r})^* v_j F^{(2)}(\mathbf{r})\,\mathrm{d}\mathbf{r} = |f|^2 \sum_{\mathbf{k},\mathbf{k}'} A^{(1)}(\mathbf{k})^* A^{(2)}(\mathbf{k}') \int_V \mathrm{e}^{-i\mathbf{k}\cdot\mathbf{r}} v_j \mathrm{e}^{i\mathbf{k}'\cdot\mathbf{r}}\,\mathrm{d}\mathbf{r}.$$

If we neglect spin–orbit interaction $v_j = p_j/m = (\hbar/im)\,\partial/\partial x_j$, so that the integral is $(\hbar/im) ik_j\,\delta_{\mathbf{k}\mathbf{k}'}V$. Hence

$$L_j = \frac{1}{m}\sum_{\mathbf{k}} A^{(1)}(\mathbf{k})^* A^{(2)}(\mathbf{k})\hbar k_j. \tag{B.17.26}$$

For calculation (b) we use superpositions of normalized Bloch functions with the Fourier coefficients $A^{(\alpha)}(\mathbf{k})$ as superposition factors, in accordance with §B.16:

$$\psi^{(\alpha)} = \sum_{\mathbf{k}} A^{(\alpha)}(\mathbf{k})\psi_{n\mathbf{k}}(\mathbf{r}) \qquad (\alpha = 1, 2). \tag{B.17.27}$$

The assumption is that one band dominates, so that there is no sum over the band number. The matrix element is

$$M_j = \sum_{\mathbf{k}} A^{(1)}(\mathbf{k})^* A^{(2)}(\mathbf{k}) \int \psi_{n\mathbf{k}}^*(\mathbf{r}) v_j \psi_{n\mathbf{k}}(\mathbf{r})\, d\mathbf{r}.$$

The $\mathbf{k}$-diagonality of v_j (problem B.15.3) has been used here.

Write $\mathbf{k} = \mathbf{k}_0 + \mathbf{q}$ and sum over $\mathbf{q}$, using (B.17.13), so that

$$M_j = \hbar^{-1} \frac{\partial E_n(\mathbf{k}_0)}{\partial k_j} \left[\sum_{\mathbf{q}} A^{(1)}(\mathbf{k})^* A^{(2)}(\mathbf{k}) \right] + \sum_{i=1}^{3} \hbar \left(\frac{1}{m} \right)_{ji}^{n\mathbf{k}_0} \sum_{\mathbf{q}} q_j A^{(1)}(\mathbf{k})^* A^{(2)}(\mathbf{k}). \tag{B.17.28}$$

We see that the simplest cases arise when $\mathbf{k}_0 = 0$ is a band extremum. Then $\mathbf{q} = \mathbf{k}$, and if only one band dominates there is in fact the simple relation

$$M_j = m \sum_{i=1}^{3} \left(\frac{1}{m} \right)_{ji}^{no} L_i \tag{B.17.29}$$

between the matrix elements (if the mass tensor is symmetrical). If the mass tensor is diagonal

$$M_j = m \left(\frac{1}{m} \right)_{jj}^{no} L_j.$$

and the matrix elements are proportional to each other[39].

Problem

(B.17.1) Investigate the sum rule (B.17.17) if only two bands are important.

$[n' = 1, n = 2:$

$$[E_1(k) - E_2(k)] |\mathbf{I}(\mathbf{k})_{2\mathbf{q}}^{1\mathbf{q}'}|^2 = \frac{\hbar^2}{2} |\mathbf{q}' - \mathbf{q}|^2 \left(\frac{1}{m} - \frac{1}{m_{2\mathbf{k}}} \right).$$

[39] W. Kohn, *Solid State Physics, Vol. 5*, Academic Press, New York, 1967 discusses this simple case.

$\underline{n' = 2, n = 1:}$

$$[E_2(\mathbf{k}) - E_1(\mathbf{k})]|\mathrm{I}(\mathbf{k})^{2\mathbf{q}}_{1\mathbf{q}'}|^2 = \frac{h^2}{2}|\mathbf{q}' - \mathbf{q}|^2\left(\frac{1}{m} - \frac{1}{m_{1\mathbf{k}}}\right).$$

Hence

$$\frac{2}{m} = \frac{1}{m_{1n}} + \frac{1}{m_{2\mathbf{k}}}.$$

Note that this is a special case of (B.15.24′).]

Further reading

J. Callaway, *Energy Band Theory*, Academic Press, New York, 1963.
C. Kittel, *Quantum Theory of Solids*, Wiley, New York, 1963.
A. B. Pippard, *The Dynamics of Conduction Electrons*, Blackie, London, 1965.
J. M. Ziman, *Principles of the Theory of Solids*, Cambridge University Press, 1964.

PART C

GROUP THEORY AND ELECTRONIC STATES IN PERFECT CRYSTALS

D. J. Morgan

CHAPTER IV

Basic Group Theory

C.1 SYMMETRY TRANSFORMATIONS

Consider a rigid body with a point P on it which has position vector $\mathbf{r} \equiv (x_1, x_2, x_3)$ with respect to some fixed origin of coordinates. If the body is rotated through an angle $-\phi$ about the x_3-axis, the point P moves to the position P' with position vector $\mathbf{r}' \equiv (x'_1, x'_2, x'_3)$, where

$$x'_1 = x_1 \cos\phi + x_2 \sin\phi, x'_2 = -x_1 \sin\phi + x_2 \cos\phi, \quad x'_3 = x_3, \tag{C.1.1}$$

$$\begin{pmatrix} x'_1 \\ x'_2 \\ x'_3 \end{pmatrix} = \begin{pmatrix} \cos\phi & \sin\phi & 0 \\ -\sin\phi & \cos\phi & 0 \\ 0 & 0 & 1 \end{pmatrix} \begin{pmatrix} x_1 \\ x_2 \\ x_3 \end{pmatrix}, \tag{C.1.2}$$

$$\mathbf{r}' = q(\phi)\mathbf{r}, \tag{C.1.3}$$

$q(\phi)$ standing for the matrix in (C.1.2). Instead of rotating the body, rotate the coordinate axes through an angle ϕ about the x_3-axis. The coordinates of the point P referred to the new coordinate system are again related to x_1, x_2 and x_3 by equations (C.1.1)–(C.1.3). The above linear transformation of the coordinates can then represent either the change in the coordinates of a moving point when the body is rotated through an angle $-\phi$ or the change in the coordinates of a fixed point when the coordinate axes are rotated through an angle ϕ.

Let the above linear transformation of the coordinates be denoted by the symbol $Q(\phi)$. The application of $Q(\phi)$ to a function $f(\mathbf{r}) \equiv f(x_1, x_2, x_3)$ of the coordinates is defined to give the function $f(q(\phi)\mathbf{r})$, i.e.

$$Q(\phi)f(\mathbf{r}) = f(q(\phi)\mathbf{r}) = f(x_1 \cos\phi + x_2 \sin\phi, -x_1 \sin\phi + x_2 \cos\phi, x_3). \tag{C.1.4}$$

This results in a function of the coordinates[1] which in general displays a different functional form from $f(\mathbf{r})$, i.e.

$$Q(\phi)f(\mathbf{r}) = f(q(\phi)\mathbf{r}) \equiv F(\mathbf{r}). \tag{C.1.5}$$

[1] Some authors use the alternative definition $Q(\phi)f(\mathbf{r}) = f(q^{-1}(\phi)\mathbf{r})$.

For example,

$$Q(\phi)(x_1 - x_2)^2 = [x_1(\cos\phi + \sin\phi) + x_2(\sin\phi - \cos\phi)]^2, \tag{C.1.6}$$

$$Q(\phi)\frac{\partial}{\partial x_1} = \cos\phi\frac{\partial}{\partial x_1} + \sin\phi\frac{\partial}{\partial x_2}, \tag{C.1.7}$$

and $Q(\phi)$ applied to the equation

$$\frac{\partial}{\partial x_1}(x_1 - x_2)^2 = 2(x_1 - x_2) \tag{C.1.8}$$

gives

$$\left(\cos\phi\frac{\partial}{\partial x_1} + \sin\phi\frac{\partial}{\partial x_2}\right)[x_1(\cos\phi + \sin\phi) + x_2(\sin\phi - \cos\phi)]^2$$
$$= 2[x_1(\cos\phi + \sin\phi) + x_2(\sin\phi - \cos\phi)], \tag{C.1.9}$$

which is still a correct equation.

It may happen, however, that the application of $Q(\phi)$ to a function of the coordinates leaves that function unchanged. $Q(\phi)$ is then said to be a symmetry transformation of that function, or $Q(\phi)$ is said to leave that function invariant. For example,

$$Q(\phi)(x_1^2 + x_2^2 - 2x_3) = (x_1^2 + x_2^2 - 2x_3), \tag{C.1.10}$$

$$Q(\phi)\nabla^2 \equiv Q(\phi)\left(\frac{\partial^2}{\partial x_1^2} + \frac{\partial^2}{\partial x_2^2} + \frac{\partial^2}{\partial x_3^2}\right) = \nabla^2, \tag{C.1.11}$$

i.e. $Q(\phi)$ is a symmetry transformation of $(x_1^2 + x_2^2 - 2x_3)$ and of ∇^2.

Consider the time-independent Schrödinger equation[2]

$$H(\mathbf{r})\psi_1(\mathbf{r}) = \epsilon\psi_1(\mathbf{r}), \tag{C.1.12}$$

and let Q be a symmetry transformation of the Hamiltonian $H(\mathbf{r})$, i.e.

$$QH(\mathbf{r}) = H(\mathbf{r}). \tag{C.1.13}$$

The application of Q to the Schrödinger equation (C.1.12) then yields

$$H(\mathbf{r})\psi_2(\mathbf{r}) = \epsilon\psi_2(\mathbf{r}), \tag{C.1.14}$$

where

$$Q\psi_1(\mathbf{r}) = \psi_2(\mathbf{r}). \tag{C.1.15}$$

$\psi_2(\mathbf{r})$ is also an eigenfunction of $H(\mathbf{r})$ belonging to the energy eigenvalue ϵ,

[2] See any Quantum Mechanics text book, e.g. L. I. Schiff, *Quantum Mechanics, 2nd ed.*, McGraw-Hill, New York, 1955.

and if ϵ is a degenerate energy level $\psi_1(\mathbf{r})$ and $\psi_2(\mathbf{r})$ are in general different functions of the coordinates.

Thus the eigenfunctions of a Hamiltonian which belong to a degenerate energy level are transformed into one another by the symmetry transformations of that Hamiltonian.

C.2 GROUPS OF SYMMETRY TRANSFORMATIONS

Let $Q_1, Q_2, \ldots, Q_g$ be a collection of g different symmetry transformations. The equation

$$Q_i Q_j = Q_p \tag{C.2.1}$$

means that for any function $f(\mathbf{r})$ of the coordinates

$$Q_i Q_j f(\mathbf{r}) = Q_i\{Q_j f(\mathbf{r})\} = Q_i F(\mathbf{r}) = G(\mathbf{r}) = Q_p f(\mathbf{r}). \tag{C.2.2}$$

In general $Q_i Q_j \neq Q_j Q_i$, the transformations do not commute and the order in which they are applied is important.

This collection is said to form a group of order g if

(i) the product of any two of the transformations in this collection, $Q_i Q_j = Q_p$, is itself contained in the collection;
(ii) the identity transformation I ($\equiv Q_1$) is included in this collection, where $IQ_j = Q_j I = Q_j$ for any transformation Q_j;
(iii) the associative law holds, $Q_i(Q_j Q_s) = (Q_i Q_j)Q_s$;
(iv) every transformation in the collection has an inverse which is itself contained in the collection. The inverse of Q_j is denoted by Q_j^{-1} and $Q_j^{-1}Q_j = Q_j Q_j^{-1} = I$, where I is the identity transformation.

The transformations which make up the group are then known as the elements of the group.

If, in addition, $Q_i Q_j = Q_j Q_i$ for all i and j, i.e. all the elements commute, the group is called an Abelian group.

As an example of a group of symmetry transformations, consider the rotations which leave an equilateral triangle in positions which are indistinguishable from the original position of that triangle. Such positions are known as equivalent positions of the triangle.

The origin, O, of coordinates is at the centre of the triangle; the x_3-axis is out of the plane of the triangle and OA, OB and OC are three other axes which are perpendicular to the sides of the triangle. It can easily be seen that the following rotations of the triangle leave it in equivalent positions, and that these rotations form a group of order six:

$$\begin{aligned}
Q_1 &\equiv I = Q_1^{-1} \text{ (the identity): no rotation} \\
Q_2 &= Q_3^{-1}: 120^\circ \text{ about the } x_3\text{-axis} \\
Q_3 &= Q_2^{-1}: 240^\circ \text{ about the } x_3\text{-axis} \\
Q_4 &= Q_4^{-1}: 180^\circ \text{ about the } OA\text{-axis} \\
Q_5 &= Q_5^{-1}: 180^\circ \text{ about the } OB\text{-axis} \\
Q_6 &= Q_6^{-1}: 180^\circ \text{ about the } OC\text{-axis}
\end{aligned} \tag{C.2.3}$$

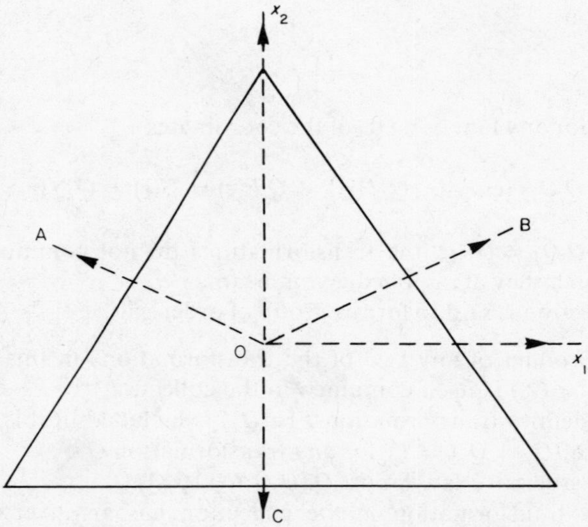

Fig. C.2.1.

For example, if the vertices of the triangle are numbered in order to show up the effects of these rotations,

$$Q_6 Q_4 \underset{2\quad 3}{\overset{1}{\triangle}} \to Q_6 \underset{1\quad 3}{\overset{2}{\triangle}} \to \underset{3\quad 1}{\overset{2}{\triangle}}, \tag{C.2.4}$$

$$Q_3 \underset{2\quad 3}{\overset{1}{\triangle}} \to \underset{3\quad 1}{\overset{2}{\triangle}}, \tag{C.2.5}$$

and hence

$$Q_6 Q_4 = Q_3. \tag{C.2.6}$$

The linear transformations of the coordinates which correspond to these operations are given below.

$$Q_1: x'_1 = x_1, \quad x'_2 = x_2, \quad x'_3 = x_3.$$

$$Q_2: x'_1 = -\frac{1}{2}x_1 + \frac{\sqrt{3}}{2}x_2, \quad x'_2 = -\frac{\sqrt{3}}{2}x_1 - \frac{1}{2}x_2, \quad x'_3 = x_3.$$

$$Q_3: x'_1 = -\frac{\sqrt{3}}{2}x_2 - \frac{1}{2}x_1, \quad x'_2 = \frac{\sqrt{3}}{2}x_1 - \frac{1}{2}x_2, \quad x'_3 = x_3.$$

$$Q_4: x'_1 = \frac{1}{2}x_1 - \frac{\sqrt{3}}{2}x_2, \quad x'_2 = -\frac{\sqrt{3}}{2}x_1 - \frac{1}{2}x_2, \quad x'_3 = -x_3. \tag{C.2.7}$$

$$Q_5: x'_1 = \frac{1}{2}x_1 + \frac{\sqrt{3}}{2}x_2, \quad x'_2 = \frac{\sqrt{3}}{2}x_1 - \frac{1}{2}x_2, \quad x'_3 = -x_3.$$

$$Q_6: x'_1 = -x_1, \quad x'_2 = x_2, \quad x'_3 = -x_3.$$

Corresponding to (C.2.4)–(C.2.6) we have

$$\begin{aligned} Q_6Q_4 f(x_1, x_2, x_3) &= Q_6 f\left(\frac{1}{2}x_1 - \frac{\sqrt{3}}{2}x_2, -\frac{\sqrt{3}}{2}x_1 - \frac{1}{2}x_2, -x_3\right) \\ &= f\left(-\frac{1}{2}x_1 - \frac{\sqrt{3}}{2}x_2, \frac{\sqrt{3}}{2}x_1 - \frac{1}{2}x_2, x_3\right) = Q_3 f(x_1, x_2, x_3). \end{aligned} \tag{C.2.8}$$

Since the product of any two elements of a group is itself an element of the group, a multiplication table for the group can be drawn up. For the above group of order six we obtain the following multiplication table:

Table C.2.1 Multiplication table for the group of order six.

Applied Second \ Applied First	Q_1	Q_2	Q_3	Q_4	Q_5	Q_6
Q_1	Q_1	Q_2	Q_3	Q_4	Q_5	Q_6
Q_2	Q_2	Q_3	Q_1	Q_5	Q_6	Q_4
Q_3	Q_3	Q_1	Q_2	Q_6	Q_4	Q_5
Q_4	Q_4	Q_6	Q_5	Q_1	Q_3	Q_2
Q_5	Q_5	Q_4	Q_6	Q_2	Q_1	Q_3
Q_6	Q_6	Q_5	Q_4	Q_3	Q_2	Q_1

In this table the element at the intersection of the ith row and the jth column is equivalent to the product Q_iQ_j. Note that each element of the group appears once only in every row and column of this table.

C.2a Point groups

The group given above, which is usually denoted by D_3, is an example of what is known as a point group. All the operations of a point group leave one point (the origin, say) unchanged.

Point groups in general contain operations which are proper rotations and improper rotations. An improper rotation may be defined as the product of a proper rotation and the inversion J. The inversion J is defined by

$$Jf(\mathbf{r}) = f(-\mathbf{r}), \tag{C.2.9}$$

where $f(\mathbf{r})$ is any function of the coordinates, and an improper rotation may be written JQ_p, where Q_p is a proper rotation. For example, the transformation of the coordinates

$$x_1' = x_1, \qquad x_2' = -x_2, \qquad x_3' = x_3 \tag{C.2.10}$$

corresponds to a reflexion in the plane $x_2 = 0$, and this reflexion can be thought of as an improper rotation JQ_6, where Q_6 is the proper rotation defined in (C.2.7).

All point group operations correspond to orthogonal transformations of the coordinates. If the point group operation Q_p corresponds to the transformation of the coordinates

$$\mathbf{r}' = q^p\mathbf{r}; \qquad x_i' = \sum_{j=1}^{3} q_{ij}^p x_j \qquad (i = 1, 2, 3), \tag{C.2.11}$$

then

$$\sum_{j=1}^{3} q_{ji}^p q_{js}^p = \delta_{is}, \qquad \sum_{j=1}^{3} q_{ij}^p q_{sj}^p = \delta_{is}, \tag{C.2.12}$$

and

$$\det|q^p| = \begin{cases} 1 & \text{if } Q_p \text{ is a proper rotation} \\ -1 & \text{if } Q_p \text{ is an improper rotation.} \end{cases} \tag{C.2.13}$$

Q_p^{-1}, the inverse of Q_p, corresponds to the transformation of the coordinates

$$\mathbf{r}' = (q^p)^{-1}\mathbf{r}; \qquad x_i' = \sum_{j=1}^{3} (q^p)_{ij}^{-1} x_j \qquad (i = 1, 2, 3). \tag{C.2.14}$$

From (C.2.11)–(C.2.13) it follows that $(q^p)^{-1} = \tilde{q}^p$, where $\tilde{q}^p$ is the transpose of the matrix q^p and is constructed by interchanging the rows and columns of q^p. Hence the transformation of the coordinates corresponding to Q_p^{-1} can also be written

$$\mathbf{r}' = \tilde{q}^p\mathbf{r}; \qquad x_i' = \sum_{j=1}^{3} q_{ji}^p x_j \qquad (i = 1, 2, 3). \tag{C.2.15}$$

C.2b Classes

Consider a group of order g with elements $Q_1 \equiv I, Q_2, \ldots, Q_g$. $Q_i^{-1}Q_jQ_i$ is itself an element of the group. Keeping Q_j fixed, allow Q_i to stand in turn for each element of the group. The elements generated in this way are said to belong to the same class as Q_j. As Q_i stands in turn for each element of the group, the set $Q_i^{-1}Q_jQ_i$ will consist of, say, h different elements, each repeated g/h times. h, the number of distinct elements in the class, is called the order of the class.

To prove this, suppose that Q_j itself appears n times in the set $Q_i^{-1}Q_jQ_i$, i.e. there are just n elements $Q_1, Q_2, \ldots, Q_n$ for which

$$Q_p^{-1}Q_jQ_p = Q_j, \qquad Q_jQ_p = Q_pQ_j \qquad (p = 1, 2, \ldots, n). \tag{C.2.16}$$

Obviously $Q_1 \equiv I$ must be one of these. Let Q_s be another element belonging to the same class as Q_j. Then for some Q_k,

$$Q_k^{-1}Q_jQ_k = Q_s. \tag{C.2.17}$$

Consider the n elements $Q_pQ_k (p = 1, 2, \ldots, n)$. Then

$$(Q_pQ_k)^{-1}Q_jQ_pQ_k = Q_k^{-1}Q_p^{-1}Q_jQ_pQ_k = Q_k^{-1}Q_jQ_k = Q_s. \tag{C.2.18}$$

This shows that Q_s occurs at least n times in the set $Q_i^{-1}Q_jQ_i$. Now let Q_t be any element of the group for which

$$Q_t^{-1}Q_jQ_t = Q_s. \tag{C.2.19}$$

Then

$$(Q_tQ_k^{-1})^{-1}Q_jQ_tQ_k^{-1} = Q_kQ_t^{-1}Q_jQ_tQ_k^{-1} = Q_kQ_sQ_k^{-1} = Q_j. \tag{C.2.20}$$

$Q_tQ_k^{-1}$ therefore commutes with Q_j and must be one of the $Q_p (p = 1, 2, \ldots, n)$, i.e. for some Q_p,

$$Q_tQ_k^{-1} = Q_p, \qquad Q_t = Q_pQ_k. \tag{C.2.21}$$

Hence the only elements Q_t which satisfy (C.2.19) are of the form Q_pQ_k $(p = 1, 2, \ldots, n)$. There are just n of these and therefore Q_s appears just n times in the set $Q_i^{-1}Q_jQ_i$. This is true for every element in the same class as Q_j, and hence

$$g = nh, \qquad \text{i.e.} \quad n = g/h. \tag{C.2.22}$$

Table C.2.2 The set $Q_i^{-1}Q_jQ_i$ for the point group D_3.

	Q_j: Q_1	Q_2	Q_3	Q_4	Q_5	Q_6
Q_i: Q_1	Q_1	Q_2	Q_3	Q_4	Q_5	Q_6
Q_2	Q_1	Q_2	Q_3	Q_5	Q_6	Q_4
Q_3	Q_1	Q_2	Q_3	Q_6	Q_4	Q_5
Q_4	Q_1	Q_3	Q_2	Q_4	Q_6	Q_5
Q_5	Q_1	Q_3	Q_2	Q_6	Q_5	Q_4
Q_6	Q_1	Q_3	Q_2	Q_5	Q_4	Q_6

Table C.2.3 Classes of the point group D_3.

Q_j	Elements in the same class as Q_j	h	g/h
Q_1	Q_1	1	6
Q_2	Q_2, Q_3	2	3
Q_3	Q_2, Q_3	2	3
Q_4	Q_4, Q_5, Q_6	3	2
Q_5	Q_4, Q_5, Q_6	3	2
Q_6	Q_4, Q_5, Q_6	3	2

In the point group D_3 there are therefore three distinct classes,

$$Q_1 \equiv I; \qquad Q_2 \text{ and } Q_3; \qquad Q_4, Q_5 \text{ and } Q_6. \tag{C.2.23}$$

The identity element $Q_1 \equiv I$ always forms a class by itself of order one. No two distinct classes have any elements in common.

The sum of elements in a distinct class will be denoted by the class symbol $\mathscr{C}$. For the point group D_3,

$$\begin{aligned} \mathscr{C}_1 &= Q_1, \qquad h_1 = 1. \\ \mathscr{C}_2 &= Q_2 + Q_3, \qquad h_2 = 2. \\ \mathscr{C}_3 &= Q_4 + Q_5 + Q_6, \qquad h_3 = 3. \end{aligned} \tag{C.2.24}$$

Obviously

$$\mathscr{C}_j = \frac{h_j}{g} \sum_{i=1}^{g} Q_i^{-1} Q_p Q_i, \tag{C.2.25}$$

where Q_p is any element in the class labelled j.

In general, a group of order g can be split up into g_c, say, distinct classes, no two of which have any elements in common.

The class symbols $\mathscr{C}_j$ commute with every element of the group. To prove this, consider

$$Q_k\mathscr{C}_j = \frac{h_j}{g}\sum_{i=1}^{g} Q_kQ_i^{-1}Q_pQ_i. \tag{C.2.26}$$

Put

$$Q_kQ_i^{-1} = Q_t^{-1}, \qquad Q_i = Q_tQ_k. \tag{C.2.27}$$

(C.2.26) then becomes

$$Q_k\mathscr{C}_j = \frac{h_j}{g}\sum_{t=1}^{g} Q_t^{-1}Q_pQ_tQ_k = \mathscr{C}_jQ_k. \tag{C.2.28}$$

It follows that the class symbols themselves commute, i.e.

$$\mathscr{C}_i\mathscr{C}_j = \mathscr{C}_j\mathscr{C}_i. \tag{C.2.29}$$

Let

$$\mathscr{C}_i = Q_1^i + Q_2^i + \cdots + Q_{h_i}^i,$$
$$\mathscr{C}_j = Q_1^j + Q_2^j + \cdots + Q_{h_j}^j. \tag{C.2.30}$$

Then

$$\mathscr{C}_i\mathscr{C}_j = \sum_{p=1}^{h_i}\sum_{k=1}^{h_j} Q_p^iQ_k^j = \sum(\text{elements of the group}). \tag{C.2.31}$$

$\mathscr{C}_i\mathscr{C}_j$ commutes with all elements of the group and must therefore consist of a sum of class symbols

$$\mathscr{C}_i\mathscr{C}_j = \sum_{p=1}^{g_c} c_{ijp}\mathscr{C}_p, \tag{C.2.32}$$

where g_c is the number of distinct classes in the group. For example, the class symbols of the point group D_3[see (C.2.24)] multiply as follows:

$$\mathscr{C}_1\mathscr{C}_1 = \mathscr{C}_1; \qquad \mathscr{C}_1\mathscr{C}_2 = \mathscr{C}_2\mathscr{C}_1 = \mathscr{C}_2; \qquad \mathscr{C}_1\mathscr{C}_3 = \mathscr{C}_3\mathscr{C}_1 = \mathscr{C}_3;$$
$$\mathscr{C}_2\mathscr{C}_2 = 2\mathscr{C}_1 + \mathscr{C}_2; \qquad \mathscr{C}_2\mathscr{C}_3 = \mathscr{C}_3\mathscr{C}_2 = 2\mathscr{C}_3;$$
$$\mathscr{C}_3\mathscr{C}_3 = 3\mathscr{C}_1 + 3\mathscr{C}_2. \tag{C.2.33}$$

Obviously, since $\mathscr{C}_i\mathscr{C}_j = \mathscr{C}_j\mathscr{C}_i$,

$$c_{ijp} = c_{jip}, \tag{C.2.34}$$

and since $\mathscr{C}_1\mathscr{C}_j = \mathscr{C}_j$, where $\mathscr{C}_1$ denotes the class of the identity,

$$c_{1jp} = c_{j1p} = \delta_{jp}. \tag{C.2.35}$$

It will be seen later that the coefficient c_{ij1} is of particular importance. Suppose that $c_{ij1} \neq 0$. Then for some p and k,

$$Q^i_p Q^j_k = Q_1, \qquad Q^j_k = (Q^i_p)^{-1}. \tag{C.2.36}$$

Hence $(Q^i_p)^{-1}$ must be in the class labelled j. Any other element Q^j_t in the class labelled j can then be written

$$Q^j_t = Q_s^{-1}(Q^i_p)^{-1}Q_s, \quad \text{i.e.} \quad (Q^j_t)^{-1} = Q_s^{-1}Q^i_pQ_s, \tag{C.2.37}$$

and $(Q^j_t)^{-1}$ must be in the class labelled i. It follows that if $c_{ij1} \neq 0$, the inverse of every element in the ith class is contained in the jth class, and the inverse of every element in the jth class is contained in the ith class. If $c_{ij1} \neq 0$ and $\mathscr{C}_i = Q^i_1 + \cdots + Q^i_{h_i}$,

$$\mathscr{C}_j = (Q^i_1)^{-1} + (Q^i_2)^{-1} + \cdots + (Q^i_{h_i})^{-1} \equiv \mathscr{C}_{i'}, \tag{C.2.38}$$

where i' labels the class containing the inverse of all the elements in the class labelled i. Clearly

$$\begin{aligned}\mathscr{C}_{i'}\mathscr{C}_i = \mathscr{C}_i\mathscr{C}_{i'} &= [(Q^i_1)^{-1} + \cdots + (Q^i_{h_i})^{-1}][Q^i_1 + \cdots + Q^i_{h_i}] \\ &= h_iQ_1 + \cdots + \cdots,\end{aligned} \tag{C.2.39}$$

and hence

$$c_{ij1} = h_i\delta_{ji'}. \tag{C.2.40}$$

This concept of a group of symmetry transformations is very important in that it can be shown that the symmetry transformations of a Hamiltonian always form a group[3]. It is the group of symmetry transformations of the Hamiltonian in the Schrödinger equation which determines the degeneracies of the energy eigenvalues and the symmetry properties of the eigenfunctions.

C.3 REPRESENTATIONS OF GROUPS

Consider a group of order g with elements $Q_1, Q_2, \ldots, Q_g$. If $M^1, M^2, \ldots, M^g$ are non-singular square matrices such that $Q_iQ_j = Q_p$ implies $M^iM^j = M^p$, where M^iM^j stands for ordinary matrix multiplication, then these matrices form a representation of the group. If the matrices are of order $n \times n$, the representation is of dimension n. Also, if M^j is the matrix representing Q_j, the matrix representing Q_j^{-1} is $M^{j^{-1}}$, where $M^{j^{-1}}$ is the inverse of the matrix M^j.

For example, two representations of the point group D_3 are

(i) $M^1 = M^2 = M^3 = M^4 = M^5 = M^6 = 1.$ (C.3.1)

[3] See, for example, V. Heine, *Group Theory in Quantum Mechanics*, Pergamon Press, London, p. 18.

This is the identity representation and is one-dimensional.

$$\text{(ii)}\ M^1 = \begin{pmatrix} 1 & 0 \\ 0 & 1 \end{pmatrix}; \qquad M^2 = \begin{pmatrix} -1/2 & -\sqrt{3}/2 \\ \sqrt{3}/2 & -1/2 \end{pmatrix};$$

$$M^3 = \begin{pmatrix} -1/2 & \sqrt{3}/2 \\ -\sqrt{3}/2 & -1/2 \end{pmatrix}; \quad M^4 = \begin{pmatrix} 1/2 & -\sqrt{3}/2 \\ -\sqrt{3}/2 & -1/2 \end{pmatrix};$$

$$M^5 = \begin{pmatrix} 1/2 & \sqrt{3}/2 \\ \sqrt{3}/2 & -1/2 \end{pmatrix}; \qquad M^6 = \begin{pmatrix} -1 & 0 \\ 0 & 1 \end{pmatrix}. \tag{C.3.2}$$

This is a two-dimensional representation.

The above matrices multiply in the same way as the elements of the point group D_3. For instance,

$$Q_6 Q_4 = Q_3, \tag{C.3.3}$$

and for the representation (i):

$$M^6 M^4 = 1 = M^3, \tag{C.3.4}$$

and for the representation (ii):

$$M^6 M^4 = \begin{pmatrix} -1 & 0 \\ 0 & 1 \end{pmatrix}\begin{pmatrix} 1/2 & -\sqrt{3}/2 \\ -\sqrt{3}/2 & -1/2 \end{pmatrix} = \begin{pmatrix} -1/2 & \sqrt{3}/2 \\ -\sqrt{3}/2 & -1/2 \end{pmatrix} = M^3. \tag{C.3.5}$$

Consider the Schrödinger equation

$$H(\mathbf{r})\psi_{nj}(\mathbf{r}) = \epsilon_n \psi_{nj}(\mathbf{r}), \tag{C.3.6}$$

and suppose that the energy level ϵ_n is d_n-fold degenerate. Corresponding to this energy level there are d_n linearly independent eigenfunctions $\psi_{nj}(\mathbf{r})$ ($j = 1, 2, 3, \ldots, d_n$). Let $Q_1, Q_2, \ldots, Q_g$ be the elements of the group of symmetry transformations of the Hamiltonian $H(\mathbf{r})$. An element Q_p of this group applied to any solution of (C.3.6) belonging to the energy ϵ_n results in another (or the same) solution of (C.3.6) belonging to the same energy. Since any solution belonging to the energy ϵ_n can be expressed as a linear combination of the functions $\psi_{nj}(\mathbf{r})$ ($j = 1, 2, 3, \ldots, d_n$), it follows that

$$Q_p \psi_{nj}(\mathbf{r}) = \sum_{s=1}^{d_n} M_{sj}^p \psi_{ns}(\mathbf{r}). \tag{C.3.7}$$

Write the d_n linearly independent solutions as a column vector $\boldsymbol{\psi}_n(\mathbf{r})$, i.e.

$$\boldsymbol{\psi}_n(\mathbf{r}) \equiv \begin{pmatrix} \psi_{n1}(\mathbf{r}) \\ \vdots \\ \psi_{nd_n}(\mathbf{r}) \end{pmatrix}. \tag{C.3.8}$$

It then follows from (C.3.7) that

$$Q_p\boldsymbol{\psi}_n(\mathbf{r}) = \tilde{M}^p\boldsymbol{\psi}_n(\mathbf{r}), \tag{C.3.9}$$

where $\tilde{M}^p$ is the transpose of the matrix M^p. By applying all the elements of the group in turn to $\boldsymbol{\psi}_n(\mathbf{r})$, g matrices $M^1, M^2, \ldots, M^g$ are generated. These matrices form a representation of the group. For instance, if we suppose that $Q_iQ_j = Q_p$, then

$$Q_iQ_j\boldsymbol{\psi}_n(\mathbf{r}) = \tilde{M}^jQ_i\boldsymbol{\psi}_n(\mathbf{r}) = \tilde{M}^j\tilde{M}^i\boldsymbol{\psi}_n(\mathbf{r}) = Q_p\boldsymbol{\psi}_n(\mathbf{r}) = \tilde{M}^p\boldsymbol{\psi}_n(\mathbf{r}). \tag{C.3.10}$$

Hence

$$\tilde{M}^j\tilde{M}^i = \tilde{M}^p \quad \text{and} \quad M^iM^j = M^p. \tag{C.3.11}$$

The linearly independent eigenfunctions belonging to a d_n-fold degenerate energy level act as basis functions for a d_n-dimensional representation of the group of the Hamiltonian.

Let $\phi_{n1}(\mathbf{r}), \phi_{n2}(\mathbf{r}), \ldots, \phi_{nd_n}(\mathbf{r})$ be linear combinations of the eigenfunctions $\psi_{nj}(\mathbf{r})$ ($j = 1, 2, \ldots, d_n$) belonging to the energy ϵ_n,

$$\phi_{np}(\mathbf{r}) = \sum_{k=1}^{d_n} S_{kp}\psi_{nk}(\mathbf{r}), \tag{C.3.12}$$

$$\boldsymbol{\phi}_n(\mathbf{r}) = \tilde{S}\boldsymbol{\psi}_n(\mathbf{r}), \tag{C.3.13}$$

where the matrix $\tilde{S}$ is non-singular so that

$$\boldsymbol{\psi}_n(\mathbf{r}) = \tilde{S}^{-1}\boldsymbol{\phi}_n(\mathbf{r}). \tag{C.3.14}$$

Then

$$Q_j\boldsymbol{\phi}_n(\mathbf{r}) = \tilde{S}Q_j\boldsymbol{\psi}_n(\mathbf{r}) = \tilde{S}\tilde{M}^j\boldsymbol{\psi}_n(\mathbf{r}) = \tilde{S}\tilde{M}^j\tilde{S}^{-1}\boldsymbol{\phi}_n(\mathbf{r}) \equiv \tilde{D}^j\boldsymbol{\phi}_n(\mathbf{r}). \tag{C.3.15}$$

The matrices $D^j = S^{-1}M^jS$ ($j = 1, 2, \ldots, g$) form another d_n-dimensional representation of the group of the Hamiltonian, and the functions $\phi_{np}(\mathbf{r})$ ($p = 1, 2, \ldots, d_n$) act as basis functions for this representation. The representation with matrices M^j and the representation with matrices D^j are equivalent representations. If two representations are equivalent, all the matrices of one of them can be changed into those of the other by the use of a single similarity transformation (the matrix S). No distinction will be made between equivalent representations. In fact, for the above representations, we write

$$M = D, \tag{C.3.16}$$

meaning that the representation with matrices M^j and the representation with matrices D^j differ at most by a similarity transformation.

C.3a Reducibility of representations

A representation with matrices $M^1, M^2, \ldots, M^g$ is reducible if, by the use of a single similarity transformation, all the matrices of this representation can be put in the form

$$\begin{pmatrix}
x & \cdot & \cdot & \cdot & \cdot & \cdot & \cdot & \cdot & \cdot & \cdot & \cdot & \cdot \\
\cdot & x & \cdot & \cdot & \cdot & \cdot & \cdot & \cdot & \cdot & \cdot & \cdot & \cdot \\
\cdot & \cdot & x & x & \cdot & \cdot & \cdot & \cdot & \cdot & \cdot & \cdot & \cdot \\
\cdot & \cdot & x & x & \cdot & \cdot & \cdot & \cdot & \cdot & \cdot & \cdot & \cdot \\
\cdot & \cdot & \cdot & \cdot & x & x & \cdot & \cdot & \cdot & \cdot & \cdot & \cdot \\
\cdot & \cdot & \cdot & \cdot & x & x & \cdot & \cdot & \cdot & \cdot & \cdot & \cdot \\
\cdot & \cdot & \cdot & \cdot & \cdot & \cdot & x & x & x & \cdot & \cdot & \cdot \\
\cdot & \cdot & \cdot & \cdot & \cdot & \cdot & x & x & x & \cdot & \cdot & \cdot \\
\cdot & \cdot & \cdot & \cdot & \cdot & \cdot & x & x & x & \cdot & \cdot & \cdot \\
\cdot & \cdot & \cdot & \cdot & \cdot & \cdot & \cdot & \cdot & \cdot & x & x & x \\
\cdot & \cdot & \cdot & \cdot & \cdot & \cdot & \cdot & \cdot & \cdot & x & x & x \\
\cdot & \cdot & \cdot & \cdot & \cdot & \cdot & \cdot & \cdot & \cdot & x & x & x
\end{pmatrix} \tag{C.3.17}$$

i.e. with single elements or sub-matrices along the diagonal and zeros elsewhere. When the matrices of this representation have been transformed so as to make these sub-matrices along the diagonal as small as possible the representation is said to be completely reduced; the sub-matrices along the diagonal then form irreducible representations.

For example, consider the following three-dimensional representation of the point group D_3:

$$M^{11} = \begin{pmatrix} 1 & 0 & 0 \\ 0 & 1 & 0 \\ 0 & 0 & 1 \end{pmatrix}, \qquad M^{12} = \begin{pmatrix} 0 & 1 & -1 \\ 1 & 0 & 1 \\ -1 & -2 & 0 \end{pmatrix},$$

$$M^{13} = \begin{pmatrix} 2 & 2 & 1 \\ -1 & -1 & -1 \\ -2 & -1 & -1 \end{pmatrix}, \qquad M^{14} = \begin{pmatrix} 0 & -1 & 1 \\ -1 & 0 & -1 \\ 0 & 0 & -1 \end{pmatrix},$$

$$M^{15} = \begin{pmatrix} -1 & 0 & 0 \\ 0 & -1 & 0 \\ 2 & 1 & 1 \end{pmatrix}, \qquad M^{16} = \begin{pmatrix} -2 & -2 & -1 \\ 1 & 1 & 1 \\ 1 & 2 & 0 \end{pmatrix}. \tag{C.3.18}$$

Let $M^{2j} = S^{-1}M^{1j}S$, with

$$S = \begin{pmatrix} 1 & -1 & 1 \\ 0 & 1 & -1 \\ -1 & 1 & 0 \end{pmatrix}, \qquad S^{-1} = \begin{pmatrix} 1 & 1 & 0 \\ 1 & 1 & 1 \\ 1 & 0 & 1 \end{pmatrix}. \tag{C.3.19}$$

Then

$$M^{21} = \begin{pmatrix} 1 & 0 & 0 \\ 0 & 1 & 0 \\ 0 & 0 & 1 \end{pmatrix}, \qquad M^{22} = \begin{pmatrix} 1 & 0 & 0 \\ 0 & -1 & 1 \\ 0 & -1 & 0 \end{pmatrix},$$

$$M^{23} = \begin{pmatrix} 1 & 0 & 0 \\ 0 & 0 & -1 \\ 0 & 1 & -1 \end{pmatrix}, \qquad M^{24} = \begin{pmatrix} -1 & 0 & 0 \\ 0 & -1 & 0 \\ 0 & -1 & 1 \end{pmatrix},$$

$$M^{25} = \begin{pmatrix} -1 & 0 & 0 \\ 0 & 0 & 1 \\ 0 & 1 & 0 \end{pmatrix}, \qquad M^{26} = \begin{pmatrix} -1 & 0 & 0 \\ 0 & 1 & -1 \\ 0 & 0 & -1 \end{pmatrix}. \tag{C.3.20}$$

Hence

$$M^{2j} = \begin{pmatrix} M^{3j} & 0 \quad 0 \\ 0 & \\ 0 & M^{4j} \end{pmatrix}, \tag{C.3.21}$$

where

$$M^{31} = M^{32} = M^{33} = 1, \qquad M^{34} = M^{35} = M^{36} = -1. \tag{C.3.22}$$

$$M^{41} = \begin{pmatrix} 1 & 0 \\ 0 & 1 \end{pmatrix}, \quad M^{42} = \begin{pmatrix} -1 & 1 \\ -1 & 0 \end{pmatrix}, \quad M^{43} = \begin{pmatrix} 0 & -1 \\ 1 & -1 \end{pmatrix},$$

$$M^{44} = \begin{pmatrix} -1 & 0 \\ -1 & 1 \end{pmatrix}, \quad M^{45} = \begin{pmatrix} 0 & 1 \\ 1 & 0 \end{pmatrix}, \quad M^{46} = \begin{pmatrix} 1 & -1 \\ 0 & -1 \end{pmatrix}. \tag{C.3.23}$$

Furthermore, no similarity transformation exists which reduces the matrices M^{11}, $M^{12}, \ldots, M^{16}$ any further. The matrices $M^{31}, \ldots, M^{36}$

and the matrices $M^{41}, \ldots, M^{46}$ form two distinct irreducible representations of the point group D_3.

The above decomposition into irreducible representations is written

$$M^{2j} = M^{3j} + M^{4j}. \tag{C.3.24}$$

The irreducible representation formed by the matrices $M^{41}, \ldots, M^{46}$ is equivalent to the representation given in (C.3.2), the transformation matrix relating the two representations being given by

$$S = \begin{pmatrix} 1/\sqrt{3} & 1 \\ 2/\sqrt{3} & 0 \end{pmatrix}; \qquad S^{-1} = \begin{pmatrix} 0 & \sqrt{3}/2 \\ 1 & -1/2 \end{pmatrix}. \tag{C.3.25}$$

(C.3.24) could then equally well have been written

$$M^{1j} = M^{3j} + M^{5j}, \tag{C.3.26}$$

where the matrices $M^{51}, \ldots, M^{56}$ are those of the representation given in (C.3.2).

In general, a completely reduced representation (M^j) contains an irreducible representation (M^α), or a representation equivalent to it, more than once along the diagonal. In this case we write

$$M^j = \sum_{\alpha=1}^{g_r} c_\alpha M^\alpha, \tag{C.3.27}$$

where g_r is the number of distinct inequivalent irreducible representations of the group and c_α is the number of times the irreducible representation M^α (or a representation equivalent to it) appears along the diagonal.

It has been shown that the linearly independent eigenfunctions belonging to a d_n-fold degenerate energy level act as basis functions for a d_n-dimensional representation of the group of symmetry transformations of the Hamiltonian. This representation is in general an irreducible representation. If it is irreducible, the degeneracy is known as 'essential degeneracy', i.e. if the d_n eigenfunctions $\psi_{n1}(\mathbf{r}), \ldots, \psi_{nd_n}(\mathbf{r})$ act as basis functions for a d_n-dimensional irreducible representation of the group of symmetry transformations of the Hamiltonian, they must of necessity belong to the same energy level.

If, however, this representation is reducible, we have what is known as 'accidental degeneracy'[4]. Suppose, for example, that the representation is reducible into two irreducible representations of dimensions m_1 and

[4] E. Wigner, *Group Theory and its Applications to the Quantum Mechanics of Atomic Spectra*, Academic Press, New York, 1959, p. 119.

m_2 $(m_1 + m_2 = d_n)$ respectively. This means that linear combinations

$$\phi_p(\mathbf{r}) = \sum_{s=1}^{d_n} B_{sp}\psi_{ns}(\mathbf{r}) \qquad (p = 1, 2, \ldots, m_1), \tag{C.3.28}$$

$$\xi_t(\mathbf{r}) = \sum_{k=1}^{d_n} C_{kt}\psi_{nk}(\mathbf{r}) \qquad (t = 1, 2, \ldots, m_2), \tag{C.3.29}$$

can be found such that $\phi_p(\mathbf{r})$ $(p = 1, 2, \ldots, m_1)$ act as basis functions for an m_1-dimensional irreducible representation, and $\xi_t(\mathbf{r})$ $(t = 1, 2, \ldots, m_2)$ act as basis functions for an m_2-dimensional irreducible representation of the group of symmetry transformations of the Hamiltonian, i.e. if Q_j is a symmetry transformation of the Hamiltonian,

$$Q_j\phi_p(\mathbf{r}) = \sum_{s=1}^{m_1} D^j_{sp}\phi_s(\mathbf{r}) \qquad (p = 1, 2, \ldots, m_1), \tag{C.3.30}$$

$$Q_j\xi_t(\mathbf{r}) = \sum_{k=1}^{m_2} E^j_{kt}\xi_k(\mathbf{r}) \qquad (t = 1, 2, \ldots, m_2). \tag{C.3.31}$$

(C.3.30) and (C.3.31) are valid even if $\phi_p(\mathbf{r})$ $(p = 1, \ldots, m_1)$ do not belong to the same energy level as $\xi_t(\mathbf{r})$ $(t = 1, \ldots, m_2)$. In other words, as far as the symmetry transformations of the Hamiltonian are concerned, the functions $\phi_p(\mathbf{r})$ and the functions $\xi_t(\mathbf{r})$ need not belong to the same energy level, i.e. this type of degeneracy is accidental.

Apart from accidental degeneracy therefore, the linearly independent eigenfunctions belonging to a d_n-fold degenerate energy level act as basis functions for a d_n-dimensional irreducible representation of the group of symmetry transformations of the Hamiltonian. Furthermore, if these linearly independent eigenfunctions are orthonormal the matrices of this irreducible representation are unitary. To prove this, we have

$$(\psi_{np}(\mathbf{r}), \psi_{nk}(\mathbf{r})) = \delta_{pk} = (Q_j\psi_{np}(\mathbf{r}), Q_j\psi_{nk}(\mathbf{r})).^5 \tag{C.3.32}$$

(C.3.32) is true because the last integral may be considered simply to be carried out in a rotated coordinate system. Hence

$$\begin{aligned}\delta_{pk} &= \left(\sum_{t=1}^{d_n} M^j_{tp}\psi_{nt}(\mathbf{r}), \sum_{s=1}^{d_n} M^j_{sk}\psi_{ns}(\mathbf{r})\right) \\ &= \sum_{t=1}^{d_n}\sum_{s=1}^{d_n} M^{j*}_{tp}M^j_{sk}(\psi_{nt}(\mathbf{r}), \psi_{ns}(\mathbf{r})) \\ &= \sum_{s=1}^{d_n} M^{j*}_{sp}M^j_{sk} = \sum_{s=1}^{d_n} M^{j\dagger}_{ps}M^j_{sk} = [M^{j\dagger}M^j]_{pk},\end{aligned} \tag{C.3.33}$$

[5] $(f(\mathbf{r}), g(\mathbf{r}))$ denotes the inner product of $f(\mathbf{r})$ and $g(\mathbf{r})$.

where $M^{j\dagger}$ is the adjoint matrix to M^j and is constructed by interchanging the rows and columns of M^j and taking the complex conjugate of each element. From (C.3.33)

$$M^{j\dagger} M^j = 1_{d_n}, \qquad M^{j\dagger} = M^{j^{-1}}, \tag{C.3.34}$$

where 1_{d_n} is the $d_n \times d_n$ unit matrix. Hence M^j is a unitary matrix.

An energy level and its eigenfunctions can therefore be labelled and described simply by naming the irreducible representation associated with the energy level. The dimension of the irreducible representation gives the degeneracy of the energy level, and the matrices of the irreducible representation tell us how the corresponding eigenfunctions act under the operations of the group of symmetry transformations of the Hamiltonian, i.e. tell us something about their symmetry properties.

C.4 ORTHOGONALITY RELATIONS

C.4a Schur's lemma

Let $M^{\alpha 1}, M^{\alpha 2}, \ldots, M^{\alpha g}$ be the matrices of an n_α-dimensional irreducible unitary representation of a group of order g, and suppose that there exists an $n_\alpha \times n_\alpha$ matrix B which commutes with all the matrices of this representation, i.e.

$$M^{\alpha j} B = B M^{\alpha j} \qquad (j = 1, 2, \ldots, g). \tag{C.4.1}$$

Taking the adjoint of both sides of (C.4.1),

$$B^\dagger M^{\alpha j\dagger} = M^{\alpha j\dagger} B^\dagger, \tag{C.4.2}$$

$$B^\dagger M^{\alpha j^{-1}} = M^{\alpha j^{-1}} B^\dagger, \tag{C.4.3}$$

since $M^{\alpha j}$ is unitary. Pre- and post-multiplying both sides of (C.4.3) by $M^{\alpha j}$ yields

$$M^{\alpha j} B^\dagger = B^\dagger M^{\alpha j}. \tag{C.4.4}$$

$B^\dagger$ thus also commutes with $M^{\alpha j}$, and hence so does the Hermitian matrix $H_1 = B + B^\dagger$, i.e.

$$H_1 M^{\alpha j} = M^{\alpha j} H_1 \qquad (j = 1, 2, \ldots, g). \tag{C.4.5}$$

By means of a unitary transformation matrix U the Hermitian matrix H_1 can be diagonalized

$$U^{-1} H_1 U = D \text{ (diagonal)}, \tag{C.4.6}$$

and, from (C.4.5),

$$D U^{-1} M^{\alpha j} U = U^{-1} M^{\alpha j} U D. \tag{C.4.7}$$

Taking the (k, p)-element of (C.4.7),

$$D_{kk}(U^{-1}M^{\alpha j}U)_{kp} = (U^{-1}M^{\alpha j}U)_{kp}D_{pp}, \tag{C.4.8}$$

$$(U^{-1}M^{\alpha j}U)_{kp}[D_{kk} - D_{pp}] = 0. \tag{C.4.9}$$

Assume that $D_{kk} \neq D_{pp}$. Then $(U^{-1}M^{\alpha j}U)_{kp}$ must be zero for all the matrices $M^{\alpha j}$ and the transformation U has brought $M^{\alpha j}$ $(j = 1, 2, \ldots, g)$ to block form, i.e. reduced it. But M^{α} is an irreducible representation and hence

$$D_{kk} = D_{pp}, \tag{C.4.10}$$

$$D = s_1 1_{n_\alpha}, \tag{C.4.11}$$

where s_1 is a scalar and 1_{n_α} is the $n_\alpha \times n_\alpha$ unit matrix. From (C.4.6),

$$H_1 = B + B^\dagger = UDU^{-1} = s_1 1_{n_\alpha}. \tag{C.4.12}$$

Similarly, by considering the Hermitian matrix $H_2 = \mathrm{i}(B - B^\dagger)$, which also commutes with $M^{\alpha j}$ $(j = 1, 2, \ldots, g)$, it can be shown that

$$H_2 = \mathrm{i}(B - B^\dagger) = s_2 1_{n_\alpha}, \tag{C.4.13}$$

and hence

$$B = \tfrac{1}{2}(H_1 - \mathrm{i}H_2) = \tfrac{1}{2}(s_1 - \mathrm{i}s_2)1_{n_\alpha}. \tag{C.4.14}$$

Any matrix which commutes with all the matrices of an irreducible representation of a group must therefore be a scalar multiple of the unit matrix. This is Schur's lemma.

Now let A be an arbitrary fixed $n_\alpha \times n_\alpha$ matrix, and consider the matrix

$$V_1 = \sum_{j=1}^{g} M^{\alpha j^{-1}} A M^{\alpha j}. \tag{C.4.15}$$

Then

$$M^{\alpha t^{-1}} V_1 M^{\alpha t} = \sum_{j=1}^{g} M^{\alpha t^{-1}} M^{\alpha j^{-1}} A M^{\alpha j} M^{\alpha t} = \sum_{p=1}^{g} M^{\alpha p^{-1}} A M^{\alpha p} = V_1, \tag{C.4.16}$$

where $M^{\alpha j}M^{\alpha t} = M^{\alpha p}$. Hence V_1 commutes with all the matrices of this irreducible representation, and by Schur's lemma

$$V_1 = m 1_{n_\alpha}, \tag{C.4.17}$$

where m is a scalar. Equating elements in (C.4.15),

$$m\delta_{st} = \sum_{j,n,p} M_{sn}^{\alpha j^{-1}} A_{np} M_{pt}^{\alpha j}, \tag{C.4.18}$$

and putting $s = t$ and summing over s,

$$mn_\alpha = \sum_{j,n,p} A_{np} \sum_{s=1}^{n_\alpha} M_{sn}^{\alpha j^{-1}} M_{ps}^{\alpha j} = \sum_{j,n,p} A_{np}\,\delta_{np}$$

$$= \sum_{j,p} A_{pp} = g \sum_{p=1}^{n_\alpha} A_{pp}. \qquad \text{(C.4.19)}$$

Hence

$$m = \frac{g}{n_\alpha} \sum_{p=1}^{n_\alpha} A_{pp}. \qquad \text{(C.4.20)}$$

Substituting (C.4.20) in (C.4.18) and equating coefficients of A_{np}, since A is an arbitrary fixed matrix,

$$\sum_{j=1}^{g} M_{sn}^{\alpha j^{-1}} M_{pt}^{\alpha j} = \frac{g}{n_\alpha} \delta_{st}\,\delta_{np}. \qquad \text{(C.4.21)}$$

This is the first of the orthogonality relations for irreducible representations.

Next let G be an arbitrary fixed $n_\alpha \times n_\beta$ matrix, and consider the matrix

$$V_2 = \sum_{j=1}^{g} M^{\beta j^{-1}} G M^{\alpha j}, \qquad \text{(C.4.22)}$$

where M^β is an n_β-dimensional irreducible unitary representation which, if $n_\alpha = n_\beta$, is not equivalent to M^α. Then

$$M^{\beta t^{-1}} V_2 M^{\alpha t} = \sum_{j=1}^{g} M^{\beta t^{-1}} M^{\beta j^{-1}} G M^{\alpha j} M^{\alpha t} = V_2, \qquad \text{(C.4.23)}$$

$$V_2 M^{\alpha t} = M^{\beta t} V_2 \qquad (t = 1, 2, \ldots, g). \qquad \text{(C.4.24)}$$

Taking the adjoint of both sides of (C.4.24),

$$M^{\alpha t^\dagger} V_2^\dagger = V_2^\dagger M^{\beta t^\dagger}, \qquad \text{(C.4.25)}$$

$$M^{\alpha t^{-1}} V_2^\dagger = V_2^\dagger M^{\beta t^{-1}}, \qquad \text{(C.4.26)}$$

since $M^{\alpha t}$ and $M^{\beta t}$ are unitary. Pre-multiplying both sides of (C.4.26) by V_2,

$$V_2 M^{\alpha t^{-1}} V_2^\dagger = V_2 V_2^\dagger M^{\beta t^{-1}}, \qquad \text{(C.4.27)}$$

or, since (C.4.24) also holds for $M^{\alpha t^{-1}}$ and $M^{\beta t^{-1}}$,

$$M^{\beta t^{-1}} V_2 V_2^\dagger = V_2 V_2^\dagger M^{\beta t^{-1}} \qquad (t = 1, 2, \ldots, g). \qquad \text{(C.4.28)}$$

Hence, by Schur's lemma,

$$V_2 V_2^\dagger = m 1_{n_\alpha} \qquad (n_\alpha \geqslant n_\beta), \tag{C.4.29}$$

where m is a scalar. We now show that $m = 0$, $V_2 = 0$ (the null matrix). Consider first the case when $n_\alpha = n_\beta$. From (C.4.29), if $m \neq 0$, $\det|V_2| \neq 0$, V_2 has an inverse and M^α and M^β are equivalent representations. But M^α and M^β are not equivalent and so $m = 0$ and $V_2 = 0$. If $n_\alpha > n_\beta$, we can fill V_2 out to a square $n_\alpha \times n_\alpha$ matrix W by inserting $(n_\alpha - n_\beta)$ columns of zeros. It then follows that

$$W W^\dagger = V_2 V_2^\dagger = m 1_{n_\alpha}. \tag{C.4.30}$$

W clearly has zero determinant and hence so does $WW^\dagger$ and $V_2 V_2^\dagger$. This again means that $m = 0$ and $V_2 = 0$. From (C.4.22), since G is an arbitrary fixed matrix,

$$\sum_{j=1}^{g} M_{sn}^{\beta j^{-1}} M_{pt}^{\alpha j} = 0 \qquad \text{for all } s, n, p \text{ and } t. \tag{C.4.31}$$

This is the second of the orthogonality relations satisfied by the matrices of irreducible representations.

Combining (C.4.21) and (C.4.31), we finally find that

$$\sum_{j=1}^{g} M_{sn}^{\beta j^{-1}} M_{pt}^{\alpha j} = \frac{g}{n_\alpha} \delta_{st}\, \delta_{pn}\, \delta_{\alpha\beta}, \tag{C.4.32}$$

or, since these are unitary matrices,

$$\sum_{j=1}^{g} M_{ns}^{\beta j *} M_{pt}^{\alpha j} = \frac{g}{n_\alpha} \delta_{st}\, \delta_{pn}\, \delta_{\alpha\beta}. \tag{C.4.33}$$

C.5 CHARACTERS

The traces (sum of the diagonal elements) of the matrices of a representation are called its characters, and the character of the matrix $M^{\alpha j}$ will be denoted by $\chi^{\alpha j}$. The trace of a matrix is unaltered by a similarity transformation and it follows that the characters of the matrices representing group elements in the same class are equal, i.e. if $Q_s = Q_k^{-1} Q_j Q_k$,

$$M^{\alpha s} = M^{\alpha k^{-1}} M^{\alpha j} M^{\alpha k} \text{ and trace } M^{\alpha s} = \text{trace}\,(M^{\alpha k^{-1}} M^{\alpha j} M^{\alpha k}) = \text{trace } M^{\alpha j}. \tag{C.5.1}$$

$\chi^\alpha(\mathscr{C}_s)$ will denote the character of all matrices $M^{\alpha j}$ which represent elements in the same class $\mathscr{C}_s$, i.e. if Q_j, Q_k, Q_p, etc. belong to the class $\mathscr{C}_s$, then

$$\chi^{\alpha j} = \chi^{\alpha k} = \chi^{\alpha p} = \cdots = \chi^\alpha(\mathscr{C}_s). \tag{C.5.2}$$

(C.4.33) with $s = n, p = t$ yields

$$\sum_{j=1}^{g} M_{ss}^{\beta j*} M_{tt}^{\alpha j} = \frac{g}{n_\alpha} \delta_{st}\, \delta_{\alpha\beta}, \tag{C.5.3}$$

and, by summing both sides over s and t,

$$\sum_{j=1}^{g} \chi^{\beta j*} \chi^{\alpha j} = g\, \delta_{\alpha\beta}. \tag{C.5.4}$$

If Q_j belongs to the class $\mathscr{C}_s$ of order h_s, (C.5.4) can be written as a sum over classes

$$\sum_{s=1}^{g_c} h_s \chi^{\beta *}(\mathscr{C}_s) \chi^{\alpha}(\mathscr{C}_s) = g\, \delta_{\alpha\beta}, \tag{C.5.5}$$

where g_c is the number of distinct classes.

Let M^j be the matrix representing Q_j in an m-dimensional reducible representation of the group of order g with elements $Q_1, Q_2, \ldots, Q_g$. The matrices of this representation can be reduced so that irreducible representations appear along the diagonal. (C.5.4) enables us to say which irreducible representations will appear. We can write

$$M^j = \sum_{\alpha=1}^{g_r} c_\alpha M^{\alpha j}, \tag{C.5.6}$$

where c_α is the number of times the αth irreducible representation appears along the diagonal and g_r is the number of distinct inequivalent irreducible representations of the group. Taking the trace of both sides of (C.5.6),

$$\chi^j = \sum_{\alpha=1}^{g_r} c_\alpha \chi^{\alpha j}, \tag{C.5.7}$$

multiplying both sides by $\chi^{\beta j*}$, summing over j and making use of (C.5.4),

$$\sum_{j=1}^{g} \chi^{\beta j*} \chi^j = \sum_{\alpha=1}^{g_r} c_\alpha \sum_{j=1}^{g} \chi^{\beta j*} \chi^{\alpha j} = \sum_{\alpha=1}^{g_r} c_\alpha g\, \delta_{\alpha\beta} = g c_\beta, \tag{C.5.8}$$

and hence

$$c_\beta = \frac{1}{g} \sum_{j=1}^{g} \chi^{\beta j*} \chi^j. \tag{C.5.9}$$

(C.5.9) tells us which irreducible representations appear along the diagonal and also how many times they appear.

One reducible representation of the group which can be written down immediately is the so-called regular representation. Write the g elements

$Q_1, \ldots, Q_g$ of the group as a column vector $\mathbf{Q}$,

$$\mathbf{Q} \equiv \begin{pmatrix} Q_1 \\ \vdots \\ Q_g \end{pmatrix}, \tag{C.5.10}$$

and operate on this column vector with some element Q_j of the group. The result is a column vector which has the elements of the group as components, but in a different order to that in the original column vector $\mathbf{Q}$. For example, for the point group D_3,

$$Q_2 \begin{pmatrix} Q_1 \\ Q_2 \\ Q_3 \\ Q_4 \\ Q_5 \\ Q_6 \end{pmatrix} = \begin{pmatrix} Q_2 \\ Q_3 \\ Q_1 \\ Q_5 \\ Q_6 \\ Q_4 \end{pmatrix} = \begin{pmatrix} 0 & 1 & 0 & 0 & 0 & 0 \\ 0 & 0 & 1 & 0 & 0 & 0 \\ 1 & 0 & 0 & 0 & 0 & 0 \\ 0 & 0 & 0 & 0 & 1 & 0 \\ 0 & 0 & 0 & 0 & 0 & 1 \\ 0 & 0 & 0 & 1 & 0 & 0 \end{pmatrix} \begin{pmatrix} Q_1 \\ Q_2 \\ Q_3 \\ Q_4 \\ Q_5 \\ Q_6 \end{pmatrix}. \tag{C.5.11}$$

Hence

$$Q_j \mathbf{Q} = \tilde{M}^{rj} \mathbf{Q}, \tag{C.5.12}$$

where the $g \times g$ matrix $\tilde{M}^{rj}$ is the transpose of M^{rj} and has $+1$ in every row and column and zeros elsewhere. The matrices $M^{r1}, \ldots, M^{rg}$ form the regular representation of the group. Obviously

$$\text{trace } M^{rj} \equiv \chi^{rj} = g\,\delta_{j1}, \tag{C.5.13}$$

where $j = 1$ corresponds to the identity element.

The regular representation is reducible and we can write

$$M^{rj} = \sum_{\alpha=1}^{g_r} c^r_\alpha M^{\alpha j}. \tag{C.5.14}$$

Using (C.5.9) and (C.5.13),

$$c^r_\alpha = \frac{1}{g} \sum_{j=1}^{g} \chi^{\alpha j *} \chi^{rj} = \frac{1}{g} \sum_{j=1}^{g} \chi^{\alpha j *} g\,\delta_{j1} = \chi^{\alpha 1 *} = n_\alpha. \tag{C.5.15}$$

Hence the number of times an irreducible representation appears in the completely reduced regular representation is equal to the dimension of that irreducible representation. All irreducible representations therefore appear in the completely reduced regular representation.

If

$$\mathscr{C}_s = Q_j + Q_k + Q_p + \cdots + \cdots, \tag{C.5.16}$$

we define

$$\mathscr{C}_s^\alpha = M^{\alpha j} + M^{\alpha k} + M^{\alpha p} + \cdots + \cdots, \tag{C.5.17}$$

where $M^{\alpha j}$ is the matrix representing Q_j in the αth irreducible representation. $\mathscr{C}_s^\alpha$ commutes with all the matrices of the αth irreducible representation and, by Schur's lemma, must be a scalar multiple of the $n_\alpha \times n_\alpha$ unit matrix,

$$\mathscr{C}_s^\alpha = \eta_s^\alpha 1_{n_\alpha}. \tag{C.5.18}$$

Taking the trace of both sides of (C.5.18),

$$\text{trace } \mathscr{C}_s^\alpha = h_s \chi^\alpha(\mathscr{C}_s) = n_\alpha \eta_s^\alpha, \tag{C.5.19}$$

and hence

$$\eta_s^\alpha = \frac{h_s}{n_\alpha} \chi^\alpha(\mathscr{C}_s). \tag{C.5.20}$$

Since

$$\mathscr{C}_s \mathscr{C}_k = \sum_{t=1}^{g_c} c_{skt} \mathscr{C}_t, \tag{C.5.21}$$

we have

$$\mathscr{C}_s^\alpha \mathscr{C}_k^\alpha = \sum_{t=1}^{g_c} c_{skt} \mathscr{C}_t^\alpha, \tag{C.5.22}$$

and, using (C.5.18) and (C.5.20),

$$h_s h_k \chi^\alpha(\mathscr{C}_s) \chi^\alpha(\mathscr{C}_k) = n_\alpha \sum_{t=1}^{g_c} c_{skt} h_t \chi^\alpha(\mathscr{C}_t). \tag{C.5.23}$$

Summing both sides of (C.5.23) over all α,

$$\begin{aligned} h_s h_k \sum_{\alpha=1}^{g_r} \chi^\alpha(\mathscr{C}_s) \chi^\alpha(\mathscr{C}_k) &= \sum_{t=1}^{g_c} c_{skt} h_t \sum_{\alpha=1}^{g_r} n_\alpha \chi^\alpha(\mathscr{C}_t) \\ &= \sum_{t=1}^{g_c} c_{skt} h_t g \delta_{t1} \quad \text{[using (C.5.13)–(C.5.15)]} \\ &= g c_{sk1} = g h_s \delta_{ks'} \quad \text{[from (C.2.40)]}. \end{aligned} \tag{C.5.24}$$

Putting $k = s'$, $\chi^\alpha(\mathscr{C}_{s'}) = \chi^{\alpha *}(\mathscr{C}_s)$ for unitary representations and summing both sides of (C.5.24) over s now yields

$$\sum_{\alpha=1}^{g_r} \sum_{s=1}^{g_c} h_s \chi^{\alpha}(\mathscr{C}_s) \chi^{\alpha *}(\mathscr{C}_s) = g g_c. \tag{C.5.25}$$

But putting $\alpha = \beta$ in (C.5.5) and summing over α gives

$$\sum_{\alpha=1}^{g_r} \sum_{s=1}^{g_c} h_s \chi^{\alpha}(\mathscr{C}_s) \chi^{\alpha *}(\mathscr{C}_s) = g g_r. \tag{C.5.26}$$

Hence

$$g_c = g_r. \tag{C.5.27}$$

The number of distinct inequivalent irreducible representations of a group is equal to the number of distinct classes in that group.

Using (C.5.13)–(C.5.15) and (C.5.27),

$$g\,\delta_{j1} = \sum_{\alpha=1}^{g_c} n_\alpha \chi^{\alpha j}, \tag{C.5.28}$$

and putting $j = 1$,

$$g = \sum_{\alpha=1}^{g_c} n_\alpha^2. \tag{C.5.29}$$

(C.5.29) is an important result. n_α is necessarily a positive integer, and (C.5.29) can be uniquely solved. For example, the point group D_3 has six elements and three classes; (C.5.29) becomes

$$\sum_{\alpha=1}^{3} n_\alpha^2 = 6. \tag{C.5.30}$$

This equation has the unique solution $n_\alpha = 1, 1, 2$. The point group D_3 therefore has two distinct inequivalent one-dimensional irreducible representations and one two-dimensional irreducible representation.

If the group under consideration is the group of symmetry transformations of a Hamiltonian, the solution of (C.5.29) gives directly the degeneracies (apart from accidental degeneracy) of the energy levels.

C.5a Calculation of character tables

To obtain the maximum amount of information from the irreducible representations of a group of symmetry transformations of a Hamiltonian it is necessary to know the matrices of these irreducible representations. For many purposes, however, it is sufficient to know the character systems of the representations. These characters can be found without knowing the actual matrices.

From (C.5.18)–(C.5.22),

$$\mathscr{C}_i^\alpha \mathscr{C}_j^\alpha = \sum_{s=1}^{g_c} c_{ijs} \mathscr{C}_s^\alpha, \tag{C.5.31}$$

$$\mathscr{C}_j^\alpha = \eta_j^\alpha 1_{n_\alpha} = \frac{h_j}{n_\alpha} \chi^\alpha(\mathscr{C}_j) 1_{n_\alpha}, \tag{C.5.32}$$

and hence

$$\sum_{s=1}^{g_c} (c_{ijs} - \eta_j^\alpha \delta_{is}) \eta_s^\alpha = 0 \qquad (i = 1, 2, \dots, g_c). \tag{C.5.33}$$

In the set of equations (C.5.33), j is fixed; the condition for the existence of finite solutions is thus

$$\det |c_{ijs} - \eta_j^\alpha \delta_{is}| = 0 \qquad (i, s = 1, 2, \dots, g_c). \tag{C.5.34}$$

The roots of (C.5.34) are the g_c values of η_j in the g_c distinct irreducible representations, i.e. j is fixed and $\alpha = 1, 2, \dots, g_c$.

The first step in the calculation of characters is therefore the determination of the coefficients c_{ijs}. Once these have been found, (C.5.34) can be used to find the g_c values of η_j^α for any fixed value of j. (C.5.20) can then be used to find the characters.

To illustrate the procedure, consider the point group D_3. From (C.2.33)

$$c_{111} = c_{122} = c_{212} = c_{133} = c_{313} = c_{222} = 1;$$

$$c_{221} = c_{233} = c_{323} = 2; \qquad c_{331} = c_{332} = 3; \tag{C.5.35}$$

and all others are zero. There are three classes in the point group D_3.

For the first class ($j = 1$), (C.5.34) yields

$$\det \begin{vmatrix} 1 - \eta_1 & 0 & 0 \\ 0 & 1 - \eta_1 & 0 \\ 0 & 0 & 1 - \eta_1 \end{vmatrix} = 0, \tag{C.5.36}$$

and hence

$$\eta_1 = 1, 1, 1. \tag{C.5.37}$$

These are the values of η_1^α for $\alpha = 1, 2, 3$, i.e. in the three distinct irreducible representations of the point group D_3.

For the second class ($j = 2$), (C.5.34) yields

$$\det \begin{vmatrix} -\eta_2 & 1 & 0 \\ 2 & 1 - \eta_2 & 0 \\ 0 & 0 & 2 - \eta_2 \end{vmatrix} = 0, \tag{C.5.38}$$

and

$$\eta_2 = -1, 2, 2.$$

These are the values of η_2^α in the three distinct irreducible representations.

For the third class ($j = 3$), (C.5.34) yields

$$\det \begin{vmatrix} -\eta_3 & 0 & 1 \\ 0 & -\eta_3 & 2 \\ 3 & 3 & -\eta_3 \end{vmatrix} = 0, \tag{C.5.40}$$

and

$$\eta_3 = -3, 3, 0. \tag{C.5.41}$$

These results are displayed in a table as follows:

Table C.5.1 η_j^α for the point group D_3.

	$j = 1$	2	3
	1	-1	-3
η_j^α	1	2	3
	1	2	0

Although the three numbers in any column of this table determine the characters of the irreducible representations, no indication has yet been given of which particular number in any column belongs to one specified irreducible representation; i.e. from the way in which these numbers have been calculated, we do not know whether the -3 in the third column (say) belongs to the same irreducible representation (same value of α) as the -1 in the second column. The simplest way to obtain this information is to make use of the orthogonality relations.

From (C.5.5)

$$\sum_{s=1}^{g_c} h_s \chi^{\beta *}(\mathscr{C}_s)\chi^\alpha(\mathscr{C}_s) = g\,\delta_{\alpha\beta}. \tag{C.5.42}$$

Let M^1 be the identity representation. Then $\chi^1(\mathscr{C}_s) = 1$ for all s, and (C.5.42) yields

$$\sum_{s=1}^{g_c} h_s \chi^\alpha(\mathscr{C}_s) = g\,\delta_{\alpha 1}. \tag{C.5.43}$$

Using (C.5.32), (C.5.43) becomes

$$\sum_{s=1}^{g_c} \eta_s^\alpha = \begin{cases} g & \text{if } \alpha \text{ labels the identity representation,} \\ 0 & \text{otherwise.} \end{cases} \tag{C.5.44}$$

The columns of table C.5.1 must therefore be rearranged so as to make the rows add up to zero, except for one row (corresponding to the identity representation) which must add up to six (the order of the group). The rearranged table is shown below.

Table C.5.2
Rearranged table of η_j^α for the point group D_3.

	$j = 1$	2	3
	1	2	3
η_j^α	1	2	-3
	1	-1	0

The first row of this table corresponds to the identity representation.

To determine which of the other two rows corresponds to the two-dimensional irreducible representation we have, from (C.5.28),

$$\sum_{\alpha=1}^{g_c} n_\alpha \chi^\alpha(\mathscr{C}_s) = g\,\delta_{s1}, \tag{C.5.45}$$

or, using (C.5.32),

$$\sum_{\alpha=1}^{g_c} n_\alpha^2 \eta_s^\alpha = \begin{cases} g & \text{if } s \text{ labels the class of the identity,} \\ 0 & \text{otherwise.} \end{cases} \tag{C.5.46}$$

The values of n_α^2 for the point group D_3 are 1,1 and 4. It can easily be seen that it is the last row of table C.5.2 which must be multiplied by four in order that (C.5.46) is satisfied, i.e. the last row of table C.5.2 corresponds to the two-dimensional irreducible representation. Knowing this, and using (C.5.20), the completed character table for the point group D_3 is as follows:

Table C.5.3
Character table for the point group D_3

	$\chi(\mathscr{C}_1)$	$\chi(\mathscr{C}_2)$	$\chi(\mathscr{C}_3)$
M^1	1	1	1
M^2	1	1	-1
M^3	2	-1	0

Note that $\chi^\alpha(\mathscr{C}_1)$ (the class of the identity) always gives the dimension of the irreducible representation.

Problems

(C.5.1) The following transformations form a point group of order twelve which is usually denoted by the symbol T:

$$Q_1(x_1, x_2, x_3), \quad Q_2(x_1, -x_2, -x_3), \quad Q_3(-x_1, x_2, -x_3),$$
$$Q_4(-x_1, -x_2, x_3), \quad Q_5(x_2, x_3, x_1), \quad Q_6(-x_2, x_3, -x_1),$$
$$Q_7(-x_2, -x_3, x_1), \quad Q_8(x_2, -x_3, -x_1), \quad Q_9(x_3, x_1, x_2),$$
$$Q_{10}(-x_3, -x_1, x_2), \quad Q_{11}(x_3, -x_1, -x_2), \quad Q_{12}(-x_3, x_1, -x_2),$$

where, for example,

$$Q_2 f(x_1, x_2, x_3) = f(x_1, -x_2, -x_3),$$
$$Q_9 f(x_1, x_2, x_3) = f(x_3, x_1, x_2),$$

$f(x_1, x_2, x_3)$ being any function of the coordinates.
Draw up the multiplication table for this group, find the classes and calculate the character table.

(C.5.2) Show that out of the transformations defined in (C.5.1) the following sets form groups in their own right:
(i) Q_1 and Q_2,
(ii) Q_1, Q_5 and Q_9,
(iii) Q_1, Q_6 and Q_{10}.
These are subgroups of the point group T. Calculate their character tables. Are there any other subgroups of the point group T?

(C.5.3) The transformations $Q_1(x_1, x_2, x_3)$, $Q_2(x_1, x_3, x_2)$, $Q_3(-x_1, x_2, x_3)$ and $Q_4(-x_1, x_3, x_2)$ form a point group of order four. Show that the following matrices form a three-dimensional representation of this group:

$$M^1 = \begin{pmatrix} 1 & 0 & 0 \\ 0 & 1 & 0 \\ 0 & 0 & 1 \end{pmatrix}, \quad M^2 = \begin{pmatrix} 1 & -2 & 2 \\ 0 & -1 & 2 \\ 0 & 0 & 1 \end{pmatrix},$$

$$M^3 = \begin{pmatrix} 1 & -2 & 2 \\ 2 & -3 & 2 \\ 2 & -2 & 1 \end{pmatrix}, \quad M^4 = \begin{pmatrix} 1 & 0 & 0 \\ 2 & -1 & 0 \\ 2 & -2 & 1 \end{pmatrix}.$$

Prove that this is a reducible representation of the group, and find the matrices of the irreducible representations which appear along the diagonal in the completely reduced form of this representation.

(C.5.4) In a three-dimensional representation of the point group T [see (C.5.1)] the matrix representing Q_4 is

$$M^4 = \begin{pmatrix} -1 & 0 & 0 \\ 0 & -1 & 0 \\ 0 & 0 & 1 \end{pmatrix},$$

the matrix representing Q_8 is

$$M^8 = \begin{pmatrix} 0 & 0 & -1 \\ 1 & 0 & 0 \\ 0 & -1 & 0 \end{pmatrix},$$

and the matrix representing Q_9 is

$$M^9 = \begin{pmatrix} 0 & 1 & 0 \\ 0 & 0 & 1 \\ 1 & 0 & 0 \end{pmatrix}.$$

Find the matrices which represent the other elements of the point group T in this representation. Is this an irreducible representation of the point group T?

(C.5.5) The following transformations form a point group of order eight which is usually denoted by the symbol C_{4v}:

$$Q_1(x_1, x_2, x_3), \qquad Q_2(x_1, -x_3, x_2),$$
$$Q_3(x_1, x_3, -x_2), \qquad Q_4(x_1, -x_2, -x_3),$$
$$Q_5(x_1, x_2, -x_3), \qquad Q_6(x_1, x_3, x_2),$$
$$Q_7(x_1, -x_3, -x_2), \qquad Q_8(x_1, -x_2, x_3).$$

Show that the functions $\phi_1 = x_2$ and $\phi_2 = x_3$ act as basis functions for a two-dimensional irreducible representation of this group. Prove that the functions $\xi_1 = x_2^2 - x_3^2 - 2x_2x_3$ and $\xi_2 = x_2^2 - x_3^2 + 2x_2x_3$ act as basis functions for a representation of this group which is reducible into two distinct one-dimensional irreducible representations. Do the functions $f_1 = x_1 + x_2 + x_3$ and $f_2 = x_1 - x_2 - x_3$ act as basis functions for a two-dimensional representation of this group?

(C.5.6) Show that the functions

$$\phi_1 = 3\,e^{ix_1} - e^{ix_2} + e^{ix_3},$$
$$\phi_2 = e^{ix_1} - e^{ix_2} - 3\,e^{ix_3},$$
$$\phi_3 = -e^{ix_1} + 3e^{ix_2} + e^{ix_3}$$

act as basis functions for a three-dimensional reducible representation of the point group of order six with elements

$$Q_1(x_1, x_2, x_3), \qquad Q_2(x_2, x_3, x_1), \qquad Q_3(x_3, x_1, x_2),$$
$$Q_4(x_1, x_3, x_2), \qquad Q_5(x_3, x_2, x_1), \qquad Q_6(x_2, x_1, x_3).$$

Is this representation reducible into three one-dimensional irreducible representations or into a one-dimensional and a two-dimensional irreducible representation?

C.6 PROJECTION OPERATORS[6]

Let $Q_1, \ldots, Q_g$ be the elements of a group of order g, and let $M^{\alpha j}$ ($j = 1, 2, \ldots, g$) be the matrices of an n_α-dimensional irreducible unitary representation of this group. Consider the operator

$$\rho_{ij}^{\alpha} \equiv \sum_{k=1}^{g} M_{ij}^{\alpha k^*} Q_k, \tag{C.6.1}$$

and let

$$f_{ij}^{\alpha} \equiv \rho_{ij}^{\alpha} f(\mathbf{r}) = \sum_{k=1}^{g} M_{ij}^{\alpha k^*} Q_k f(\mathbf{r}), \tag{C.6.2}$$

where $f(\mathbf{r})$ is an arbitrary function of the coordinates. Operating on (C.6.2) with an element Q_s of the group yields

$$Q_s f_{ij}^{\alpha} = \sum_{k=1}^{g} M_{ij}^{\alpha k^*} Q_s Q_k f(\mathbf{r}). \tag{C.6.3}$$

If $Q_s Q_k = Q_t$,

$$Q_k = Q_s^{-1} Q_t, \tag{C.6.4}$$

$$M^{\alpha k} = M^{\alpha s^{-1}} M^{\alpha t} = \tilde{M}^{\alpha s^*} M^{\alpha t}, \tag{C.6.5}$$

since these are unitary matrices, and (C.6.3) becomes

$$Q_s f_{ij}^{\alpha} = \sum_{t=1}^{g} \sum_{v=1}^{n_\alpha} \tilde{M}_{iv}^{\alpha s} M_{vj}^{\alpha t^*} Q_t f(\mathbf{r}) = \sum_{v=1}^{n_\alpha} M_{vi}^{\alpha s} f_{vj}^{\alpha}. \tag{C.6.6}$$

For fixed j, let

$$\mathbf{f}_j^{\alpha} \equiv \begin{pmatrix} f_{1j}^{\alpha} \\ f_{2j}^{\alpha} \\ \vdots \\ f_{n_\alpha j}^{\alpha} \end{pmatrix}. \tag{C.6.7}$$

(C.6.6) then becomes

$$Q_s \mathbf{f}_j^{\alpha} = \tilde{M}^{\alpha s} \mathbf{f}_j^{\alpha}. \tag{C.6.8}$$

This last equation shows that, for given j, the functions $f_{1j}(\mathbf{r}), f_{2j}(\mathbf{r}), \ldots, f_{n_\alpha j}(\mathbf{r})$ act as basis functions for the αth irreducible representation of the group.

[6] E. Wigner, *Group Theory and its Application to the Quantum Mechanics of Atomic Spectra*, Academic Press, New York, 1959, p. 112.

The operators ρ_{ij}^{α} take an arbitrary function and project out of it those portions which transform as partners in a basis for the αth irreducible representation.

Let $\xi_1^{\beta}, \xi_2^{\beta}, \ldots, \xi_{n_\beta}^{\beta}$ be basis functions for the βth irreducible representation of the group. Then

$$\rho_{ij}^{\alpha}\xi_k^{\beta} = \sum_{s=1}^{g} M_{ij}^{\alpha s^*} Q_s \xi_k^{\beta} = \sum_{s=1}^{g} M_{ij}^{\alpha s^*} \sum_{t=1}^{n_\beta} M_{tk}^{\beta s}\xi_t^{\beta}. \tag{C.6.9}$$

From the orthogonality relations for irreducible representations (C.4.33)

$$\sum_{s=1}^{g} M_{ij}^{\alpha s^*} M_{tk}^{\beta s} = \frac{g}{n_\alpha}\delta_{it}\,\delta_{jk}\,\delta_{\alpha\beta}, \tag{C.6.10}$$

and (C.6.9) becomes

$$\rho_{ij}^{\alpha}\xi_k^{\beta} = \frac{g}{n_\alpha}\delta_{jk}\,\delta_{\alpha\beta}\sum_{t=1}^{n_\beta}\delta_{it}\xi_t^{\beta} = \frac{g}{n_\alpha}\delta_{jk}\,\delta_{\alpha\beta}\xi_i^{\beta}. \tag{C.6.11}$$

In particular, (C.6.11) yields

$$\rho_{ij}^{\alpha}\xi_j^{\alpha} = \text{const. }\xi_i^{\alpha}. \tag{C.6.12}$$

ρ_{ij}^{α} can therefore be regarded as a step operator, taking us from one basis function to another.

Operating on (C.6.11) with ρ_{kl}^{β},

$$\begin{aligned}\rho_{kl}^{\beta}\rho_{ij}^{\alpha}\xi_k^{\beta} &= \frac{g}{n_\alpha}\delta_{jk}\,\delta_{\alpha\beta}\rho_{kl}^{\beta}\xi_i^{\beta} \\ &= \frac{g^2}{n_\alpha^2}\delta_{jk}\,\delta_{\alpha\beta}\,\delta_{il}\xi_k^{\beta}.\end{aligned} \tag{C.6.13}$$

Hence

$$\rho_{kl}^{\beta}\rho_{ij}^{\alpha} = \frac{g^2}{n_\alpha^2}\delta_{jk}\,\delta_{il}\,\delta_{\alpha\beta}. \tag{C.6.14}$$

This is the orthogonality relation for projection operators.

Problems

(C.6.1) There are four distinct one-dimensional irreducible representations and one two-dimensional irreducible representation of the point group C_{4v} [see problem (C.5.5)]. The matrices of these representations are as follows:

$M^1 : M^{11} = M^{12} = M^{13} = M^{14} = M^{15} = M^{16} = M^{17} = M^{18} = 1.$

$M^2 : M^{21} = M^{22} = M^{23} = M^{24} = 1,\ M^{25} = M^{26} = M^{27} = M^{28} = -1.$

$M^3 : M^{31} = M^{34} = M^{35} = M^{38} = 1,\ M^{32} = M^{33} = M^{36} = M^{37} = -1.$

$M^4 : M^{41} = M^{44} = M^{46} = M^{47} = 1,\ M^{42} = M^{43} = M^{45} = M^{48} = -1.$

$$M^5 : M^{51} = \begin{pmatrix} 1 & 0 \\ 0 & 1 \end{pmatrix}, \quad M^{52} = \begin{pmatrix} 0 & 1 \\ -1 & 0 \end{pmatrix},$$

$$M^{53} = \begin{pmatrix} 0 & -1 \\ 1 & 0 \end{pmatrix}, \quad M^{54} = \begin{pmatrix} -1 & 0 \\ 0 & -1 \end{pmatrix},$$

$$M^{55} = \begin{pmatrix} 1 & 0 \\ 0 & -1 \end{pmatrix}, \quad M^{56} = \begin{pmatrix} 0 & 1 \\ 1 & 0 \end{pmatrix},$$

$$M^{57} = \begin{pmatrix} 0 & -1 \\ -1 & 0 \end{pmatrix}, \quad M^{58} = \begin{pmatrix} -1 & 0 \\ 0 & 1 \end{pmatrix}.$$

By operating on the function

$$\begin{aligned} \phi = {} & a_1 \exp[\mathrm{i}(x_2 + 2x_3)] + a_2 \exp[\mathrm{i}(x_2 - 2x_3)] + a_3 \exp[\mathrm{i}(2x_3 - x_2)] \\ & + a_4 \exp[\mathrm{i}(-x_2 - 2x_3)] + a_5 \exp[\mathrm{i}(2x_2 + x_3)] \\ & + a_6 \exp[\mathrm{i}(2x_2 - x_3)] + a_7 \exp[\mathrm{i}(x_3 - 2x_2)] \\ & + a_8 \exp[\mathrm{i}(-2x_2 - x_3)], \end{aligned}$$

where the a_j $(j = 1, 2, \ldots, 8)$ are constants, with the projection operators of this group, find basis functions for all the irreducible representations of the point group C_{4v}.

(C.6.2) Consider the function

$$\xi_{b_1 b_2 b_3}(\mathbf{r}) = \exp[\mathrm{i}(b_1 x_1 + b_2 x_2 + b_3 x_3)],$$

where b_1, b_2 and b_3 are non-zero constants. By operating on this function with the projection operators of the point group C_{4v}, show that

$$\phi^1_{b_1 b_2 b_3} \equiv \exp[ib_1 x_1]\{\cos b_2 x_2 \cos b_3 x_3 + \cos b_3 x_2 \cos b_2 x_3\}$$

acts as basis function for the irreducible representation M^1 [see problem (C.6.1)];

$$\phi^2_{b_1 b_2 b_3} \equiv \exp[ib_1 x_1]\{\sin b_2 x_2 \sin b_3 x_3 - \sin b_3 x_2 \sin b_2 x_3\}$$

acts as basis function for the irreducible representation M^2;

$$\phi^3_{b_1 b_2 b_3} \equiv \exp[ib_1 x_1]\{\cos b_2 x_2 \cos b_3 x_3 - \cos b_3 x_2 \cos b_2 x_3\}$$

acts as basis function for the irreducible representation M^3;

$$\phi^4_{b_1 b_2 b_3} \equiv \exp[ib_1 x_1]\{\sin b_2 x_2 \sin b_3 x_3 + \sin b_3 x_2 \sin b_2 x_3\}$$

acts as basis function for the irreducible representation M^4; and

$$\phi^{51}_{b_1 b_2 b_3} \equiv \exp[ib_1 x_1] \sin b_2 x_2 \cos b_3 x_3,$$

$$\phi^{52}_{b_1 b_2 b_3} \equiv \exp[ib_1 x_1] \sin b_3 x_2 \cos b_2 x_3$$

act as basis functions for the irreducible representation M^5 of the point group C_{4v}.

CHAPTER V

Solid State Applications

C.7 TRANSLATION GROUPS AND SPACE GROUPS

The time-independent Schrödinger equation for an electron in a crystalline solid is[7]

$$H(\mathbf{r})\psi(\mathbf{r}) \equiv \left[-\frac{\hbar^2}{2m}\nabla^2 + V(\mathbf{r}) \right]\psi(\mathbf{r}) = E\psi(\mathbf{r}), \tag{C.7.1}$$

where the Hamiltonian $H(\mathbf{r})$ is invariant under all the operations of the space group[8] of the particular crystalline structure to which it refers. A space group contains operations which are translations, rotations (proper and improper), reflexions, glide reflexions (translation + reflexion) and screw displacements (translation + rotation).

The distinguishing feature of crystalline structure is the existence of unit cells in the crystal[9]. These are small volumes, all of identical shape, size, orientation and constitution, which taken together fill all the crystal, i.e. the crystal consists of a regular periodic array of identical unit cells. The unit cell may, in the simplest cases, contain only one atom or, in more complicated cases, many atoms. The size of the unit cell so defined is not unique. If a cell twice as large as the original one is taken, for example, it would still have all the properties described as defining a unit cell but would contain twice as many atoms as the smaller cell. The smallest possible cell, the primitive unit cell, is generally used, but sometimes, for reasons of symmetry, a larger unit cell is more convenient.

Let $\mathbf{R}_s$, a lattice translation vector, be a vector joining a point of one unit cell to the equivalent point of another unit cell. $\mathbf{R}_s$ can be written as a linear combination of three noncoplanar primitive translation vectors $\mathbf{a}_1, \mathbf{a}_2$ and $\mathbf{a}_3$,

$$\mathbf{R}_s = s_1\mathbf{a}_1 + s_2\mathbf{a}_2 + s_3\mathbf{a}_3, \tag{C.7.2}$$

s_1, s_2 and s_3 being integers (positive, negative or zero). The primitive translation vectors $\mathbf{a}_1$, $\mathbf{a}_2$ and $\mathbf{a}_3$ define the unit cell of the crystalline structure. The unit cell is the parallelepiped with $\mathbf{a}_1$, $\mathbf{a}_2$ and $\mathbf{a}_3$ as edges,

[7] See §A.1.
[8] L. P. Bouckaert, R. Smoluchowski and E. Wigner, *Phys. Rev.*, **50**, 58 (1936); C. Herring, *J. Franklin Inst.*, **233**, 525 (1942); G. F. Koster, *Solid State Phys.*, **5**, 173 (1957); F. Herman, *Rev. Mod. Phys.*, **30**, 102 (1958).
[9] See §B.2.

and it has volume Ω given by

$$\Omega = \mathbf{a}_1 \cdot (\mathbf{a}_2 \times \mathbf{a}_3). \qquad \text{(C.7.3)}$$

As s_1, s_2 and s_3 take on all positive and negative integer values including zero, the set of points (C.7.2) map out a translation lattice, i.e. the translation lattice points have position vectors $\mathbf{R_s} = s_1\mathbf{a}_1 + s_2\mathbf{a}_2 + s_3\mathbf{a}_3$.

Let the operation corresponding to the lattice translation $\mathbf{R_s}$ be denoted by $\{Q_1|\mathbf{R_s}\}$, where Q_1 is the identity element of some point group, i.e.

$$\{Q_1|\mathbf{R_s}\}f(\mathbf{r}) = f(\mathbf{r} + \mathbf{R_s}) = f(\mathbf{r} + s_1\mathbf{a}_1 + s_2\mathbf{a}_2 + s_3\mathbf{a}_3). \qquad \text{(C.7.4)}$$

If the crystal is infinite in extent, $\{Q_1|\mathbf{R_s}\}$ is a symmetry operation and

$$\{Q_1|\mathbf{R_s}\}V(\mathbf{r}) = V(\mathbf{r} + \mathbf{R_s}) = V(\mathbf{r}),$$
$$\{Q_1|\mathbf{R_s}\}H(\mathbf{r}) = H(\mathbf{r} + \mathbf{R_s}) = H(\mathbf{r}). \qquad \text{(C.7.5)}$$

These translation operators form a group, the translation group, which is a subgroup of the space group of the crystalline structure. If the crystal is infinite in extent, as it must be for $\{Q_1|\mathbf{R_s}\}$ to be a symmetry operation, the translation group contains an infinite number of elements. A finite crystal is not really invariant under any of the $\{Q_1|\mathbf{R_s}\}$, and in order to make use of this translational symmetry we must modify our problem slightly and introduce periodic boundary conditions[10].

Take as a sample a crystal of finite size $N_1\mathbf{a}_1 \times N_2\mathbf{a}_2 \times N_3\mathbf{a}_3$, i.e. made up of $N \equiv N_1N_2N_3$ unit cells, where N_1, N_2 and N_3 are large whole numbers (as large as we like), and consider an infinite number of such samples stacked together. We then regard the three translations $N_1\mathbf{a}_1$, $N_2\mathbf{a}_2$ and $N_3\mathbf{a}_3$, which take us from a position in one sample to an equivalent position in a neighbouring sample, as effecting no change at all. In other words, we put these translations equal to zero translation. Expressed in terms of the wavefunction $\psi(\mathbf{r})$, this procedure gives the usual periodic boundary conditions

$$\psi(\mathbf{r}) = \psi(\mathbf{r} + N_1\mathbf{a}_1) = \psi(\mathbf{r} + N_2\mathbf{a}_2) = \psi(\mathbf{r} + N_3\mathbf{a}_3), \qquad \text{(C.7.6)}$$

or

$$\{Q_1|N_1\mathbf{a}_1\}\psi(\mathbf{r}) = \{Q_1|N_2\mathbf{a}_2\}\psi(\mathbf{r}) = \{Q_1|N_3\mathbf{a}_3\}\psi(\mathbf{r})$$
$$= \{Q_1|0\}\psi(\mathbf{r}) = \psi(\mathbf{r}). \qquad \text{(C.7.7)}$$

If the volume ($= N\Omega$) of this sample crystal is so large that it has ordinary macroscopic dimensions, rather than being on an atomic scale, any physical results derived from the model should be independent of N.

[10] See §B.6.

With this sample crystal and periodic boundary conditions, there are only $N \equiv N_1 N_2 N_3$ independent lattice translations $\mathbf{R_s}$. The corresponding operators $\{Q_1|\mathbf{R_s}\}$ all commute, and hence the translation group is a finite Abelian group of order N.

The space group of a crystalline structure may now be set up by combining the translation operations with the operations of a point group. An element of the space group will be denoted by the symbol $\{Q_j|\mathbf{v}_j + \mathbf{R_s}\}$, and the corresponding operation defined by

$$\{Q_j|\mathbf{v}_j + \mathbf{R_s}\} f(\mathbf{r}) = f(q^j\mathbf{r} + \mathbf{v}_j + \mathbf{R_s}), \tag{C.7.8}$$

where $\mathbf{r}' = q^j\mathbf{r}$ is the transformation of the coordinates corresponding to the point group element Q_j, and $\mathbf{v}_j$ is a so-called nonprimitive translation. $\mathbf{v}_j$ can be different for each operation of the point group but does not depend upon $\mathbf{s}$. Space groups which have $\mathbf{v}_j = 0$ for all j are called **symmorphic** space groups; the others are called **nonsymmorphic**. The significance of these nonprimitive translations will be seen later.

From (C.7.8),

$$\begin{aligned}\{Q_p|\mathbf{v}_p + \mathbf{R_m}\}\{Q_j|\mathbf{v}_j + \mathbf{R_s}\} f(\mathbf{r}) &= \{Q_p|\mathbf{v}_p + \mathbf{R_m}\} f(q^j\mathbf{r} + \mathbf{v}_j + \mathbf{R_s}) \\ &= f(q^j(q^p\mathbf{r} + \mathbf{v}_p + \mathbf{R_m}) + \mathbf{v}_j + \mathbf{R_s}) \\ &= f(q^j q^p\mathbf{r} + q^j\mathbf{v}_p + q^j\mathbf{R_m} + \mathbf{v}_j + \mathbf{R_s}).\end{aligned} \tag{C.7.9}$$

Since these operations form a group,

$$\{Q_p|\mathbf{v}_p + \mathbf{R_m}\}\{Q_j|\mathbf{v}_j + \mathbf{R_s}\} = \{Q_t|\mathbf{v}_t + \mathbf{R_n}\}, \tag{C.7.10}$$

and hence, using (C.7.9),

$$q^j q^p = q^t, \qquad Q_p Q_j = Q_t, \tag{C.7.11}$$

$$q^j\mathbf{v}_p + q^j\mathbf{R_m} + \mathbf{v}_j + \mathbf{R_s} = \mathbf{v}_t + \mathbf{R_n}. \tag{C.7.12}$$

(C.7.12) is not automatically fulfilled for any arbitrary sets of primitive and nonprimitive translations. It puts great restrictions on the combinations of point groups, primitive translations and nonprimitive translations which can occur in space groups. For instance, from (C.7.12), $q^j\mathbf{R_m}$ must be a lattice translation vector. This is one of the fundamental requirements of a space group: the operations of the point group must transform the lattice translation vectors into the same set of vectors. Similarly, $q^j\mathbf{v}_p + \mathbf{v}_j$ must be a nonprimitive translation or, at most, differ from a nonprimitive translation by a lattice translation.

These restrictions on the primitive and nonprimitive translations which can be associated with a given point group greatly limit the number of space groups which can be set up. In fact there are just 230 possible space groups permitted by these restrictions, 73 of which are symmorphic. These are made up by combining 14 different types of translation lattices[11], called Bravais lattices, with 32 possible point groups[12].

To illustrate the above concepts, we shall now discuss the crystalline structures of the metal aluminium (Al) and the semiconductors germanium (Ge) and indium antimonide (InSb). When setting up the space groups for these structures it is necessary to consider (i) the Bravais lattice, (ii) the point group and (iii) the nonprimitive translations associated with each structure.

(i) *The Bravais lattices*

The Al, Ge and InSb structures all have the same Bravais lattice, namely the face-centred cubic lattice. A portion of this lattice is shown in fig. B.2.4a. This is a cube of edge-length a, with lattice points at the vertices and at the centre of each face. O is the origin of the primitive unit cell which is shown (dashed lines) together with the primitive translation vectors. Each lattice point is at the origin of a primitive unit cell.

If O is taken as the origin of coordinates and the coordinate axes are as shown in fig. B.2.4a,

$$\mathbf{a}_1 = \frac{a}{2}(0, 1, 1), \qquad \mathbf{a}_2 = \frac{a}{2}(1, 0, 1), \qquad \mathbf{a}_3 = \frac{a}{2}(1, 1, 0), \tag{C.7.13}$$

and the volume of the unit cell is

$$\Omega = \mathbf{a}_1 \cdot (\mathbf{a}_2 \times \mathbf{a}_3) = a^3/4, \tag{C.7.14}$$

i.e. one quarter of the volume of the cube. The lattice translation vectors are given by

$$\mathbf{R}_s = \frac{a}{2}(s_2 + s_3, s_1 + s_3, s_1 + s_2), \tag{C.7.15}$$

s_1, s_2 and s_3 being integers.

The Al structure (face-centred cubic structure) is obtained by putting an Al atom at the origin of each of the unit cells, i.e. at every lattice point. For this structure, therefore, each unit cell contains one atom.

The InSb structure (zinc-blende structure) is obtained by putting an In atom at the origin of each of the unit cells and an Sb atom at the point $\frac{1}{4}(\mathbf{a}_1 + \mathbf{a}_2 + \mathbf{a}_3) = a(\frac{1}{4}, \frac{1}{4}, \frac{1}{4})$ inside each unit cell. For this structure, therefore, the unit cell contains two atoms, one In atom and one Sb atom.

[11] See §B.2, also J. C. Slater, *Quantum Theory of Molecules and Solids, Vol. 2—Symmetry and Energy Bands in Crystals*, McGraw-Hill, New York, 1965, p. 10.
[12] G. F. Koster, J. O. Dimmock, R. G. Wheeler and H. Statz, *Properties of the Thirty-two Point Groups*, Massachusetts Institute of Technology Press, Cambridge, Mass., 1963.

The Ge structure (diamond structure) is obtained by putting a Ge atom at the origin of each of the unit cells and a Ge atom at the point

$$\tfrac{1}{4}(\mathbf{a}_1 + \mathbf{a}_2 + \mathbf{a}_3) = \frac{a}{4}(1, 1, 1)$$

inside each unit cell. The unit cell therefore contains two Ge atoms.

The lattice parameter a has, of course, a different value for each of these substances. For Al, $a = 4{\cdot}05$ Å; for Ge, $a = 5{\cdot}66$ Å; and for InSb, $a = 6{\cdot}48$ Å.

(ii) *The point groups*

The point group of order 48 which consists of all operations which take a cube into itself[13], i.e. leave it in equivalent positions, is denoted by O_h. Let the elements of this point group be denoted by $Q_1, Q_2, \ldots, Q_{24}$; $Q'_1, Q'_2, \ldots, Q'_{24}$. If the centre of the cube is taken as the origin of coordinates, the first 24 operations of this point group, which themselves form a point group denoted by T_d, are as given in table C.7.1.

Table C.7.1 Elements of the point group T_d.

$Q_1(x_1x_2x_3)$	$Q_2(x_1\bar{x}_2\bar{x}_3)$	$Q_3(\bar{x}_1x_2\bar{x}_3)$	$Q_4(\bar{x}_1\bar{x}_2x_3)$	$Q_5(x_2x_3x_1)$	$Q_6(\bar{x}_2x_3\bar{x}_1)$
$Q_7(\bar{x}_2\bar{x}_3x_1)$	$Q_8(x_2\bar{x}_3\bar{x}_1)$	$Q_9(x_3x_1x_2)$	$Q_{10}(\bar{x}_3\bar{x}_1x_2)$	$Q_{11}(x_3\bar{x}_1\bar{x}_2)$	$Q_{12}(\bar{x}_3x_1\bar{x}_2)$
$Q_{13}(\bar{x}_1x_3\bar{x}_2)$	$Q_{14}(\bar{x}_1\bar{x}_3x_2)$	$Q_{15}(\bar{x}_3\bar{x}_2x_1)$	$Q_{16}(x_3\bar{x}_2\bar{x}_1)$	$Q_{17}(x_2\bar{x}_1\bar{x}_3)$	$Q_{18}(\bar{x}_2x_1\bar{x}_3)$
$Q_{19}(x_1x_3x_2)$	$Q_{20}(x_1\bar{x}_3\bar{x}_2)$	$Q_{21}(x_3x_2x_1)$	$Q_{22}(\bar{x}_3x_2\bar{x}_1)$	$Q_{23}(x_2x_1x_3)$	$Q_{24}(\bar{x}_2\bar{x}_1x_3)$

For example,

$$\begin{aligned} Q_6 f(x_1, x_2, x_3) &= f(-x_2, x_3, -x_1), \\ Q_{24} f(x_1, x_2, x_3) &= f(-x_2, -x_1, x_3), \end{aligned} \tag{C.7.16}$$

$f(x_1, x_2, x_3)$ being any function of the coordinates. The remaining operations of the point group O_h, i.e. $Q'_1, \ldots, Q'_{24}$, are obtained from table C.7.1 by multiplying the corresponding unprimed operation by the inversion J. For example,

$$\begin{aligned} Q'_{10} f(x_1, x_2, x_3) &= J Q_{10} f(x_1, x_2, x_3) \\ &= J f(-x_3, -x_1, x_2) = f(x_3, x_1, -x_2). \end{aligned} \tag{C.7.17}$$

[13] G. F. Koster, J. O. Dimmock, R. G. Wheeler and H. Statz, *Properties of the Thirty-two Point Groups*, Massachusetts Institute of Technology Press, Cambridge, Mass., 1963; J. C. Slater, *Quantum Theory of Molecules and Solids, Vol 2—Symmetry and Energy Bands in Crystals*, McGraw-Hill, New York, 1965, p. 28.

The point group O_h appears in the space group of the Al structure and in the space group of the Ge structure. The point group T_d appears in the space group of the InSb structure.

(iii) *The nonprimitive translations and the space groups*

The Al and InSb structures do not have any nonprimitive translations associated with them ($\mathbf{v}_j = 0$ for all j) and the corresponding space groups are symmorphic.

The space group of the Al structure is denoted by O_h^5. If the origin of coordinates is taken at the origin of a unit cell, a typical operation of this space group is

$$\{Q_6'|\mathbf{R_s}\}f(x_1, x_2, x_3) = f\left(x_2 + \frac{a}{2}(s_2 + s_3), -x_3 + \frac{a}{2}(s_1 + s_3), x_1 + \frac{a}{2}(s_1 + s_2)\right). \tag{C.7.18}$$

The space group of the InSb structure is denoted by T_d^2. If the origin of coordinates is taken at the origin of a unit cell, a typical operation of this space group is

$$\{Q_{10}|\mathbf{R_s}\}f(x_1, x_2, x_3) = f\left(-x_3 + \frac{a}{2}(s_2 + s_3), -x_1 + \frac{a}{2}(s_1 + s_3), x_2 + \frac{a}{2}(s_1 + s_2)\right). \tag{C.7.19}$$

The Ge structure does have nonprimitive translations associated with it. Taking the origin of coordinates at the origin of a unit cell, the first 24 operations (those in table C.7.1) of the point group O_h have no nonprimitive translations associated with them, but the remaining elements $Q_1', \ldots, Q_{24}'$ have the same nonprimitive translation $\mathbf{v} = a(\frac{1}{4}, \frac{1}{4}, \frac{1}{4})$. This nonprimitive translation carries us from the atom at the origin to the other atom in the unit cell. The space group of the Ge structure is denoted by O_h^7 and has elements $\{Q_t|\mathbf{R_s}\}$ and $\{Q_t'|\mathbf{v} + \mathbf{R_s}\}$ $(t = 1, 2, \ldots, 24)$. Typical operations of this space group are

$$\{Q_6|\mathbf{R_s}\}f(x_1, x_2, x_3) = f\left(-x_2 + \frac{a}{2}(s_2 + s_3), x_3 + \frac{a}{2}(s_1 + s_3), -x_1 + \frac{a}{2}(s_1 + s_2)\right). \tag{C.7.20}$$

$$\{Q_6'|\mathbf{v} + \mathbf{R_s}\} f(x_1, x_2, x_3)$$

$$= f\left(x_2 + \frac{a}{2}(s_2 + s_3) + \frac{a}{4},\right.$$

$$\left. -x_3 + \frac{a}{2}(s_1 + s_3) + \frac{a}{4}, x_1 + \frac{a}{2}(s_1 + s_2) + \frac{a}{4}\right). \quad \text{(C.7.21)}$$

C.8 RECIPROCAL LATTICES, BLOCH FUNCTIONS AND BRILLOUIN ZONES[14]

Consider the Bravais lattice defined by the primitive translations $\mathbf{a}_1$, $\mathbf{a}_2$ and $\mathbf{a}_3$, and let $\mathbf{b}_1$, $\mathbf{b}_2$ and $\mathbf{b}_3$ be noncoplanar vectors defined by

$$\mathbf{a}_i \cdot \mathbf{b}_j = 2\pi\,\delta_{ij} \qquad (i, j = 1, 2, 3). \qquad \text{(C.8.1)}$$

If $\Omega \equiv \mathbf{a}_1 \cdot (\mathbf{a}_2 \times \mathbf{a}_3)$, it follows from this definition that

$$\mathbf{b}_1 = \frac{2\pi}{\Omega}\mathbf{a}_2 \times \mathbf{a}_3, \qquad \mathbf{b}_2 = \frac{2\pi}{\Omega}\mathbf{a}_3 \times \mathbf{a}_1, \qquad \mathbf{b}_3 = \frac{2\pi}{\Omega}\mathbf{a}_1 \times \mathbf{a}_2, \qquad \text{(C.8.2)}$$

as shown in §B.2. It was also explained there that the reciprocal lattice to the lattice with primitive translations $\mathbf{a}_1, \mathbf{a}_2$ and $\mathbf{a}_3$ has lattice points with position vectors $\mathbf{K}_l$ given by

$$\mathbf{K}_l = l_1\mathbf{b}_1 + l_2\mathbf{b}_2 + l_3\mathbf{b}_3, \qquad \text{(C.8.3)}$$

l_1, l_2 and l_3 being integers (positive, negative or zero). $\mathbf{K}_l$ is known as a reciprocal lattice vector. The unit cell of the reciprocal lattice, defined by the vectors $\mathbf{b}_1, \mathbf{b}_2$ and $\mathbf{b}_3$, has volume $\mathbf{b}_1 \cdot (\mathbf{b}_2 \times \mathbf{b}_3)$ which, from (C.8.2), is equal to $8\pi^3/\Omega$.

For example, the face-centred cubic Bravais lattice has the primitive translations

$$\mathbf{a}_1 = \frac{a}{2}(0, 1, 1), \qquad \mathbf{a}_2 = \frac{a}{2}(1, 0, 1), \qquad \mathbf{a}_3 = \frac{a}{2}(1, 1, 0), \qquad \text{(C.8.4)}$$

and, from (C.8.2), the vectors $\mathbf{b}_1, \mathbf{b}_2$ and $\mathbf{b}_3$ of the corresponding reciprocal lattice are given by

$$\mathbf{b}_1 = \frac{2\pi}{a}(-1, 1, 1), \qquad \mathbf{b}_2 = \frac{2\pi}{a}(1, -1, 1), \qquad \mathbf{b}_3 = \frac{2\pi}{a}(1, 1, -1). \qquad \text{(C.8.5)}$$

[14] See also §B.2, §B.5 and §B.11.

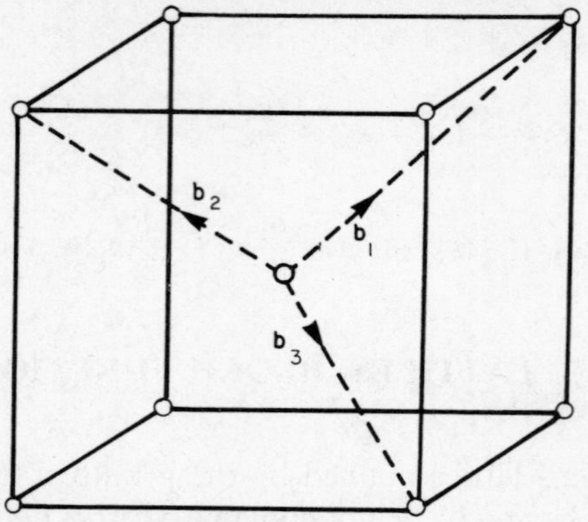

Fig. C.8.1. Primitive translations of the body-centred cubic lattice.

These are in fact the primitive translations for a body-centred cubic Bravais lattice with cube edge-length $4\pi/a$ (fig. C.8.1). The body-centred cubic lattice is the reciprocal lattice to the face-centred cubic lattice.

We now begin the classification of electronic states in crystals by seeing what effect the operators $\{Q_1|\mathbf{R_s}\}$ of the translation group have on the eigenfunctions of the Hamiltonian $H(\mathbf{r})$.

The translation group is Abelian and hence the number of classes equals the number of elements $N \equiv N_1 N_2 N_3$, i.e.

$$\{Q_1|\mathbf{R_s}\}^{-1}\{Q_1|\mathbf{R_m}\}\{Q_1|\mathbf{R_s}\} = \{Q_1|\mathbf{R_m}\}$$

for all $\mathbf{R_s}$ and each element forms a class by itself. The number of distinct irreducible representations equals the number of classes and hence the translation group has N distinct irreducible representations. From

$$\sum_{\alpha=1}^{N} n_\alpha^2 = N, \tag{C.8.6}$$

where n_α is the dimension of the αth irreducible representation, all of these irreducible representations are one-dimensional.

The Hamiltonian $H(\mathbf{r})$ is invariant under the operations of the translation group and the eigenfunctions of $H(\mathbf{r})$ act as basis functions for the irreducible representations of this group. If $\psi(\mathbf{r})$ is a basis function for a one-dimensional irreducible representation of the translation group

$$\{Q_1|\mathbf{a}_j\}\psi(\mathbf{r}) = \psi(\mathbf{r} + \mathbf{a}_j) = t\psi(\mathbf{r}),$$

$$\{Q_1|2\mathbf{a}_j\}\psi(\mathbf{r}) = \psi(\mathbf{r} + 2\mathbf{a}_j) = t^2\psi(\mathbf{r}),$$

$$\{Q_1|s_j\mathbf{a}_j\}\psi(\mathbf{r}) = \psi(\mathbf{r} + s_j\mathbf{a}_j) = t^{s_j}\psi(\mathbf{r}),$$
$$\{Q_1|N_j\mathbf{a}_j\}\psi(\mathbf{r}) = \psi(\mathbf{r} + N_j\mathbf{a}_j) = t^{N_j}\psi(\mathbf{r}) = \psi(\mathbf{r}), \tag{C.8.7}$$

where periodic boundary conditions have been applied. Hence

$$t^{N_j} = 1, \qquad t = \exp[2\pi i h_j/N_j], \tag{C.8.8}$$

where $h_j (j = 1, 2, 3)$ can be any integer (positive, negative or zero). It now follows that

$$\{Q_1|\mathbf{R}_s\}\psi(\mathbf{r}) = \psi(\mathbf{r} + \mathbf{R}_s) = \psi(\mathbf{r} + s_1\mathbf{a}_1 + s_2\mathbf{a}_2 + s_3\mathbf{a}_3)$$

$$= \exp[2\pi i(s_1h_1/N_1 + s_2h_2/N_2 + s_3h_3/N_3)]\psi(\mathbf{r}), \tag{C.8.9}$$

h_1, h_2 and h_3 being integers. Defining a vector $\mathbf{k}$, the wave-vector in reciprocal lattice space, by

$$\mathbf{k} \equiv \frac{h_1}{N_1}\mathbf{b}_1 + \frac{h_2}{N_2}\mathbf{b}_2 + \frac{h_3}{N_3}\mathbf{b}_3, \tag{C.8.10}$$

(C.8.9) can then be written

$$\{Q_1|\mathbf{R}_s\}\psi(\mathbf{r}) = \psi(\mathbf{r} + \mathbf{R}_s) = e^{i\mathbf{k}\cdot\mathbf{R}_s}\psi(\mathbf{r}). \tag{C.8.11}$$

For alternative approaches to the last two important results, see equations (B.6.4) and (B.11.10).

The wave-vector $\mathbf{k}$ labels the irreducible representations of the translation group. The element $\{Q_1|\mathbf{R}_s\}$ is represented in the $\mathbf{k}$th irreducible representation by the one-dimensional 'matrix' $e^{i\mathbf{k}\cdot\mathbf{R}_s}$.

Let an eigenfunction belonging to (acting as basis function for) the $\mathbf{k}$th irreducible representation of the translation group be denoted by $\psi_\mathbf{k}(\mathbf{r})$. It has the property

$$\{Q_1|\mathbf{R}_s\}\psi_\mathbf{k}(\mathbf{r}) = \psi_\mathbf{k}(\mathbf{r} + \mathbf{R}_s) = e^{i\mathbf{k}\cdot\mathbf{R}_s}\psi_\mathbf{k}(\mathbf{r}). \tag{C.8.12}$$

Such functions are known as Bloch functions[15]. If we write

$$\psi_\mathbf{k}(\mathbf{r}) = e^{i\mathbf{k}\cdot\mathbf{r}}u_\mathbf{k}(\mathbf{r}), \tag{C.8.13}$$

then

$$\{Q_1|\mathbf{R}_s\}\psi_\mathbf{k}(\mathbf{r}) = e^{i\mathbf{k}\cdot\mathbf{R}_s}[e^{i\mathbf{k}\cdot\mathbf{r}}u_\mathbf{k}(\mathbf{r} + \mathbf{R}_s)], \tag{C.8.14}$$

and, using (C.8.12), it follows that

$$u_\mathbf{k}(\mathbf{r} + \mathbf{R}_s) = u_k(\mathbf{r}). \tag{C.8.15}$$

[15] F. Bloch, *Z. Physik*, **52**, 555 (1928); see also §B.11b.

The Bloch function $\psi_{\mathbf{k}}(\mathbf{r})$ can be written as $e^{i\mathbf{k}\cdot\mathbf{r}}u_{\mathbf{k}}(\mathbf{r})$, where $u_{\mathbf{k}}(\mathbf{r})$, the modulating part of this Bloch function, has the basic periodicity of the crystal lattice, i.e.

$$u_{\mathbf{k}}(\mathbf{r} + \mathbf{R}_s) = u_{\mathbf{k}}(\mathbf{r}) \text{ for all lattice translations } \mathbf{R}_s. \qquad \text{(C.8.16)}$$

There are just N distinct irreducible representations of the translation group and hence just N different wave-vectors are needed to label these irreducible representations. We note that if two wave-vectors, $\mathbf{k}$ and $\mathbf{k}'$, differ by a reciprocal lattice vector $\mathbf{K}_l$, then they label the same irreducible representation of the translation group, i.e. if $\mathbf{k} - \mathbf{k}' = \mathbf{K}_l$,

$$e^{i\mathbf{k}\cdot\mathbf{R}_s} = e^{i\mathbf{k}'\cdot\mathbf{R}_s}\, e^{i\mathbf{K}_l\cdot\mathbf{R}_s} = e^{i\mathbf{k}'\cdot\mathbf{R}_s} \qquad \text{(C.8.17)}$$

since

$$\mathbf{K}_l \cdot \mathbf{R}_s = 2\pi(s_1 l_1 + s_1 l_2 + s_3 l_3) = 2\pi \times \text{integer}. \qquad \text{(C.8.18)}$$

We must therefore choose the N wave-vectors in such a way that no two of them differ by a reciprocal lattice vector. These N wave-vectors will then label the irreducible representations uniquely.

The N wave-vectors chosen for this purpose are those with the smallest magnitudes. Drawn from the origin ($\mathbf{k} = 0$) in reciprocal lattice space, their end-points lie in a region of reciprocal lattice space which is symmetrical about the origin $\mathbf{k} = 0$. This region of reciprocal lattice space is called the first Brillouin zone[16], and the N wave-vectors whose end-points lie in this region are called reduced wave-vectors.

To illustrate these concepts, we now consider the two-dimensional square lattice with lattice parameter a (fig. C.8.2). The reciprocal lattice is also a square lattice but with lattice parameter $2\pi/a$ (fig. C.8.3). The first Brillouin zone in this case is the square ABCD.

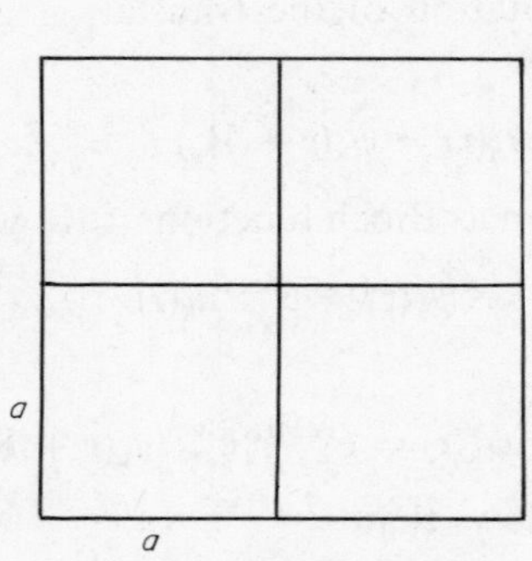

Fig. C.8.2. Square lattice.

[16] See §B.5.

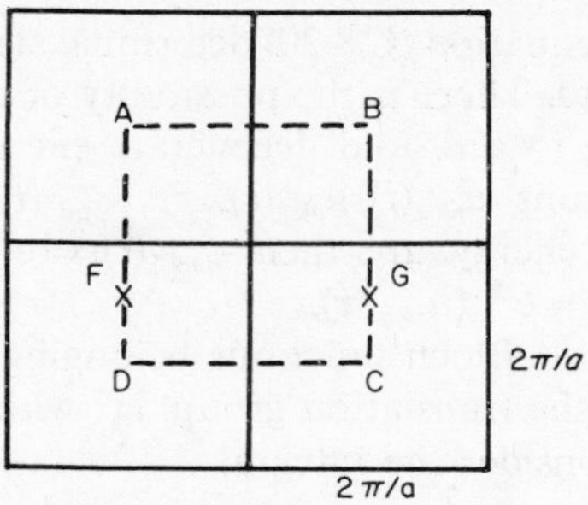

Fig. C.8.3. Reciprocal lattice.

In confining the wave-vectors to the first Brillouin zone, we have only partially fulfilled the requirement that no two wave-vectors should differ by a reciprocal lattice vector. There is still some ambiguity about wave-vectors on the surface of the first Brillouin zone. Consider, for example, the points F and G (fig. C.8.3). $\mathbf{k}_F$ and $\mathbf{k}_G$ differ by a reciprocal lattice vector and hence refer to the same irreducible representation of the translation group. Either of them can be used to label the representation. Likewise, we can without discrimination use any one of $\mathbf{k}_A$, $\mathbf{k}_B$, $\mathbf{k}_C$ or $\mathbf{k}_D$ (fig. C.8.3) to label the corresponding irreducible representation. A Bloch state, defined by the irreducible representation of the translation group to which it belongs, on the surface of the first Brillouin zone is thus represented by at least two points on the surface (see also problem B.11.8).

If the energy eigenvalue corresponding to the Bloch function $\psi_{\mathbf{k}}(\mathbf{r}) = e^{i\mathbf{k}\cdot\mathbf{r}}u_{\mathbf{k}}(\mathbf{r})$, where $\mathbf{k}$ is a vector in the first Brillouin zone, is denoted by $E(\mathbf{k})$,

$$H(\mathbf{r})\psi_{\mathbf{k}}(\mathbf{r}) = E(\mathbf{k})\psi_{\mathbf{k}}(\mathbf{r}). \qquad \text{(C.8.19)}$$

Multiplying both sides of (C.8.19) by $e^{-i\mathbf{k}\cdot\mathbf{r}}$, we obtain

$$H_{\mathbf{k}}(\mathbf{r})u_{\mathbf{k}}(\mathbf{r}) = E(\mathbf{k})u_{\mathbf{k}}(\mathbf{r}), \qquad \text{(C.8.20)}$$

where we have again used the transformation (B.15.10)

$$H_{\mathbf{k}}(\mathbf{r}) \equiv e^{-i\mathbf{k}\cdot\mathbf{r}}H(\mathbf{r})\,e^{i\mathbf{k}\cdot\mathbf{r}} = H(\mathbf{r}) + \frac{\hbar^2k^2}{2m} - \frac{i\hbar^2}{m}\mathbf{k}\cdot\nabla. \qquad \text{(C.8.21)}$$

For a fixed wave-vector $\mathbf{k}$, equation (C.8.20) will have an infinite number of eigenvalues which we denote by $E_1(\mathbf{k}), E_2(\mathbf{k}), \ldots, E_n(\mathbf{k}), \ldots$, numbering them in order of ascending energies. The energy can thus be regarded as a many-valued function of $\mathbf{k}$. The subscript n labels the energy band, and the eigenvalues of equation (C.8.20) determine single states at the point **k** in each energy band. There is the possibility of degenerate energy levels and corresponding to a d_n-fold degenerate energy level $E_n(\mathbf{k})$ there will be the eigenfunctions $u_{n\mathbf{k}1}(\mathbf{r}), u_{n\mathbf{k}2}(\mathbf{r}), \ldots, u_{n\mathbf{k}d_n}(\mathbf{r})$. The Bloch functions belonging to this energy are then $\psi_{n\mathbf{k}1}(\mathbf{r}) = e^{i\mathbf{k}\cdot\mathbf{r}}u_{n\mathbf{k}1}(\mathbf{r}), \psi_{n\mathbf{k}2}(\mathbf{r}) = e^{i\mathbf{k}\cdot\mathbf{r}} u_{n\mathbf{k}2}(\mathbf{r}), \ldots, \psi_{n\mathbf{k}d_n}(\mathbf{r}) = e^{i\mathbf{k}\cdot\mathbf{r}}u_{n\mathbf{k}d_n}(\mathbf{r})$.

We now show that Bloch functions belonging to different irreducible representations of the translation group, i.e. with different wave-vectors, are orthogonal. Consider the integral

$$
\begin{aligned}
I^{n\mathbf{k}j}_{n'\mathbf{k}'j'} &\equiv \int_{N\Omega} \psi^*_{n'\mathbf{k}'j'}(\mathbf{r})\psi_{n\mathbf{k}j}(\mathbf{r})\,d\mathbf{r} \\
&= \int_{N\Omega} e^{i(\mathbf{k}-\mathbf{k}')\cdot\mathbf{r}} u^*_{n'\mathbf{k}'j'}(\mathbf{r})u_{n\mathbf{k}j}(\mathbf{r})\,d\mathbf{r}.
\end{aligned}
\tag{C.8.22}
$$

Let $\mathbf{r} = \mathbf{r}' + \mathbf{R_s}$. The volume ($= N\Omega$) of the crystal can then be spanned by allowing $\mathbf{r}'$ to vary over a unit cell and summing over all $\mathbf{R_s}$. Hence

$$
\begin{aligned}
I^{n\mathbf{k}j}_{n'\mathbf{k}'j'} &= \sum_{\mathbf{R_s}} \int_{\Omega} e^{i(\mathbf{k}-\mathbf{k}')\cdot(\mathbf{r}'+\mathbf{R_s})} u^*_{n'\mathbf{k}'j'}(\mathbf{r}' + \mathbf{R_s})u_{n\mathbf{k}j}(\mathbf{r}' + \mathbf{R_s})\,d\mathbf{r}' \\
&= \left\{ \sum_{\mathbf{R_s}} e^{i(\mathbf{k}-\mathbf{k}')\cdot\mathbf{R_s}} \right\} \int_{\Omega} \psi^*_{n'\mathbf{k}'j'}(\mathbf{r})\psi_{n\mathbf{k}j}(\mathbf{r})\,d\mathbf{r}.
\end{aligned}
\tag{C.8.23}
$$

We have shown previously [equation (C.4.33)] that for two irreducible unitary representations M^α, M^β, of a group of order g

$$
\sum_{j=1}^{g} M^{\beta j*}_{ls} M^{\alpha j}_{pt} = \frac{g}{n_\alpha}\,\delta_{st}\,\delta_{lp}\,\delta_{\alpha\beta}.
\tag{C.8.24}
$$

Applying this result to the Abelian group of translations ($\alpha \to \mathbf{k}, \beta \to \mathbf{k}'$),

$$
\sum_{\mathbf{R_s}} e^{i(\mathbf{k}-\mathbf{k}')\cdot\mathbf{R_s}} = N\,\delta_{\mathbf{k},\mathbf{k}'}.
\tag{C.8.25}
$$

Thus if $\mathbf{k} \neq \mathbf{k}'$, $I^{n\mathbf{k}j}_{n'\mathbf{k}'j'} = 0$ and $\psi_{n'\mathbf{k}'j'}(\mathbf{r})$ and $\psi_{n\mathbf{k}j}(\mathbf{r})$ are orthogonal.

The relation [equation (C.5.24)]

$$
h_l h_s \sum_{\alpha=1}^{g_c} \chi^\alpha(\mathscr{C}_l)\chi^\alpha(\mathscr{C}_s) = g h_l\,\delta_{sl'},
\tag{C.8.26}
$$

when applied to the Abelian group of translations, yields another useful result[17], namely

$$\sum_{\substack{\mathbf{k} \\ \text{(in the first} \\ \text{Brillouin zone)}}} e^{i\mathbf{k}\cdot(\mathbf{R}_l+\mathbf{R}_s)} = N\,\delta_{\mathbf{R}_s,-\mathbf{R}_l}. \tag{C.8.27}$$

C.9 SYMMETRY PROPERTIES OF $E_n(\mathbf{k})$ IN THE FIRST BRILLOUIN ZONE AND THE GROUP OF THE WAVE VECTOR k

Consider the Bloch function $\psi_{n\mathbf{k}j}(\mathbf{r})$ belonging to the energy eigenvalue $E_n(\mathbf{k})$ and satisfying the equation

$$H(\mathbf{r})\psi_{n\mathbf{k}j}(\mathbf{r}) = E_n(\mathbf{k})\psi_{n\mathbf{k}j}(\mathbf{r}). \tag{C.9.1}$$

The Hamiltonian $H(\mathbf{r})$ is invariant under the operations of the space group of the particular crystalline structure to which it refers, and an element $\{Q_l|\mathbf{v}_l + \mathbf{R}_s\}$ of this space group applied to the Bloch function $\psi_{n\mathbf{k}j}(\mathbf{r})$ must result in another Bloch function with energy equal to $E_n(\mathbf{k})$. Now

$$\begin{aligned}\{Q_l|\mathbf{v}_l + \mathbf{R}_s\}\psi_{n\mathbf{k}j}(\mathbf{r}) &= \{Q_l|\mathbf{v}_l + \mathbf{R}_s\}\, e^{i\mathbf{k}\cdot\mathbf{r}}u_{n\mathbf{k}j}(\mathbf{r}) \\ &= e^{i\mathbf{k}\cdot(\mathbf{v}_l+\mathbf{R}_s)}\, e^{i\mathbf{k}\cdot q^l\mathbf{r}}\xi(\mathbf{r}),\end{aligned} \tag{C.9.2}$$

where

$$\xi(\mathbf{r}) \equiv u_{n\mathbf{k}j}(q^l\mathbf{r} + \mathbf{v}_l + \mathbf{R}_s) = u_{n\mathbf{k}j}(q^l\mathbf{r} + \mathbf{v}_l). \tag{C.9.3}$$

Furthermore

$$\mathbf{k}\cdot q^l\mathbf{r} = \sum_{p=1}^{3} k_p \sum_{t=1}^{3} q^l_{pt}x_t = \sum_{t=1}^{3}\left[\sum_{p=1}^{3} q^l_{pt}k_p\right]x_t = [(q^l)^{-1}\mathbf{k}]\cdot\mathbf{r}, \tag{C.9.4}$$

since q^l is an orthogonal matrix. Hence (C.9.2) can be written

$$\{Q_l|\mathbf{v}_l + \mathbf{R}_s\}\psi_{n\mathbf{k}j}(\mathbf{r}) = e^{i\mathbf{k}\cdot(\mathbf{v}_l+\mathbf{R}_s)}\, e^{i(q^{l^{-1}}\mathbf{k})\cdot\mathbf{r}}\,\xi(\mathbf{r}). \tag{C.9.5}$$

From (C.9.3)

$$\xi(\mathbf{r} + \mathbf{R}_\mathbf{m}) = \xi(\mathbf{r}) \qquad \text{for all } \mathbf{R}_\mathbf{m}, \tag{C.9.6}$$

and the right-hand side of (C.9.5) is therefore a Bloch function with wave-vector $q^{l^{-1}}\mathbf{k}$.

The element $\{Q_l|\mathbf{v}_l + \mathbf{R}_s\}$ of the space group transforms a Bloch function with wave-vector $\mathbf{k}$ and energy $E_n(\mathbf{k})$ into another Bloch function with wave-vector $q^{l^{-1}}\mathbf{k}$ and equal energy. By considering all energy bands, it can be seen that if the energy eigenvalues at the point $\mathbf{k}$ in the first

[17] See equation (B.6.20) and (B.6.22) for a different derivation.

Brillouin zone are $E_1(\mathbf{k}), E_2(\mathbf{k}), \ldots, E_n(\mathbf{k}), \ldots$ and the energy eigenvalues at the point $q^{l^{-1}}\mathbf{k}$ are $E_1(q^{l^{-1}}\mathbf{k}), E_2(q^{l^{-1}}\mathbf{k}), \ldots, E_n(q^{l^{-1}}\mathbf{k}), \ldots$, then

$$E_n(\mathbf{k}) = E_n(q^{l^{-1}}\mathbf{k}) = Q_l^{-1}E_n(\mathbf{k}). \tag{C.9.7}$$

(C.9.7) is true for all elements of the point group.

In every band the energy, regarded as a function of position within the first Brillouin zone (i.e. as a function of **k**), possesses the full point group symmetry of the crystal.

Amongst the elements $\{Q_l|\mathbf{v}_l + \mathbf{R}_s\}$ of the space group there will be some which, when applied to $\psi_{n\mathbf{k}j}(\mathbf{r})$, will result in Bloch functions with the same wave-vector $\mathbf{k}$ ($q^{l^{-1}}\mathbf{k} = \mathbf{k}$) or with an equivalent wave-vector ($q^{l^{-1}}\mathbf{k} = \mathbf{k} + \mathbf{K}_\mathbf{p}$). All such elements form a subgroup of the space group, called the group of the wave-vector **k**. It is this subgroup which determines the degeneracies and symmetry properties of the energy states belonging to the point **k** in the first Brillouin zone, i.e. any element of the group of the wave-vector **k** transforms a Bloch function with wave-vector **k** and energy $E_n(\mathbf{k})$ into another Bloch function with the same (or equivalent) wave-vector and energy.

To illustrate these concepts, we shall now consider the groups of wave-vectors at certain points of high symmetry in the first Brillouin zone of the face-centred cubic lattice, taking as specific substances Ge and InSb.

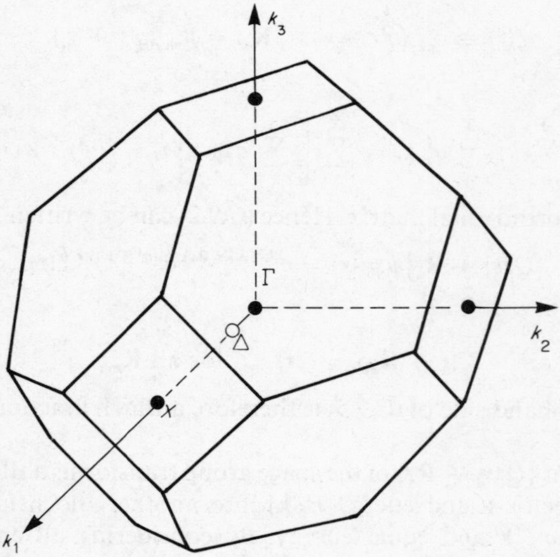

Fig. C.9.1. First Brillouin zone of the face-centred cubic lattice.

Figure C.9.1. shows the first Brillouin zone (a truncated octahedron) of the face-centred cubic lattice. The centre of the first Brillouin zone ($\mathbf{k} = 0$) is always labelled Γ. The points of high symmetry which we shall consider are Γ, Δ and X (fig. C.9.1). Δ is any point on the line joining Γ to X (the Δ-axis), but does not coincide with either Γ or X. The coordinates of these symmetry points are as follows:

$$\Gamma : \mathbf{k}_\Gamma = \frac{2\pi}{a}(0, 0, 0),$$

$$\Delta : \mathbf{k}_\Delta = \frac{2\pi}{a}(\eta, 0, 0), \qquad (0 < \eta < 1)$$

$$X : \mathbf{k}_X = \frac{2\pi}{a}(1, 0, 0). \tag{C.9.8}$$

The irreducible representations of symmorphic space groups are somewhat easier to handle than those of nonsymmorphic space groups, and we shall therefore first consider the InSb structure.

C.9a Groups of wave-vectors in the InSb structure[18]

The point group associated with the InSb structure is T_d (see table C.7.1); table C.9.1 shows the elements of this point group which transform the wave vectors at Γ, Δ and X into themselves or into equivalent wave vectors.

Table C.9.1 Point group operations belonging to the groups of wave-vectors in the InSb structure.

Γ:	All the elements of the point group T_d.
Δ:	The elements Q_1, Q_2, Q_{19}, Q_{20} of the point group T_d.
X:	Those belonging to Δ plus the elements Q_3, Q_4, Q_{13}, Q_{14} of the point group T_d.

At Δ, for example, the operations (applied in $\mathbf{k}$-space) are

$$Q_1(k_1, k_2, k_3), \quad Q_2(k_1, -k_2, -k_3), \quad Q_{19}(k_1, k_3, k_2), \quad Q_{20}(k_1, -k_3, -k_2) \tag{C.9.9}$$

These four operations leave the wave-vector $\mathbf{k}_\Delta = (2\pi/a)(\eta, 0, 0)$ invariant. At X, these four operations leave $\mathbf{k}_X = (2\pi/a)(1, 0, 0)$ invariant, and the

[18] G. Dresselhaus, *Phys. Rev.*, **100**, 580 (1955); R. H. Parmenter, *Phys. Rev.*, **100**, 573 (1955).

additional operations

$$Q_3(-k_1, k_2, -k_3), \qquad Q_4(-k_1, -k_2, k_3),$$
$$Q_{13}(-k_1, k_3, -k_2), \qquad Q_{14}(-k_1, -k_3, k_2) \tag{C.9.10}$$

transform $\mathbf{k}_X$ into $-\mathbf{k}_X$, an equivalent wave-vector.

The group of the wave-vector at Γ is therefore the full space group T_d^2 of the InSb structure. Let $E_n(0)$ be a d_n-fold (essentially) degenerate energy level with wave-vector $\mathbf{k}_\Gamma = (0, 0, 0)$ and Bloch functions $\psi_{n0j}(\mathbf{r})$ $(j = 1, 2, \ldots, d_n)$. These functions act as basis functions for a d_n-dimensional irreducible representation of the group of the wave-vector at Γ. Hence if $\{Q_l|\mathbf{R_m}\}$ is an element of the symmorphic space group T_d^2,

$$\{Q_l|\mathbf{R_m}\}\psi_{n0j}(\mathbf{r}) = \sum_{s=1}^{d_n} M_{sj}^{lm}\psi_{n0s}(\mathbf{r}). \tag{C.9.11}$$

Now

$$\{Q_l|\mathbf{R_m}\}\psi_{n0j}(\mathbf{r}) = e^{i\mathbf{k}_\Gamma\cdot\mathbf{R_m}}\psi_{n0j}(q^l\mathbf{r}) = Q_l\psi_{n0j}(\mathbf{r}), \tag{C.9.12}$$

and (C.9.11) can be written

$$Q_l\psi_{n0j}(\mathbf{r}) = \sum_{s=1}^{d_n} M_{sj}^{lm}\psi_{n0s}(\mathbf{r}). \tag{C.9.13}$$

By considering all elements $\{Q_l|\mathbf{R_m}\}$ with l fixed and $\mathbf{m}$ varying, we see that

$$M^{lm} = M^{lp} = \cdots = M^l(\text{say}). \tag{C.9.14}$$

All space group elements with the same point group operation Q_l are therefore represented by the same matrix M^l. (C.9.13) now becomes

$$Q_l\psi_{n0j}(\mathbf{r}) = \sum_{s=1}^{d_n} M_{sj}^{l}\psi_{n0s}(\mathbf{r}) \qquad (l = 1, 2, \ldots, 24). \tag{C.9.15}$$

The Bloch functions $\psi_{n0j}(\mathbf{r})$ $(j = 1, 2, \ldots, d_n)$ therefore act as basis functions for a d_n-dimensional irreducible representation of the point group T_d, the point group element Q_l being represented by the matrix M^l. To find the irreducible representations of the group of the wave-vector at Γ, we need only find the irreducible representations of the point group T_d. To each irreducible representation of the point group T_d will correspond an irreducible representation of the group of the wave-vector at Γ.

The character table for the point group T_d, with the notation for the irreducible representations, is given below. This is then also the character table for the group of the wave-vector at Γ.

Table C.9.2 Character table for the point group T_d.

	Q_1	Q_2–Q_4	Q_5–Q_{12}	Q_{13}–Q_{18}	Q_{19}–Q_{24}
Γ_1	1	1	1	1	1
Γ_2	1	1	1	−1	−1
Γ_{12}	2	2	−1	0	0
Γ_{15}	3	−1	0	−1	1
Γ_{25}	3	−1	0	1	−1

The numbers in this table are the characters of the matrices M^l. At the point Γ in the first Brillouin zone there are therefore nondegenerate states (two types: Γ_1 and Γ_2), two-fold degenerate states (one type: Γ_{12}) and three-fold degenerate states (two types: Γ_{15} and Γ_{25}).

Next consider the point Δ. The elements of the group of the wave-vector $\mathbf{k}_\Delta$ are $\{Q_1|\mathbf{R_m}\}$, $\{Q_2|\mathbf{R_m}\}$, $\{Q_{19}|\mathbf{R_m}\}$ and $\{Q_{20}|\mathbf{R_m}\}$. If $E_n(\mathbf{k}_\Delta)$ is a d_n-fold degenerate energy level at Δ with Bloch functions $\psi_{n\mathbf{k}_\Delta j}(\mathbf{r})$ $(j = 1, 2, \ldots, d_n)$,

$$\{Q_l|\mathbf{R_m}\}\psi_{n\mathbf{k}_\Delta j}(\mathbf{r}) = \mathrm{e}^{i\mathbf{k}_\Delta\cdot\mathbf{R_m}}Q_l\psi_{n\mathbf{k}_\Delta j}(\mathbf{r})$$

$$= \sum_{s=1}^{d_n} M_{sj}^{lm}\psi_{n\mathbf{k}_\Delta s}(\mathbf{r}) \qquad (l = 1, 2, 19, 20), \qquad \text{(C.9.16)}$$

and hence

$$Q_l\psi_{n\mathbf{k}_\Delta j}(\mathbf{r}) = \sum_{s=1}^{d_n} \mathrm{e}^{-i\mathbf{k}_\Delta\cdot\mathbf{R_m}}M_{sj}^{lm}\psi_{n\mathbf{k}_\Delta s}(\mathbf{r}). \qquad \text{(C.9.17)}$$

By considering all elements with the same point group operation Q_l but different translations $\mathbf{R_m}$, we see that

$$\mathrm{e}^{-i\mathbf{k}_\Delta\cdot\mathbf{R_m}}M^{lm} = \mathrm{e}^{-i\mathbf{k}_\Delta\cdot\mathbf{R_s}}M^{ls} = \cdots = M^l(\text{say}), \qquad \text{(C.9.18)}$$

and (C.9.17) can be written

$$Q_l\psi_{n\mathbf{k}_\Delta j}(\mathbf{r}) = \sum_{s=1}^{d_n} M_{sj}^{l}\psi_{n\mathbf{k}_\Delta s}(\mathbf{r}) \qquad (l = 1, 2, 19, 20). \qquad \text{(C.9.19)}$$

The Bloch functions $\psi_{n\mathbf{k}_\Delta j}(\mathbf{r})$ $(j = 1, 2, \ldots, d_n)$ therefore act as basis functions for a d_n-dimensional irreducible representation of the point group of order four with elements Q_1, Q_2, Q_{19} and Q_{20}, the element Q_l being represented by the matrix M^l. To find the irreducible representations of the group of the wave-vector at Δ, we need only find the irreducible representations of the point group of order four with elements Q_1, Q_2, Q_{19} and Q_{20}. To each irreducible representation of this point group will correspond an irreducible representation of the group of the wave-vector at Δ. The character table for this point group, with the notation for the irreducible representations, is given below. This is then also the character table for the group of the wave-vector at Δ.

Table C.9.3 Character table for the point group at Δ.

	Q_1	Q_2	Q_{19}	Q_{20}
Δ_1	1	1	1	1
Δ_2	1	1	−1	−1
Δ_3	1	−1	1	−1
Δ_4	1	−1	−1	1

The numbers in this table are the characters of the matrices M^l. At the point Δ in the first Brillouin zone there are only nondegenerate states (four types: $\Delta_1, \Delta_2, \Delta_3$ and Δ_4).

Finally, for InSb, we consider the irreducible representations of the group of the wave-vector at X. These irreducible representations are found in the same way as those for the point Δ. The point group elements belonging to the wave-vector at X are $Q_1, Q_2, Q_3, Q_4, Q_{13}, Q_{14}, Q_{19}$ and Q_{20}. These form a point group of order eight. The Bloch functions at X belonging to a d_n-fold degenerate energy level act as basis functions for a d_n-dimensional irreducible representation of this point group. To each irreducible representation of this point group corresponds an irreducible representation of the group of the wave-vector at X. If M^l ($l = 1, 2, 3, 4, 13, 14, 19, 20$) are the matrices of an irreducible representation of this point group, the matrix representing the element $\{Q_l|\mathbf{R_m}\}$ in the corresponding irreducible representation of the group of the wave-vector at X is $e^{i\mathbf{k}_X \cdot \mathbf{R_m}} M^l$. The character table for this point group, with the notation for the irreducible representations, is given below.

Table C.9.4 Character table for the point group at X.

	Q_1	Q_2	Q_3, Q_4	Q_{13}, Q_{14}	Q_{19}, Q_{20}
X_1	1	1	1	1	1
X_2	1	1	1	−1	−1
X_3	1	1	−1	−1	1
X_4	1	1	−1	1	−1
X_5	2	−2	0	0	0

At the point X in the first Brillouin zone there are therefore nondegenerate states (four types: X_1, X_2, X_3, X_4) and two-fold degenerate states (one type: X_5).

C.9b Groups of wave-vectors in the Ge Structure[19]

We now consider the nonsymmorphic space group O_h^7 belonging to the Ge structure. The point group associated with this structure is O_h, and the elements of this point group which transform the wave-vectors at Γ, Δ and X into themselves or into equivalent wave-vectors are given in table C.9.5.

Table C.9.5 Point group operations belonging to the groups of wave-vectors in the Ge structure (see table C.7.1).

Γ:	All the elements of the point group O_h.
Δ:	The elements $Q_1, Q_2, Q_{19}, Q_{20}, Q_3', Q_4', Q_{13}', Q_{14}'$ of the point group O_h.
X:	Those elements belonging to Δ plus the elements $Q_3, Q_4, Q_{13}, Q_{14}, Q_1', Q_2', Q_{19}', Q_{20}'$ of the point group O_h.

The group of the wave-vector at Γ is the full space group O_h^7 with elements $\{Q_l|\mathbf{R_m}\}$ and $\{Q_l'|\mathbf{v} + \mathbf{R_m}\}$ $(l = 1, 2, \ldots, 24)$, where $\mathbf{v} \equiv a(\frac{1}{4}, \frac{1}{4}, \frac{1}{4})$. If $E_n(0)$ is a d_n-fold degenerate energy level at Γ with Bloch functions $\psi_{n0j}(\mathbf{r})$ $(j = 1, 2, \ldots, d_n)$,

$$\{Q_l|\mathbf{R_m}\}\psi_{n0j}(\mathbf{r}) = Q_l\psi_{n0j}(\mathbf{r}) = \sum_{s=1}^{d_n} M_{sj}^{lm}\psi_{n0s}(\mathbf{r}), \tag{C.9.20}$$

$$\{Q_l'|\mathbf{v} + \mathbf{R_m}\}\psi_{n0j}(\mathbf{r}) = \{Q_l'|\mathbf{v}\}\psi_{n0j}(\mathbf{r}) = \sum_{s=1}^{d_n} M_{sj}^{l'm}\psi_{n0s}(\mathbf{r}). \tag{C.9.21}$$

By considering all space group elements with the same point group operation but different lattice translations, we see that

$$M^{lm} = M^{ls} = M^{lp} = \cdots = M^l \text{ (say)}, \tag{C.9.22}$$

$$M^{l'm} = M^{l's} = M^{l'p} = \cdots = M^{l'} \text{ (say)}, \tag{C.9.23}$$

and (C.9.20) and (C.9.21) can be written

$$Q_l\psi_{n0j}(\mathbf{r}) = \sum_{s=1}^{d_n} M_{sj}^{l}\psi_{n0s}(\mathbf{r}), \tag{C.9.24}$$

$$\{Q_l'|\mathbf{v}\}\psi_{n0j}(\mathbf{r}) = \sum_{s=1}^{d_n} M_{sj}^{l'}\psi_{n0s}(\mathbf{r}). \tag{C.9.25}$$

[19] C. Herring, *J. Franklin Inst.*, **233**, 525 (1942); R. J. Elliott, *Phys. Rev.*, **96**, 280 (1954); G. Dresselhaus, A. F. Kip and C. Kittel, *Phys. Rev.*, **98**, 368 (1955); H. Jones, *The Theory of Brillouin Zones and Electronic States in Crystals,* North Holland, Amsterdam, 1962.

The 48 operations Q_l and $\{Q_l'|\mathbf{v}\}$ ($l = 1, 2, \ldots, 24$) do not form a group. For example,

$$\{Q_2'|\mathbf{v}\}\{Q_2'|\mathbf{v}\} = \{Q_1|\mathbf{R}\}, \tag{C.9.26}$$

where $\mathbf{R}$ is a lattice translation $[= a(0, \frac{1}{2}, \frac{1}{2})]$ and this is not one of the same set of operations. However, if we consider the 48 matrices M^l and $M^{l'}$ ($l = 1, 2, \ldots, 24$), we see that the product of any two of these matrices is itself one of the same set of matrices and that the inverse of every one of these matrices is itself contained in the set. These matrices therefore form an irreducible representation of a group and, as far as the Bloch functions $\psi_{n0j}(\mathbf{r})$ ($j = 1, 2, \ldots, d_n$) are concerned, the 48 operations Q_l and $\{Q_l'|\mathbf{v}\}$ may be regarded as forming a group.

The Bloch functions $\psi_{n0j}(\mathbf{r})$ ($j = 1, 2, \ldots, d_n$) act as basis functions for a d_n-dimensional irreducible representation of the 'group' with elements Q_l and $\{Q_l'|\mathbf{v}\}$ ($l = 1, 2, \ldots, 24$). To find the irreducible representations of the group of the wave-vector at Γ, we need only find the irreducible representations of the above 'group' of order 48. To each irreducible representation of this 'group' of order 48 corresponds an irreducible representation of the group of the wave-vector at Γ.

The product

$$\{Q_l|\mathbf{R_s}\}\{Q_p'|\mathbf{v} + \mathbf{R_m}\} = \{Q_t'|\mathbf{v} + \mathbf{R_w}\} \tag{C.9.27}$$

implies

$$Q_lQ_p' = Q_t', \qquad M^lM^{p'} = M^{t'}. \tag{C.9.28}$$

By considering products of space group elements, we see that the matrices M^l and $M^{l'}$ ($l = 1, 2, \ldots, 24$) multiply in the same way as the elements Q_l and Q_l' of the point group O_h. These matrices therefore form an irreducible representation of the point group O_h.

The 'group' with elements Q_l and $\{Q_l'|\mathbf{v}\}$ ($l = 1, 2, \ldots, 24$) is simply isomorphic with the point group O_h with elements Q_l and Q_l' ($l = 1, 2, \ldots, 24$). These two groups have the same irreducible representations and the same character table, i.e. if the matrices M^l and $M^{l'}$ form an irreducible representation of the point group O_h then they also form an irreducible representation of the 'group' with elements Q_l and $\{Q_l'|\mathbf{v}\}$ in which the element Q_l is represented by the matrix M^l and the element $\{Q_l'|\mathbf{v}\}$ is represented by the matrix $M^{l'}$.

From the character table for the point group O_h we therefore obtain the following character table for the 'group' of order 48 with elements Q_l and $\{Q_l'|\mathbf{v}\}$. The numbers in this table are the characters of the matrices M^l and $M^{l'}$ defined in (C.9.22) and (C.9.23).

Table C.9.6 Character table for the group Γ in the Ge structure.

	Q_1	$Q_2 - Q_4$	$Q_5 - Q_{12}$	$Q_{13} - Q_{18}$	$Q_{19} - Q_{24}$	$\{Q'_1\|\mathbf{v}\}$	$\{Q'_2\|\mathbf{v}\} - \{Q'_4\|\mathbf{v}\}$	$\{Q'_5\|\mathbf{v}\} - \{Q'_{12}\|\mathbf{v}\}$	$\{Q'_{13}\|\mathbf{v}\} - \{Q'_{18}\|\mathbf{v}\}$	$\{Q'_{19}\|\mathbf{v}\} - \{Q'_{24}\|\mathbf{v}\}$
Γ_1	1	1	1	1	1	1	1	1	1	1
Γ'_1	1	1	1	−1	−1	−1	−1	−1	1	1
Γ_2	1	1	1	−1	−1	1	1	1	−1	−1
Γ'_2	1	1	1	1	1	−1	−1	−1	−1	−1
Γ_{12}	2	2	−1	0	0	2	2	−1	0	0
Γ'_{12}	2	2	−1	0	0	−2	−2	1	0	0
Γ_{15}	3	−1	0	−1	1	−3	1	0	1	−1
Γ'_{15}	3	−1	0	1	−1	3	−1	0	1	−1
Γ_{25}	3	−1	0	1	−1	−3	1	0	−1	1
Γ'_{25}	3	−1	0	−1	1	3	−1	0	−1	1

At the point Γ in the first Brillouin zone of the Ge structure there are therefore nondegenerate states (four types: $\Gamma_1, \Gamma'_1, \Gamma_2$ and Γ'_2), two-fold degenerate states (two types: Γ_{12} and Γ'_{12}), and three-fold degenerate states (four types: $\Gamma_{15}, \Gamma'_{15}, \Gamma_{25}$ and Γ'_{25}).

The group of the wave-vector at Δ has elements $\{Q_l|\mathbf{R_m}\}$ ($l = 1, 2, 19, 20$) and $\{Q'_t|\mathbf{v} + \mathbf{R_m}\}$ ($t = 3, 4, 13, 14$). If $E_n(\mathbf{k}_\Delta)$ is a d_n-fold degenerate energy level at Δ with Bloch functions $\psi_{n\mathbf{k}_\Delta j}(\mathbf{r})$ ($j = 1, 2, \ldots, d_n$),

$$\{Q_l|\mathbf{R_m}\}\psi_{n\mathbf{k}_\Delta j}(\mathbf{r}) = e^{i\mathbf{k}_\Delta \cdot \mathbf{R_m}} Q_l \psi_{n\mathbf{k}_\Delta j}(\mathbf{r}) = \sum_{s=1}^{d_n} M_{sj}^{lm} \psi_{n\mathbf{k}_\Delta s}(\mathbf{r}), \quad (C.9.29)$$

$$\{Q'_t|\mathbf{v} + \mathbf{R_m}\}\psi_{n\mathbf{k}_\Delta j}(\mathbf{r}) = e^{i\mathbf{k}_\Delta \cdot \mathbf{R_m}} \{Q'_t|\mathbf{v}\}\psi_{n\mathbf{k}_\Delta j}(\mathbf{r}) = \sum_{s=1}^{d_n} M_{sj}^{t'm} \psi_{n\mathbf{k}_\Delta s}(\mathbf{r}), \quad (C.9.30)$$

and hence

$$e^{-i\mathbf{k}_\Delta \cdot \mathbf{R_m}} M^{lm} = e^{-i\mathbf{k}_\Delta \cdot \mathbf{R_p}} M^{lp} = \cdots = M^l \text{ (say)}, \quad (C.9.31)$$

$$e^{-i\mathbf{k}_\Delta \cdot \mathbf{R_m}} M^{t'm} = e^{-i\mathbf{k}_\Delta \cdot \mathbf{R_p}} M^{t'p} = \cdots = M^{t'} \text{ (say)}. \quad (C.9.32)$$

(C.9.29) and (C.9.30) can therefore be written

$$Q_l \psi_{n\mathbf{k}_\Delta j}(\mathbf{r}) = \sum_{s=1}^{d_n} M_{sj}^{l} \psi_{n\mathbf{k}_\Delta s}(\mathbf{r}), \quad (C.9.33)$$

$$\{Q'_t|\mathbf{v}\}\psi_{n\mathbf{k}_\Delta j}(\mathbf{r}) = \sum_{s=1}^{d_n} M_{sj}^{t'} \psi_{n\mathbf{k}_\Delta s}(\mathbf{r}). \quad (C.9.34)$$

The operations Q_l ($l = 1, 2, 19, 20$) and $\{Q'_t|\mathbf{v}\}$ ($t = 3, 4, 13, 14$) do not form a group, and the product of two of the matrices M^l and $M^{t'}$ is not in general one of the same set of matrices. Suppose that

$$\{Q'_t|\mathbf{v} + \mathbf{R_s}\}\{Q'_l|\mathbf{v} + \mathbf{R_m}\} = \{Q_p|\mathbf{R_w}\}. \quad (C.9.35)$$

This implies

$$Q'_t Q'_l = Q_p, \qquad \mathbf{R_w} - \mathbf{R_m} = q^{l'}(\mathbf{v} + \mathbf{R_s}) + \mathbf{v}, \quad (C.9.36)$$

and, from (C.9.29)–(C.9.32),

$$M^{t's}M^{l'm} = M^{pw}, \quad (C.9.37)$$

$$M^{t'}M^{l'} = e^{i\mathbf{k}_\Delta \cdot (\mathbf{R_w} - \mathbf{R_m} - \mathbf{R_s})} M^p = e^{i\mathbf{k}_\Delta \cdot [q^{l'}(\mathbf{v} + \mathbf{R_s}) + \mathbf{v} - \mathbf{R_s}]} M^p. \quad (C.9.38)$$

Since Q'_l is an operation of the point group belonging to the wave-vector at Δ,

$$e^{i\mathbf{k}_\Delta \cdot [q^{l'}(\mathbf{v} + \mathbf{R_s})]} = e^{i(q^{l'-1}\mathbf{k}_\Delta) \cdot (\mathbf{v} + \mathbf{R_s})} = e^{i\mathbf{k}_\Delta \cdot (\mathbf{v} + \mathbf{R_s})}, \quad (C.9.39)$$

and hence (C.9.38) becomes

$$M^{t'}M^{l'} = e^{2i\mathbf{k}_\Delta \cdot \mathbf{v}}M^p. \tag{C.9.40}$$

(C.9.36) and (C.9.40) show that the eight matrices M^l ($l = 1, 2, 19, 20$) and $M^{t'}$ ($t = 3, 4, 13, 14$) do not multiply in the same way as the point group elements Q_l and Q'_t. However, if we consider the matrices M^l ($l = 1, 2, 19, 20$) and $L^{t'} \equiv e^{-i\mathbf{k}_\Delta \cdot \mathbf{v}}M^{t'}$ ($t = 3, 4, 13, 14$) we see that these matrices do multiply in the same way as the elements Q_l and Q'_t, e.g. (C.9.36) and (C.9.40) now yield

$$Q'_tQ'_l = Q_p, \qquad L^{t'}L^{l'} = M^p. \tag{C.9.41}$$

The matrices M^l and $L^{t'}$ form an irreducible representation of the point group with elements Q_l ($l = 1, 2, 19, 20$) and Q'_t ($t = 3, 4, 13, 14$), and to find the irreducible representations of the group of the wave-vector at Δ we need only find the irreducible representations of this point group of order eight.

The character table for this point group, with the notation for the irreducible representations, is given below. The numbers in this table are the characters of the matrices M^l and $L^{t'}$.

Table C.9.7 Character table for the point group at Δ.

	Q_1	Q_2	Q_{19}, Q_{20}	Q'_3, Q'_4	Q'_{13}, Q'_{14}
Δ_1	1	1	1	1	1
Δ'_1	1	1	-1	-1	1
Δ_2	1	1	-1	1	-1
Δ'_2	1	1	1	-1	-1
Δ_5	2	-2	0	0	0

From this character table we then obtain the following character table for the group of the wave-vector at Δ. The numbers in this table are the characters of the matrices M^l and $M^{t'}$ defined in (C.9.31) and (C.9.32).

Table C.9.8 Character table for the group of the wave-vector at Δ in the first Brillouin zone of the Ge structure.

	Q_1	Q_2	Q_{19}, Q_{20}	$\{Q'_3\|\mathbf{v}\}, \{Q'_4\|\mathbf{v}\}$	$\{Q'_{13}\|\mathbf{v}\}, \{Q'_{14}\|\mathbf{v}\}$
Δ_1	1	1	1	$e^{i\mathbf{k}_\Delta \cdot \mathbf{v}}$	$e^{i\mathbf{k}_\Delta \cdot \mathbf{v}}$
Δ'_1	1	1	-1	$-e^{i\mathbf{k}_\Delta \cdot \mathbf{v}}$	$e^{i\mathbf{k}_\Delta \cdot \mathbf{v}}$
Δ_2	1	1	-1	$e^{i\mathbf{k}_\Delta \cdot \mathbf{v}}$	$-e^{i\mathbf{k}_\Delta \cdot \mathbf{v}}$
Δ'_2	1	1	1	$-e^{i\mathbf{k}_\Delta \cdot \mathbf{v}}$	$-e^{i\mathbf{k}_\Delta \cdot \mathbf{v}}$
Δ_5	2	-2	0	0	0

At the point Δ in the first Brillouin zone of the Ge structure there are therefore nondegenerate states (four types: Δ_1, Δ_1', Δ_2 and Δ_2') and two-fold degenerate states (one type: Δ_5). For instance, a Bloch function $\psi_{\Delta_2'}(\mathbf{r})$ belonging to a Δ_2'-type state satisfies

$$\{Q_l|\mathbf{R_s}\}\psi_{\Delta_2'}(\mathbf{r}) = \mathrm{e}^{\mathrm{i}\mathbf{k}_\Delta\cdot\mathbf{R_s}}\psi_{\Delta_2'}(\mathbf{r}) \qquad (l = 1, 2, 19, 20), \tag{C.9.42}$$

$$\{Q_t'|\mathbf{v} + \mathbf{R_s}\}\psi_{\Delta_2'}(\mathbf{r}) = -\mathrm{e}^{\mathrm{i}\mathbf{k}_\Delta\cdot(\mathbf{v}+\mathbf{R_s})}\psi_{\Delta_2'}(\mathbf{r}) \qquad (t = 3, 4, 13, 14). \tag{C.9.43}$$

Finally, we come to the point X on the surface of the first Brillouin zone. The group of the wave vector at X has elements $\{Q_t|\mathbf{R_m}\}$ and $\{Q_t'|\mathbf{v} + \mathbf{R_m}\}$ $(t = 1, 2, 3, 4, 13, 14, 19, 20)$, and if $E_n(\mathbf{k}_X)$ is a d_n-fold degenerate energy level at X with Bloch functions $\psi_{n\mathbf{k}_Xj}(\mathbf{r})$ $(j = 1, 2, \ldots, d_n)$,

$$\{Q_t|\mathbf{R_m}\}\psi_{n\mathbf{k}_Xj}(\mathbf{r}) = \mathrm{e}^{\mathrm{i}\mathbf{k}_X\cdot\mathbf{R_m}}Q_t\psi_{n\mathbf{k}_Xj}(\mathbf{r}) = \sum_{s=1}^{d_n} M_{sj}^{tm}\psi_{n\mathbf{k}_Xs}(\mathbf{r}), \tag{C.9.44}$$

$$\{Q_t'|\mathbf{v} + \mathbf{R_m}\}\psi_{n\mathbf{k}_Xj}(\mathbf{r}) = \mathrm{e}^{\mathrm{i}\mathbf{k}_X\cdot\mathbf{R_m}}\{Q_t'|\mathbf{v}\}\psi_{n\mathbf{k}_Xj}(\mathbf{r}) = \sum_{s=1}^{d_n} M_{sj}^{t'm}\psi_{n\mathbf{k}_Xs}(\mathbf{r}), \tag{C.9.45}$$

and hence

$$\mathrm{e}^{-\mathrm{i}\mathbf{k}_X\cdot\mathbf{R_m}}M^{tm} = \mathrm{e}^{-\mathrm{i}\mathbf{k}_X\cdot\mathbf{R_p}}M^{tp} = \cdots = M^t \text{ (say)}, \tag{C.9.46}$$

$$\mathrm{e}^{-\mathrm{i}\mathbf{k}_X\cdot\mathbf{R_m}}M^{t'm} = \mathrm{e}^{-\mathrm{i}\mathbf{k}_X\cdot\mathbf{R_p}}M^{t'p} = \cdots = M^{t'} \text{ (say)}. \tag{C.9.47}$$

(C.9.44) and (C.9.45) can therefore be written

$$Q_t\psi_{n\mathbf{k}_Xj}(\mathbf{r}) = \sum_{s=1}^{d_n} M_{sj}^{t}\psi_{n\mathbf{k}_Xs}(\mathbf{r}), \tag{C.9.48}$$

$$\{Q_t'|\mathbf{v}\}\psi_{n\mathbf{k}_Xj}(\mathbf{r}) = \sum_{s=1}^{d_n} M_{sj}^{t'}\psi_{n\mathbf{k}_Xs}(\mathbf{r}). \tag{C.9.49}$$

The operations Q_t and $\{Q_t'|\mathbf{v}\}$ $(t = 1, 2, 3, 4, 13, 14, 19, 20)$ do not form a group.

If $\mathbf{R_m} = \dfrac{a}{2}(m_2 + m_3, m_1 + m_3, m_1 + m_2)$, where m_1, m_2 and m_3 are integers, and $\mathbf{k}_X = \dfrac{2\pi}{a}(1, 0, 0)$,

$$\mathrm{e}^{\mathrm{i}\mathbf{k}_x\cdot\mathbf{R_m}} = \mathrm{e}^{\mathrm{i}\pi(m_2+m_3)} = \begin{cases} 1 \text{ if } m_2 + m_3 \text{ is even;} \\ -1 \text{ if } m_2 + m_3 \text{ is odd.} \end{cases} \tag{C.9.50}$$

Let $\mathbf{R}_e$ denote any lattice translation for which $\mathrm{e}^{\mathrm{i}\mathbf{k}_X\cdot\mathbf{R}_e} = 1$, and $\mathbf{R}_0$ denote any lattice translation for which $\mathrm{e}^{\mathrm{i}\mathbf{k}_X\cdot\mathbf{R}_0} = -1$. From (C.9.44)–(C.9.47),

$\{Q_t|\mathbf{R}_e\}$ is represented by the matrix M^t,

$\{Q_t|\mathbf{R}_0\}$ is represented by the matrix $-M^t \equiv L^t$,

$\{Q_t'|\mathbf{v} + \mathbf{R}_e\}$ is represented by the matrix $M^{t'}$, and

$$\{Q_t'|\mathbf{v} + \mathbf{R}_0\} \text{ is represented by the matrix } -M^{t'} \equiv L^{t'}. \quad \text{(C.9.51)}$$

It is obvious that the 32 matrices M^t, $M^{t'}$, L^t, $L^{t'}$ ($t = 1, 2, 3, 4, 13, 14, 19, 20$) form an irreducible representation of some group of order 32. Suppose that we denote the elements of this group by A_t (represented by M^t), B_t (represented by L^t), C_t' (represented by $M^{t'}$) and D_t' (represented by $L^{t'}$). We can draw up the multiplication table for this group and hence find its classes. To do this, we make use of the fact that the matrices M^t, L^t, $M^{t'}$ and $L^{t'}$ multiply in the same way as the space group elements. For example,

$$\{Q_2|\mathbf{R}_e\}\{Q_3'|\mathbf{v} + \mathbf{R}_0\} = \{Q_4'|\mathbf{v} + \mathbf{R}_0\}, \quad \text{(C.9.52)}$$

and hence

$$M^2 L^{3'} = L^{4'}, \qquad A_2 D_3' = D_4'. \quad \text{(C.9.53)}$$

Once the multiplication table has been obtained the classes can be found. These are given below.

$$\mathscr{C}_1 = A_1; \quad \mathscr{C}_2 = A_3 + A_4 + B_3 + B_4; \quad \mathscr{C}_3 = A_2;$$

$$\mathscr{C}_4 = C_{19}' + D_{19}'; \quad \mathscr{C}_5 = C_{13}' + C_{14}' + D_{13}' + D_{14}'; \quad \mathscr{C}_6 = C_1' + D_1';$$

$$\mathscr{C}_7 = C_3' + C_4' + D_3' + D_4'; \quad \mathscr{C}_8 = C_2' + D_2'; \quad \mathscr{C}_9 = A_{19} + A_{20};$$

$$\mathscr{C}_{10} = A_{13} + A_{14} + B_{13} + B_{14}; \quad \mathscr{C}_{11} = B_{19} + B_{20};$$

$$\mathscr{C}_{12} = C_{20}' + D_{20}'; \quad \mathscr{C}_{13} = B_2; \quad \mathscr{C}_{14} = B_1. \quad \text{(C.9.54)}$$

The calculation of the character table, by the methods described earlier, is then straightforward.

Table C.9.9 Character table for the group of order 32 at X.

	$\mathscr{C}_1$	$\mathscr{C}_2$	$\mathscr{C}_3$	$\mathscr{C}_4$	$\mathscr{C}_5$	$\mathscr{C}_6$	$\mathscr{C}_7$	$\mathscr{C}_8$	$\mathscr{C}_9$	$\mathscr{C}_{10}$	$\mathscr{C}_{11}$	$\mathscr{C}_{12}$	$\mathscr{C}_{13}$	$\mathscr{C}_{14}$
	1	1	1	1	1	1	1	1	1	1	1	1	1	1
	1	1	1	−1	−1	1	1	1	−1	−1	−1	−1	1	1
	1	−1	1	−1	1	1	−1	1	−1	1	−1	−1	1	1
	1	−1	1	1	−1	1	−1	1	1	−1	1	1	1	1
	2	0	−2	0	0	2	0	−2	0	0	0	0	−2	2
	1	1	1	1	1	−1	−1	−1	−1	−1	−1	1	1	1
	1	1	1	−1	−1	−1	−1	−1	1	1	1	−1	1	1
	1	−1	1	−1	1	−1	1	−1	1	−1	1	−1	1	1
	1	−1	1	1	−1	−1	1	−1	−1	1	−1	1	1	1
	2	0	−2	0	0	−2	0	2	0	0	0	0	−2	2
X_1	2	0	2	0	0	0	0	0	2	0	−2	0	−2	−2
X_2	2	0	2	0	0	0	0	0	−2	0	2	0	−2	−2
X_3	2	0	−2	2	0	0	0	0	0	0	0	2	2	−2
X_4	2	0	−2	−2	0	0	0	0	0	0	0	2	2	−2

We next note that in the irreducible representation of this group of order 32 formed by the matrices M^t, L^t, $M^{t'}$ and $L^{t'}$, the trace of the matrix representing the element A_t is the negative of the trace of the matrix representing the element B_t, and the trace of the matrix representing the element C_t' is the negative of the trace of the matrix representing the element D_t'. The only irreducible representations of this group which satisfy these requirements are those denoted by X_1, X_2, X_3 and X_4 in table C.9.9. From this table, therefore, we obtain the following character table for the group of the wave-vector at X:

Table C.9.10 Character table for the group of the wave vector at the point X in the first Brillouin zone of the Ge structure.

	Q_1	Q_2	Q_3, Q_4	Q_{13}, Q_{14}	Q_{19}, Q_{20}	$\{Q_1'\|\mathbf{v}\}$	$\{Q_2'\|\mathbf{v}\}$
X_1	2	2	0	0	2	0	0
X_2	2	2	0	0	-2	0	0
X_3	2	-2	0	0	0	0	0
X_4	2	-2	0	0	0	0	0

	$\{Q_3'\|\mathbf{v}\}, \{Q_4'\|\mathbf{v}\}$	$\{Q_{13}'\|\mathbf{v}\}, \{Q_{14}'\|\mathbf{v}\}$	$\{Q_{19}'\|\mathbf{v}\}$	$\{Q_{20}'\|\mathbf{v}\}$
X_1	0	0	0	0
X_2	0	0	0	0
X_3	0	0	2	-2
X_4	0	0	-2	2

The numbers in table C.9.10 are the characters of the matrices M^t and $M^{t'}$ defined in (C.9.46) and (C.9.47). At the point X in the first Brillouin zone of the Ge structure there are therefore two-fold degenerate states only (four types: X_1, X_2, X_3 and X_4).

We can now summarize the method that we have used for finding the irreducible representations of the group of the wave-vector **k** where

(i) for symmorphic space groups, **k** is *any point* (including surface points) in the first Brillouin zone;

(ii) for nonsymmorphic space groups, **k** is an *interior point* (*not* on the surface) of the first Brillouin zone.

Suppose that the group of the wave-vector **k** (satisfying the above requirements) has elements $\{Q_t|\mathbf{v}_t + \mathbf{R_s}\}$ $(t = 1, 2, \ldots, g)$, where for symmorphic space groups $\mathbf{v}_t = 0$ for all t. The point group operations Q_t $(t = 1, 2, \ldots, g)$ then form a point group of order g. First find the irreducible representations of this point group. To each irreducible representation of this point group will then correspond an irreducible representation of the group of the wave-vector $\mathbf{k}$. If the matrices of an irreducible representation of this point group are M^t $(t = 1, 2, \ldots, g)$, the space group element $\{Q_t|\mathbf{v}_t + \mathbf{R}_s\}$ is represented in the corresponding irreducible representation of the group of the wave-vector $\mathbf{k}$ by the matrix $e^{i\mathbf{k}\cdot(\mathbf{v}_t+\mathbf{R}_s)}M^t$. In this way the character tables of the groups of wave-vectors can be written down immediately the character tables of the associated point groups are known.

For nonsymmorphic space groups the above method fails for wave-vectors on the surface of the first Brillouin zone. These points can, however, be handled by the same method[20] as that described for the point X in the first Brillouin zone of the Ge structure.

Problem

(C.9.1) The group of the wave-vector $\mathbf{k}_\Delta = \dfrac{2\pi}{a}(\eta, 0, 0)(0 < \eta < \frac{1}{2})$ in the first Brillouin zone of the garnet structure (space group O_h^{10}) consists of the following symmetry transformations:

$$\left\{x_1x_2x_3\middle|\frac{a}{2}(-n_1 + n_2 + n_3, n_1 - n_2 + n_3, n_1 + n_2 - n_3)\right\},$$

$$\left\{x_1\bar{x}_2\bar{x}_3\middle|\frac{a}{2}(-n_1 + n_2 + n_3, n_1 - n_2 + n_3, n_1 + n_2 - n_3 + 1)\right\},$$

$$\left\{x_1x_3x_2\middle|\frac{a}{4}(-2n_1 + 2\eta_2 + 2n_3 + 1, 2n_1 - 2n_2 + 2n_3 + 1, 2n_1 + 2n_2 - 2n_3 + 1)\right\},$$

$$\left\{x_1\bar{x}_3\bar{x}_2\middle|\frac{a}{4}(-2n_1 + 2n_2 + 2n_3 + 1, 2n_1 - 2n_2 + 2n_3 - 1, 2n_1 + 2n_2 - 2n_3 + 1)\right\},$$

$$\left\{x_1\bar{x}_2x_3\middle|\frac{a}{2}(-n_1 + n_2 + n_3 + 1, n_1 - n_2 + n_3, n_1 + n_2 - n_3)\right\},$$

$$\left\{x_1x_2\bar{x}_3\middle|\frac{a}{2}(-n_1 + n_2 + n_3, n_1 - n_2 + n_3 + 1, n_1 + n_2 - n_3)\right\},$$

$$\left\{x_1\bar{x}_3x_2\middle|\frac{a}{4}(-2n_1 + 2n_2 + 2n_3 + 1, 2n_1 - 2n_2 + 2n_3 + 1, 2n_1 + 2n_2 - 2n_3 - 1)\right\},$$

[20] H. Jones, *The Theory of Brillouin Zones and Electronic States in Crystals*, North Holland, Amsterdam, 1962, Chapter 4.

$$\left\{ x_1 x_3 \bar{x}_2 \middle| \frac{a}{4}(-2n_1 + 2n_2 + 2n_3 - 1, 2n_1 - 2n_2 + 2n_3 + 1, 2n_1 + 2n_2 - 2n_3 + 1) \right\},$$

where a is a lattice parameter and n_1, n_2 and n_3 take on positive and negative integer values including zero.

What is the Bravais lattice associated with this structure? Calculate the appropriate character table for this group, and describe the electronic states which can occur at the point Δ in the first Brillouin zone of the garnet structure.

C.10 COMPATIBILITY RELATIONS AND TIME-REVERSAL SYMMETRY

C.10a Compatibility relations

Consider the Δ-axis which terminates at the symmetry point Γ in the first Brillouin zone of the Ge structure. From the appropriate character tables we obtain the following table for group elements which are common to both Γ and Δ:

Table C.10.1 Characters of group elements common to both Γ and Δ (Ge).

	Q_1	Q_2	Q_{19}, Q_{20}	$\{Q'_3\|\mathbf{v}\}, \{Q'_4\|\mathbf{v}\}$	$\{Q'_{13}\|\mathbf{v}\}, \{Q'_{14}\|\mathbf{v}\}$
Δ_1	1	1	1	$e^{i\mathbf{k}_\Delta\cdot\mathbf{v}}$	$e^{i\mathbf{k}_\Delta\cdot\mathbf{v}}$
Δ'_1	1	1	-1	$-e^{i\mathbf{k}_\Delta\cdot\mathbf{v}}$	$e^{i\mathbf{k}_\Delta\cdot\mathbf{v}}$
Δ_2	1	1	-1	$e^{i\mathbf{k}_\Delta\cdot\mathbf{v}}$	$-e^{i\mathbf{k}_\Delta\cdot\mathbf{v}}$
Δ'_2	1	1	1	$-e^{i\mathbf{k}_\Delta\cdot\mathbf{v}}$	$-e^{i\mathbf{k}_\Delta\cdot\mathbf{v}}$
Δ_5	2	-2	0	0	0
Γ_1	1	1	1	1	1
Γ'_1	1	1	-1	-1	1
Γ_2	1	1	-1	1	-1
Γ'_2	1	1	1	-1	-1
Γ_{12}	2	2	0	2	0
Γ'_{12}	2	2	0	-2	0
Γ_{15}	3	-1	1	1	1
Γ'_{15}	3	-1	-1	-1	1
Γ_{25}	3	-1	-1	1	-1
Γ'_{25}	3	-1	1	-1	-1

As we approach the symmetry point Γ along the Δ-axis, $e^{i\mathbf{k}_\Delta\cdot\mathbf{v}} \to 1$, and we see, for example, that a wavefunction belonging to a Δ_1-type state then behaves under these common operations in the same way as a wavefunction belonging to a Γ_1-type state. In other words, as we approach Γ along the Δ-axis a Δ_1-type state joins smoothly on to a Γ_1-type state. Γ_1 is said to be compatible[21] with Δ_1 and, in a similar way, we see that Γ_1' is compatible with Δ_1', Γ_2 is compatible with Δ_2 and Γ_2' is compatible with Δ_2'.

Now consider a Γ_{12}-type state, and let $\psi_1^{\Gamma_{12}}(\mathbf{r})$ and $\psi_2^{\Gamma_{12}}(\mathbf{r})$ be Bloch functions at Γ belonging to this state. The matrices representing the elements common to both Γ and Δ in the irreducible representation denoted by Γ_{12} are given in table C.10.2.

Table C.10.2 Matrices of the irreducible representation Γ_{12}.

$$Q_1:M^1 = \begin{pmatrix}1 & 0\\0 & 1\end{pmatrix}; \quad Q_2:M^2 = \begin{pmatrix}1 & 0\\0 & 1\end{pmatrix}; \quad Q_{19}:M^{19} = \begin{pmatrix}1 & 0\\0 & -1\end{pmatrix};$$

$$Q_{20}:M^{20} = \begin{pmatrix}1 & 0\\0 & -1\end{pmatrix}; \quad \{Q_3'|\mathbf{v}\}:M^{3'} = \begin{pmatrix}1 & 0\\0 & 1\end{pmatrix}; \quad \{Q_4'|\mathbf{v}\}:M^{4'} = \begin{pmatrix}1 & 0\\0 & 1\end{pmatrix};$$

$$\{Q_{13}'|\mathbf{v}\}:M^{13'} = \begin{pmatrix}1 & 0\\0 & -1\end{pmatrix}; \quad \{Q_{14}'|\mathbf{v}\}:M^{14'} = \begin{pmatrix}1 & 0\\0 & -1\end{pmatrix}.$$

The Bloch functions $\psi_1^{\Gamma_{12}}(\mathbf{r})$ and $\psi_2^{\Gamma_{12}}(\mathbf{r})$ act as basis functions for these matrices, e.g.

$$\{Q_{13}'|\mathbf{v}\}\psi_1^{\Gamma_{12}}(\mathbf{r}) = \psi_1^{\Gamma_{12}}(\mathbf{r}), \qquad \{Q_{13}'|\mathbf{v}\}\psi_2^{\Gamma_{12}}(\mathbf{r}) = -\psi_2^{\Gamma_{12}}(\mathbf{r}). \quad \text{(C.10.1)}$$

As we approach Γ along the Δ-axis, $e^{i\mathbf{k}_\Delta\cdot\mathbf{v}} \to 1$, and from tables C.10.1 and C.10.2 we see that a wavefunction belonging to a Δ_1-type state then behaves under these common operations in the same way as $\psi_1^{\Gamma_{12}}(\mathbf{r})$, and a wavefunction belonging to a Δ_2-type state then behaves in the same way as $\psi_2^{\Gamma_{12}}(\mathbf{r})$. Δ_1- and Δ_2-type states therefore join smoothly on to a Γ_{12}-type state at Γ. Γ_{12} is compatible with Δ_1 and Δ_2.

In the above manner we obtain the following compatibility relations for states at Γ and states at Δ in the Ge structure:

Table C.10.3 Compatibility relations for states at Γ and at Δ(Ge).

Γ_1	Γ_2	Γ_1'	Γ_2'	Γ_{12}	Γ_{12}'	Γ_{15}	Γ_{15}'	Γ_{25}	Γ_{25}'
Δ_1	Δ_2	Δ_1'	Δ_2'	$\Delta_1\Delta_2$	$\Delta_1'\Delta_2'$	$\Delta_1\Delta_5$	$\Delta_1'\Delta_5$	$\Delta_2\Delta_5$	$\Delta_2'\Delta_5$

[21] L. P. Bouckaert, R. Smoluchowski, and E. Wigner, *Phys. Rev.*, **50**, 58 (1936).

At the other end of the Δ-axis is the symmetry point X and, in a similar way to that above, we obtain the following compatibility relations for states at X and states at Δ in the Ge structure:

Table C.10.4 Compatibility relations for states at X and at Δ (Ge).

X_1	X_2	X_3	X_4
$\Delta_1\Delta_2'$	$\Delta_1'\Delta_2$	Δ_5	Δ_5

The continuity of the energy and wavefunctions in the first Brillouin zone implies that only compatible states can exist together in the same energy band. From the above tables, for example, we see that a Γ_{15}-type state and a Δ_2-type state cannot exist together in the same energy band, and if at the point Γ in an energy band there is a Γ_1-type state, at the point Δ in the same band there must be a Δ_1-type state and at the point X in the same band there must be an X_1-type state.

There is a rule which determines the symmetry types along an axis which are compatible with a given symmetry type at the end of that axis. This rule states that, for every class, as the end-point is approached the sum of the characters of the compatible representations along the axis equals the character of the representation at the end of the axis. For example, for the Ge structure we obtain tables C.10.5 and C.10.6.

Table C.10.5 Example of the rule for determining compatibility relations (Ge).

	Q_1	Q_2	Q_{19}, Q_{20}	$\{Q_3'\|\mathbf{v}\}, \{Q_4'\|\mathbf{v}\}$	$\{Q_{13}'\|\mathbf{v}\}, \{Q_{14}'\|\mathbf{v}\}$
Δ_1	1	1	1	$e^{i\mathbf{k}_\Delta\cdot\mathbf{v}} \to 1$	$e^{i\mathbf{k}_\Delta\cdot\mathbf{v}} \to 1$
Δ_2	1	1	−1	$e^{i\mathbf{k}_\Delta\cdot\mathbf{v}} \to 1$	$-e^{i\mathbf{k}_\Delta\cdot\mathbf{v}} \to -1$
Γ_{12}	2	2	0	2	0

Table C.10.6 Example of the rule for determining compatibility relations (Ge).

	Q_1	Q_2	Q_{19}, Q_{20}	$\{Q_3'\|\mathbf{v}\}, \{Q_4'\|\mathbf{v}\}$	$\{Q_{13}'\|\mathbf{v}\}, \{Q_{14}'\|\mathbf{v}\}$
Δ_1	1	1	1	$e^{i\mathbf{k}_\Delta\cdot\mathbf{v}} \to e^{i\mathbf{k}_X\cdot\mathbf{v}}$	$e^{i\mathbf{k}_\Delta\cdot\mathbf{v}} \to e^{i\mathbf{k}_X\cdot\mathbf{v}}$
Δ_2'	1	1	1	$-e^{i\mathbf{k}_\Delta\cdot\mathbf{v}} \to -e^{i\mathbf{k}_X\cdot\mathbf{v}}$	$-e^{i\mathbf{k}_\Delta\cdot\mathbf{v}} \to -e^{i\mathbf{k}_X\cdot\mathbf{v}}$
X_1	2	2	2	0	0

Using this rule, we obtain the following compatibility relations for Γ, Δ and X states in the InSb structure.

Table C.10.7 Compatibility relations for states at Γ and at Δ (InSb).

Γ_1	Γ_2	Γ_{12}	Γ_{15}	Γ_{25}
Δ_1	Δ_2	$\Delta_1\Delta_2$	$\Delta_1\Delta_3\Delta_4$	$\Delta_2\Delta_3\Delta_4$

Table C.10.8 Compatibility relations for states at X and at Δ (InSb).

X_1	X_2	X_3	X_4	X_5
Δ_1	Δ_2	Δ_1	Δ_2	$\Delta_3\Delta_4$

C.10b Time-reversal symmetry

Consider the time-independent Schrödinger equation

$$H(\mathbf{r})\psi_{n\mathbf{k}}(\mathbf{r}) \equiv \left[\frac{-\hbar^2}{2m}\nabla^2 + V(\mathbf{r})\right]\psi_{n\mathbf{k}}(\mathbf{r}) = E_n(\mathbf{k})\psi_{n\mathbf{k}}(\mathbf{r}). \qquad \text{(C.10.2)}$$

Taking the complex conjugate of this equation,

$$H(\mathbf{r})\psi^*_{n\mathbf{k}}(\mathbf{r}) = E_n(\mathbf{k})\psi^*_{n\mathbf{k}}(\mathbf{r}), \qquad \text{(C.10.3)}$$

and hence $\psi^*_{n\mathbf{k}}(\mathbf{r})$ is also an eigenfunction of $H(\mathbf{r})$ and belongs to the same energy eigenvalue $E_n(\mathbf{k})$ as $\psi_{n\mathbf{k}}(\mathbf{r})$. Now

$$\psi^*_{n\mathbf{k}}(\mathbf{r}) = \mathrm{e}^{-\mathrm{i}\mathbf{k}\cdot\mathbf{r}}u^*_{n\mathbf{k}}(\mathbf{r}) \qquad \text{(C.10.4)}$$

and so $\psi^*_{n\mathbf{k}}(\mathbf{r})$ is a Bloch function with wave-vector -**k**. By considering all energy bands, we now see that in all cases, whether the point group contains the inversion or not,

$$E_n(\mathbf{k}) = E_n(-\mathbf{k}). \qquad \text{(C.10.5)}$$

This result is referred to as time-reversal symmetry, because if we started with the time-dependent Schrödinger equation

$$H(\mathbf{r})\psi_{n\mathbf{k}}(\mathbf{r},t) = \mathrm{i}\hbar\frac{\partial\psi_{n\mathbf{k}}(\mathbf{r},t)}{\partial t}, \qquad \text{(C.10.6)}$$

where

$$\psi_{n\mathbf{k}}(\mathbf{r},t) = \mathrm{e}^{-\mathrm{i}E_n(\mathbf{k})t/\hbar}\psi_{n\mathbf{k}}(\mathbf{r}), \qquad \text{(C.10.7)}$$

then $\psi^*_{n\mathbf{k}}(\mathbf{r},-t) = \mathrm{e}^{-\mathrm{i}E_n(\mathbf{k})t/\hbar}\psi^*_{n\mathbf{k}}(\mathbf{r})$ satisfies the same equation (C.10.6) as $\psi_{n\mathbf{k}}(\mathbf{r},t)$ and hence

$$H(\mathbf{r})\psi^*_{n\mathbf{k}}(\mathbf{r}) = E_n(\mathbf{k})\psi^*_{n\mathbf{k}}(\mathbf{r}). \qquad \text{(C.10.8)}$$

We now show that time-reversal symmetry can introduce additional degeneracies[22] to those already present and due to space-like symmetry properties.

Consider the InSb structure and let $\psi^3_{\mathbf{k}_\Delta}(\mathbf{r}) = e^{i\mathbf{k}_\Delta\cdot\mathbf{r}}u^3_{\mathbf{k}_\Delta}(\mathbf{r})$ be a Δ_3-type Bloch function belonging to the energy $E(\mathbf{k}_\Delta)$ and satisfying the equation

$$H(\mathbf{r})\psi^3_{\mathbf{k}_\Delta}(\mathbf{r}) = E(\mathbf{k}_\Delta)\psi^3_{\mathbf{k}_\Delta}(\mathbf{r}). \tag{C.10.9}$$

Application of the element $Q_{13}(-x_1, x_3, -x_2)$ of the space group T^2_d to this equation yields

$$H(\mathbf{r})\phi_{-\mathbf{k}_\Delta}(\mathbf{r}) = E(\mathbf{k}_\Delta)\phi_{-\mathbf{k}_\Delta}(\mathbf{r}), \tag{C.10.10}$$

where

$$\phi_{-\mathbf{k}_\Delta}(\mathbf{r}) = Q_{13}\psi^3_{\mathbf{k}_\Delta}(\mathbf{r}) = e^{-i\mathbf{k}_\Delta\cdot\mathbf{r}}u_{\mathbf{k}_\Delta}(-x_1, x_3, -x_2). \tag{C.10.11}$$

Taking the complex conjugate of equation (C.10.10), we now obtain the equation

$$H(\mathbf{r})\psi_{\mathbf{k}_\Delta}(\mathbf{r}) = E(\mathbf{k}_\Delta)\psi_{\mathbf{k}_\Delta}(\mathbf{r}), \tag{C.10.12}$$

where

$$\psi_{\mathbf{k}_\Delta}(\mathbf{r}) = \phi^*_{-\mathbf{k}_\Delta}(\mathbf{r}) = e^{i\mathbf{k}_\Delta\cdot\mathbf{r}}u^{3*}_{\mathbf{k}_\Delta}(-x_1, x_3, -x_2) = Q_{13}\psi^{3*}_{\mathbf{k}_\Delta}(\mathbf{r}). \tag{C.10.13}$$

$\psi_{\mathbf{k}_\Delta}(\mathbf{r})$ is a Bloch function with wave-vector $\mathbf{k}_\Delta$ and belongs to the energy eigenvalue $E(\mathbf{k}_\Delta)$. It must therefore belong to one of the irreducible representations of the group of the wave-vector at Δ. From the character table for the group of the wave-vector at Δ and from the multiplication table for the point group T_d, we see that

$$\begin{aligned}
Q_1\psi_{\mathbf{k}_\Delta}(\mathbf{r}) &= \psi_{\mathbf{k}_\Delta}(\mathbf{r}),\\
Q_2\psi_{\mathbf{k}_\Delta}(\mathbf{r}) &= Q_2Q_{13}\psi^{3*}_{\mathbf{k}_\Delta}(\mathbf{r}) = Q_{13}Q_2\psi^{3*}_{\mathbf{k}_\Delta}(\mathbf{r}) = -\psi_{\mathbf{k}_\Delta}(\mathbf{r}),\\
Q_{19}\psi_{\mathbf{k}_\Delta}(\mathbf{r}) &= Q_{19}Q_{13}\psi^{3*}_{\mathbf{k}_\Delta}(\mathbf{r}) = Q_{13}Q_{20}\psi^{3*}_{\mathbf{k}_\Delta}(\mathbf{r}) = -\psi_{\mathbf{k}_\Delta}(\mathbf{r}),\\
Q_{20}\psi_{\mathbf{k}_\Delta}(\mathbf{r}) &= Q_{20}Q_{13}\psi^{3*}_{\mathbf{k}_\Delta}(\mathbf{r}) = Q_{13}Q_{19}\psi^{3*}_{\mathbf{k}_\Delta}(\mathbf{r}) = \psi_{\mathbf{k}_\Delta}(\mathbf{r}).
\end{aligned} \tag{C.10.14}$$

Hence $\psi_{\mathbf{k}_\Delta}(\mathbf{r})$ is a Δ_4-type Bloch function belonging to the same energy eigenvalue $E(\mathbf{k}_\Delta)$ as the Δ_3-type Bloch function $\psi^3_{\mathbf{k}_\Delta}(\mathbf{r})$.

If there is a Δ_3-type state with energy $E(\mathbf{k}_\Delta)$ there must also be a Δ_4-type state with this energy, or Δ_3- and Δ_4-type states always stick together.

[22] For further discussion of time-reversal symmetry and description of the tests used to determine the additional degeneracies see C. Herring, *Phys. Rev.*, **52**, 361 (1937); R. J. Elliott, *Phys. Rev.*, **96**, 280 (1954); D. F. Johnston, *Proc. Roy. Soc. (Lond.)*, **A243**, 546 (1958); E. Wigner, *Group Theory and its Applications to the Quantum Mechanics of Atomic Spectra*, Academic Press, New York, 1959. See also Appendix for time reversal with spin.

If we plot the energy as a function of **k** along the Δ-axis then from the compatibility relations and time-reversal symmetry a possible energy band structure is that shown in fig. C.10.1, while an impossible one is that shown in fig. C.10.2.

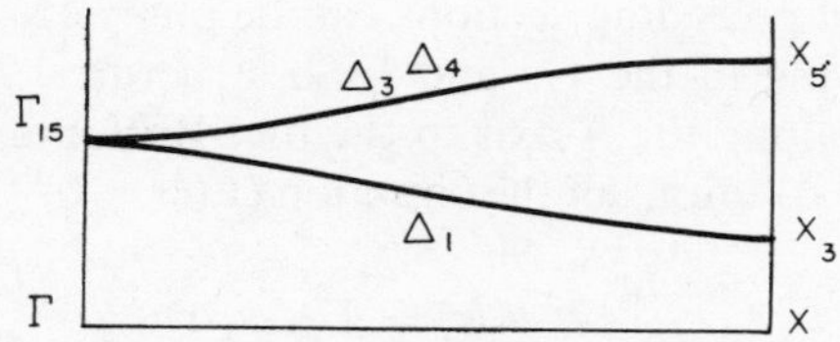

Fig. C.10.1. Possible band structure for InSb.

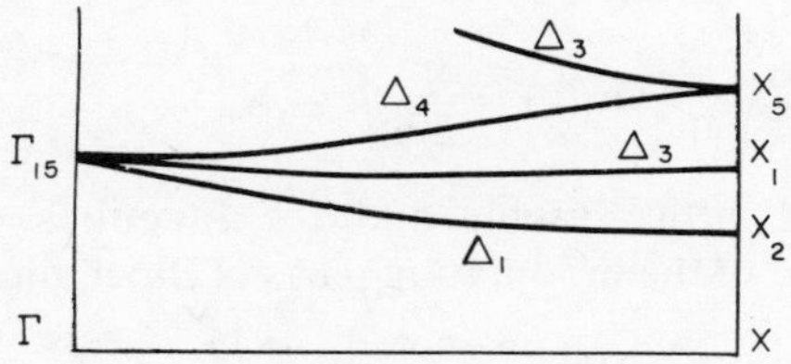

Fig. C.10.2. Impossible band structure for InSb.

Problem

(C.10.1) The character table for the point group of order three associated with the group of the wave-vector $\mathbf{k}_\Lambda = \dfrac{2\pi}{a}(\eta, \eta, \eta)(0 < \eta < \frac{1}{2})$ in the first Brillouin zone of the pyrite structure (space group T_h^6) is given below.

	$Q_1(x_1x_2x_3)$	$Q_2(x_2x_3x_1)$	$Q_3(x_3x_1x_2)$
Λ_1	1	1	1
Λ_2	1	ω	ω^2
Λ_3	1	ω^2	ω

Here ω stands for $e^{2\pi i/3}$.

Using the fact that the space group of the pyrite structure contains the inversion operator $J(-x_1, -x_2, -x_3)$, prove that along the Λ-axis in the first Brillouin zone Λ_2-type states and Λ_3-type states always stick together.

C.11 FREE-ELECTRON ENERGY BANDS[23]

If it is assumed that an electron moves in a potential which is constant (= 0) everywhere the energy band problem can be solved exactly. This is the free-electron approximation.

As an example of free-electron energy bands, and to illustrate the use to which the results of preceding sections can be put, we shall now consider free electrons moving in the Ge and InSb structures. Attention will be confined to states along the Δ-axis in the first Brillouin zone.

The appropriate solutions of the equation ($V(\mathbf{r}) = 0$)

$$-\frac{\hbar^2}{2m}\nabla^2\psi_{\mathbf{k}}(\mathbf{r}) = E(\mathbf{k})\psi_{\mathbf{k}}(\mathbf{r}) \qquad \text{(C.11.1)}$$

are of the form

$$\psi_{\mathbf{k}\mathbf{l}}(\mathbf{r}) = e^{i(\mathbf{k}-\mathbf{K}_\mathbf{l})\cdot\mathbf{r}} \qquad \text{(C.11.2)}$$

with corresponding eigenvalues

$$E_\mathbf{l}(\mathbf{k}) = \frac{\hbar^2}{2m}|\mathbf{k} - \mathbf{K}_\mathbf{l}|^2. \qquad \text{(C.11.3)}$$

$\mathbf{K}_\mathbf{l}$ is a reciprocal lattice vector and the functions $e^{-i\mathbf{K}_\mathbf{l}\cdot\mathbf{r}}$ have the periodicity required of the modulating parts of Bloch functions, i.e.

$$e^{-i\mathbf{K}_\mathbf{l}\cdot(\mathbf{r}+\mathbf{R}_s)} = e^{-i\mathbf{K}_\mathbf{l}\cdot\mathbf{r}}, \qquad \text{(C.11.4)}$$

where $\mathbf{R}_s$ is a translation lattice vector.

For the Ge and InSb structures

$$\mathbf{K}_\mathbf{l} = l_1\mathbf{b}_1 + l_2\mathbf{b}_2 + l_3\mathbf{b}_3 = \frac{2\pi}{a}(-l_1 + l_2 + l_3, l_1 - l_2 + l_3, l_1 + l_2 - l_3)$$

$$\equiv \frac{2\pi}{a}(L_1, L_2, L_3) = \frac{2\pi}{a}\mathbf{L}, \qquad \text{(C.11.5)}$$

and a wave-vector $\mathbf{k}$ in the first Brillouin zone may be written

$$\mathbf{k} = \frac{2\pi}{a}(\xi, \eta, \zeta). \qquad \text{(C.11.6)}$$

The unnormalized wavefunctions and the energies are then given by

$$\psi_{\mathbf{k}\mathbf{L}}(\mathbf{r}) = \exp\left\{\frac{2\pi i}{a}[(\xi - L_1)x_1 + (\eta - L_2)x_2 + (\zeta - L_3)x_3]\right\}, \qquad \text{(C.11.7)}$$

[23] F. Herman, *Rev. Mod. Phys.*, **30**, 102, (1958); H. Jones, *The Theory of Brillouin Zones and Electronic States in Crystals*, North Holland, Amsterdam, 1962, Sections 30, 33, 34 and 35; J. C. Slater, *Quantum Theory of Molecules and Solids, Vol. 2*, McGraw-Hill, New York, 1965, Section 10.2 and Appendix 7.

$$W_{\mathbf{L}}(\mathbf{k}) \equiv \frac{2ma^2}{h^2} E_{\mathbf{L}}(\mathbf{k}) = (\xi - L_1)^2 + (\eta - L_2)^2 + (\zeta - L_3)^2. \quad \text{(C.11.8)}$$

Note that only those values of L_j ($j = 1, 2, 3$) are allowed which make the $l_j (j = 1, 2, 3)$ integers, e.g. $\mathbf{L} = (1,0,0)$ is not allowed since this corresponds to $l_2 = l_3 = \frac{1}{2}$.

For given $\mathbf{L} = (L_1, L_2, L_3)$, the energies and wavefunctions at Γ, Δ and X in the first Brillouin zone are as follows:

$$\mathbf{k}_\Gamma = \frac{2\pi}{a}(0, 0, 0)$$

$$W_{\mathbf{L}}(\Gamma) = L_1^2 + L_2^2 + L_3^2 \quad \text{(C.11.9)}$$

$$\psi_{\mathbf{L}\Gamma}(\mathbf{r}) = \exp\left[-\frac{2\pi i}{a}(L_1 x_1 + L_2 x_2 + L_3 x_3)\right] \quad \text{(C.11.10)}$$

$$\mathbf{k}_\Delta = \frac{2\pi}{a}(\xi, 0, 0), \qquad (0 < \xi < 1)$$

$$W_{\mathbf{L}}(\Delta) = (\xi - L_1)^2 + L_2^2 + L_3^2 \quad \text{(C.11.11)}$$

$$\psi_{\mathbf{L}\Delta}(\mathbf{r}) = \exp\left\{\frac{2\pi i}{a}\left[(\xi - L_1)x_1 - L_2 x_2 - L_3 x_3\right]\right\} \quad \text{(C.11.12)}$$

$$\mathbf{k}_X = \frac{2\pi}{a}(1, 0, 0)$$

$$W_{\mathbf{L}}(X) = (1 - L_1)^2 + L_2^2 + L_3^2 \quad \text{(C.11.13)}$$

$$\psi_{\mathbf{L}X}(\mathbf{r}) = \exp\left\{\frac{2\pi i}{a}[(1 - L_1)x_1 - L_2 x_2 - L_3 x_3]\right\}. \quad \text{(C.11.14)}$$

The three lowest energies at Γ, with the associated **L**-vectors, are given in table C.11.1.

Table C.11.1 Energies of free electrons at Γ.

Energy	Associated **L**-vectors
$W(\Gamma) = 0$	$(0, 0, 0)$
$W(\Gamma) = 3$	$(1, 1, 1)$, $(-1, 1, 1)$, $(1, -1, 1)$, $(1, 1, -1)$, $(-1, 1, -1)$, $(-1, -1, 1)$, $(1, -1, -1)$, $(-1, -1, -1)$
$W(\Gamma) = 4$	$(2, 0, 0)$, $(0, 2, 0)$, $(0, 0, 2)$, $(-2, 0, 0)$, $(0, -2, 0)$, $(0, 0, -2)$

Thus $W(\Gamma) = 0$ is a nondegenerate level, $W(\Gamma) = 3$ is an eight-fold degenerate level and $W(\Gamma) = 4$ is a six-fold degenerate level.

The three lowest energies at X, with the associated **L**-vectors, are given in table C.11.2.

Table C.11.2 Energies of free electrons at X.

Energy	Associated **L**-vectors			
$W(X) = 1$	$(0, 0, 0)$,	$(2, 0, 0)$		
$W(X) = 2$	$(1, 1, 1)$,	$(1, -1, 1)$,	$(1, 1, -1)$,	
	$(1, -1, -1)$			
$W(X) = 5$	$(0, 2, 0)$,	$(0, 0, 2)$,	$(0, -2, 0)$,	$(0, 0, -2)$
	$(2, 0, 2)$,	$(2, 0, -2)$,	$(2, 2, 0)$,	$(2, -2, 0)$

Thus $W(X) = 1$ is a two-fold degenerate level, $W(X) = 2$ is a four-fold degenerate level and $W(X) = 5$ is an eight-fold degenerate level.

We next note that the **L**-vector $(0, 0, 0)$ is associated with $W(\Gamma) = 0$ and with $W(X) = 1$. This vector, when put into (C.11.11), gives the energy band $W(\Delta) = \xi^2$ joining $W(\Gamma) = 0$ to $W(X) = 1$. Similarly, the **L**-vectors $(1, 1, 1), (1, -1, 1), (1, 1, -1)$ and $(1, -1, -1)$ are associated with $W(\Gamma) = 3$ and with $W(X) = 2$. These vectors, when put into (C.11.11), give the energy band $W(\Delta) = (\xi - 1)^2 + 2$ joining $W(\Gamma) = 3$ to $W(X) = 2$. The Δ-states in this band are therefore four-fold degenerate. In this way table C.11.3 may be constructed.

Table C.11.3

Energies	Common **L**-vectors	Energy Bands
$W(\Gamma) = 0$, $W(X) = 1$	$(0, 0, 0)$	ξ^2
$W(\Gamma) = 3$, $W(X) = 2$	$(1, 1, 1)$, $(1, -1, 1)$, $(1, 1, -1)$, $(1, -1, -1)$	$(\xi - 1)^2 + 2$
$W(\Gamma) = 4$, $W(X) = 1$	$(2, 0, 0)$	$(\xi - 2)^2$
$W(\Gamma) = 4$, $W(X) = 5$	$(0, 2, 0)$, $(0, 0, 2)$, $(0, -2, 0)$, $(0, 0, -2)$	$\xi^2 + 4$

Using table C.11.3, we can now draw a portion of the energy band diagram. This is shown in fig. C.11.1.

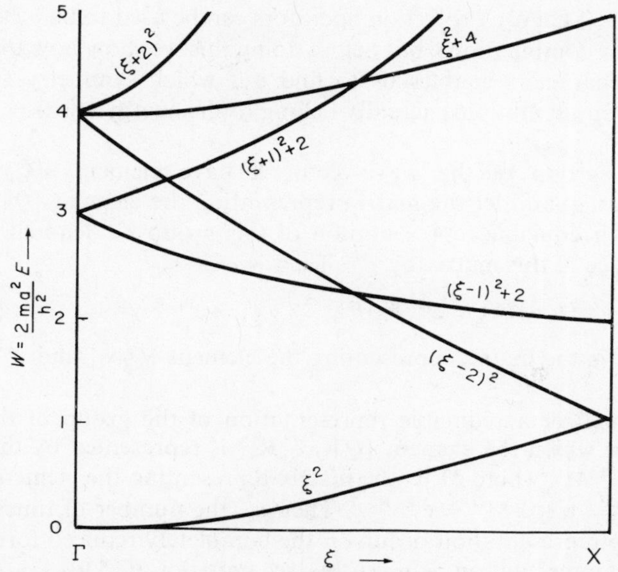

Fig. C.11.1. Free-electron energy bands—Ge and InSb structures.

Now consider, for example, the energy $W(\Delta) = (\xi - 1)^2 + 2$ $(0 < \xi < 1)$, which from table C.11.3 and equation (C.11.12) is four-fold degenerate with Bloch functions

$$\psi_1 = \exp\left\{\frac{2\pi i}{a}[(\xi - 1)x_1 - x_2 - x_3]\right\},$$

$$\psi_2 = \exp\left\{\frac{2\pi i}{a}[(\xi - 1)x_1 + x_2 - x_3]\right\},$$

$$\psi_3 = \exp\left\{\frac{2\pi i}{a}[(\xi - 1)x_1 - x_2 + x_3]\right\},$$

$$\psi_4 = \exp\left\{\frac{2\pi i}{a}[(\xi - 1)x_1 + x_2 + x_3]\right\}. \tag{C.11.15}$$

Any linear combination of these four functions is again a Bloch function belonging to the energy $W(\Delta) = (\xi - 1)^2 + 2$, and the problem is to find those linear combinations which act as basis functions for one or more of the irreducible representations of the group of the wave-vector at Δ. Such linear combinations[24] are known as symmetrized combinations of

[24] L. Mariot, *Group Theory and Solid State Physics*, transl. A. Nussbaum, Prentice-Hall, New Jersey, 1962, p. 94; H. C. Schlosser, *J. Phys. Chem. Solids*, **23**, 963 (1962).

plane waves (SCPW). Projection operators can be used to find the appropriate linear combinations, but before doing this we show how the theory of group characters enables us to find out which symmetry types (irreducible representations) actually belong to the energy $W(\Delta) = (\xi - 1)^2 + 2$.

Let the group of the wave-vector $\mathbf{k}$ have elements $\{Q_l|\mathbf{v}_l + \mathbf{R}_s\}$ $(l = 1, 2, \ldots, g)$ and let the matrix representing the element $\{Q_l|\mathbf{v}_l + \mathbf{R}_s\}$ in the αth irreducible representation of this group be denoted by $M^{\alpha ls}$, and the trace of this matrix by $\chi^{\alpha ls}$. Then

$$M^{\alpha ls} = e^{i\mathbf{k}\cdot\mathbf{R}_s} M^{\alpha l}, \qquad \chi^{\alpha ls} = e^{i\mathbf{k}\cdot\mathbf{R}_s}\chi^{\alpha l}, \tag{C.11.16}$$

where $M^{\alpha l}$ is the matrix representing the element $\{Q_l|\mathbf{v}_l\}$ and $\chi^{\alpha l}$ = trace $M^{\alpha l}$.

Now consider a reducible representation of the group of the wave-vector $\mathbf{k}$ in which the element $\{Q_l|\mathbf{v}_l + \mathbf{R}_s\}$ is represented by the matrix $M^{ls} = e^{i\mathbf{k}\cdot\mathbf{R}_s} M^l$, where M^l is the matrix representing the element $\{Q_l|\mathbf{v}_l\}$, and let χ^{ls} = trace $M^{ls} = e^{i\mathbf{k}\cdot\mathbf{R}_s}\chi^l$. Then c_α, the number of times the αth irreducible representation occurs in the completely reduced form of this reducible representation, is given by [see equation (C.5.9)]

$$c_\alpha = \frac{1}{gN}\sum_{l,s} \chi^{\alpha ls *}\chi^{ls} = \frac{1}{g}\sum_{l} \chi^{\alpha l *}\chi^{l}. \tag{C.11.17}$$

The four functions in (C.11.15) act as basis functions for such a reducible representation of the group of the wave-vector at Δ, and hence (C.11.17) can be used to find out which symmetry types belong to the energy $W(\Delta) = (\xi - 1)^2 + 2$.

Consider first the InSb structure. Writing the four functions in (C.11.15) as a column vector and operating on this column vector with the elements $Q_t (t = 1, 2, 19, 20)$ of the point group T_d gives four matrices, the traces of which are

$$\chi^1 = 4, \qquad \chi^2 = 0, \qquad \chi^{19} = 2, \qquad \chi^{20} = 2. \tag{C.11.18}$$

Using these traces and the appropriate character table (table C.9.3), (C.11.17) yields

$$c_{\Delta_1} = \tfrac{1}{4}(4 + 2 + 2) = 2, \qquad c_{\Delta_2} = \tfrac{1}{4}(4 - 2 - 2) = 0,$$

$$c_{\Delta_3} = \tfrac{1}{4}(4 + 2 - 2) = 1, \qquad c_{\Delta_4} = \tfrac{1}{4}(4 - 2 + 2) = 1. \tag{C.11.19}$$

Hence belonging to the free-electron energy $W(\Delta) = (\xi - 1)^2 + 2$ there are two Δ_1-type states, a Δ_3-type state and a Δ_4-type state. Here we have an example of accidental degeneracy.

Next consider the Ge structure. Writing the four functions in (C.11.15) as a column vector and operating on this column vector with the elements $Q_l (l = 1, 2, 19, 20)$ and $\{Q'_t|\mathbf{v}\}$ $(t = 3, 4, 13, 14)$ of the space group O_h^7 gives eight matrices, the traces of which are

$$\chi^1 = 4, \qquad \chi^2 = 0, \qquad \chi^{19} = 2, \qquad \chi^{20} = 2,$$
$$\chi^{3'} = 0, \qquad \chi^{4'} = 0, \qquad \chi^{13'} = 0, \qquad \chi^{14'} = 0. \qquad \text{(C.11.20)}$$

Using these traces and the appropriate character table (table C.9.8), (C.11.17) yields

$$c_{\Delta_1} = \tfrac{1}{8}(4 + 2 + 2) = 1, \qquad c_{\Delta_2'} = \tfrac{1}{8}(4 + 2 + 2) = 1,$$
$$c_{\Delta_2} = \tfrac{1}{8}(4 - 2 - 2) = 0, \qquad c_{\Delta_1'} = \tfrac{1}{8}(4 - 2 - 2) = 0,$$
$$c_{\Delta_5} = \tfrac{1}{8}(4 + 2 + 2) = 1. \qquad \text{(C.11.21)}$$

Hence belonging to the free-electron energy $W(\Delta) = (\xi - 1)^2 + 2$ there are Δ_1-, Δ_2'- and Δ_5-type states. Again there is accidental degeneracy.

All the free-electron energy levels can be examined in this way, and figs. C.11.2 and C.11.3 show the results of such calculations for the states along the Δ-axis in the first Brillouin zone.

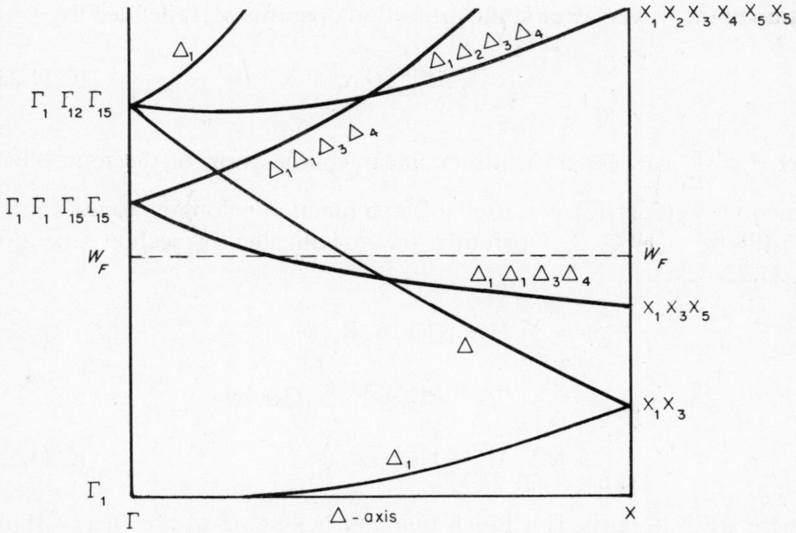

Fig. C.11.2. Free-electron energy bands for InSb.

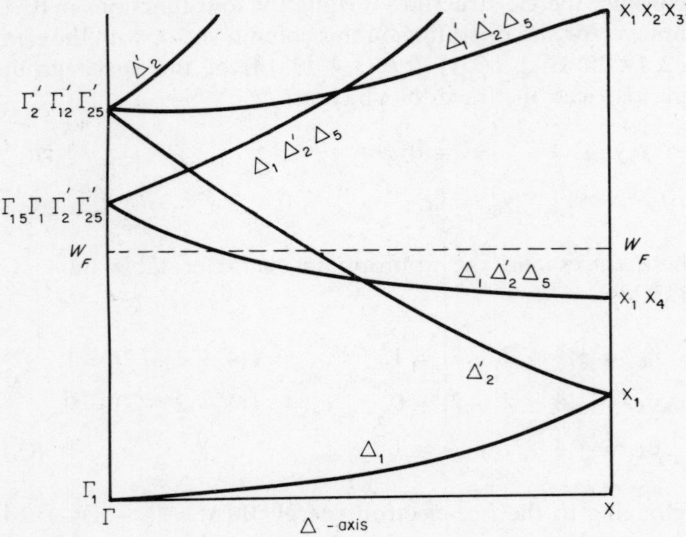

Fig. C.11.3. Free-electron energy bands for Ge.

Now we return to the wavefunctions. If $M^{\alpha ls}$ is the matrix representing the element $\{Q_l|\mathbf{v}_l + \mathbf{R}_\mathbf{s}\}$ in the αth irreducible representation of the group of the wave-vector **k**, the projection operator ρ^α_{jp} is defined by

$$\rho^\alpha_{jp} = \sum_{l,\mathbf{s}} M^{\alpha ls*}_{jp}\{Q_l|\mathbf{v}_l + \mathbf{R}_\mathbf{s}\}. \tag{C.11.22}$$

Let $\phi = \sum_{t=1}^{4} b_t\psi_t$ be an arbitrary linear combination of the four Bloch functions in (C.11.15). ϕ is itself a Bloch function belonging to the energy $W(\Delta) = (\xi - 1)^2 + 2$. Operating on ϕ with the projection operator (C.11.22) gives

$$\begin{aligned}\rho^\alpha_{jp}\phi &= \sum_{l,\mathbf{s}} M^{\alpha ls*}_{jp}\{Q_l|\mathbf{v}_l + \mathbf{R}_\mathbf{s}\}\phi \\ &= \sum_{l,\mathbf{s}} \mathrm{e}^{-\mathrm{i}\mathbf{k}_\Delta\cdot\mathbf{R}_\mathbf{s}} M^{\alpha l*}_{jp}\, \mathrm{e}^{\mathrm{i}\mathbf{k}_\Delta\cdot\mathbf{R}_\mathbf{s}}\{Q_l|\mathbf{v}_l\}\phi \\ &= N\sum_{l} M^{\alpha l*}_{jp}\{Q_l|\mathbf{v}_l\}\phi = \psi^\alpha_j,\end{aligned} \tag{C.11.23}$$

where ψ^α_j, if it exists, is a Bloch function belonging to the energy $W(\Delta) = (\xi - 1)^2 + 2$ and transforming according to the jth row of the αth irreducible representation of the group of the wave-vector at Δ, i.e. it is one of the required symmetrized combinations of plane waves.

For the InSb structure the appropriate space group elements are $Q_l (l = 1, 2, 19, 20)$. These have the following effects on $\phi = \Sigma b_t \psi_t$:

$$
\begin{aligned}
Q_1\phi &= \phi, \\
Q_2\phi &= b_1\psi_4 + b_2\psi_3 + b_3\psi_2 + b_4\psi_1, \\
Q_{19}\phi &= b_1\psi_1 + b_2\psi_3 + b_3\psi_2 + b_4\psi_4, \\
Q_{20}\phi &= b_1\psi_4 + b_2\psi_2 + b_3\psi_3 + b_4\psi_1.
\end{aligned} \tag{C.11.24}
$$

From table C.9.3:

$$
\begin{aligned}
M^{\Delta_1 l} &= 1 \qquad \text{for} \quad l = 1, 2, 19, 20 \\
M^{\Delta_2 l} &= \begin{cases} 1 & \text{for} \quad l = 1, 2 \\ -1 & \text{for} \quad l = 19, 20 \end{cases} \\
M^{\Delta_3 l} &= \begin{cases} 1 & \text{for} \quad l = 1, 19 \\ -1 & \text{for} \quad l = 2, 20 \end{cases} \\
M^{\Delta_4 l} &= \begin{cases} 1 & \text{for} \quad l = 1, 20 \\ -1 & \text{for} \quad l = 2, 19 \end{cases}
\end{aligned} \tag{C.11.25}
$$

Using these results in (C.11.23), we now find that

$$
\begin{aligned}
\rho_{11}^{\Delta_1}\phi &= 2N[(b_1 + b_4)(\psi_1 + \psi_4) + (b_2 + b_3)(\psi_2 + \psi_3)], \\
\rho_{11}^{\Delta_2}\phi &= 0, \\
\rho_{11}^{\Delta_3}\phi &= 2N(b_1 - b_4)(\psi_1 - \psi_4), \\
\rho_{11}^{\Delta_4}\phi &= 2N(b_2 - b_3)(\psi_2 - \psi_3).
\end{aligned} \tag{C.11.26}
$$

Hence, apart from constant factors, the symmetrized Bloch functions belonging to the free-electron energy $W(\Delta) = (\xi - 1)^2 + 2$ in InSb are

$$
\begin{aligned}
\Delta_1 : \psi_1 + \psi_4 &= \mathrm{e}^{2\pi i(\xi - 1)x_1/a} \cos 2\pi(x_2 + x_3)/a, \\
\Delta_1 : \psi_2 + \psi_3 &= \mathrm{e}^{2\pi i(\xi - 1)x_1/a} \cos 2\pi(x_2 - x_3)/a, \\
\Delta_3 : \psi_1 - \psi_4 &= \mathrm{e}^{2\pi i(\xi - 1)x_1/a} \sin 2\pi(x_2 + x_3)/a, \\
\Delta_4 : \psi_2 - \psi_3 &= \mathrm{e}^{2\pi i(\xi - 1)x_1/a} \sin 2\pi(x_2 - x_3)/a.
\end{aligned} \tag{C.11.27}
$$

For the Ge structure the appropriate elements of the space group O_h^7 are $Q_l (l = 1, 2, 19, 20)$ and $\{Q'_t|\mathbf{v}\} (t = 3, 4, 13, 14)$. These additional elements have the following effects on $\phi = \Sigma b_t \psi_t$:

$$\{Q'_3|\mathbf{v}\}\phi = \mathrm{i}\, e^{\mathrm{i}\mathbf{k}_\Delta \cdot \mathbf{v}}(b_1\psi_2 - b_2\psi_1 - b_3\psi_4 + b_4\psi_3),$$

$$\{Q'_4|\mathbf{v}\}\phi = \mathrm{i}\, e^{\mathrm{i}\mathbf{k}_\Delta \cdot \mathbf{v}}(b_1\psi_3 - b_2\psi_4 - b_3\psi_1 + b_4\psi_2),$$

$$\{Q'_{13}|\mathbf{v}\}\phi = \mathrm{i}\, e^{\mathrm{i}\mathbf{k}_\Delta \cdot \mathbf{v}}(b_1\psi_3 - b_2\psi_1 - b_3\psi_4 + b_4\psi_2),$$

$$\{Q'_{14}|\mathbf{v}\}\phi = \mathrm{i}\, e^{\mathrm{i}\mathbf{k}_\Delta \cdot \mathbf{v}}(b_1\psi_2 - b_2\psi_4 - b_3\psi_1 + b_4\psi_3). \qquad \text{(C.12.28)}$$

Using (C.11.24), (C.11.28) and known matrix elements[25], (C.11.23) now yields

$$\rho_{11}^{\Delta_1}\phi = 2N(b_1 + b_4 - \mathrm{i}b_2 - \mathrm{i}b_3)(\psi_1 + \psi_4 + \mathrm{i}\psi_2 + \mathrm{i}\psi_3),$$

$$\rho_{11}^{\Delta'_2}\phi = 2N(b_1 + b_4 + \mathrm{i}b_2 + \mathrm{i}b_3)(\psi_1 + \psi_4 - \mathrm{i}\psi_2 - \mathrm{i}\psi_3),$$

$$\rho_{11}^{\Delta_2}\phi = 0, \qquad \rho_{11}^{\Delta'_1}\phi = 0,$$

$$\rho_{11}^{\Delta_5}\phi = N(b_1 - b_4 + \mathrm{i}b_2 - \mathrm{i}b_3)(\psi_1 - \psi_4 - \mathrm{i}\psi_2 + \mathrm{i}\psi_3),$$

$$\rho_{21}^{\Delta_5}\phi = N(b_1 - b_4 + \mathrm{i}b_2 - \mathrm{i}b_3)(\psi_1 - \psi_4 + \mathrm{i}\psi_2 - \mathrm{i}\psi_3). \qquad \text{(C.11.29)}$$

Hence, apart from constant factors, the symmetrized Bloch functions belonging to the free-electron energy $W(\Delta) = (\xi - 1)^2 + 2$ in Ge are

$$\Delta_1 : \psi_1 + \psi_4 + \mathrm{i}\psi_2 + \mathrm{i}\psi_3 = e^{2\pi \mathrm{i}(\xi - 1)x_1/a}[\cos 2\pi(x_2 + x_3)/a + \mathrm{i}\cos 2\pi(x_2 - x_3)/a],$$

$$\Delta'_2 : \psi_1 + \psi_4 - \mathrm{i}\psi_2 - \mathrm{i}\psi_3 = e^{2\pi \mathrm{i}(\xi - 1)x_1/a}[\cos 2\pi(x_2 + x_3)/a - \mathrm{i}\cos 2\pi(x_2 - x_3)/a],$$

$$\Delta_5 : \psi_1 - \psi_4 - \mathrm{i}\psi_2 + \mathrm{i}\psi_3 = e^{2\pi \mathrm{i}(\xi - 1)x_1/a}[\sin 2\pi(x_2 + x_3)/a + \mathrm{i}\sin 2\pi(x_2 - x_3)/a]$$

$$\psi_1 - \psi_4 + \mathrm{i}\psi_2 - \mathrm{i}\psi_3 = e^{2\pi \mathrm{i}(\xi - 1)x_1/a}[\sin 2\pi(x_2 + x_3)/a - \mathrm{i}\sin 2\pi(x_2 - x_3)/a]$$

$$\text{(C.11.30)}$$

In this way the appropriate symmetrized combinations of plane waves can be obtained for all the free-electron energy levels.

The Fermi Level

The Ge atom has four outer electrons which correspond, in the solid, to the valence and conduction electrons. Each unit cell of the Ge structure contains two atoms, and hence in a crystal of N unit cells there is a total of $8N$ valence and conduction electrons. The In atom has three outer

[25] J. C. Slater, *Quantum Theory of Molecules and Solids, Vol. 2*, McGraw-Hill, New York, 1965, p. 376.

electrons and the Sb atom has five (a III–V compound). Each unit cell of the InSb structure contains one In atom and one Sb atom, and once again the crystal contains $8N$ valence and conduction electrons.

We expect energy levels to be filled up, starting at the bottom, until all the electrons are accounted for. The highest energy level occupied in this way at the absolute zero of temperature is the Fermi level, and it is of interest to calculate the position of the Fermi level in the free-electron approximation.

Consider a free-electron energy level which is d_n-fold degenerate with wavefunctions $\psi_1 = \mathrm{e}^{\mathrm{i}\boldsymbol{\gamma}_1 \cdot \mathbf{r}}$, $\psi_2 = \mathrm{e}^{\mathrm{i}\boldsymbol{\gamma}_2 \cdot \mathbf{r}}, \ldots, \psi_{d_n} = \mathrm{e}^{\mathrm{i}\boldsymbol{\gamma}_{d_n} \cdot \mathbf{r}}$, and energy (in non-dimensional units)

$$W = \frac{a^2}{4\pi^2}\gamma_1^2 = \frac{a^2}{4\pi^2}\gamma_2^2 = \cdots = \frac{a^2}{4\pi^2}\gamma_{d_n}^2, \qquad \text{(C.11.31)}$$

where

$$\boldsymbol{\gamma}_j = \mathbf{k} - \mathbf{K}_{\mathbf{l}_j} \qquad (j = 1, 2, \ldots, d_n), \qquad \text{(C.11.32)}$$

$\mathbf{K}_{\mathbf{l}_j}$ being a reciprocal lattice vector. When filling up the energy levels with electrons, this level can accommodate $2d_n$ electrons, the factor 2 appearing because of the electron spin. Suppose we write the Fermi level as $W_F = a^2\gamma_F^2/4\pi^2$, γ_F being the radius of the Fermi sphere in reciprocal lattice space. The total number of electrons ($=8N$) filling the free-electron energy levels up to and including the energy W_F is then equal to twice the number of vectors $\boldsymbol{\gamma}$ which satisfy the condition

$$\gamma^2 \leqslant \gamma_F^2, \qquad \text{(C.11.33)}$$

and this is just twice the number of vectors $\boldsymbol{\gamma}$ which are contained in the Fermi sphere of volume $4\pi\gamma_F^3/3$. Now in the first Brillouin zone of volume $8\pi^3/\Omega$ there are N $\mathbf{k}$-vectors which are uniformly distributed with density $N\Omega/8\pi^3$ per unit volume. The vectors $\boldsymbol{\gamma}$ are also uniformly distributed with this density, and hence

$$8N = 2 \cdot \frac{N\Omega}{8\pi^3} \cdot \frac{4}{3}\pi\gamma_F^3,$$

$$\gamma_F^3 = 24\pi^2/\Omega. \qquad \text{(C.11.34)}$$

For the Ge and InSb structures $\Omega = a^3/4$, and hence

$$W_F = a^2\gamma_F^2/4\pi^2 = (12/\pi)^{2/3} = 2{\cdot}44. \qquad \text{(C.11.35)}$$

This energy is shown as a dashed line in figs. C.11.2 and C.11.3.

Problem

(C.11.1) The character table and symmetry transformations of the group of the wave-vector $\mathbf{k}_\Sigma = (2\pi/a)(\eta, \eta, 0)(0 < \eta < \frac{3}{4})$ in the first Brillouin zone of the diamond structure are as follows:

	$\{x_1x_2x_3\|0\}$	$\{x_2x_1x_3\|0\}$	$\{x_1x_2\bar{x}_3\|\mathbf{v}\}$	$\{x_2x_1\bar{x}_3\|\mathbf{v}\}$
Σ_1	1	1	α	α
Σ_2	1	-1	$-\alpha$	α
Σ_3	1	1	$-\alpha$	$-\alpha$
Σ_4	1	-1	α	$-\alpha$

Here a is a lattice parameter, the nonprimitive translation $\mathbf{v} \equiv a(\frac{1}{4}, \frac{1}{4}, \frac{1}{4})$ and $\alpha \equiv e^{i\pi\eta}$.

Defining $W_\Sigma \equiv (2ma^2/h^2)E_\Sigma$, where E_Σ is the electron energy at the point Σ in the first Brillouin zone, m is the electron mass and h is Planck's constant, some typical energy bands along the Σ-axis for the case of free electrons are as follows:

$$W_\Sigma = (\eta + 1)^2 + (\eta - 1)^2 + 1,$$

$$W_\Sigma = 2(\eta - 1)^2 + 1,$$

$$W_\Sigma = 2\eta^2.$$

By using projection operators, or otherwise, find the symmetrized free-electron wavefunctions belonging to these energy bands and indicate on an energy band diagram the symmetry types in each band and the degeneracy of each band. Do not discuss the symmetry points at the ends of the Σ-axis.

C.12 SPIN–ORBIT COUPLING AND DOUBLE GROUPS

In the discussion of the electronic states in crystals spin–orbit interaction and other relativistic effects have so far been neglected. This is justified to a first approximation. In many cases, however, these effects become important and the energy band structure is modified in ways that cannot be ignored[26].

Spin–orbit interaction is an essentially relativistic effect and the Hamiltonian generally used is obtained as an approximation from Dirac's equation[27]. It is as follows:

$$H(\mathbf{s}, \mathbf{r}) \equiv H(\mathbf{r}) - \frac{i\hbar}{2m^2c^2}\mathbf{s} \cdot [\nabla V(\mathbf{r}) \times \nabla] - \frac{\hbar^4\nabla^4}{8m^3c^2} - \frac{\hbar^2}{4m^2c^2}\nabla V(\mathbf{r}) \cdot \nabla, \tag{C.12.1}$$

[26] R. J. Elliott, *Phys. Rev.*, **96**, 266, 280 (1954).

[27] J. C. Slater, *Quantum Theory of Molecules and Solids, Vol. 2*, McGraw-Hill, New York, 1965, Appendix 9.

where

$$H(\mathbf{r}) \equiv -\frac{\hbar^2}{2m}\nabla^2 + V(\mathbf{r}). \tag{C.12.2}$$

$V(\mathbf{r})$ is the periodic potential, c is the velocity of light and $\mathbf{s} \equiv (s_1, s_2, s_3)$ is the spin angular momentum operator. The second term on the right-hand side of (C.12.1) is the spin–orbit interaction and the third term is a relativistic term which arises from the change of mass with velocity.

If we let

$$\mathscr{H}(\mathbf{r}) \equiv H(\mathbf{r}) - \frac{\hbar^4\nabla^4}{8m^3c^2} - \frac{\hbar^2}{4m^2c^2}\nabla V(\mathbf{r}) \cdot \nabla, \tag{C.12.3}$$

and define a vector operator $\mathbf{P}$ with components

$$P_j = \frac{\hbar}{2m^2c^2}[\nabla V(\mathbf{r}) \times \nabla]_j \qquad (j = 1, 2, 3), \tag{C.12.4}$$

then (C.12.1) can be written

$$H(\mathbf{s}, \mathbf{r}) \equiv \mathscr{H}(\mathbf{r}) - \mathrm{i} \sum_{j=1}^{3} s_j P_j. \tag{C.12.5}$$

If σ is a spin coordinate which can take on the values ± 1, the corresponding values of the x_3-component of the spin angular momentum being $\pm\hbar/2$, the energy eigenvalue equation can be written

$$H(\mathbf{s}, \mathbf{r})\Psi(\sigma, \mathbf{r}) = E\Psi(\sigma, \mathbf{r}) \tag{C.12.6}$$

with

$$\Psi(\sigma, \mathbf{r}) \equiv \psi^{(1)}(\mathbf{r})v_1(\sigma) + \psi^{(2)}(\mathbf{r})v_2(\sigma). \tag{C.12.7}$$

$v_1(\sigma)$ and $v_2(\sigma)$ are Pauli spin functions defined by[28]

$$v_1(+1) = 1, v_1(-1) = 0, v_2(+1) = 0, v_2(-1) = 1, \tag{C.12.8}$$

and satisfying

$$\begin{aligned}
&s_1 v_1(\sigma) = \tfrac{1}{2}\hbar v_2(\sigma), && s_1 v_2(\sigma) = \tfrac{1}{2}\hbar v_1(\sigma),\\
&s_2 v_1(\sigma) = \frac{\mathrm{i}}{2}\hbar v_2(\sigma), && s_2 v_2(\sigma) = -\frac{\mathrm{i}}{2}\hbar v_1(\sigma),\\
&s_3 v_1(\sigma) = \tfrac{1}{2}\hbar v_1(\sigma), && s_3 v_2(\sigma) = -\tfrac{1}{2}\hbar v_2(\sigma),\\
&\mathbf{s}^2 v_1(\sigma) = (s_1^2 + s_2^2 + s_3^2)v_1(\sigma) = \tfrac{3}{4}\hbar^2 v_1(\sigma), &&\\
&\mathbf{s}^2 v_2(\sigma) = \tfrac{3}{4}\hbar^2 v_2(\sigma). &&
\end{aligned} \tag{C.12.9}$$

[28] V. Heine, *Group Theory in Quantum Mechanics*, Pergamon Press, London, 1960, p. 78.

The normalization of the wavefunction (C.12.7) is then

$$\sum_{\sigma=\pm 1} \int \Psi^*(\sigma, \mathbf{r})\Psi(\sigma, \mathbf{r})\, d\mathbf{r} = 1$$

$$= \int [\psi^{*(1)}(\mathbf{r})\psi^{(1)}(\mathbf{r}) + \psi^{*(2)}(\mathbf{r})\psi^{(2)}(\mathbf{r})]\, d\mathbf{r}. \qquad \text{(C.12.10)}$$

The Hamiltonian $H(\mathbf{s}, \mathbf{r})$ is invariant under lattice translations and once again the wavefunctions and energies can be described in terms of wave-vectors lying in the first Brillouin zone. If $E_n(\mathbf{k})$ is a d_n-fold degenerate energy level with eigenfunctions $\Psi_{n\mathbf{k}j}(\sigma, \mathbf{r})$ $(j = 1, 2, \ldots, d_n)$,

$$H(\mathbf{s}, \mathbf{r})\, \Psi_{n\mathbf{k}j}(\sigma, \mathbf{r}) = E_n(\mathbf{k})\Psi_{n\mathbf{k}j}(\sigma, \mathbf{r}), \qquad \text{(C.12.11)}$$

and

$$\begin{aligned}\Psi_{n\mathbf{k}j}(\sigma, \mathbf{r}) &= \psi^{(1)}_{n\mathbf{k}j}(\mathbf{r})v_1(\sigma) + \psi^{(2)}_{n\mathbf{k}j}(\mathbf{r})v_2(\sigma)\\ &= e^{i\mathbf{k}\cdot\mathbf{r}}[u^{(1)}_{n\mathbf{k}j}(\mathbf{r})v_1(\sigma) + u^{(2)}_{n\mathbf{k}j}(\mathbf{r})v_2(\sigma)]\\ &= e^{i\mathbf{k}\cdot\mathbf{r}}u_{n\mathbf{k}j}(\sigma, \mathbf{r}),\end{aligned} \qquad \text{(C.12.12)}$$

where $u^{(1)}_{n\mathbf{k}j}(\mathbf{r})$, $u^{(2)}_{n\mathbf{k}j}(\mathbf{r})$ and $u_{n\mathbf{k}j}(\sigma, \mathbf{r})$ are lattice-periodic, i.e.

$$u_{n\mathbf{k}j}(\sigma, \mathbf{r} + \mathbf{R}_t) = u_{n\mathbf{k}j}(\sigma, \mathbf{r}) \qquad \text{(C.12.13)}$$

for any lattice translation $\mathbf{R_t}$.

Let $\{Q_l|\mathbf{v}_l + \mathbf{R_t}\}$ be an element of the group of the wave-vector $\mathbf{k}$; consider the effect of applying this element to the Hamiltonian $H(\mathbf{s}, \mathbf{r})$ and to the wavefunction $\Psi_{n\mathbf{k}j}(\sigma, \mathbf{r})$. The spin coordinate σ is defined with respect to the x_3-axis, which means that the spin functions $v_1(\sigma)$ and $v_2(\sigma)$ and also the spin operator $\mathbf{s} = (s_1, s_2, s_3)$ change under the application of an element of the group of the wave-vector $\mathbf{k}$, i.e. in a rotated coordinate system the spin no longer 'points' along the x_3-axis. With the help of (C.12.4), (C.12.5) and (C.12.7),

$$\{Q_l|\mathbf{v}_l + \mathbf{R_t}\}H(\mathbf{s}, \mathbf{r}) = \mathscr{H}(\mathbf{r}) - i\sum_{j=1}^{3} [Q_l s_j][Q_l P_j], \qquad \text{(C.12.14)}$$

$$\{Q_l|\mathbf{v}_l + \mathbf{R_t}\}\Psi_{n\mathbf{k}j}(\sigma, \mathbf{r}) = \sum_{p=1}^{2} [\{Q_l|\mathbf{v}_l + \mathbf{R_t}\}\psi^{(p)}_{n\mathbf{k}j}(\mathbf{r})][Q_l v_p(\sigma)]. \qquad \text{(C.12.15)}$$

We must now consider the effect of rotations (proper and improper) on the spin angular momentum operator $\mathbf{s}$ and on the spin functions $v_1(\sigma)$ and $v_2(\sigma)$. The results can be derived by analogy from the corresponding results for the orbital angular momentum operator.

Let $\mathbf{L} \equiv (L_1, L_2, L_3)$ denote the orbital angular momentum operator, i.e.

$$L \equiv -i\hbar[\mathbf{r} \times \nabla], \qquad L_1 = -i\hbar\left[x_2\frac{\partial}{\partial x_3} - x_3\frac{\partial}{\partial x_2}\right], \text{etc.,} \qquad \text{(C.12.16)}$$

and suppose that the point group element Q_l corresponds to the orthogonal transformation of the coordinates

$$x_j' = \sum_{p=1}^{3} q_{jp}^l x_p, \qquad \det|q_{jp}^l| = \pm 1. \qquad \text{(C.12.17)}$$

The application of Q_l to a component L_j of the orbital angular momentum operator then yields

$$Q_l L_j = \pm \sum_{p=1}^{3} q_{jp}^l L_p \qquad (j = 1, 2, 3), \qquad \text{(C.12.18)}$$

where the plus sign appears if $\det|q_{jp}^l| = 1$ and the minus sign appears if $\det|q_{jp}^l| = -1$.

We now assume that the components of the spin angular momentum operator behave in the same way under rotations, i.e.

$$Q_l s_j = \begin{cases} \displaystyle\sum_{p=1}^{3} q_{jp}^l s_p & \text{if } Q_l \text{ is a proper rotation} \\ \displaystyle -\sum_{p=1}^{3} q_{jp}^l s_p & \text{if } Q_l \text{ is an improper rotation.} \end{cases} \qquad \text{(C.12.19)}$$

This is the first of the required results.

Using (C.12.4) we find that the operators P_j $(j = 1, 2, 3)$ also behave in the same way as the orbital angular momentum operators under rotations, and hence

$$\sum_{j=1}^{3} [Q_l s_j][Q_l P_j] = \sum_{j=1}^{3} \sum_{p=1}^{3} \sum_{t=1}^{3} q_{jp}^l q_{jt}^l s_p P_t = \sum_{t=1}^{3} s_t P_t, \qquad \text{(C.12.20)}$$

where we have made use of the fact that

$$\sum_{j=1}^{3} q_{jp}^l q_{jt}^l = \delta_{tp}. \qquad \text{(C.12.21)}$$

Using (C.12.14) it now follows that

$$\{Q_l|\mathbf{v}_l + \mathbf{R}_t\}H(\mathbf{s}, \mathbf{r}) = H(\mathbf{s}, \mathbf{r}), \qquad \text{(C.12.22)}$$

i.e. the elements of the group of the wave-vector $\mathbf{k}$ when applied simultaneously to space and spin coordinates leave the Hamiltonian $H(\mathbf{s}, \mathbf{r})$ invariant.

We next consider the effect of a *proper* rotation Q_l on the spin functions $v_1(\sigma)$ and $v_2(\sigma)$, where Q_l corresponds to the orthogonal transformation of the coordinates

$$\mathbf{r}' = q^l\mathbf{r}, \qquad \det |q^l_{jp}| = 1. \tag{C.12.23}$$

Any proper rotation can be expressed in terms of three angles known as the Eulerian angles of rotation[29]. If the above proper rotation corresponds to the Eulerian angles θ_1, θ_2 and θ_3, the matrix q^l is given by

$$\begin{pmatrix} \begin{array}{c}\cos\theta_1\cos\theta_2\cos\theta_3\\ -\sin\theta_1\sin\theta_3\end{array}, & \begin{array}{c}\cos\theta_1\cos\theta_2\sin\theta_3\\ +\sin\theta_1\cos\theta_3\end{array}, & -\cos\theta_1\sin\theta_2 \\ \begin{array}{c}-\sin\theta_1\cos\theta_2\cos\theta_3\\ -\cos\theta_1\sin\theta_3\end{array}, & \begin{array}{c}-\sin\theta_1\cos\theta_2\sin\theta_3\\ +\cos\theta_1\cos\theta_3\end{array}, & \sin\theta_1\sin\theta_2 \\ \sin\theta_2\cos\theta_3, & \sin\theta_2\sin\theta_3, & \cos\theta_2 \end{pmatrix} \tag{C.12.24}$$

and the proper rotation $Q_l(\theta_1, \theta_2, \theta_3)$ can be written

$$Q_l(\theta_1, \theta_2, \theta_3) = Q(x_3, \theta_3)Q(x_2, \theta_2)Q(x_3, \theta_1), \tag{C.12.25}$$

where $Q(x_j, \theta)$ denotes a rotation of the coordinate system through an angle θ about the x_j-axis, e.g.

$$Q(x_3, \theta)f(x_1, x_2, x_3) = f(x_1 \cos\theta + x_2 \sin\theta, -x_1 \sin\theta + x_2 \cos\theta, x_3). \tag{C.12.26}$$

Let $Q(x_3, \delta\theta)$ denote a rotation of the coordinate system through an infinitesimal angle $\delta\theta$ about the x_3-axis. Then

$$\begin{aligned} Q(x_3, \delta\theta)f(x_1, x_2, x_3) &= f(x_1 + x_2\,\delta\theta, x_2 - x_1\,\delta\theta, x_3) \\ &= f(x_1, x_2, x_3) + \delta\theta x_2\,\partial f(x_1, x_2, x_3)/\partial x_1 \\ &\quad - \delta\theta x_1\,\partial f(x_1, x_2, x_3)/\partial x_2 \\ &= \left[1 + \delta\theta\left(x_2\frac{\partial}{\partial x_1} - x_1\frac{\partial}{\partial x_2}\right)\right] f(x_1, x_2, x_3) \\ &= \left[1 + \frac{\delta\theta}{i\hbar}L_3\right] f(x_1, x_2, x_3), \end{aligned} \tag{C.12.27}$$

where $\mathbf{L} \equiv (L_1, L_2, L_3)$ is the orbital angular momentum operator. A rotation through an angle θ about the x_3-axis is equal to n successive

[29] E. Wigner, *Group Theory and its Applications to the Quantum Mechanics of Atomic Spectra*, Academic Press, New York, 1959, p. 90.

rotations through θ/n, and hence, from (C.12.27),

$$Q(x_3, \theta)f(x_1, x_2, x_3) = \lim_{n\to\infty}\left[1 + \frac{\theta}{ni\hbar}L_3\right]^n f(x_1, x_2, x_3)$$

$$= [1 + (\theta L_3/i\hbar) + (\theta L_3/i\hbar)^2/2! + \cdots + \cdots]f(x_1, x_2, x_3), \qquad \text{(C.12.28)}$$

which we can sum in a purely formal manner to give

$$Q(x_3, \theta)f(x_1, x_2, x_3) = e^{-i\theta L_3/\hbar}f(x_1, x_2, x_3). \qquad \text{(C.12.29)}$$

As far as functions of the coordinates x_1, x_2 and x_3 are concerned therefore, rotations can be expressed explicitly in terms of orbital angular momentum operators.

We now assume that corresponding to (C.12.29) we can write

$$Q(x_j, \theta)v_p(\sigma) = e^{-i\theta s_j/\hbar}v_p(\sigma) = [1 + (\theta s_j/i\hbar) + \cdots + \cdots]v_p(\sigma) \quad (p = 1, 2), \qquad \text{(C.12.30)}$$

where $\mathbf{s} = (s_1, s_2, s_3)$ is the spin angular momentum operator. Using (C.12.9), (C.12.25) and (C.12.30), we then find that

$$Q_l(\theta_1, \theta_2, \theta_3)v_1(\sigma) = \cos\tfrac{1}{2}\theta_2\, e^{-\frac{1}{2}i(\theta_1+\theta_3)}v_1(\sigma) + \sin\tfrac{1}{2}\theta_2\, e^{\frac{1}{2}i(\theta_3-\theta_1)}v_2(\sigma),$$
$$Q_l(\theta_1, \theta_2, \theta_3)v_2(\sigma) = -\sin\tfrac{1}{2}\theta_2\, e^{\frac{1}{2}i(\theta_1-\theta_3)}v_1(\sigma) + \cos\tfrac{1}{2}\theta_2\, e^{\frac{1}{2}i(\theta_1+\theta_3)}v_2(\sigma). \qquad \text{(C.12.31)}$$

These are the standard equations for the effect of a rotation on the spin functions $v_1(\sigma)$ and $v_2(\sigma)$. In matrix notation

$$Q_l(\theta_1, \theta_2, \theta_3)\begin{pmatrix} v_1(\sigma) \\ v_2(\sigma) \end{pmatrix} = \tilde{D}^l(\theta_1, \theta_2, \theta_3)\begin{pmatrix} v_1(\sigma) \\ v_2(\sigma) \end{pmatrix}, \qquad \text{(C.12.32)}$$

where $\tilde{D}^l(\theta_1, \theta_2, \theta_3)$ is the transpose of the matrix $D^l(\theta_1, \theta_2, \theta_3)$ given by

$$D^l(\theta_1, \theta_2, \theta_3) = \begin{pmatrix} \cos\frac{1}{2}\theta_2\, e^{-\frac{1}{2}i(\theta_1+\theta_3)}, & -\sin\frac{1}{2}\theta_2\, e^{\frac{1}{2}i(\theta_1-\theta_3)} \\ \sin\frac{1}{2}\theta_2\, e^{\frac{1}{2}i(\theta_3-\theta_1)}, & \cos\frac{1}{2}\theta_2\, e^{\frac{1}{2}i(\theta_1+\theta_3)} \end{pmatrix} \qquad \text{(C.12.33)}$$

Let $\bar{Q}_1$ denote a rotation through 2π about any given axis and consider the rotation $\bar{Q}_l \equiv \bar{Q}_1 Q_l$, where Q_l is a proper rotation characterized by the Eulerian angles θ_1, θ_2 and θ_3. $\bar{Q}_l$ has the same effect as Q_l when applied to a function of the coordinates x_1, x_2, x_3 and we would ordinarily write $\bar{Q}_l = Q_l$. Because of the half-angles in (C.12.31), however, we find that

$$\bar{Q}_1 v_1(\sigma) = -v_1(\sigma), \qquad \bar{Q}_1 v_2(\sigma) = -v_2(\sigma), \qquad \text{(C.12.34)}$$

i.e. the spin functions change sign under a rotation through 2π. It follows that if

$$Q_l\begin{pmatrix} v_1(\sigma) \\ v_2(\sigma) \end{pmatrix} = \tilde{D}^l(\theta_1, \theta_2, \theta_3)\begin{pmatrix} v_1(\sigma) \\ v_2(\sigma) \end{pmatrix}, \tag{C.12.35}$$

then

$$\bar{Q}_l\begin{pmatrix} v_1(\sigma) \\ v_2(\sigma) \end{pmatrix} = -\tilde{D}^l(\theta_1, \theta_2, \theta_3)\begin{pmatrix} v_1(\sigma) \\ v_2(\sigma) \end{pmatrix}. \tag{C.12.36}$$

As far as the spin functions are concerned therefore, $\bar{Q}_l$ is not the same operation as Q_l. $\{Q_l|\mathbf{v}_l + \mathbf{R}_\mathbf{t}\}$ and $\{\bar{Q}_l|\mathbf{v}_l + \mathbf{R}_\mathbf{t}\}$ both leave the Hamiltonian $H(\mathbf{s}, \mathbf{r})$ invariant and $\{\bar{Q}_l|\mathbf{v}_l + \mathbf{R}_\mathbf{t}\}$ does not have the same effect as $\{Q_l|\mathbf{v}_l + \mathbf{R}_\mathbf{t}\}$ on the spin-dependent wavefunctions. Both of these operations must therefore be included in the group of symmetry transformations of the Hamiltonian $H(\mathbf{s}, \mathbf{r})$.

An improper rotation can be considered as the product of the inversion and a proper rotation. Since we have already found the effect of a proper rotation on the spin functions, we need now only consider the effect of the inversion J.

The application of the inversion J to the equation

$$s_3 v_p(\sigma) = (\tfrac{1}{2} - \delta_{2p})\hbar v_p(\sigma) \qquad (p = 1, 2) \tag{C.12.37}$$

yields

$$s_3[J v_p(\sigma)] = (\tfrac{1}{2} - \delta_{2p})\hbar[J v_p(\sigma)], \tag{C.12.38}$$

since, from (C.12.19), J leaves the spin operator invariant. It follows that

$$J v_p(\sigma) = A v_p(\sigma), \tag{C.12.39}$$

where A is a constant. A further application of the inversion now yields

$$J^2 v_p(\sigma) = A^2 v_p(\sigma). \tag{C.12.40}$$

J^2 can be put equal to either Q_1 or $\bar{Q}_1$, and it follows from (C.12.40) that

$$A = 1, -1, \mathrm{i} \quad \text{or} -\mathrm{i}. \tag{C.12.41}$$

For simplicity, the value $A = 1$ is generally used, and we therefore take

$$J v_1(\sigma) = v_1(\sigma), \qquad J v_2(\sigma) = v_2(\sigma). \tag{C.12.42}$$

The improper rotation $Q_m = JQ_l$ therefore has the same effect on the spin functions as the proper rotation Q_l, and the improper rotation $\bar{Q}_m = J\bar{Q}_l$ has the same effect as the proper rotation $\bar{Q}_l$.

The elements $\{Q_p|\mathbf{v}_p + \mathbf{R}_t\}$ and $\{\bar{Q}_p|\mathbf{v}_p + \mathbf{R}_t\}$ $(p = 1, 2, \ldots, g)$, all of which transform a wavefunction with wave-vector **k** into another wavefunction with the same or equivalent wave-vector, form what is known as the double group of the wave-vector **k**. This double group has twice as many elements as the single group of the wave-vector **k** with elements $\{Q_p|\mathbf{v}_p + \mathbf{R}_t\}$ $(p = 1, 2, \ldots, g)$. In the presence of spin–orbit coupling the eigenfunctions $\Psi_{n\mathbf{k}j}(\sigma, \mathbf{r})$ $(j = 1, 2, \ldots, d_n)$ belonging to a d_n-fold degenerate energy level $E_n(\mathbf{k})$ act as basis functions for a d_n-dimensional irreducible representation of the double group of the wave-vector **k**.

Double groups in the InSb and Ge structures

In many cases (cf. § C.9) the character table of the double group of the wave-vector **k** follows directly from the character table of the associated double point group. We now, for the purpose of illustration, carry out the calculation of the character table for the point Δ in the first Brillouin zone of the InSb structure[30].

The elements (see tables C.7.1 and C.9.1) of the double group of the wave-vector at the point Δ in the first Brillouin zone of the InSb structure are $\{Q_p|\mathbf{R}_t\}$ and $\{\bar{Q}_p|\mathbf{R}_t\}$ $(p = 1, 2, 19, 20)$, where the point group elements correspond to the following orthogonal transformations of the coordinates x_1, x_2 and x_3:

$$Q_1, \bar{Q}_1: \begin{pmatrix} x'_1 \\ x'_2 \\ x'_3 \end{pmatrix} = \begin{pmatrix} 1 & 0 & 0 \\ 0 & 1 & 0 \\ 0 & 0 & 1 \end{pmatrix} \begin{pmatrix} x_1 \\ x_2 \\ x_3 \end{pmatrix};$$

$$Q_2, \bar{Q}_2: \begin{pmatrix} x'_1 \\ x'_2 \\ x'_3 \end{pmatrix} = \begin{pmatrix} 1 & 0 & 0 \\ 0 & -1 & 0 \\ 0 & 0 & -1 \end{pmatrix} \begin{pmatrix} x_1 \\ x_2 \\ x_3 \end{pmatrix};$$

$$Q_{19}, \bar{Q}_{19}: \begin{pmatrix} x'_1 \\ x'_2 \\ x'_3 \end{pmatrix} = \begin{pmatrix} 1 & 0 & 0 \\ 0 & 0 & 1 \\ 0 & 1 & 0 \end{pmatrix} \begin{pmatrix} x_1 \\ x_2 \\ x_3 \end{pmatrix};$$

$$Q_{20}, \bar{Q}_{20}: \begin{pmatrix} x'_1 \\ x'_2 \\ x'_3 \end{pmatrix} = \begin{pmatrix} 1 & 0 & 0 \\ 0 & 0 & -1 \\ 0 & -1 & 0 \end{pmatrix} \begin{pmatrix} x_1 \\ x_2 \\ x_3 \end{pmatrix}. \qquad \text{(C.12.43)}$$

[30] R. H. Parmenter, *Phys. Rev.*, **100**, 573 (1955); G. Dresselhaus, *Phys. Rev.*, **100**, 580 (1955).

We now note that $Q_{19}, \bar{Q}_{19}, Q_{20}$ and $\bar{Q}_{20}$ are improper rotations and so we write $Q_{19} = JQ'_{19}$, $\bar{Q}_{19} = J\bar{Q}'_{19}$, $Q_{20} = JQ'_{20}$ and $\bar{Q}_{20} = J\bar{Q}'_{20}$, where $Q'_{19}, \bar{Q}'_{19}, Q'_{20}$ and $\bar{Q}'_{20}$ are proper rotations corresponding to the following orthogonal transformations of the coordinates x_1, x_2 and x_3:

$$Q'_{19}, \bar{Q}'_{19}: \begin{pmatrix} x'_1 \\ x'_2 \\ x'_3 \end{pmatrix} = \begin{pmatrix} -1 & 0 & 0 \\ 0 & 0 & -1 \\ 0 & -1 & 0 \end{pmatrix} \begin{pmatrix} x_1 \\ x_2 \\ x_3 \end{pmatrix};$$

$$Q'_{20}, \bar{Q}'_{20}: \begin{pmatrix} x'_1 \\ x'_2 \\ x'_3 \end{pmatrix} = \begin{pmatrix} -1 & 0 & 0 \\ 0 & 0 & 1 \\ 0 & 1 & 0 \end{pmatrix} \begin{pmatrix} x_1 \\ x_2 \\ x_3 \end{pmatrix}. \qquad \text{(C.12.44)}$$

If $E_n(\mathbf{k}_\Delta)$ is a d_n-fold degenerate energy level with spin-dependent eigenfunctions $\Psi_{n\mathbf{k}_\Delta j}(\sigma, \mathbf{r})$ $(j = 1, 2, \ldots, d_n)$,

$$\{Q_p|\mathbf{R}_t\}\Psi_{n\mathbf{k}_\Delta j}(\sigma, \mathbf{r}) = e^{i\mathbf{k}_\Delta \cdot \mathbf{R}_t} Q_p \Psi_{n\mathbf{k}_\Delta j}(\sigma, \mathbf{r})$$

$$= \sum_{l=1}^{d_n} M^{pt}_{lj} \Psi_{n\mathbf{k}_\Delta l}(\sigma, \mathbf{r}), \qquad \text{(C.12.45)}$$

$$\{\bar{Q}_p|\mathbf{R}_t\}\Psi_{n\mathbf{k}_\Delta j}(\sigma, r) = e^{i\mathbf{k}_\Delta \cdot \mathbf{R}_t} \bar{Q}_p \Psi_{n\mathbf{k}_\Delta j}(\sigma, \mathbf{r})$$

$$= \sum_{l=1}^{d_n} (-M^{pt}_{lj}) \Psi_{n\mathbf{k}_\Delta l}(\sigma, \mathbf{r}), \qquad \text{(C.12.46)}$$

from which

$$e^{-i\mathbf{k}_\Delta \cdot \mathbf{R}_t} M^{pt} = e^{-i\mathbf{k}_\Delta \cdot \mathbf{R}_m} M^{pm} = \cdots = M^p \text{ (say)}, \qquad \text{(C.12.47)}$$

and

$$Q_p \Psi_{n\mathbf{k}_\Delta j}(\sigma, \mathbf{r}) = \sum_{l=1}^{d_n} M^p_{lj} \Psi_{n\mathbf{k}_\Delta l}(\sigma, \mathbf{r}), \qquad \text{(C.17.48)}$$

$$\bar{Q}_p \Psi_{n\mathbf{k}_\Delta j}(\sigma, \mathbf{r}) = \sum_{l=1}^{d_n} [-M^p_{lj}] \Psi_{n\mathbf{k}_\Delta l}(\sigma, \mathbf{r}). \qquad \text{(C.12.49)}$$

The functions $\Psi_{n\mathbf{k}_\Delta j}(\sigma, \mathbf{r})$ $(j = 1, 2, \ldots, d_n)$ therefore act as basis functions for a d_n-dimensional irreducible representation of the double point group with elements Q_p and $\bar{Q}_p$ $(p = 1, 2, 19, 20)$, the element Q_p being represented by the matrix M^p and the element $\bar{Q}_p$ by the matrix $-M^p$. To find the irreducible representations of the double group with elements $\{Q_p|\mathbf{R}_t\}$ and $\{\bar{Q}_p|\mathbf{R}_t\}$ $(p = 1, 2, 19, 20)$ we need only find the irreducible representations of the double point group with elements Q_p and $\bar{Q}_p$ $(p = 1, 2, 19, 20)$.

The spin functions $v_1(\sigma)$ and $v_2(\sigma)$ themselves act as basis functions for a two-dimensional irreducible representation of a double point group, in which the element Q_p is represented by the matrix D^p[see (C.12.33)] and the element $\bar{Q}_p$ by the matrix $-D^p$. We can therefore use these matrices to draw up the multiplication table of the double point group and hence calculate its character table.

By comparing the first two matrices in (C.12.43) and the matrices in (C.12.44) with the matrix (C.12.24) we can find the Eulerian angles corresponding to these rotations and hence find the effect of these rotations on the spin functions. In this way we obtain the matrices

$$Q_1, \bar{Q}_1 : D^1 = \pm\begin{pmatrix} 1 & 0 \\ 0 & 1 \end{pmatrix}; \qquad Q_2, \bar{Q}_2 : D^2 = \pm\begin{pmatrix} 0 & -i \\ -i & 0 \end{pmatrix};$$

$$Q'_{19}, \bar{Q}'_{19} : D^{19} = \pm\begin{pmatrix} i/\sqrt{2} & -1/\sqrt{2} \\ 1/\sqrt{2} & -i/\sqrt{2} \end{pmatrix};$$

$$Q'_{20}, \bar{Q}'_{20} : D^{20} = \pm\begin{pmatrix} -i/\sqrt{2} & -1/\sqrt{2} \\ 1/\sqrt{2} & i/\sqrt{2} \end{pmatrix}. \tag{C.12.50}$$

We have written these results in the above manner because, for example, we can either associate the matrix $+\begin{pmatrix} 0 & -i \\ -i & 0 \end{pmatrix}$ with Q_2 and the matrix $-\begin{pmatrix} 0 & -i \\ -i & 0 \end{pmatrix}$ with $\bar{Q}_2$ *or* the matrix $-\begin{pmatrix} 0 & -i \\ -i & 0 \end{pmatrix}$ with Q_2 and the matrix $\begin{pmatrix} 0 & -i \\ -i & 0 \end{pmatrix}$ with $\bar{Q}_2$. The choice is open and depends simply on which Eulerian angles are used in the first place to characterize the proper rotation $Q_2(\theta_1, \theta_2, \theta_3)$; for example, two acceptable sets of Eulerian angles for the proper rotation Q_2 are $\theta_1 = \pi, \theta_2 = \pi, \theta_3 = 0$ and $\theta_1 = 0, \theta_2 = \pi$, $Q_3 = \pi$, the first set leading to $\begin{pmatrix} 0 & -i \\ -i & 0 \end{pmatrix}$ being associated with Q_2 and $-\begin{pmatrix} 0 & -i \\ -i & 0 \end{pmatrix}$ being associated with $\bar{Q}_2$, and the second set leading to $-\begin{pmatrix} 0 & -i \\ -i & 0 \end{pmatrix}$ being associated with Q_2 and $\begin{pmatrix} 0 & -i \\ -i & 0 \end{pmatrix}$ being associated with $\bar{Q}_2$. In the present calculation we choose the matrices with the plus signs to go with the unbarred elements and the matrices with the minus signs to go with the barred elements. We then arrive at the following two-dimensional irreducible representation of the double point group:

$$Q_1:\begin{pmatrix}1 & 0\\ 0 & 1\end{pmatrix};\quad \bar{Q}_1:\begin{pmatrix}-1 & 0\\ 0 & -1\end{pmatrix};\quad Q_2:\begin{pmatrix}0 & -\mathrm{i}\\ -\mathrm{i} & 0\end{pmatrix};$$

$$\bar{Q}_2:\begin{pmatrix}0 & \mathrm{i}\\ \mathrm{i} & 0\end{pmatrix};\quad Q_{19}:\frac{1}{\sqrt{2}}\begin{pmatrix}\mathrm{i} & -1\\ 1 & -\mathrm{i}\end{pmatrix};\quad \bar{Q}_{19}:\frac{1}{\sqrt{2}}\begin{pmatrix}-\mathrm{i} & 1\\ -1 & \mathrm{i}\end{pmatrix};$$

$$Q_{20}:\frac{1}{\sqrt{2}}\begin{pmatrix}-\mathrm{i} & -1\\ 1 & \mathrm{i}\end{pmatrix};\quad \bar{Q}_{20}:\frac{1}{\sqrt{2}}\begin{pmatrix}\mathrm{i} & 1\\ -1 & -\mathrm{i}\end{pmatrix}. \tag{C.12.51}$$

The calculation of the multiplication table for the double point group is now a simpler matter. For example,

$$\frac{1}{\sqrt{2}}\begin{pmatrix}-\mathrm{i} & 1\\ -1 & \mathrm{i}\end{pmatrix}\frac{1}{\sqrt{2}}\begin{pmatrix}-\mathrm{i} & -1\\ 1 & \mathrm{i}\end{pmatrix}=\begin{pmatrix}0 & \mathrm{i}\\ \mathrm{i} & 0\end{pmatrix} \tag{C.12.52}$$

and hence

$$\bar{Q}_{19}Q_{20}=\bar{Q}_2. \tag{C.12.53}$$

Table C.12.1 shows, for example, that

$$Q_2^{-1}=\bar{Q}_2,\qquad Q_{19}^{-1}=\bar{Q}_{19},\qquad \bar{Q}_2\bar{Q}_{19}=Q_{20}. \tag{C.12.54}$$

Table C.12.1 Multiplication table for the double point group at the point Δ in the first Brillouin zone of the InSb structure.

	Q_1	$\bar{Q}_1$	Q_2	$\bar{Q}_2$.	Q_{19}	$\bar{Q}_{19}$	Q_{20}	$\bar{Q}_{20}$
Q_1	Q_1	$\bar{Q}_1$	Q_2	$\bar{Q}_2$	Q_{19}	$\bar{Q}_{19}$	Q_{20}	$\bar{Q}_{20}$
$\bar{Q}_1$	$\bar{Q}_1$	Q_1	$\bar{Q}_2$	Q_2	$\bar{Q}_{19}$	Q_{19}	$\bar{Q}_{20}$	Q_{20}
Q_2	Q_2	$\bar{Q}_2$	$\bar{Q}_1$	Q_1	Q_{20}	$\bar{Q}_{20}$	$\bar{Q}_{19}$	Q_{19}
$\bar{Q}_2$	$\bar{Q}_2$	Q_2	Q_1	$\bar{Q}_1$	$\bar{Q}_{20}$	Q_{20}	Q_{19}	$\bar{Q}_{19}$
Q_{19}	Q_{19}	$\bar{Q}_{19}$	$\bar{Q}_{20}$	Q_{20}	$\bar{Q}_1$	Q_1	Q_2	$\bar{Q}_2$
$\bar{Q}_{19}$	$\bar{Q}_{19}$	Q_{19}	Q_{20}	$\bar{Q}_{20}$	Q_1	$\bar{Q}_1$	$\bar{Q}_2$	Q_2
Q_{20}	Q_{20}	$\bar{Q}_{20}$	Q_{19}	$\bar{Q}_{19}$	$\bar{Q}_2$	Q_2	$\bar{Q}_1$	Q_1
$\bar{Q}_{20}$	$\bar{Q}_{20}$	Q_{20}	$\bar{Q}_{19}$	Q_{19}	Q_2	$\bar{Q}_2$	Q_1	$\bar{Q}_1$

The classes of the double point group can now be obtained. These are given below.

$$\mathscr{C}_1=Q_1;\qquad \mathscr{C}_2=\bar{Q}_1;\qquad \mathscr{C}_3=Q_2+\bar{Q}_2;$$

$$\mathscr{C}_4=Q_{19}+\bar{Q}_{19};\qquad \mathscr{C}_5=Q_{20}+\bar{Q}_{20}. \tag{C.12.55}$$

There are five classes and eight elements and hence, using (C.5.29), this double point group has four distinct one-dimensional irreducible representations and one two-dimensional irreducible representation. The character table can now be obtained by the same method used for the single group.

Table C.12.2 Character table for the double point group at Δ (InSb).

	$\mathscr{C}_1$	$\mathscr{C}_2$	$\mathscr{C}_3$	$\mathscr{C}_4$	$\mathscr{C}_5$
Δ_1	1	1	1	1	1
Δ_2	1	1	1	-1	-1
Δ_3	1	1	-1	1	-1
Δ_4	1	1	-1	-1	1
Δ_5	2	-2	0	0	0

The first four rows of table C.12.2 could in fact have been written down immediately. An irreducible representation of the single group automatically furnishes an irreducible representation of the double group, i.e. if we have an irreducible representation of the single group in which the element Q_j is represented by the matrix M^j, then by associating the matrix M^j with both Q_j and $\bar{Q}_j$ we obtain an irreducible representation of the double group. In this way the first four rows of table C.12.2 are obtainable from the character table of the single group (table C.9.3). The remaining representation, Δ_5, cannot be obtained in this way and is known as an extra representation of the double group. The matrices of this extra representation are in fact those given in (C.12.51).

From (C.12.45) and (C.12.46) we see that in the d_n-dimensional irreducible representation for which the functions $\psi_{n\mathbf{k}_\Delta j}(\sigma, \mathbf{r})$ $(j = 1, 2, \ldots, d_n)$ act as basis functions the element Q_p is represented by the matrix M^p and the element $\bar{Q}_p$ is represented by the matrix $-M^p$. If Q_p and $\bar{Q}_p$ are in the same class it follows that trace $M^p = -$trace $M^p = 0$. Now in table C.12.2, Q_2 and $\bar{Q}_2$, for example, are in the same class and the matrices representing these elements in the d_n-dimensional irreducible representation must have zero trace. The only irreducible representation which satisfies this requirement is Δ_5. Hence spin-dependent wavefunctions at the point Δ in the first Brillouin zone of the InSb structure can act as basis functions only for the extra representation Δ_5, and it is this representation only which is appropriate to the present theory[31].

[31] W. Opechowski, *Physica*, **7**, 552 (1940).

In the presence of spin–orbit coupling the spin-dependent wavefunctions with wave-vector **k** can act as basis functions only for the extra irreducible representations of the double group of the wave-vector **k**.

The part of the character table of the double group of the wave-vector at the point Δ in the first Brillouin zone of the InSb structure which is appropriate to the theory of spin–orbit coupling is therefore that given in table C.12.3.

Table C.12.3 The extra irreducible representation at the point Δ (InSb).

	$\mathscr{C}_1$	$\mathscr{C}_2$	$\mathscr{C}_3$	$\mathscr{C}_4$	$\mathscr{C}_5$
Δ_5	2	-2	0	0	0

With spin–orbit interaction, therefore, there are only two-fold degenerate states (one type only: Δ_5) at the point Δ in the first Brillouin zone of the InSb structure.

We now give the character tables appropriate to the double groups at the points Γ, Δ and X in the first Brillouin zone of the InSb structure and in the first Brillouin zone of the Ge structure[32]. These tables describe the extra irreducible representations of the double groups and can be derived in ways similar to those used for the single groups.

Table C.12.4 The extra representations at the point Γ in the first Brillouin zone of the InSb structure.

	Q_1	$\bar{Q}_1$	$Q_2 - Q_4$, $\bar{Q}_2 - \bar{Q}_4$	$Q_5 - Q_{12}$	$\bar{Q}_5 - \bar{Q}_{12}$	$Q_{13} - Q_{18}$	$\bar{Q}_{13} - \bar{Q}_{18}$	$Q_{19} - Q_{24}$, $\bar{Q}_{19} - \bar{Q}_{24}$
Γ_6	2	-2	0	1	-1	$\sqrt{2}$	$-\sqrt{2}$	0
Γ_7	2	-2	0	1	-1	$-\sqrt{2}$	$\sqrt{2}$	0
Γ_8	4	-4	0	-1	1	0	0	0

There are therefore two-fold degenerate states (two types: Γ_6 and Γ_7) and four-fold degenerate states (one type: Γ_8) at the point Γ in the first Brillouin zone of the InSb structure.

Table C.12.5 The extra representations at the point X in the first Brillouin zone of the InSb structure.

	Q_1	$\bar{Q}_1$	$Q_2, \bar{Q}_2$	$Q_3, Q_4, \bar{Q}_3, \bar{Q}_4$	Q_{13}, Q_{14}	$\bar{Q}_{13}, \bar{Q}_{14}$	$Q_{19}, Q_{20}, \bar{Q}_{19}, \bar{Q}_{20}$
X_6	2	-2	0	0	$\sqrt{2}$	$-\sqrt{2}$	0
X_7	2	-2	0	0	$-\sqrt{2}$	$\sqrt{2}$	0

[32] R. J. Elliott, *Phys. Rev.*, **96**, 280 (1954); R. H. Parmenter, *Phys. Rev.*, **100**, 573 (1955); G. Dresselhaus, *Phys. Rev.*, **100**, 580 (1955).

Table C.12.6 The extra representations at the point Γ in the first Brillouin zone of the Ge structure.

	Q_1	$\bar{Q}_1$	$Q_2 - Q_4, \bar{Q}_2 - \bar{Q}_4$	$Q_5 - Q_{12}$	$\bar{Q}_5 - \bar{Q}_{12}$	$Q_{13} - Q_{18}$	$\bar{Q}_{13} - \bar{Q}_{18}$	$Q_{19} - Q_{24}, \bar{Q}_{19} - \bar{Q}_{24}$
Γ_6^+	2	−2	0	1	−1	$\sqrt{2}$	$-\sqrt{2}$	0
Γ_6^-	2	−2	0	1	−1	$-\sqrt{2}$	$\sqrt{2}$	0
Γ_7^+	2	−2	0	1	−1	$-\sqrt{2}$	$\sqrt{2}$	0
Γ_7^-	2	−2	0	1	−1	$\sqrt{2}$	$-\sqrt{2}$	0
Γ_8^+	4	−4	0	−1	1	0	0	0
Γ_8^-	4	−4	0	−1	1	0	0	0

	$\{Q'_1\|\mathbf{v}\}$	$\{\bar{Q}'_1\|\mathbf{v}\}$	$\{Q'_2 - Q'_4, \bar{Q}'_2 - \bar{Q}'_4\|\mathbf{v}\}$	$\{Q'_5 - Q'_{12}\|\mathbf{v}\}$	$\{\bar{Q}'_5 - \bar{Q}'_{12}\|\mathbf{v}\}$	$\{Q'_{13} - Q'_{18}\|\mathbf{v}\}$	$\{\bar{Q}'_{13} - \bar{Q}'_{18}\|\mathbf{v}\}$	$\{Q'_{19} - Q'_{24}, \bar{Q}'_{19} - \bar{Q}'_{24}\|\mathbf{v}\}$
Γ_6^+	2	−2	0	1	−1	$\sqrt{2}$	$-\sqrt{2}$	0
Γ_6^-	−2	2	0	−1	1	$\sqrt{2}$	$-\sqrt{2}$	0
Γ_7^+	2	−2	0	1	−1	$-\sqrt{2}$	$\sqrt{2}$	0
Γ_7^-	−2	2	0	−1	1	$-\sqrt{2}$	$\sqrt{2}$	0
Γ_8^+	4	−4	0	−1	1	0	0	0
Γ_8^-	−4	4	0	1	−1	0	0	0

In the presence of spin–orbit interaction there are therefore two-fold degenerate states only (two types: X_6 and X_7) at the point X in the first Brillouin zone of the InSb structure.

There are two-fold degenerate states (four types: Γ_6^+, Γ_6^-, Γ_7^+ and Γ_7^-) and four-fold degenerate states (two types: Γ_8^+ and Γ_8^-) at the point Γ in the first Brillouin zone of the Ge structure.

Table C.12.7 The extra representations at the point Δ in the first Brillouin zone of the Ge structure.

	Q_1	$\bar{Q}_1$	$Q_2, \bar{Q}_2$	$Q_{19}, Q_{20}, \bar{Q}_{19}, \bar{Q}_{20}$	$\{Q'_3, \bar{Q}'_3, Q'_4, \bar{Q}'_4 \mid \mathbf{v}\}$	$\{Q'_{13}, Q'_{14} \mid \mathbf{v}\}$	$\{\bar{Q}'_{13}, \bar{Q}'_{14} \mid \mathbf{v}\}$
Δ_6	2	-2	0	0	0	$\sqrt{2}e^{i\mathbf{k}_\Delta \cdot \mathbf{v}}$	$-\sqrt{2}e^{i\mathbf{k}_\Delta \cdot \mathbf{v}}$
Δ_7	2	-2	0	0	0	$-\sqrt{2}e^{i\mathbf{k}_\Delta \cdot \mathbf{v}}$	$\sqrt{2}e^{i\mathbf{k}_\Delta \cdot \mathbf{v}}$

At the point Δ in the first Brillouin zone of the Ge structure there are therefore two-fold degenerate states only (two types: Δ_6 and Δ_7).

Table C.12.8 The extra representations at the point X in the first Brillouin zone of the Ge structure.

	Q_1	$\bar{Q}_1$	$Q_2, \bar{Q}_2$	$Q_3, Q_4, \bar{Q}_3, \bar{Q}_4$	Q_{13}, Q_{14}	$\bar{Q}_{13}, \bar{Q}_{14}$	$Q_{19}, Q_{20}, \bar{Q}_{19}, \bar{Q}_{20}$	$\{Q'_1 \mid \mathbf{v}\}$	$\{\bar{Q}'_1 \mid \mathbf{v}\}$
X_5	4	-4	0	0	0	0	0	-4	4

$\{Q'_2, \bar{Q}'_2 \mid \mathbf{v}\}$	$\{Q'_3, Q'_4, \bar{Q}'_3, \bar{Q}'_4 \mid \mathbf{v}\}$	$\{Q'_{13}, Q'_{14} \mid \mathbf{v}\}$	$\{\bar{Q}'_{13}, \bar{Q}'_{14} \mid \mathbf{v}\}$	$\{Q'_{19}, \bar{Q}'_{19} \mid \mathbf{v}\}$	$\{Q'_{20}, \bar{Q}'_{20} \mid \mathbf{v}\}$
0	0	0	0	0	0

At the point X in the first Brillouin zone of the Ge structure there are therefore four-fold degenerate states only (one type only: X_5).

The compatibility relations, which tell us the symmetry types that can exist together in the same energy band, are easily obtained from the above character tables. These relations are given in table C.12.9 for InSb and in table C.12.10 for Ge.

Table C.12.9 Compatibility relations for double groups (InSb).

Γ_6	Γ_7	Γ_8	X_6	X_7
Δ_5	Δ_5	$\Delta_5\Delta_5$	Δ_5	Δ_5

Table C.12.10 Compatibility relations for double groups (Ge).

Γ_6^+	Γ_6^-	Γ_7^+	Γ_7^-	Γ_8^+	Γ_8^-	X_5
Δ_6	Δ_6	Δ_7	Δ_7	$\Delta_6\Delta_7$	$\Delta_6\Delta_7$	$\Delta_6\Delta_7$

Problems

(C.12.1) Using table C.12.7, calculate the character table appropriate to the extra irreducible representations of the double group of the wave-vector $\mathbf{k}_\Delta$ in the first Brillouin zone of the garnet structure [see problem (C.9.1)].

(C.12.2) Calculate the character table appropriate to the extra irreducible representations of the double group of the wave-vector $\mathbf{k}_\Sigma$ in the first Brillouin zone of the diamond structure [see problem (C.11.1)].

(C.12.3) Calculate the character table appropriate to the extra irreducible representations of the double group of the wave-vector $\mathbf{k}_\Lambda$ in the first Brillouin zone of the pyrite structure [see problem (C.10.1)].

(C.12.4) Verify tables C.12.4–C.12.10 of the text.

C.13 SPIN–ORBIT SPLITTING IN InSb AND Ge

Consider the spin-independent Hamiltonian

$$H(\mathbf{r}) \equiv -\frac{\hbar^2}{2m}\nabla^2 + V(\mathbf{r}), \qquad \text{(C.13.1)}$$

and let $E_n(\mathbf{k})$ be an n_α-fold degenerate energy level with Bloch functions $\psi_{n\mathbf{k}j}(\mathbf{r})$ $(j = 1, 2, \ldots, n_\alpha)$. Suppose that these functions act as basis functions for the αth irreducible representation of the single group of the wave-vector $\mathbf{k}$, i.e.

$$\{Q_l|\mathbf{v}_l + \mathbf{R}_\mathbf{t}\}\psi_{n\mathbf{k}j}(\mathbf{r}) = \sum_{p=1}^{n_\alpha} M_{pj}^{\alpha l t}\psi_{n\mathbf{k}p}(\mathbf{r}), \qquad \text{(C.13.2)}$$

where $\{Q_l|\mathbf{v}_l + \mathbf{R}_\mathbf{t}\}$ $(l = 1, 2, \ldots, g)$ are the elements of the single group of the wave-vector $\mathbf{k}$.

If $v_1(\sigma)$ and $v_2(\sigma)$ are the spin functions defined in (C.12.8) and (C.12.9), the spin-dependent Bloch functions belonging to this energy level $E_n(\mathbf{k})$ are

$$\begin{aligned} &v_1(\sigma)\psi_{n\mathbf{k}1}(\mathbf{r}), v_1(\sigma)\psi_{n\mathbf{k}2}(\mathbf{r}), \ldots, v_1(\sigma)\psi_{n\mathbf{k}n_\alpha}(\mathbf{r}), \\ &v_2(\sigma)\psi_{n\mathbf{k}1}(\mathbf{r}), v_2(\sigma)\psi_{n\mathbf{k}2}(\mathbf{r}), \ldots, v_2(\sigma)\psi_{n\mathbf{k}n_\alpha}(\mathbf{r}), \end{aligned} \qquad \text{(C.13.3)}$$

i.e. with spin degeneracy included the total degeneracy of the energy level is $2n_\alpha$. The spin-dependent Bloch functions (C.13.3) act as basis functions for a $2n_\alpha$-dimensional representation of the double group of the wave-vector **k**. This representation may or may not be reducible. The fact that this representation may be reducible has important physical consequences.

If the rotation Q_l is characterized by the Eulerian angles θ_1, θ_2 and θ_3,

$$\{Q_l|\mathbf{v}_l + \mathbf{R}_t\}[v_f(\sigma)\psi_{n\mathbf{k}j}(\mathbf{r})] = \sum_{d=1}^{2}\sum_{p=1}^{n_\alpha} D^l_{df}(\theta_1, \theta_2, \theta_3)M^{\alpha lt}_{pj}v_d(\sigma)\psi_{n\mathbf{k}p}(\mathbf{r})$$

$$= \sum_{d=1}^{2}\sum_{p=1}^{n_\alpha} [D^l(\theta_1, \theta_2, \theta_3) \times M^{\alpha lt}]_{dp,\,fj}v_d(\sigma)\psi_{n\mathbf{k}p}(\mathbf{r}), \qquad \text{(C.13.4)}$$

where $D^l(\theta_1, \theta_2, \theta_3)$ is the matrix given in (C.12.33) and $[D^l(\theta_1, \theta_2, \theta_3) \times M^{\alpha lt}]_{dp,fj}$ denotes the (dp, fj) element of the direct product $D^l(\theta_1, \theta_2, \theta_3) \times M^{\alpha lt}$, d and p labelling the rows and f and j labelling the columns. $D^l \times M^{\alpha lt}$ is a $2n_\alpha \times 2n_\alpha$ matrix given by

$$D^l \times M^{\alpha lt} = \begin{pmatrix} D^l_{11} & D^l_{12} \\ D^l_{21} & D^l_{22} \end{pmatrix} \times M^{\alpha lt} = \begin{pmatrix} D^l_{11}M^{\alpha lt} & D^l_{12}M^{\alpha lt} \\ D^l_{21}M^{\alpha lt} & D^l_{22}M^{\alpha lt} \end{pmatrix}$$

$$= \begin{pmatrix} D^l_{11}M^{\alpha lt}_{11} \ldots D^l_{11}M^{\alpha lt}_{1n_\alpha} D^l_{12}M^{\alpha lt}_{11} \ldots D^l_{12}M^{\alpha lt}_{1n_\alpha} \\ \vdots \qquad\qquad \vdots \qquad\qquad \vdots \\ D^l_{11}M^{\alpha lt}_{n_\alpha 1} \ldots D^l_{11}M^{\alpha lt}_{n_\alpha n_\alpha} D^l_{12}M^{\alpha lt}_{n_\alpha 1} \ldots D^l_{12}M^{\alpha lt}_{n_\alpha n_\alpha} \\ D^l_{21}M^{\alpha lt}_{11} \ldots D^l_{21}M^{\alpha lt}_{1n_\alpha} D^l_{22}M^{\alpha lt}_{11} \ldots D^l_{22}M^{\alpha lt}_{1n_\alpha} \\ \vdots \qquad\qquad \vdots \qquad\qquad \vdots \\ D^l_{21}M^{\alpha lt}_{n_\alpha 1} \ldots D^l_{21}M^{\alpha lt}_{n_\alpha n_\alpha} D^l_{22}M^{\alpha lt}_{n_\alpha 1} \ldots D^l_{22}M^{\alpha lt}_{n_\alpha n_\alpha} \end{pmatrix} \qquad \text{(C.13.5)}$$

If the spin-dependent Bloch functions (C.13.3) are written as a column vector

$$\boldsymbol{\psi}_{n\mathbf{k}}(\sigma, \mathbf{r}) \equiv \begin{pmatrix} v_1(\sigma)\psi_{n\mathbf{k}1}(\mathbf{r}) \\ v_1(\sigma)\psi_{n\mathbf{k}2}(\mathbf{r}) \\ \vdots \\ v_1(\sigma)\psi_{n\mathbf{k}n_\alpha}(\mathbf{r}) \\ v_2(\sigma)\psi_{n\mathbf{k}1}(\mathbf{r}) \\ v_2(\sigma)\psi_{n\mathbf{k}2}(\mathbf{r}) \\ \vdots \\ v_2(\sigma)\psi_{n\mathbf{k}n_\alpha}(\mathbf{r}) \end{pmatrix} \qquad \text{(C.13.6)}$$

then

$$\{Q_l|\mathbf{v}_l + \mathbf{R}_t\}\psi_{n\mathbf{k}}(\sigma, \mathbf{r}) = \widetilde{D^l \times M^{\alpha l t}}\psi_{n\mathbf{k}}(\sigma, \mathbf{r}), \qquad \text{(C.13.7)}$$

where $\widetilde{D^l \times M^{\alpha l t}}$ denotes the transpose of the matrix $D^l \times M^{\alpha l t}$. Obviously

$$\{\bar{Q}_l|\mathbf{v}_l + \mathbf{R}_t\}\psi_{n\mathbf{k}}(\sigma, \mathbf{r}) = -\widetilde{D^l \times M^{\alpha l t}}\psi_{n\mathbf{k}}(\sigma, \mathbf{r}). \qquad \text{(C.13.8)}$$

The spin-dependent Bloch functions (C.13.3) therefore act as basis functions for a $2n_\alpha$-dimensional representation of the double group of the wave-vector $\mathbf{k}$ in which the element $\{Q_l|\mathbf{v}_l + \mathbf{R}_t\}$ is represented by the matrix $D^l \times M^{\alpha l t}$ and the element $\{\bar{Q}_l|\mathbf{v}_l + \mathbf{R}_t\}$ is represented by the matrix $-D^l \times M^{\alpha l t}$.

From (C.12.33) and (C.13.5),

$$\begin{aligned}\text{trace }[D^l(\theta_1, \theta_2, \theta_3) \times M^{\alpha l t}] &= \{\text{trace }[D^l(\theta_1, \theta_2, \theta_3]\}\{\text{trace }[M^{\alpha l t}]\} \\ &= 2\cos\tfrac{1}{2}\theta_2 \cos\tfrac{1}{2}(\theta_1 + \theta_3)\,\chi^{\alpha l t}.\end{aligned} \qquad \text{(C.13.9)}$$

The traces of these matrices can therefore easily be found and, using the theory of group characters (§C.5), we can find the irreducible representations of the double group which appear along the diagonals in the completely reduced forms of these matrices.

The results obtained in this way for the point Γ in the first Brillouin zone of the InSb structure are given in table C.13.1.

Table C.13.1 Double group irreducible representations in the absence of spin-dependent forces at the point Γ in the first Brillouin zone of the InSb structure.

$D \times \Gamma_1 = \Gamma_6$,	$D \times \Gamma_2 = \Gamma_7$,	$D \times \Gamma_{12} = \Gamma_8$,
$D \times \Gamma_{15} = \Gamma_7 + \Gamma_8$,	$D \times \Gamma_{25} = \Gamma_6 + \Gamma_8$.	

Table C.13.1 tells us, for example, that in the absence of spin–orbit coupling the six-fold degenerate (spin degeneracy included) Γ_{15} state is in fact made up of a two-fold degenerate Γ_7 state and a four-fold degenerate Γ_8 state which are accidentally degenerate with one another. This means in turn that spin–orbit coupling will split the six-fold degenerate Γ_{15} state into a two-fold degenerate Γ_7 state and a four-fold degenerate Γ_8 state. This effect is shown schematically in fig. C.13.1 where the conduction and valence bands in InSb are shown along the Δ-axis (a) before and (b) after the inclusion of spin–orbit coupling. Use has also been made of the compatibility relations (table C.12.9).

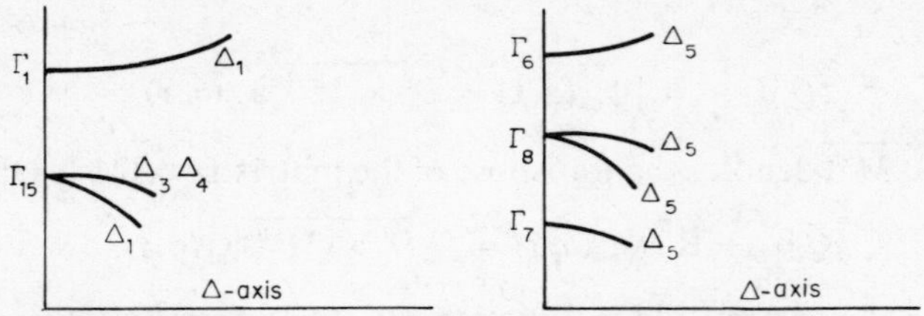

Fig. C.13.1. Spin–orbit splitting in InSb (schematic). (a) Valence and conduction bands without spin–orbit coupling. (b) Valence and conduction bands with spin–orbit coupling.

The difference between the energy of the Γ_8 state and the energy of the Γ_7 state is the spin–orbit splitting at $\mathbf{k} = 0$ in InSb.

The results obtained for the point Γ in the first Brillouin zone of the Ge structure are given in table C.13.2.

Table C.13.2 Double group irreducible representations in the absence of spin-dependent forces at the point Γ in the first Brillouin zone of the Ge structure.

$D \times \Gamma_1 = \Gamma_6^+$,	$D \times \Gamma_1' = \Gamma_6^-$,	$D \times \Gamma_2 = \Gamma_7^+$,	$D \times \Gamma_2' = \Gamma_7^-$,
$D \times \Gamma_{12} = \Gamma_8^+$,	$D \times \Gamma_{12}' = \Gamma_8^-$,	$D \times \Gamma_{15} = \Gamma_6^- + \Gamma_8^-$,	
$D \times \Gamma_{15}' = \Gamma_6^+ + \Gamma_8^+$,	$D \times \Gamma_{25} = \Gamma_7^- + \Gamma_8^-$,	$D \times \Gamma_{25}' = \Gamma_7^+ + \Gamma_8^+$.	

Table C.13.2 tells us, for example, that the six-fold degenerate (spin degeneracy included) Γ_{25}' state will be split by spin–orbit coupling into a two-fold degenerate Γ_7^+ state and a four-fold degenerate Γ_8^+ state. This effect is shown schematically in fig. C.13.2, where the conduction and valence bands in Ge are shown along the Δ-axis (a) before and (b) after the inclusion of spin–orbit coupling. Use has also been made of the compatibility relations (table C.12.10).

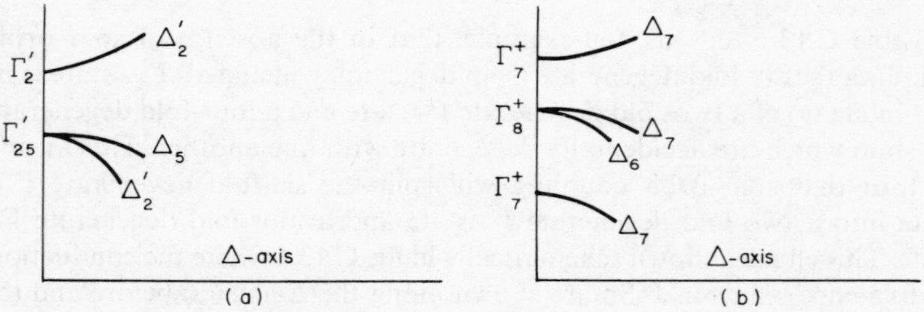

Fig. C.13.2. Spin–orbit splitting in Ge (schematic). (a) Valence and conduction bands without spin–orbit coupling. (b) Valence and conduction bands with spin–orbit coupling.

The difference between the energy of the Γ_8^+ state and the energy of the lower Γ_7^+ state is the spin–orbit splitting at $\mathbf{k} = 0$ in Ge.

C.13a Perturbation calculation of the spin-orbit splitting in InSb and Ge

(i) *InSb*

In the absence of spin–orbit interaction the Γ_{15} state in InSb, with energy $E(\Gamma_{15})$, is six-fold degenerate (spin degeneracy included) with wave-functions

$$\begin{aligned}
\psi_1(\sigma, \mathbf{r}) &= \nu_1(\sigma)\psi_1^{\Gamma_{15}}(\mathbf{r}), \qquad \psi_2(\sigma, \mathbf{r}) = \nu_2(\sigma)\psi_1^{\Gamma_{15}}(\mathbf{r}),\\
\psi_3(\sigma, \mathbf{r}) &= \nu_1(\sigma)\psi_2^{\Gamma_{15}}(\mathbf{r}), \qquad \psi_4(\sigma, \mathbf{r}) = \nu_2(\sigma)\psi_2^{\Gamma_{15}}(\mathbf{r}),\\
\psi_5(\sigma, \mathbf{r}) &= \nu_1(\sigma)\psi_3^{\Gamma_{15}}(\mathbf{r}), \qquad \psi_6(\sigma, \mathbf{r}) = \nu_2(\sigma)\psi_3^{\Gamma_{15}}(\mathbf{r}), \qquad \text{(C.13.10)}
\end{aligned}$$

where the functions $\psi_1^{\Gamma_{15}}(\mathbf{r})$, $\psi_2^{\Gamma_{15}}(\mathbf{r})$ and $\psi_3^{\Gamma_{15}}(\mathbf{r})$ act as basis functions for the Γ_{15} irreducible representation of the single group of the wave-vector at the point Γ in the first Brillouin zone of the InSb structure.

We take as the perturbation the spin–orbit interaction term

$$H^{(1)} \equiv -\frac{i\hbar}{2m^2c^2}\mathbf{s}\cdot(\nabla V(\mathbf{r}) \times \nabla) \equiv -\mathrm{i}\sum_{j=1}^{3} s_j P_j, \qquad \text{(C.13.11)}$$

where the spin angular momentum operator $\mathbf{s} = (s_1, s_2, s_3)$ acts only on the spin-dependent parts of the wavefunctions. If

$$H_{jq}^{(1)} \equiv (\psi_j(\sigma, \mathbf{r}), H^{(1)}\psi_q(\sigma, \mathbf{r})) \equiv \sum_{\sigma=\pm 1} \int \psi_j^*(\sigma, \mathbf{r})H^{(1)}\psi_q(\sigma, \mathbf{r})\, d\mathbf{r}, \qquad \text{(C.13.12)}$$

the perturbation splits the energy level $E(\Gamma_{15})$ into levels $E(\Gamma_{15}) + \epsilon_j$, where the ϵ_j are given by the roots of the equation

$$\det|H_{jq}^{(1)} - \epsilon\delta_{jq}| = 0. \qquad \text{(C.13.13)}$$

This is a standard result of perturbation theory[33].

Using (C.12.9), we see that

$$\begin{aligned}
H^{(1)}\psi_1(\sigma, \mathbf{r}) &= -\mathrm{i}\hbar/2[P_1\psi_2(\sigma, \mathbf{r}) + \mathrm{i}P_2\psi_2(\sigma, \mathbf{r}) + P_3\psi_1(\sigma, \mathbf{r})],\\
H^{(1)}\psi_2(\sigma, \mathbf{r}) &= -\mathrm{i}\hbar/2[P_1\psi_1(\sigma, \mathbf{r}) - \mathrm{i}P_2\psi_1(\sigma, \mathbf{r}) - P_3\psi_2(\sigma, \mathbf{r})],\\
H^{(1)}\psi_3(\sigma, \mathbf{r}) &= -\mathrm{i}\hbar/2[P_1\psi_4(\sigma, \mathbf{r}) + \mathrm{i}P_2\psi_4(\sigma, \mathbf{r}) + P_3\psi_3(\sigma, \mathbf{r})],\\
H^{(1)}\psi_4(\sigma, \mathbf{r}) &= -\mathrm{i}\hbar/2[P_1\psi_3(\sigma, \mathbf{r}) - \mathrm{i}P_2\psi_3(\sigma, \mathbf{r}) - P_3\psi_4(\sigma, \mathbf{r})],\\
H^{(1)}\psi_5(\sigma, \mathbf{r}) &= -\mathrm{i}\hbar/2[P_1\psi_6(\sigma, \mathbf{r}) + \mathrm{i}P_2\psi_6(\sigma, \mathbf{r}) + P_3\psi_5(\sigma, \mathbf{r})],\\
H^{(1)}\psi_6(\sigma, \mathbf{r}) &= -\mathrm{i}\hbar/2[P_1\psi_5(\sigma, \mathbf{r}) - \mathrm{i}P_2\psi_5(\sigma, \mathbf{r}) - P_3\psi_6(\sigma, \mathbf{r})].
\end{aligned} \qquad \text{(C.13.14)}$$

[33] L. I. Schiff, *Quantum Mechanics, 2nd ed.*, McGraw-Hill, New York, 1955, p. 156.

Because of the orthogonality of the spin functions $v_1(\sigma)$ and $v_2(\sigma)$, it now follows that

$$H_{11}^{(1)} \equiv \sum_{\sigma=\pm 1} \int \psi_1^*(\sigma, \mathbf{r}) H^{(1)} \psi_1(\sigma, \mathbf{r})\, d\mathbf{r} = -i\hbar/2 \int \psi_1^{\Gamma_{15}*}(\mathbf{r}) P_3 \psi_1^{\Gamma_{15}}(\mathbf{r})\, d\mathbf{r},$$

$$H_{12}^{(1)} = -i\hbar/2 \int \psi_1^{\Gamma_{15}*}(\mathbf{r})[P_1 - iP_2]\psi_1^{\Gamma_{15}}(\mathbf{r})\, d\mathbf{r},$$

$$H_{13}^{(1)} = -i\hbar/2 \int \psi_1^{\Gamma_{15}*}(\mathbf{r}) P_3 \psi_2^{\Gamma_{15}}(\mathbf{r})\, d\mathbf{r}, \text{ etc.} \tag{C.13.15}$$

The matrix elements $H_{jq}^{(1)}$ are therefore expressible in terms of the matrix elements of the components P_j $(j = 1, 2, 3)$ of the operator $\mathbf{P}$ taken between the unperturbed spin-independent functions $\psi_1^{\Gamma_{15}}(\mathbf{r})$, $\psi_2^{\Gamma_{15}}(\mathbf{r})$ and $\psi_3^{\Gamma_{15}}(\mathbf{r})$. These functions act as basis functions for the Γ_{15} irreducible representation of the single group of the wave-vector at the point Γ and, in fact, behave under the operations of the point group T_d in exactly the same way as the coordinates x_1, x_2 and x_3, e.g. one element of the point group T_d is $Q_{11}(x_3, -x_1, -x_2)$ and

$$Q_{11}\psi_1^{\Gamma_{15}}(\mathbf{r}) = \psi_3^{\Gamma_{15}}(\mathbf{r}), \quad Q_{11}\psi_2^{\Gamma_{15}}(\mathbf{r}) = -\psi_1^{\Gamma_{15}}(\mathbf{r}), \quad Q_{11}\psi_3^{\Gamma_{15}}(\mathbf{r}) = -\psi_2^{\Gamma_{15}}(\mathbf{r}). \tag{C.13.16}$$

If Q_t is an element of the point group T_d which corresponds to the orthogonal transformations of the coordinates

$$x'_j = \sum_{w=1}^{3} q_{jw}^t x_w \qquad (j = 1, 2, 3), \tag{C.13.17}$$

then

$$Q_t \psi_j^{\Gamma_{15}}(\mathbf{r}) = \sum_{w=1}^{3} q_{jw}^t \psi_w^{\Gamma_{15}}(\mathbf{r}) \qquad (j = 1, 2, 3). \tag{C.13.18}$$

Also, from (C.12.18),

$$Q_t P_j = \pm \sum_{w=1}^{3} q_{jw}^t P_w \qquad (j = 1, 2, 3), \tag{C.13.19}$$

where the plus sign appears if $\det|q_{jw}^t| = 1$ and the minus sign appears if $\det|q_{jw}^t| = -1$. Since the point group operations leave inner products invariant, it follows from (C.13.18) and (C.13.19) that

$$\int \psi_j^{\Gamma_{15}*}(\mathbf{r}) P_w \psi_v^{\Gamma_{15}}(\mathbf{r})\, d\mathbf{r} = \int [Q_t\psi_j^{\Gamma_{15}}(\mathbf{r})]^* [Q_t P_w][Q_t \psi_v^{\Gamma_{15}}(\mathbf{r})]\, d\mathbf{r}$$

$$= \pm \sum_{l=1}^{3} \sum_{u=1}^{3} \sum_{d=1}^{3} q_{jl}^t q_{wu}^t q_{vd}^t \int \psi_l^{\Gamma_{15}*}(\mathbf{r}) P_u \psi_d^{\Gamma_{15}}(\mathbf{r})\, d\mathbf{r}. \tag{C.13.20}$$

(C.13.20) enables us to find relationships between the matrix elements of the components P_j ($j = 1, 2, 3$) of the operator $\mathbf{P}$. In fact, using (C.13.20), one can show that there are only six non-zero matrix elements[34] of the components P_j of the operator $\mathbf{P}$ and that these six matrix elements are related as follows:

$$\begin{aligned}\int \psi_1^{\Gamma_{15}*}(\mathbf{r})P_3\psi_2^{\Gamma_{15}}(\mathbf{r})\,d\mathbf{r} &= \int \psi_2^{\Gamma_{15}*}(\mathbf{r})P_1\psi_3^{\Gamma_{15}}(\mathbf{r})\,d\mathbf{r} \\ &= \int \psi_3^{\Gamma_{15}*}(\mathbf{r})P_2\psi_1^{\Gamma_{15}}(\mathbf{r})\,d\mathbf{r} \\ &= -\int \psi_1^{\Gamma_{15}*}(\mathbf{r})P_2\psi_3^{\Gamma_{15}}(\mathbf{r})\,d\mathbf{r} \\ &= -\int \psi_2^{\Gamma_{15}*}(\mathbf{r})P_3\psi_1^{\Gamma_{15}}(\mathbf{r})\,d\mathbf{r} \\ &= -\int \psi_3^{\Gamma_{15}*}(\mathbf{r})P_1\psi_2^{\Gamma_{15}}(\mathbf{r})\,d\mathbf{r} = W \text{ (say).}\end{aligned} \tag{C.13.21}$$

Using (C.13.14), (C.13.15) and (C.13.21) it now follows that

$$\begin{aligned}H_{11}^{(1)} &= H_{12}^{(1)} = H_{14}^{(1)} = H_{15}^{(1)} = H_{21}^{(1)} = H_{22}^{(1)} = H_{23}^{(1)} = H_{26}^{(1)} = H_{32}^{(1)} = H_{33}^{(1)} \\ &= H_{34}^{(1)} = H_{35}^{(1)} = H_{41}^{(1)} = H_{43}^{(1)} = H_{44}^{(1)} = H_{46}^{(1)} = H_{51}^{(1)} = H_{53}^{(1)} = H_{55}^{(1)} \\ &= H_{56}^{(1)} = H_{62}^{(1)} = H_{64}^{(1)} = H_{65}^{(1)} = H_{66}^{(1)} = 0,\\ H_{13}^{(1)} &= H_{24}^{(1)*} = H_{31}^{(1)*} = H_{36}^{(1)} = H_{42}^{(1)} = H_{45}^{(1)} = H_{54}^{(1)*} = H_{63}^{(1)*} = -i\hbar W/2,\\ H_{16}^{(1)} &= -H_{25}^{(1)} = -H_{52}^{(1)} = H_{61}^{(1)} = \hbar W/2.\end{aligned} \tag{C.13.22}$$

The secular equation (C.13.13) then reduces to

$$(\epsilon - \hbar W/2)^4(\epsilon + \hbar W)^2 = 0 \tag{C.13.23}$$

and it follows that

$$E(\Gamma_8) = E(\Gamma_{15}) + \hbar W/2, \tag{C.13.24}$$

$$E(\Gamma_7) = E(\Gamma_{15}) - \hbar W. \tag{C.13.25}$$

Hence, to this approximation, the spin–orbit splitting $E(\Gamma_8) - E(\Gamma_7)$ at $\mathbf{k} = 0$ is $3\hbar W/2$, where W is defined in (C.13.21).

[34] See also the discussion in §E.14.

(ii) *Ge*

In the absence of spin–orbit interaction the Γ'_{25} state in Ge, with energy $E(\Gamma'_{25})$, is six-fold degenerate with wavefunctions

$$\psi_1(\sigma,\mathbf{r}) = v_1(\sigma)\psi_1^{\Gamma'_{25}}(\mathbf{r}), \qquad \psi_2(\sigma,\mathbf{r}) = v_2(\sigma)\psi_1^{\Gamma'_{25}}(\mathbf{r}),$$
$$\psi_3(\sigma,\mathbf{r}) = v_1(\sigma)\psi_2^{\Gamma'_{25}}(\mathbf{r}), \qquad \psi_4(\sigma,\mathbf{r}) = v_2(\sigma)\psi_2^{\Gamma'_{25}}(\mathbf{r}),$$
$$\psi_5(\sigma,\mathbf{r}) = v_1(\sigma)\psi_3^{\Gamma'_{25}}(\mathbf{r}), \qquad \psi_6(\sigma,\mathbf{r}) = v_2(\sigma)\psi_3^{\Gamma'_{25}}(\mathbf{r}), \qquad \text{(C.13.26)}$$

where the functions $\psi_1^{\Gamma'_{25}}(\mathbf{r})$, $\psi_2^{\Gamma'_{25}}(\mathbf{r})$ and $\psi_3^{\Gamma'_{25}}(\mathbf{r})$ act as basis functions for the Γ'_{25} irreducible representation of the single group of the wave-vector at the point Γ in the first Brillouin zone of the Ge structure.

Once again the matrix elements $H_{jq}^{(1)} \equiv (\psi_j(\sigma,\mathbf{r}),\ H^{(1)}\psi_q(\sigma,\mathbf{r}))$ are expressible in terms of the matrix elements of the components P_j ($j = 1, 2, 3$) of the operator $\mathbf{P}$ taken between the unperturbed spin-independent functions $\psi_1^{\Gamma'_{25}}(\mathbf{r})$, $\psi_2^{\Gamma'_{25}}(\mathbf{r})$ and $\psi_3^{\Gamma'_{25}}(\mathbf{r})$. These functions behave under the operations of the point group O_h in exactly the same way as the functions x_2x_3, x_1x_3 and x_1x_2. Using this fact, we find that there are again only six non-zero matrix elements of the components P_j of the operator $\mathbf{P}$ and that these six non-zero matrix elements are related as follows:

$$\begin{aligned}\int \psi_1^{\Gamma'_{25}*}(\mathbf{r})P_3\psi_2^{\Gamma'_{25}}(\mathbf{r})\,d\mathbf{r} &= \int \psi_2^{\Gamma'_{25}*}(\mathbf{r})P_1\psi_3^{\Gamma'_{25}}(\mathbf{r})\,d\mathbf{r}\\ &= \int \psi_3^{\Gamma'_{25}}(\mathbf{r})P_2\psi_1^{\Gamma'_{25}}(\mathbf{r})\,d\mathbf{r}\\ &= -\int \psi_1^{\Gamma'_{25}*}(\mathbf{r})P_2\psi_3^{\Gamma'_{25}}(\mathbf{r})\,d\mathbf{r}\\ &= -\int \psi_2^{\Gamma'_{25}*}(\mathbf{r})P_3\psi_1^{\Gamma'_{25}}(\mathbf{r})\,d\mathbf{r}\\ &= -\int \psi_3^{\Gamma'_{25}*}(\mathbf{r})P_1\psi_2^{\Gamma'_{25}}(\mathbf{r})\,d\mathbf{r} = Z \text{ (say)}.\end{aligned} \qquad \text{(C.13.27)}$$

Using these results we find that the only non-zero matrix elements $H_{jq}^{(1)}$ which appear in the secular determinant are

$$H_{13}^{(1)} = H_{31}^{(1)*} = H_{24}^{(1)*} = H_{42}^{(1)} = H_{36}^{(1)} = H_{63}^{(1)*} = H_{45}^{(1)} = H_{54}^{(1)*} = -i\hbar Z/2,$$
$$H_{16}^{(1)} = H_{61}^{(1)} = -H_{25}^{(1)} = -H_{52}^{(1)} = \hbar Z/2, \qquad \text{(C.13.28)}$$

and the secular equation reduces to

$$(\epsilon - \hbar Z/2)^4(\epsilon + \hbar Z)^2 = 0. \qquad \text{(C.13.29)}$$

It follows that

$$E(\Gamma_8^+) = E(\Gamma'_{25}) + \hbar Z/2, \tag{C.13.30}$$

$$E(\Gamma_7^+) = E(\Gamma'_{25}) - \hbar Z, \tag{C.13.31}$$

and hence to this approximation the spin–orbit splitting $E(\Gamma_8^+) - E(\Gamma_7^+)$ at $\mathbf{k} = 0$ is $3\hbar Z/2$, where Z is defined in (C.13.27).

Further reading

Altmann, S. L., *Group Theory*, *Quantum Theory*, Academic Press, New York, 1962.
Hall, G. G., *Applied Group Theory*, Longmans, London, 1967.
Hamermesh, M., *Group Theory*, Addison-Wesley, Reading, Massachusetts, 1962.
Lomont, J. S., *Applications of Finite Groups*, Academic Press, New York, 1959.
McWeeney, R. *Symmetry—An Introduction to Group Theory and its Applications*, Pergamon Press, London, 1963.
Meijer, P. H. E. and Bauer, E., *Group Theory: The Application to Quantum Mechanics*, North Holland, Amsterdam, 1962.
Murnaghan, F. D., *The Theory of Group Representations*, Johns Hopkins, Baltimore, Maryland, 1938.
Tinkham, M. *Group Theory and Quantum Mechanics*, McGraw-Hill, New York, 1964.

PART D

STATISTICS OF ELECTRONS AND APPLICATIONS TO DEVICES

D. A. Evans

CHAPTER VI

Equilibrium Statistics

D.1 QUANTUM STATES FOR ELECTRONS IN A PURE CRYSTAL

The purpose of this chapter is to discuss what quantum states for electrons, arising either from the structure of the crystal or from impurities and defects in the structure, exist in a semiconductor, and how the electrons are distributed among these states in equilibrium. To facilitate this, we restate in this section in a simple way and in an appropriate notation the results obtained earlier in this book.

Since electrons moving in a periodic crystal structure have some properties in common with electrons moving freely in a constant potential, it is convenient to start with a discussion of such free electrons. Consider, then, an electron moving with constant potential energy U in a cubical box of side L and volume L^3. The time-independent Schrödinger equation for the wave function ψ is

$$-\frac{\hbar^2}{2m}\nabla^2\psi + [E - U]\psi = 0, \qquad \text{(D.1.1)}$$

where E is the electron's total energy and m its mass. Since U is a constant, the solutions of (D.1.1) are plane waves,

$$\psi(\mathbf{r}) = C\,\mathrm{e}^{\mathrm{i}\mathbf{k}\cdot\mathbf{r}} = C\exp[\mathrm{i}(xk_x + yk_y + zk_z)], \qquad \text{(D.1.2)}$$

where $\mathbf{k}$ is known as the wave-vector and is related to the energy and to the momentum $\mathbf{p}$ by

$$\frac{\hbar^2k^2}{2m} = E - U, \qquad k \equiv |\mathbf{k}|, \qquad \mathbf{p} = \hbar\mathbf{k}. \qquad \text{(D.1.3)}$$

The constant energy surfaces $E = E_1$ are spheres in $\mathbf{k}$-space, the radius of a typical sphere being

$$k_1 = \left[\frac{2m(E_1 - U)}{\hbar^2}\right]^{\frac{1}{2}} \qquad \text{(D.1.4)}$$

and its volume in $\mathbf{k}$-space being

$$V_{\mathbf{k}}(E_1) = \frac{4\pi}{3}k_1^3 = \frac{4\pi}{3}\left[\frac{2m(E_1 - U)}{\hbar^2}\right]^{\frac{3}{2}}. \qquad \text{(D.1.5)}$$

However, not all values of **k** are allowed. This may be seen by considering the x component of the momentum, $p_x = \hbar k_x$. Since the position x of the electron is known to be between 0 and L, the momentum must be uncertain to an extent

$$\Delta p_x = \frac{h}{L}, \qquad \Delta k_x = \frac{h}{\hbar L} = \frac{2\pi}{L}. \tag{D.1.6}$$

Thus allowed values of k_x are separated by a distance $2\pi/L$. The same is clearly true of k_y and k_z, so that the allowed states can be thought of as cubes of side $2\pi/L$, the actual value of **k** for each state being somewhere in the relevant cube. A more rigorous approach (see §B.6) in terms of periodic boundary conditions leads to the same result.

We conclude that the number of allowed values of **k** in any volume $V_{\mathbf{k}}$ of **k**-space is the number of cubes of side $2\pi/L$ contained in that volume. Since *two* electron states of different spin correspond to each value of **k**, the number of electron states in a volume $V_{\mathbf{k}}$ of **k**-space is

$$n(V_{\mathbf{k}}) = \frac{2V_{\mathbf{k}}}{(2\pi/L)^3} = \frac{L^3 V_{\mathbf{k}}}{4\pi^3}. \tag{D.1.7}$$

In particular, the number of electron states with energies less than E_1 is (D.1.7) with $V_{\mathbf{k}}$ identified as the volume of the constant-energy surface $E = E_1$ and given by (D.1.5). Thus, denoting this number by $n(E_1)$,

$$n(E_1) = \frac{L^3}{3\pi^2}\left[\frac{2m(E_1 - U)}{\hbar^2}\right]^{\frac{3}{2}}. \tag{D.1.8}$$

It is usual to define a density of states $g(E)$ so that $g(E)\,\mathrm{d}E$ is the number of states per unit sample volume in the energy range $E \to E + \mathrm{d}E$. In this case

$$g(E) \equiv \frac{1}{L^3}\frac{\mathrm{d}n(E)}{\mathrm{d}E} = \frac{1}{2\pi^2}\left(\frac{2m}{\hbar^2}\right)^{\frac{3}{2}}(E_1 - U)^{\frac{1}{2}}. \tag{D.1.9}$$

Returning now to the motion of electrons in a crystal lattice, the potential energy of an electron can be expressed as the sum of a potential due to the relatively fixed atoms of the lattice and a potential due to all the other electrons. To make the problem manageable it is assumed that the electrical potential due to the electrons can be replaced by its time-averaged value at any point, and that the resulting potential energy $U(\mathbf{r})$ has lattice periodicity in the sense that if $\mathbf{R_n}$ is any vector of the direct lattice as defined in §B.2, then

$$U(\mathbf{r} + \mathbf{R_n}) = U(\mathbf{r}). \tag{D.1.10}$$

The Schrödinger equation is now

$$-\frac{\hbar^2}{2m}\nabla^2\psi + [E - V(\mathbf{r})]\psi = 0, \tag{D.1.11}$$

and according to the Bloch theorem (see §B.11), solutions exist of the form

$$\psi_{\mathbf{k}}(\mathbf{r}) = u_{\mathbf{k}}(\mathbf{r})\, e^{i\mathbf{k}\cdot\mathbf{r}}, \tag{D.1.12}$$

where $u_{\mathbf{k}}(\mathbf{r})$ has lattice periodicity as a function of $\mathbf{r}$. $\psi_{\mathbf{k}}(\mathbf{r})$ is called a Bloch function and $u_{\mathbf{k}}(\mathbf{r})$ the modulating part of the Bloch function.

The functional dependence of E on $\mathbf{k}$ depends, of course, on the particular crystal concerned. It is often true however that some ranges of E correspond to real $\mathbf{k}$ and others to complex $\mathbf{k}$. Since a solution of the form (D.1.12) with complex $\mathbf{k}$ contains a factor $\exp(ax + by + cz)$ with a, b, c not all zero, it cannot satisfy periodic boundary conditions. Therefore no physically real states exist in the ranges of energy corresponding to complex $\mathbf{k}$. Such a range is called a forbidden energy band, and the ranges corresponding to real $\mathbf{k}$ are called allowed energy bands.

There exist (see §B.2, §B.3 and §B.5) vectors $\mathbf{K}_{\mathbf{m}}$ in $\mathbf{k}$-space with the property that if $\mathbf{R}_{\mathbf{n}}$ is any vector of the direct lattice, $\mathbf{K}_{\mathbf{m}}\cdot\mathbf{R}_{\mathbf{n}}$ is 2π times an integer. It follows that a wave function such as (D.1.12) can equally well be expressed as

$$\psi_{\mathbf{k}}(\mathbf{r}) = [u_{\mathbf{k}}(\mathbf{r})\, e^{i\mathbf{K}_{\mathbf{m}}\cdot\mathbf{r}}]\, e^{i(\mathbf{k}-\mathbf{K}_{\mathbf{m}})\cdot\mathbf{r}}, \tag{D.1.13}$$

the factor in square brackets having lattice periodicity, so that (D.1.13) is an acceptable Bloch function with wave-vector $(\mathbf{k} - \mathbf{K}_{\mathbf{m}})$. By the use of such transformations any Bloch function can be assigned a wave-vector in a finite region of $\mathbf{k}$-space containing the origin. Within this region, known as the first Brillouin zone, E is a multi-valued function of $\mathbf{k}$, the different branches of the function corresponding to different allowed energy bands.

In a semiconductor, attention is usually focussed on a particular forbidden band and the two adjacent allowed bands. The upper allowed band is called the conduction band and is defined in energy by $E \geqslant E_c$, the upper limit of the band being irrelevant. The lower band, called the valence band, is defined by $E \leqslant E_v$, the lower limit being irrelevant. The width of the forbidden gap is denoted by

$$E_g \equiv E_c - E_v. \tag{D.1.14}$$

The energy as a function of $\mathbf{k}$ in the conduction band, $E_c(\mathbf{k})$, evidently has a minimum value E_c at some value $\mathbf{k}_c$ of $\mathbf{k}$. If the energy is expanded in a Taylor series about $\mathbf{k} = \mathbf{k}_c$ and only second order terms are retained, one finds in general that

$$\begin{aligned} E_c(\mathbf{k}) = E_c &+ a_{xx}(k_x - k_{cx})^2 + a_{yy}(k_y - k_{cy})^2 + a_{zz}(k_z - k_{cz})^2 \\ &+ 2a_{xy}(k_x - k_{cx})(k_y - k_{cy}) + 2a_{yz}(k_y - k_{cy})(k_z - k_{cz}) \\ &+ 2a_{zx}(k_z - k_{cz})(k_x - k_{cx}), \end{aligned} \tag{D.1.15}$$

where $a_{xx} \equiv \frac{1}{2}\left(\frac{\partial^2 E_c}{\partial k_x^2}\right)_{\mathbf{k}=\mathbf{k}_c}$ and the other a's are similarly defined. (D.1.15) is the equation of an ellipsoid, and if the coordinate axes are chosen parallel to the principal axes of the ellipsoid the cross terms a_{xy}, a_{yz}, a_{zx} vanish. The resulting expression can be rewritten as

$$\mathrm{E}_c(\mathbf{k}) = E_c + \frac{\hbar^2}{2}\left[\frac{(k_x - k_{cx})^2}{m_x} + \frac{(k_y - k_{cy})^2}{m_y} + \frac{(k_z - k_{cz})^2}{m_z}\right], \tag{D.1.16}$$

where $m_x \equiv \hbar^2/2a_{xx}$ etc. By analogy with (D.1.3), m_x, m_y, m_z are called the effective masses in the three principal directions. The constant energy surface $E_c(\mathbf{k}) = E_1$ is therefore an ellipsoid in $\mathbf{k}$-space, whose volume can be shown to be

$$V_{\mathbf{k}}(E_1) = \frac{4\pi}{3}(m_x m_y m_z)^{\frac{1}{2}}\left(\frac{2(E_1 - E_c)}{\hbar^2}\right)^{\frac{3}{2}}. \tag{D.1.17}$$

It will be assumed that the argument based on (D.1.6) remains valid. The calculation of a density of states $g(E)$ for the conduction band then follows exactly the pattern of (D.1.7)–(D.1.9), with the exception that, because of the crystal symmetry, there will in general be more than one conduction band minimum with energy E_c. Denoting the number of such minima by n_c, the counterpart of (D.1.9) is

$$g(E) = \frac{n_c(m_x m_y m_z)^{\frac{1}{2}}}{2\pi^2}\left(\frac{2}{\hbar^2}\right)^{\frac{3}{2}}(E - E_c)^{\frac{1}{2}} \tag{D.1.18}$$

or

$$g(E) = \frac{1}{2\pi^2}\left(\frac{2m_c}{\hbar^2}\right)^{\frac{3}{2}}(E - E_c)^{\frac{1}{2}}, \tag{D.1.19}$$

where the density-of-states effective mass for the conduction band is defined by

$$m_c \equiv \{n_c^2 m_x m_y m_z\}^{\frac{1}{3}}. \tag{D.1.20}$$

(D.1.19), being based on the neglect of third order terms in the Taylor expansion of E, is valid only for energies fairly close to E_c. It will appear (in §D.3) that most conduction band electrons have energies near E_c, so it will often be reasonable to use (D.1.19). Band structures may, however, be encountered for which the second derivatives a_{xx} etc. are all small compared to higher derivatives, and (D.1.19) will then not be accurate.

A very similar argument, starting with the expansion of the valence band energy $E_v(\mathbf{k})$ about a *maximum* value E_v at $\mathbf{k} = \mathbf{k}_v$,

$$E_v(\mathbf{k}) = E_v - \frac{\hbar^2}{2}\left[\frac{(k_x - k_{vx})^2}{m'_x} + \frac{(k_y - k_{vy})^2}{m'_y} + \frac{(k_z - k_{vz})^2}{m'_z}\right] \qquad \text{(D.1.21)}$$

leads, if there are n_v such maxima, to a density of states

$$g(E) = \frac{1}{2\pi^2}\left(\frac{2m_v}{\hbar^2}\right)^{\frac{3}{2}}(E_v - E)^{\frac{1}{2}} \qquad \text{(D.1.22)}$$

with a density-of-states effective mass for the valence band defined by

$$m_v \equiv \{n_v^2 m'_x m'_y m'_z\}^{\frac{1}{3}}. \qquad \text{(D.1.23)}$$

The simplest possible form for the conduction band would be one in which

$$n_c = 1, \qquad \mathbf{k}_c = 0, \qquad m_x = m_y = m_z = m_c, \qquad E = E_c + \frac{\hbar^2 k^2}{2m_c}. \qquad \text{(D.1.24)}$$

Such a conduction band is said to have a scalar effective mass, and comparison with (D.1.3) shows the close analogy between an electron moving in a crystal with such a conduction band and a free electron. It appears plausible that, for instance, the dynamical response of an electron to an external electric field or to the attraction of an impurity atom could then be discussed by regarding the electron as a free particle with an effective mass m_c, and a scalar effective mass is often assumed in considering such problems. Actual band structures, however, are often very far from the form (D.1.24). In germanium, for example, the conduction band has four minima, the effective masses in the principal directions being ($1{\cdot}64\,m$, $0{\cdot}082\,m$, $0{\cdot}082\,m$), so that $m_c \approx 0{\cdot}56\,m$. The valence band has two branches, for each of which $\mathbf{k}_v = 0$ and having the same value of E_v, for which the density-of-states effective masses are $m_{v1} = 0{\cdot}043\,m$ and $m_{v2} = 0{\cdot}36\,m$. The final density-of-states effective mass for this band is $m_v = (m_{v1}^{\frac{3}{2}} + m_{v2}^{\frac{3}{2}})^{\frac{2}{3}} \approx 0{\cdot}37\,m$. Further details of actual band structures are given in Blakemore[1]. For a more rigorous discussion of the effective mass concept, see Chapter III above.

[1] J. S. Blakemore, *Semiconductor Statistics*, Pergamon Press, London, 1962; see also discussion in §A.3. and problem (A.3.1).

D.2 QUANTUM STATES DUE TO IMPURITIES AND DEFECTS

In the previous section we considered a pure semiconductor, in which each atom has just enough electrons in its outer shell to form the valence bonds linking it to other atoms. If a single atom is replaced by an impurity atom which has a different number of outer shell electrons, a new situation is created. For definiteness we consider a pure semiconductor such as germanium or silicon which is in Group IV of the periodic table and therefore has four outer shell electrons.

Suppose first that a single semiconductor atom is replaced by an atom of a Group V element—nickel, phosphorus, arsenic, etc. The impurity atom has five outer shell electrons, only four of which can take part in valence bonds. The fifth electron is lightly bound to the atom and a small amount of energy supplied from the thermal motions of atoms will allow it to be given up to the conduction band. An impurity atom of this type is therefore called a *donor*.

In order to obtain a rough first estimate for the energy of an electron bound to such a donor, consider that the donor without this electron has a net positive charge equal to the electronic charge e. There is therefore an analogy with the hydrogen atom, which also consists of an electron bound to a relatively fixed positively charged nucleus. However, while the binding energy of a hydrogen atom in its lowest state is about 13·6 eV, the binding energy for an electron bound to a donor must be less than the energy gap between valence and conduction bands, which is of the order of one electron volt. The radius of the first Bohr orbit for the donor–electron system, a_{0d}, must therefore be considerably larger than the corresponding figure for hydrogen, $a_0 = 0{\cdot}53$ Å. Since the distance between nearest-neighbour atoms in, for example, germanium is about 2·5 Å, it is at least possible that the first Bohr orbit for the electron bound to a donor contains a large number of other atoms. If this is so, two other assumptions suggest themselves: (i) that the electron acts dynamically with a mass equal to an effective mass for the conduction band, m_c^*, rather than the free electron mass m, and (ii) that the dielectric polarization of the atoms inside the orbit reduces the electrostatic potential by a factor equal to the relative dielectric constant ϵ of the bulk material.

The Schrödinger equation for a hydrogen-like donor, i.e. a donor for which the assumptions of the previous paragraph are valid, is discussed in §B.13f. The relevant results are equations (B.13.36)–(B.13.39) with $n = 1$ and $q = e$, which, in the notation of the present section, become

$$a_{0d} = \frac{\hbar^2 \epsilon}{m_c^* e^2} = \left(\frac{\epsilon m}{m_c^*}\right) a_0 \tag{D.2.1}$$

and

$$E_d = E_c - E_{id}, E_{id} = \frac{m_c^* e^4}{2\hbar^2 \epsilon^2} = \frac{m_c^*}{m\epsilon^2} E_{ih}. \quad \text{(D.2.2)}$$

Here E_{ih} is the ionization energy of the hydrogen atom in its lowest energy state, i.e. the energy required to remove the electron from the atom entirely, and $E_{ih} = 13{\cdot}6$ eV. Similarly E_{id}, the energy required to remove the bound electron from the donor and place it in the lowest state of the conduction band, is called the ionization energy of the donor. The notation of this section is connected with that of §B.13f by $a_{0d} \equiv a_1$, $E_{id} \equiv -E_1$.

For germanium, using $\epsilon = 16$ and assuming for simplicity that m_c^* can be approximated by the density-of-states effective mass $m_c = 0{\cdot}56\,m$, (D.2.1) and (D.2.2) give

$$a_{0d} \approx 29a_0 \approx 15\text{Å}, \; E_{id} \approx 0{\cdot}0022\, E_{ih} \approx 0{\cdot}03 \text{ eV}. \quad \text{(D.2.3)}$$

The fact that the calculated radius is six times the distance between nearest-neighbour atoms suggests that the basic assumptions (i) and (ii) above are reasonable. In the case of germanium, however, as discussed at the end of the last section, the conduction band is very far from a simple scalar effective mass shape. It is therefore not surprising that measured values of E_{id} are considerably different from (D.2.3), being of the order 0·0096 eV (antimony) to 0·0127 eV (arsenic). Better agreement is obtained by using for m_c^* the geometric mean of the effective masses m_x, m_y, m_z for the principal axes of each ellipsoid, so that $m_c^* = [1{\cdot}64\,m \times (0{\cdot}082\,m)^2]^{1/3} \approx 0{\cdot}22\,m_0$. (D.2.3) then becomes

$$a_{0d} \approx 38 \text{ Å}, \qquad E_{id} \approx 0{\cdot}0117 \text{ eV},$$

but there is no simple way of justifying the use of this value.

Suppose next that a single semiconductor atom is replaced by an atom of a Group III element—boron, indium, gallium, etc. The impurity atom has only three outer shell electrons, so that a fourth electron is required to make up the system of valence bonds. Since, however, the system (impurity atom + four valence electrons) is not electrically neutral but has a net electrical charge $-e$, the fourth electron will be less tightly bound to its atom than normal valence electrons. Such an impurity atom therefore gives rise to an electron energy level above the top of the valence band.

If the extra electron required is taken from the valence band, a hole is created. Two possibilities now exist—the hole may be in a true valence band state or (considering it as a positively charged particle of positive mass) it may enter an orbit around the impurity. In the first case the hole is free to contribute to conduction, and the atom is said to be ionized. In the second place the hole is bound to the atom, which is effectively neutral. Because of its ability to accept an electron from the valence band, thus creating a free hole, an atom of this type is called an *acceptor*. The energy required to create a hole in this way, which is the ionization energy E_{ia} of the acceptor, is also the difference between the energy of the fourth electron referred to above, which will be called E_a, and the maximum energy of an electron in the valence band, E_v.

The order of magnitude of E_{ia} can be worked out in the same way as for E_{id}, m_c^* being replaced by a valence band effective mass m_v^*. Use of the density-of-states effective mass $m_v = 0{\cdot}37\,m$ for germanium again results in an overestimate for the ionization energy, the calculated value being $E_{ia} = 0{\cdot}02$ eV while measured values are of order 0·010 eV to 0·011 eV. For a more detailed treatment of hydrogen-like donors and acceptors see §B.13 and §B.16.

The donors and acceptors referred to above are monovalent in the sense that each can provide at most one charge carrier. Impurity atoms from other groups of the periodic table can act as multivalent donors or acceptors. For example, zinc in germanium is a divalent acceptor, being capable of accepting two valence band electrons, while gold in germanium can act as a trivalent acceptor and also as a monovalent donor. As well as impurity atoms, other forms of lattice defect can give rise to states with energies in the forbidden band, such as vacancies (lattice sites not occupied by an atom), interstitial atoms (atoms between the normal lattice sites) and dislocations (distortions of the lattice extending over several atoms). Such defects are present to some extent in almost all crystals, and their number can be greatly increased by radiation damage. A realistic model of a semiconductor should therefore include (i) conduction and valence band states, (ii) deliberately introduced donors and acceptors, and (iii) an indefinite number of other defects giving rise to states throughout the forbidden gap.

D.3 THE FERMI–DIRAC STATISTICS FOR ELECTRONS IN BANDS AND LOCALIZED STATES

The occupation statistics for all electronic quantum states in a semiconductor in equilibrium can be derived from a single result of statistical mechanics. Consider any system of identical particles which can interact with surrounding systems in such a way that the number of particles and the total energy are free to fluctuate, but the mean number of particles, mean energy and volume remain fixed. Let suffix i label quantum states of the whole system, and let n_i be the number of particles and U_i the total energy in quantum state i. The required result is that the probability of finding the system in quantum state i is

$$P_i = \frac{1}{\Xi}\exp[(\mu n_i - U_i)/k_0T] \equiv \frac{t_i}{\Xi} \tag{D.3.1}$$

where

$$\Xi \equiv \sum_{\text{all } i} \exp[(\mu n_i - U_i)/k_0T] \equiv \sum_{\text{all } i} t_i. \tag{D.3.2}$$

Here μ is a parameter known as the *chemical potential* or (for electrons only) the *Fermi level* for the system, k_0 is Boltzmann's constant and T is the absolute temperature. Ξ is known as the grand partition function for the system. (D.3.1) is proved[2] by expressing the entropy as

$$S = -k_0 \sum_i P_i \ln P_i$$

and then maximizing the entropy subject to the conditions that the mean number $\bar{n} = \sum_i n_i P_i$ and the mean energy $\bar{U} = \sum_i P_i U_i$ are fixed.

One further general result is required. The mean number of particles in the system is

$$\bar{n} = \sum_i n_i P_i = \frac{1}{\Xi}\sum_i n_i t_i = \frac{k_0T}{\Xi}\sum_i \frac{(\partial \ln t_i)}{\partial \mu} t_i$$

$$= \frac{k_0T}{\Xi}\sum_i \frac{\partial t_i}{\partial \mu} = \frac{k_0T}{\Xi}\frac{\partial \Xi}{\partial \mu}. \tag{D.3.3}$$

Consider first the system of all electrons in the conduction and valence bands. Let suffix j label single-electron quantum states in the bands, a typical state having energy E_j. Since electrons are indistinguishable, a quantum state i of the whole system is completely specified by the number n_j of electrons in each single-electron quantum state, and because of the Pauli exclusion principle $n_j = 0$ or 1 for all j. Assuming no interaction between different single-electron quantum states, the total energy and number of particles in the system are

$$U_i = \sum_{j=1}^{N} n_j E_j, \qquad n_i = \sum_{j=1}^{N} n_j, \tag{D.3.4}$$

N being the total number of single-electron quantum states. If we further define

$$t_j \equiv \exp[(\mu - E_j)/k_0T], \tag{D.3.5}$$

[2] P. T. Landsberg, *Thermodynamics with Quantum Statistical Illustrations*, Interscience, New York, 1961, p. 409.

then (D.3.1) can be rewritten as

$$P_i \equiv P(n_1, n_2, \ldots n_N) = \frac{1}{\Xi} \cdot t_1^{n_1} t_2^{n_2} \ldots t_N^{n_N} \tag{D.3.6}$$

and (D.3.2) as

$$\Xi = \sum_{n_1, n_2 \cdots n_N} t_1^{n_1} t_2^{n_2} \ldots t_N^{n_N} = \left(\sum_{n_1} t_1^{n_1} \right) \left(\sum_{n_2 \cdots n_N} t_2^{n_2} \ldots t_N^{n_N} \right). \tag{D.3.7}$$

Now (D.3.6) represents the probability of finding specified values of all the occupation numbers n_j. The probability of finding a particular state (say the first) occupied, i.e. of finding $n_1 = 1$, is obtained by summing over all other occupation numbers,

$$\begin{aligned} P(n_1 = 1) &= \sum_{n_2 \cdots n_N} P(1, n_2, \ldots n_N) \\ &= \frac{t_1}{\Xi} \sum_{n_2 \cdots n_N} t_2^{n_2} \cdots t_N^{n_N} \\ &= \frac{t_1}{\sum_{n_1} t_1^{n_1}}, \end{aligned} \tag{D.3.8}$$

where (D.3.7) has been used. Since n_1 can only be 0 or 1 the sum has only two terms,

$$P(n_1 = 1) = \frac{t_1}{1 + t_1} = \frac{1}{1 + t_1^{-1}} = \frac{1}{1 + \exp[(E_1 - \mu)/k_0 T]}. \tag{D.3.9}$$

Finally, dropping the suffix 1, we conclude that the probability of finding an electron in any single-electron state with energy E is given by the Fermi-Dirac distribution function

$$f_e(E) = \frac{1}{1 + \exp[(E - \mu)/k_0 T]}. \tag{D.3.10}$$

The form of $f_e(E)$ at absolute zero and at a finite temperature is illustrated in fig. D.3.1. It will be seen that $f_e(E)$ becomes very small for energies several $k_0 T$ above the Fermi level, and approaches unity for energies several $k_0 T$ below the Fermi level.

We next consider an impurity atom or other defect in the pure crystal structure, capable of trapping up to M electrons ($M \geqslant 1$). The system referred to in equation (D.3.1) is now identified with the set of all electrons trapped at the defect. It is assumed that a defect having trapped r electrons ($0 \leqslant r \leqslant M$) can be in a number of quantum states s (depending on r), the total energy of (defect + electrons) in state s being E_{rs}. The grand partition function for the defect is then

$$\Xi = \sum_{r=0}^{M} \left\{ \sum_{s} \exp[(r\mu - E_{rs})/k_0 T] \right\}$$

$$= \sum_{r=0}^{M} Z_r \exp(r\mu/k_0 T), \qquad \text{(D.3.11)}$$

where

$$Z_r \equiv \sum_{s} \exp(-E_{rs}/k_0 T), \qquad \text{(D.3.12)}$$

the sum being over all quantum states of a defect with r electrons. Z_r is called the partition function for a defect with r electrons. From (D.3.3) and (D.3.11) the mean number of electrons trapped at a single defect (or 'trap') is

$$\bar{r} = \frac{\sum_{r=1}^{M} r Z_r \exp(r\mu/k_0 T)}{\sum_{r=0}^{M} Z_r \exp(r\mu/k_0 T)}, \qquad \text{(D.3.13)}$$

and if there are N_t identical defects, the total number of electrons trapped at them is

$$n_t = N_t \bar{r}. \qquad \text{(D.3.14)}$$

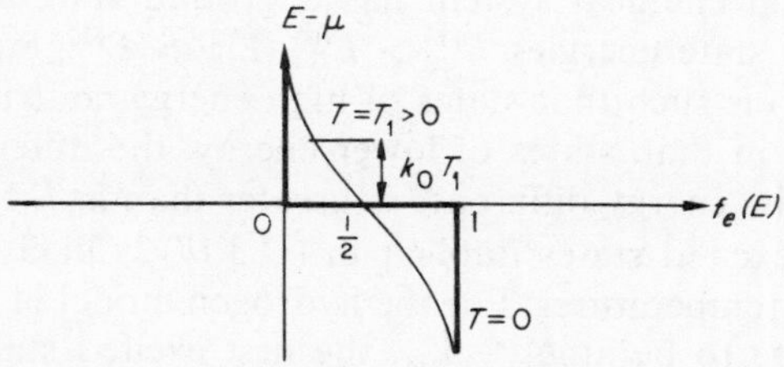

Fig. D.3.1. The Fermi–Dirac function.

An important special case occurs when the defect, like most donors and acceptors, is capable of trapping at most one electron ($r = 0$ or 1, $M = 1$). For this case $\bar{r}$ is simply the probability that the defect will have trapped an electron at any given time, and can be put into a form identical to that of the Fermi-Dirac probability function,

$$\bar{r} = \frac{Z_1 \exp(\mu/k_0 T)}{Z_0 + Z_1 \exp(\mu/k_0 T)} = \frac{1}{1 + \exp[(E_t^* - \mu)/k_0 T]}, \qquad \text{(D.3.15)}$$

where E_t^* is an effective energy level for the trapped electron, defined by

$$E_t^* \equiv k_0 T \log_e(Z_0/Z_1). \quad \text{(D.3.16)}$$

A further simplification is often made. Assume that the defect with no electrons has only a single energy level E_0, with degeneracy g_0, while the defect with one electron has a single energy level $E_0 + E_t$ with degeneracy g_1, so that E_t can reasonably be identified with the energy level of the trapped electron. The partition functions (D.3.12) are

$$Z_0 = g_0 \exp(-E_0/k_0 T), \qquad Z_1 = g_1 \exp[-(E_0 + E_t)/k_0 T] \quad \text{(D.3.17)}$$

so that from (D.3.16) the effective energy level is

$$E_t^* = E_t + k_0 T \log_e(g_0/g_1) \quad \text{(D.3.18)}$$

and from (D.3.15) the occupation probability for the defect is

$$\bar{r} = \frac{1}{1 + (g_0/g_1)\exp[(E_t - \mu)/k_0 T]}. \quad \text{(D.3.19)}$$

For a donor, in which the state $r = 1$ is the neutral state with an *unpaired* electron loosely bound to the atom, the two spin states available to this electron make it plausible that $g_0/g_1 = \frac{1}{2}$. Similarly for an acceptor, in which the state $r = 0$ has an *unpaired* valence electron, $g_0/g_1 = 2$ would be expected.

The assumption leading to (D.3.17), that for each charge state of the centre there is only one energy level, is less restrictive than it appears. Any quantum mechanical system has a ground state energy $E^{(0)}$ and various excited state energies $E^{(1)} > E^{(0)}$, $E^{(2)} > E^{(1)}$, etc. However the form of (D.3.12) is such that states of high energy contribute less to the partition function than states of lower energy, the difference becoming marked when the energy difference is greater than $k_0 T$. In general, then, the neglect of excited states implicit in (D.3.17) is likely to be valid at sufficiently low temperatures. For the hydrogen model of a donor, taking the ground state to be at $E_c - E_{id}$, the first excited state will be at $E_c - (E_{id}/4)$, the energy difference being $3E_{id}/4 \approx 0{\cdot}008$ eV in germanium. This will be greater than $2k_0 T$ at temperatures below about 50°K, and for such temperatures the first excited state can be ignored in the present context. This is, however, not obviously true at room temperature.

D.4 EQUILIBRIUM ELECTRON AND HOLE CONCENTRATIONS FOR INTRINSIC AND EXTRINSIC SEMICONDUCTORS

The concentration of electrons $n(E)\,\mathrm{d}E$ in any small energy range $\mathrm{d}E$ is the product of the number of quantum states per unit crystal volume in that range, $g(E)\,dE$, and the Fermi–Dirac occupation probability $f_e(E)$. The total concentration of electrons in the conduction band is obtained by integrating this product between the lower and upper limits of the band. In practice, however, the upper limit of the conduction band usually has an energy at least 20 k_0T above the Fermi level, so that at the upper limit $f_e(E) \sim e^{-20}$ and is negligible. It is therefore justifiable to replace the upper limit of integration formally by infinity, giving the concentration of conduction band electrons as

$$n = \int_{E_c}^{\infty} g(E) f_e(E)\,dE. \tag{D.4.1}$$

Similar arguments apply to the valence band. The probability of finding a hole in a valence band state of energy E is $1 - f_e(E)$, and at the lower edge of the valence band this probability is negligible, so that the lower limit of integration can be replaced formally by minus infinity. The concentration of valence band holes is therefore

$$p = \int_{-\infty}^{E_v} g(E)[1 - f_e(E)]\,dE. \tag{D.4.2}$$

Substituting (D.1.19) for $g(E)$ and (D.3.10) for $f_e(E)$ into (D.4.1) gives directly

$$n = \frac{1}{2\pi^2}\left(\frac{2m_c}{\hbar^2}\right)^{\frac{3}{2}} \int_{E_c}^{\infty} \frac{(E - E_c)^{\frac{1}{2}}\,dE}{1 + \exp[(E - \mu)/k_0T]}. \tag{D.4.3}$$

It is convenient to define dimensionless variables by

$$\eta \equiv (E - E_c)/k_0T, \qquad \eta_c \equiv E_c/k_0T, \qquad F \equiv \mu/k_0T, \tag{D.4.4}$$

and in terms of these (D.4.3) is

$$n = \frac{1}{2\pi^2}\left(\frac{2m_ck_0T}{\hbar^2}\right)^{\frac{3}{2}} \int_0^{\infty} \frac{\eta^{\frac{1}{2}}\,d\eta}{1 + \exp[\eta - (F - \eta_c)]}. \tag{D.4.5}$$

(D.4.5) can finally be written in the form

$$n = N_cI_{\frac{1}{2}}(F - \eta_c) = N_cI_{\frac{1}{2}}[(\mu - E_c)/k_0T], \tag{D.4.6}$$

where

$$N_c \equiv \frac{1}{4\pi^{\frac{3}{2}}}\left(\frac{2m_ckT}{\hbar^2}\right)^{\frac{3}{2}}, \tag{D.4.7}$$

$$I_{\frac{1}{2}}(x) \equiv \frac{2}{\sqrt{\pi}} \int_0^{\infty} \frac{\eta^{\frac{1}{2}}\,d\eta}{1 + \exp(\eta - x)}. \tag{D.4.8}$$

$I_{\frac{1}{2}}(x)$ is known as the Fermi–Dirac integral of order $\frac{1}{2}$, and has the important property that, for $x < -2$,

$$I_{\frac{1}{2}}(x) \approx \exp(-x), \tag{D.4.9}$$

with a fractional error of about 4 % at $x = -2$ and 0·7 % at $x = -4$. If an accuracy of 4 % is acceptable, as is very often the case in semiconductor work, (D.4.6) can be approximated by

$$n \approx N_c \exp(F - \eta_c) = N_c \exp[(\mu - E_c)/k_0 T] \qquad \text{for } \mu < E_c - 2k_0 T. \tag{D.4.10}$$

A semiconductor for which (D.4.10) is an adequate approximation is said to have a *non-degenerate* conduction band.

An argument similar to that leading from (D.4.1) to (D.4.6)–(D.4.8) can be applied to (D.4.2). With the additional variables $\eta' \equiv (E_v - E)/k_0 T$ and $\eta_v \equiv E_v/k_0 T$ one finds

$$p = N_v I_{\frac{1}{2}}(\eta_v - F) = N_v I_{\frac{1}{2}}[(E_v - \mu)/k_0 T], \tag{D.4.11}$$

where

$$N_v \equiv \frac{1}{4\pi^{\frac{3}{2}}} \left(\frac{2m_v k_0 T}{\hbar^2} \right)^{\frac{3}{2}}. \tag{D.4.12}$$

The approximation derived from (D.4.9) is in this case

$$p \approx N_v \exp(\eta_v - F) = N_v \exp[(E_v - \mu)/k_0 T] \qquad \text{for } \mu > E_v + 2k_0 T, \tag{D.4.13}$$

and when this holds the valence band is said to be *non-degenerate.*

Many problems in semiconductor theory are capable of analytic solution only if both bands are assumed to be non-degenerate. The conditions for this assumption to be valid are that the Fermi level should be within the forbidden band and at least $mk_0 T$ from the conduction band edges ($m = 2$ or more depending on the required accuracy), or equivalently that $n \ll N_c$ and $p \ll N_v$. Non-degeneracy of the bands will be assumed in the remainder of this chapter. Also, the suffix 0 will be used to denote equilibrium, so that n_0, p_0, μ_0, F_0 are the equilibrium values of n, p, μ, F. This notation is necessary because in the next chapter equations such as (D.3.10) and (D.4.6) will be assumed to hold in certain non-equilibrium situations.

We consider first a completely pure or *intrinsic* semiconductor. There are no quantum states having energies in the forbidden gap, and at absolute zero the conduction band is empty and the valence band full. At any finite temperature T, however, there are present photons of black body radiation and quanta of lattice vibration, known as phonons, having average energies of the order of k_0T. There will always be a few photons with energies greater than the energy gap $E_g \equiv E_c - E_v$, and such a photon is capable, possibly in cooperation with a phonon, of giving up its energy to a valence band electron which is thereby transferred to the conduction band. Since this process leaves a hole in the valence band it is referred to as the generation of an electron-hole pair. The balance between this process and its converse leads at a given temperature to definite equilibrium concentrations of electrons and holes. Evidently these concentrations are equal, i.e

$$n_0(\text{intrinsic}) = p_0(\text{intrinsic}) = n_i. \qquad \text{(D.4.14)}$$

This equation, together with (D.4.10) and (D.4.13), enables us to calculate the intrinsic concentration n_i for non-degenerate bands. Multiplying (D.4.10) and (D.4.13) together gives

$$n_i^2 = n_0 p_0 = N_c N_v \exp[(E_v - E_c)/k_0 T] = N_c N_v \exp(-E_g/k_0 T) \qquad \text{(D.4.15)}$$

or

$$n_i = (N_c N_v)^{\frac{1}{2}} \exp(-E_g/2k_0 T). \qquad \text{(D.4.16)}$$

The position of the Fermi level in intrinsic material is found by solving the equation

$$n_0 = N_c \exp[(\mu_0 - E_c)/k_0 T] = N_v \exp[(E_v - \mu_0)/k_0 T] = p_0 \qquad \text{(D.4.17)}$$

to be at $\mu_0 = E_i$, where

$$E_i \equiv \tfrac{1}{2}(E_c + E_v) - \frac{k_0 T}{2} \log_e\left(\frac{N_c}{N_v}\right) = \tfrac{1}{2}(E_c + E_v) - \frac{3k_0 T}{4} \log_e\left(\frac{m_c}{m_v}\right). \qquad \text{(D.4.18)}$$

E_i is called the intrinsic energy level. For germanium, $m_c/m_v \approx 1{\cdot}5$ and $E_i \approx \frac{1}{2}(E_c + E_v) - 0{\cdot}3\, k_0 T$ and so is very near the middle of the forbidden gap.

The temperature dependence of (D.4.16) is complicated by the possible variation of E_g with temperature. If a linear variation is assumed, of the form

$$E_g = E_{g0} - \beta T, \qquad \text{(D.4.19)}$$

then together with (D.4.7) and (D.4.12), (D.4.16) gives

$$n_i = \frac{(m_c m_v)^{\frac{3}{4}}}{4} \left(\frac{2k_0 T}{\pi \hbar^2}\right)^{\frac{3}{2}} \exp(\beta/2k_0) \exp(-E_{g0}/2k_0 T). \qquad \text{(D.4.20)}$$

A plot of $\log_e(n_i/T^{\frac{3}{2}})$ against T^{-1} should therefore be linear with a slope related not to the actual energy gap E_g but to the extrapolated energy gap E_{g0}. Excellent agreement with the experimental data[3] for germanium in the range 250°–500°K is obtained with $E_{g0} = 0{\cdot}785$ eV, and other evidence indicates that in germanium at $T > 200$°K

$$E_g \approx (0{\cdot}785 - 0{\cdot}0004T)\,\text{eV}. \tag{D.4.21}$$

Since $kT \approx 0{\cdot}026$ eV at room temperature, the exponential in (D.4.20) is very small and varies with temperature much more rapidly than $T^{\frac{3}{2}}$. In germanium, for example, n_i increases by a factor of 10 as the temperature increases from 286°K to 330°K.

We consider next a semiconductor which contains a concentration N_d of identical donors, with an energy level E_d a short distance below E_c. Such a semiconductor always contains more electrons than holes, and is called extrinsic *n*-type. An effective donor energy level E_d^* can then be defined as in (D.3.16) and will probably be lower than E_d by an amount of order k_0T [compare (D.3.18) with $g_1/g_0 = 2$]. Defining a dimensionless energy $\eta_d^* \equiv E_d^*/k_0T$, and recalling that (D.3.15) represents the probability f_d that a defect has trapped an electron, the concentration of *neutral* donors is

$$N_d^0 = N_d f_d = \frac{N_d}{1 + \exp(\eta_d^* - F_0)} \tag{D.4.22}$$

and the concentration of *ionized* (positively charged) donors is

$$N_d^+ = N_d(1 - f_d) = \frac{N_d}{1 + \exp(F_0 - \eta_d^*)}. \tag{D.4.23}$$

Now each electron in the conduction band must have come either from a now vacant state in the valence band or from a now ionized donor, so that

$$n_0 = p_0 + N_d^+, \tag{D.4.24}$$

which can also be regarded as an overall charge neutrality condition.

In order to solve (D.4.24) we first recall that from (D.4.10) and (D.4.13) $n_0 \propto \exp(F_0)$ and $p_0 \propto \exp(-F_0)$, and that $n_0 = p_0 = n_i$ when $F_0 = E_i/k_0T \equiv \eta_i$. It follows that

$$n_0 = n_i \exp(F_0 - \eta_i), \qquad p_0 = n_i \exp(\eta_i - F_0), \qquad n_0 p_0 = n_i^2. \tag{D.4.25}$$

[3] F. J. Morin and J. P. Maita, *Phys. Rev.*, **94**, 1525 (1954); see also E. G. S. Paige, *Progress in Semiconductors 8*, Temple Press, London, 1964, p.134.

Now (D.4.25) can be used to express F_0 as a function of n_0 and the result substituted in (D.4.23) for N_d^+. When this is done, (D.4.24) becomes

$$n_0 - \frac{n_i^2}{n_0} - \frac{N_d}{1 + cn_0} = 0, \qquad c \equiv (n_i)^{-1} \exp(\eta_i - \eta_d^*). \quad \text{(D.4.26)}$$

This is in general a cubic equation for n_0 and has one real root. In almost all practical cases, however, one of two approximations, each of which effectively reduces (D.4.26) to a quadratic, is valid.

The first approximation referred to consists in assuming that the donors are fully ionized in the sense that $N_d^+ \approx N_d$. From (D.4.23) this will be the case if the Fermi level μ_0 is several k_0T below the donor effective energy E_d^*, or equivalently if $cn_0 \ll 1$ in (D.4.26). The latter then becomes

$$n_0^2 - n_i^2 - N_d n_0 = 0, \qquad n_0 = \tfrac{1}{2}[N_d + \sqrt{(N_d^2 + 4n_i^2)}]. \quad \text{(D.4.27)}$$

The solution has two limiting forms. If the number of donors is very small, $N_d \ll n_i$, then to the first order in N_d/n_i

$$n_0 \approx n_i + N_d/2, \qquad p_0 \approx n_i - N_d/2, \quad \text{(D.4.28)}$$

while if the number of donors is large, $N_d \gg n_i$, then

$$n_0 \approx N_d, \qquad p_0 \approx n_i^2/N_d \ll n_0. \quad \text{(D.4.29)}$$

As N_d increases a point will be reached at which the Fermi level is near the effective energy of the donors and these will cease to be fully ionized. The second approximation referred to is then appropriate. It consists in assuming that $p_0 \ll n_0$, so that $n_i^2 \ll n_0^2$. (D.4.26) then becomes

$$n_0(1 + cn_0) - N_d = 0, \quad \text{(D.4.30)}$$

with general solution

$$n_0 = \frac{1}{2c}[\sqrt{(1 + 4cN_d)} - 1]. \quad \text{(D.4.31)}$$

The limiting case of $cN_d \ll 1$ leads back to (D.4.29), so that the donors are then fully ionized. The limit $cN_d \gg 1$ leads to

$$n_0 \approx (N_d/c)^{\frac{1}{2}} = (N_d n_i)^{\frac{1}{2}} \exp[(\eta_d^* - \eta_i)/2]. \quad \text{(D.4.32)}$$

(D.4.32) is in fact a low-temperature limiting case. This becomes clearer if n_i is replaced in (D.4.26) by $N_c \exp(\eta_i - \eta_c)$, and η_d^* is replaced by $\eta_d + \log_e(g_0/g_1)$, so that

$$c = N_c^{-1}(g_1/g_0) \exp[(E_c - E_d)/k_0T]$$

and increases rapidly with decreasing temperature. Thus for any N_d the condition $cN_d \gg 1$ is reached at a sufficiently low temperature. (D.4.32) can be written

$$n_0 \text{ (low temperature)} = (N_d N_c g_0/g_1)^{\frac{1}{2}} \exp[-(E_c - E_d)/2k_0T], \tag{D.4.33}$$

a result analogous to (D.4.16). Since in practice $N_d \lesssim N_c$, a semiconductor containing only donors will become non-degenerate ($n_0 \ll N_c$) at a sufficiently low temperature.

A situation very similar to that just considered arises in the case of a semiconductor which contains a concentration N_a of acceptors, with an effective energy level $E_a^* \equiv k_0T\eta_a^*$ a short distance above the top of the valence band. Such a semiconductor always contains more holes than electrons and is called extrinsic *p*-type. The important factor is now the concentration of ionized (negatively charged) acceptors

$$N_a^- = N_a f_a \equiv \frac{N_a}{1 + \exp(\eta_a^* - F_0)}. \tag{D.4.34}$$

Since the electrons in the conduction band and those on ionized acceptors have all come from the valence band, the charge neutrality condition is

$$p_0 = n_0 + N_a^-. \tag{D.4.35}$$

Provided that the Fermi level is several k_0T above the acceptor effective energy level, the acceptors can be regarded as fully ionized, and results analogous to (D.4.27)–(D.4.29) apply. In general

$$p_0 = \tfrac{1}{2}[N_a + \sqrt{(N_a^2 + 4n_i^2)}], \tag{D.4.36}$$

while for $N_a \ll n_i$,

$$p_0 \approx n_i + N_a/2, \qquad n_0 \approx n_i - N_a/2, \tag{D.4.37}$$

and for $N_a \gg n_i$,

$$p_0 \approx N_a, \qquad n_0 \approx n_i^2/N_a \ll p_0. \tag{D.4.38}$$

When the acceptors are not fully ionized, a result similar to (D.4.31) is found.

In practice it is rare for a semiconductor to contain only one type of impurity. We therefore briefly consider the case of a semiconductor containing N_d donors *and* N_a acceptors per unit volume, but no other impurities or defects. The charge conservation equation is

$$n_0 = p_0 + N_d^+ - N_a^-, \tag{D.4.39}$$

where all the symbols have the same meanings as previously. It is convenient to define effective concentrations of donors and acceptors by

$$N_d^* \equiv N_d^+ - N_a^-, \qquad N_a^* \equiv N_a^- - N_d^+. \tag{D.4.40}$$

It will be seen that, if $N_d^* > 0$, (D.4.39) is of the same form as (D.4.24), while if $N_a^* > 0$, (D.4.39) is of the same form as (D.4.35). Accordingly, if $N_d^* > 0$ the material is n-type and n_0 satisfies equations of the form (D.4.27)–(D.4.29), while if $N_a^* > 0$ the material is p-type and p_0 satisfies equations of the form (D.4.36)–(D.4.38), with N_d replaced by N_d^* and N_a by N_a^*. In general these equations still require solution, since N_d^* and N_a^* are themselves functions of the Fermi level. It is, however, often the case that donors *and* acceptors are fully ionized, so that

$$N_d^* \approx N_d - N_a, \qquad N_a^* \approx N_a - N_d. \tag{D.4.41}$$

Thus if $N_d > N_a$ the effect of the acceptors is simply to neutralize an equal number of the donors, and conversely if $N_a > N_d$. The important results are now effectively (D.4.27) and (D.4.36),

$$n_0\ (n\text{-type, impurities fully ionized}) = \tfrac{1}{2}[N_d^* + \sqrt{(N_d^{*2} + 4n_i^2)}],$$

$$p_0\ (p\text{-type, impurities fully ionized}) = \tfrac{1}{2}[N_a^* + \sqrt{(N_a^{*2} + 4n_i^2)}], \tag{D.4.42}$$

where N_d^* and N_a^* are determined by (D.4.41).

The form of (D.4.40) is an instance of the more general phenomenon of *compensation*, whereby the effect on carrier concentration of one type of impurity is reduced by the presence of other impurities or defects. To discuss this qualitatively consider a strongly n-type semiconductor, containing N_d donors, for which the Fermi level is near the conduction band. As discussed at the end of §D.2, unavoidable lattice defects give rise to a number of electron states in the forbidden gap. It will be assumed that these states are distributed more or less uniformly in energy, and also that some are neutral when they contain an electron (donor-type) and others are neutral when empty (acceptor type). Then donor-type states several k_0T above the Fermi level will be positively charged, and acceptor-type states several k_0T below the Fermi level will be negatively charged. The charge neutrality condition for the whole semiconductor can be written

$$n_0 - p_0 = N_d^+ + n_d^+ - n_a^- = N_d^+ - n_c, \qquad n_c \equiv n_a^- - n_d^+, \tag{D.4.43}$$

where n_d^+ is the concentration of positively charged donor-type states and n_a^- the concentration of negatively charged acceptor states, while n_c can be regarded as the effective concentration of compensating defects. The effective concentration of ionized donors is evidently $N_d^+ - n_c$.

For a Fermi level near the conduction band $n_a^- \gg n_d^+$ and $n_c > 0$. Further, as more donors are added the Fermi level rises, n_a^- increases, n_d^+ decreases and n_c rises, so that the amount of compensation increases automatically with the number of donors. An alternative assumption is that the defect states are all below the Fermi level so that n_c is effectively constant. This model is considered in detail by Blakemore[1].

CHAPTER VII

Recombination and Charge Transport Statistics

D.5 TYPES OF RECOMBINATION MECHANISM

As mentioned briefly in Chapter VI, the equilibrium electron and hole concentrations result from a balance between processes which generate electron–hole pairs and processes which lead to the disappearance of electron–hole pairs. Equivalently, the balance is between generation and recombination processes. These processes involve a considerable change in energy of at least one electron, and it is convenient to classify them according to the means by which this energy is supplied or removed. A division can also be made[4] between processes which can occur in a perfect crystal lattice ('*unavoidable*' processes) and those which require the presence of an impurity or defect ('*avoidable*' processes). The description will concentrate on recombination processes, the generation process being the exact inverse.

The most commonly considered unavoidable recombination mechanisms are those in which an electron is transferred from the conduction band to the valence band without any intermediate state. The energy lost by the electron, which is at least equal to the band gap E_g, can in principle be absorbed by emitting a photon (radiative recombination), by emitting a number of phonons (multi-phonon recombination), or by exciting a second electron to a higher energy level (electron collision recombination or Auger recombination). The multi-phonon process is normally considered unlikely because the number of phonons involved is large. It is also possible for the recombining electron and hole to first become loosely bound together, forming an exciton [see problem (B.13.3)] of binding energy E_b. The remaining energy $E_g - E_b$ could then be lost by any of the mechanisms just mentioned. The possibility of exciton formation will not be considered further.

Radiative recombination may occur in two forms, depending on the band structure of the material. Since a photon is effectively a particle travelling at the speed of light, its momentum–energy ratio is much smaller than that of a conduction or valence electron. Momentum conservation therefore requires that the recombining electron and hole should

[4] P. T. Landsberg, *Proc. Inst. Elec. Engrs.*, (*London*) *Pt. B*, **106**, *Suppl. 17*, 908 (1959).

have nearly equal momenta and therefore nearly equal values of $\mathbf{k}$. Also, the form of the Fermi distribution means that most electrons and holes are found near the band edges, with wave vectors near $\mathbf{k}_c$ and $\mathbf{k}_v$ respectively. Most recombination processes therefore involve a change in wave vector of amount $(\mathbf{k}_v - \mathbf{k}_c)$. If $\mathbf{k}_c = \mathbf{k}_v$, momentum conservation can be satisfied in a purely radiative process. Such a process is also called direct radiative recombination and the material is said to have a direct band gap. If on the other hand $\mathbf{k}_c \neq \mathbf{k}_v$, some other entity must supply the difference in momentum. This is usually a phonon, whose energy is so low that essentially all the energy is still taken up by the emitted photon. The process is referred to as indirect radiative recombination and the material as having an indirect band gap. Other things being equal, the direct process would be expected to be more probable than the indirect, so that radiative processes would be expected to be more important in direct-gap materials. Also, since the generation process requires the presence in the black body radiation of a photon with an energy of slightly over E_g, its probability should vary with the band gap approximately as $\exp(-E_g/k_0T)$. Unavoidable radiative recombination is therefore expected to be more important in materials with a small energy gap.

In the electron collision process, known as an Auger process because of its similarity with processes occurring within atoms and discovered by Pierre Auger, the second electron can absorb any required change of momentum as well as of energy, so that the details of band structure are less important. Since Auger recombination requires the presence of a second electron, free to change state (i.e. the presence of a conduction electron or hole), its relative importance increases when electron and hole concentrations increase, in particular at high temperatures. It is known[5] to become dominant over direct radiative recombination in tellurium above 350°K.

The avoidable generation-recombination processes can be considered as occurring in at least two steps. Considering only recombination for the moment, the first step would be the transition of an electron from the conduction band to a localised state within the forbidden gap ('electron capture'). The electron may then transfer to another localised state, and finally to the valence band where it annihilates a hole ('hole capture'). Each step in the process may in principle involve photons, phonons or Auger effects. The existence of excited states of impurity atoms opens the possibility of a multi-phonon cascade process[6,7] in which an electron

[5] J. S. Blakemore, *Proc. Intern. Conf. Semiconductor Physics, Prague*, 981 (1960). For the case of InSb see A. R. Beattie and G. Smith, *Phys. Stat. Solidi*, **19**, 577 (1967) and A. R. Beattie and P. T. Landsberg, *Proc. Roy. Soc.*, **A249**, 16 (1959).

[6] M. Lax, *Phys. Rev.*, **119**, 1952 (1960); *J. Phys. Chem. Solids*, **8**, 66 (1959).

[7] E. F. Smith and P. T. Landsberg, *J. Phys. Chem. Solids*, **27**, 1727 (1966); D. R. Hamann and A. L. McWhorter, *Phys. Rev.*, **134**, A250 (1964); F. Beleznay and G. Pataki, *Phys. Stat. Solidi*, **13**, 499 (1966); M. Nagae, *J. Phys. Soc. Japan*, **18**, 207 (1963).

captured into a highly excited state can reach the ground state by a series of transitions, in each of which the energy lost is of order k_0T and can be taken up by a single phonon. A radiative cascade is also possible[8]. The distinction between direct and indirect band-gap materials is unimportant as the impurity atom has sufficient mass to absorb any momentum difference. These considerations, allied with the extreme difficulty of removing all defects and impurities from a semiconductor, mean that in most practical problems the avoidable processes are predominant. In the same way as for unavoidable processes, Auger effects are expected to become relatively more important as electron and hole concentrations increase. In *n*-type germanium containing copper recombination centres at 300°K, for example[9], hole capture is predominantly an Auger process at electron concentrations exceeding 10^{17} cm^{-3}.

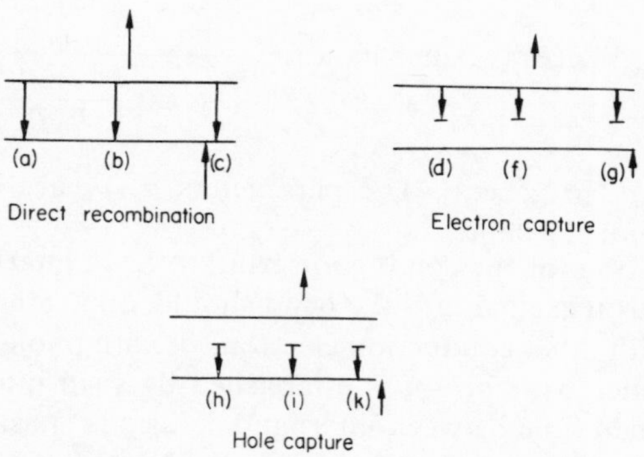

Fig. D.5.1. Recombination Mechanisms. Arrows denote electron transitions.

In many problems of semiconductor statistics one is concerned only with the effect of a given mechanism on the electrons involved in it. Fig. D.5.1 therefore represents all recombination mechanisms involving at most one recombination centre, and excluding exciton formation, classified according to the number and initial and final states of the electrons. For example (a) includes all band–band recombination processes involving one electron, whether radiative (direct or indirect) or multiphonon. It will appear in the next section that this classification is sufficiently detailed to allow calculation of the recombination rates.

[8] P. Lal and P. T. Landsberg, *Phys. Rev.*, **140**, A46 (1965).

[9] P. T. Landsberg, D. A. Evans and C. Rhys-Roberts, *Proc. Phys. Soc.*, **83**, 325 (1964). See also M. Sheinkman, *Soviet Phys.: Solid State*, **7**, 18 (1965); P. T. Landsberg, C. Rhys-Roberts and P. Lal, *Proc. Phys. Soc.*, **84**, 915 (1964); V. I. Bonch-Bruevich and Y. V. Gulyaev, *Soviet Phys.: Solid State*, **2**, 431 (1960). For a recent theoretical discussion of impact ionization see M. E. Cohen and P. T. Landsberg, *Phys. Rev.*, **154**, 683 (1967).

D.6 QUASI-FERMI LEVELS AND MASS-ACTION LAWS

In equilibrium, the occupation probability for an electron state anywhere in the bands is determined through (D.3.10) by a single parameter, the Fermi level μ. When the semiconductor as a whole is no longer in equilibrium, it may still be legitimate to consider the conduction and valence band as being in *internal* equilibrium, so that the occupation probability for a conduction band state is determined by a parameter μ_e,

$$f_e(E)\ \text{(conduction band)} = \frac{1}{1+\exp\left(\dfrac{E-\mu_e}{k_0 T}\right)} \tag{D.6.1}$$

and the occupation probability for a valence band state by a parameter μ_h,

$$f_e(E)\ \text{(valence band)} = \frac{1}{1+\exp\left(\dfrac{E-\mu_h}{k_0 T}\right)}, \tag{D.6.2}$$

where $\mu_e \neq \mu_h$ in general. The parameters μ_e, μ_h are known as quasi-Fermi levels for the bands.

The condition for the conduction band to be in internal equilibrium is that an electron excited into the band should, on average, undergo many collisions with other conduction electrons or with phonons before leaving the band, since equilibrium is established by such interband collisions. Since the mean time between interband collisions is usually about 10^{-12} sec while the mean time before leaving the band can vary from 10^{-3} sec to 10^{-11} sec, the assumption of internal equilibrium is often valid. It may break down in materials containing many recombination centres, or when very high frequency external signals are present. Similar considerations apply of course to the valence band. Assuming that (D.6.1) and (D.6.2) apply and that the bands are non-degenerate, (D.4.25) is replaced for the non-equilibrium case by

$$n = n_i \exp(F_e - \eta_i), \qquad p = n_i \exp(\eta_i - F_h), \qquad np = n_i^2 \exp(F_e - F_h), \tag{D.6.3}$$

where, in conformity with previous notation, $F_{e,h} \equiv \mu_{e,h}/k_0 T$.

To illustrate the calculation of generation and recombination rates, consider the recombination process symbolized by fig. D.5.1a. Supposing that a conduction band state of energy E_1 is known to be full and that a valence band state of energy E_2 is known to be empty, let the probability of a transition from E_1 to E_2 in the next small time-interval δt be $p(E_1,$

$E_2)\,\delta t$. In a unit crystal volume the number of full conduction band states in a small energy range dE_1 will be $f_e(E_1)g(E_1)\,dE_1$ and the number of empty valence band states in a small energy range dE_2 will be $[1 - f_e(E_2)]g(E_2)\,dE_2$. The total recombination rate per unit crystal volume per unit time due to process (a) is therefore

$$R_r(a) = \int_{E_c}^{\infty} dE_1 \int_{-\infty}^{E_v} dE_2\, f_e(E_1)[1 - f_e(E_2)]g(E_1)g(E_2)p(E_1, E_2). \qquad \text{(D.6.4)}$$

This can be placed in a more useful form by recalling the definitions (D.4.1), (D.4.2) of n and p, and noting that

$$np = \int_{E_c}^{\infty} dE_1 \int_{-\infty}^{E_v} dE_2\, f_e(E_1)[1 - f_e(E_2)]g(E_1)g(E_2)$$

so that

$$R_r(a) = \langle p(E_1, E_2)\rangle np, \qquad \text{(D.6.5)}$$

where $\langle p(E_1, E_2)\rangle$ is an average of the transition probability $p(E_1, E_2)$ over the actual distribution of electrons and holes,

$$\langle p(E_1, E_2)\rangle \equiv \frac{\displaystyle\int_{E_c}^{\infty} dE_1 \int_{-\infty}^{E_v} dE_2\, f_e(E_1)[1 - f_e(E_2)]g(E_1)g(E_2)p(E_1, E_2)}{\displaystyle\int_{E_c}^{\infty} dE_1 \int_{-\infty}^{E_v} dE_2\, f_e(E_1)[1 - f_e(E_2)]g(E_1)g(E_2)}. \qquad \text{(D.6.6)}$$

We now make the assumption of non-degeneracy in the form

$$f_e(E_1) \approx \exp[(\mu_e - E_1)/k_0T], \qquad 1 - f_e(E_2) \approx \exp[(E_2 - \mu_h)/k_0T]. \qquad \text{(D.6.7)}$$

For $\mu_e = E_c - 4k_0T$ the greatest fractional error in $f_e(E_1)$ is about 2%, but if $p(E_1, E_2)$ does not vary too violently with energy the overall error introduced in (D.6.6) should be less, as is the case in (D.4.9). With the approximation (D.6.7), we have

$$f_e(E_1)[1 - f_e(E_2)] \approx \exp[(\mu_e - \mu_h)/k_0T]\exp[(E_2 - E_1)/k_0T]$$

and the dependence of both numerator and denominator of (D.6.6) on quasi-Fermi levels is contained in the factor $\exp[(\mu_e - \mu_h)/k_0T]$, which can be taken outside the integrals and cancelled. We conclude that, within the non-degenerate approximation, $\langle p(E_1, E_2)\rangle$ is independent of

n and p, so that $R_r(\mathrm{a}) \propto np$. It is convenient to rename the proportionality factor B^s, so that

$$R_r(\mathrm{a}) \approx B^s np \qquad \text{(non-degenerate).} \tag{D.6.8}$$

The total generation rate per unit volume and per unit time due to the reverse of process (a) is calculated in the same way as (D.6.4) and is found to be

$$R_g(\mathrm{a}) = \int_{E_c}^{\infty} \mathrm{d}E_1 \int_{-\infty}^{E_v} \mathrm{d}E_2 [1 - f_e(E_1)] f_e(E_2) g(E_1) g(E_2) p(E_2, E_1), \tag{D.6.9}$$

where $p(E_2, E_1)$ is the probability per unit time for the upward transition. However, to the same accuracy as (D.6.7),

$$1 - f_e(E_1) \approx 1, \qquad f_e(E_2) \approx 1$$

so that

$$R_g(\mathrm{a}) \approx \text{constant (non-degenerate).} \tag{D.6.10}$$

The constant can be evaluated by using the *principle of detailed balance.* This states that in equilibrium the generation and recombination rates due to any process and its converse are equal, so that

$$R_g(\mathrm{a}) = B^s n_0 p_0 = B^s n_i^2 . \tag{D.6.11}$$

It is usual to define a net recombination rate for process (a) and its converse by

$$U(\mathrm{a}) \equiv R_r(\mathrm{a}) - R_g(\mathrm{a}) = B^s(np - n_i^2). \tag{D.6.12}$$

Similar arguments can be used to calculate the net rates for all the processes occurring in fig. D.5.1, with an additional integration over electron energy for the Auger processes. However, the results of such calculations may be anticipated by noting that each conduction electron participating in any process leads to a factor $f_e(E_1)$ in the integral and thus to a factor n in the rate. Similarly each valence band hole leads to a factor p in the rate. Now in process (b) and its converse one more conduction electron participates than in (a), so that one expects a net rate

$$U(\mathrm{b}) = B_1 n(np - n_i^2) \tag{D.6.13}$$

and similarly for process (c)

$$U(\mathrm{c}) = B_2 p(np - n_i^2) \tag{D.6.14}$$

where B_1, B_2 are averages of transition probabilities for Auger processes.

In discussing the rates of recombination via centres (or 'traps') it will be assumed for the sake of simplicity that a concentration N_t of identical centres is present, each of which can exist in only two distinct states of charge described as full and empty. Thus if f_t is the probability of finding a trap full, n_t the concentration of full traps and p_t the concentration of empty traps,

$$n_t = N_t f_t, \qquad p_t = N_t(1 - f_t) = N_t - n_t. \tag{D.6.15}$$

The rate of electron capture by process (d) per unit volume and time can be written

$$R_r(\mathrm{d}) = N_t \sum_s \int_{E_c}^{\infty} \mathrm{d}E_1 \, f_e(E_1) f_t^*(E_s) g(E_1) p(E_1, E_s), \tag{D.6.16}$$

where the sum is over all possible quantum states of an empty trap with energies E_s, and $f_t^*(E_s)$ denotes the probability of finding a trap empty and in the particular quantum state referred to. It will be assumed that $f_t^*(E_s) = (1 - f_t)\phi(E_s)$, where $\phi(E_s)$ does not depend on quasi-Fermi levels. Then $f_t^*(E_s)$ gives rise to a factor $1 - f_t$, while as in the previous calculations $f_e(E_1)$ gives rise to a factor n for a non-degenerate conduction band. Thus

$$R_r(\mathrm{d}) = T_1^s n N_t(1 - f_t) = T_1^s n p_t, \tag{D.6.17}$$

where T_1^s is independent of n and p_t.

The rate of the reverse process can be seen to be proportional to n_t, say $R_g(\mathrm{d}) = X_1^s n_t$. The principle of detailed balance requires that the two rates be equal in equilibrium, $T_1^s n_0 p_{t0} = X_1^s n_{t0}$. This leads to the net rate of electron capture by this process as

$$U(\mathrm{d}) = R_r(\mathrm{d}) - R_g(\mathrm{d}) = T_1^s(np_t - n_1 n_t) \tag{D.6.18}$$

where

$$n_1 \equiv \frac{n_0 p_{t0}}{n_{t0}} = \frac{n_0(1 - f_{t0})}{f_{t0}} = n_i \exp[(E_t^* - E_i)/k_0 T]. \tag{D.6.19}$$

Here E_t^* is defined as in (D.3.16) and f_{t0} has been identified with $\bar{r}$ of (D.3.15). Evidently n_1 is the value of n_0 corresponding to a Fermi level for which the traps are half full.

The net rates of electron capture by processes (f) and (g) are derived from that of (d) by noting that in each case an additional electron or hole participates:

$$U(\mathrm{f}) = T_1 n(np_t - n_1 n_t) \tag{D.6.20}$$

$$U(\mathrm{g}) = T_2 p(np_t - n_1 n_t). \tag{D.6.21}$$

For the hole capture processes (h), (j) and (k) a similar argument leads to net rates

$$U(\mathrm{h}) = T_2^s(pn_t - p_1 p_t) \qquad \text{(D.6.22)}$$

$$U(\mathrm{j}) = T_3 n(pn_t - p_1 p_t) \qquad \text{(D.6.23)}$$

$$U(\mathrm{k}) = T_4 p(pn_t - p_1 p_t), \qquad \text{(D.6.24)}$$

where

$$p_1 \equiv \frac{p_0 n_{t0}}{p_{t0}} = n_i \exp[(E_i - E_t^*)/k_0 T], \qquad n_1 p_1 = n_i^2. \qquad \text{(D.6.25)}$$

It is convenient to collect together the three electron–hole recombination processes to give a net direct recombination rate

$$U_{\text{direct}} = U(\mathrm{a}) + U(\mathrm{b}) + U(\mathrm{c}) = F(np - n_i^2) \qquad \text{(D.6.26)}$$

and similarly to define a net electron capture rate

$$U_{ec} = U(\mathrm{d}) + U(\mathrm{f}) + U(\mathrm{g}) = G(np_t - n_1 n_t) \qquad \text{(D.6.27)}$$

and a net hole capture rate

$$U_{hc} = U(\mathrm{h}) + U(\mathrm{j}) + U(\mathrm{k}) = H(pn_t - p_1 p_t), \qquad \text{(D.6.28)}$$

where

$$F \equiv B^s + B_1 n + B_2 p, \qquad G \equiv T_1^s + T_1 n + T_2 p,$$
$$H \equiv T_2^s + T_3 n + T_4 p. \qquad \text{(D.6.29)}$$

Expressions such as (D.6.27)–(D.6.29) are sometimes called 'mass-action laws' by analogy with the laws of reaction kinetics in chemistry, where the forward rate of a chemical reaction is proportional to the product of the concentrations of the reactants. They are valid only to the extent to which the electron and hole distribution functions have the quasi-equilibrium form (D.6.1) and (D.6.2), and are non-degenerate. A more detailed account of recombination statistics is found elsewhere[10].

D.7 STEADY-STATE AND TRANSIENT RECOMBINATION RATES AND LIFETIMES

D.7a Steady-state conditions

When a semiconductor device is run under operating conditions which do not vary with time the values of n and p at any point are determined by

[10] P. T. Landsberg in *Festkörperprobleme*, **VI**, 174 (1967).

differential equations of a type discussed in the following section, but the concentration of trapped electrons, n_t, is subject to no such equation. To evaluate the recombination rates (D.6.27) and (D.6.28) it is therefore necessary to eliminate n_t, which is done by means of the assumptions

(a) that electrons are not transferred from one trap to another,

(b) that there are no externally induced transitions involving the traps, and

(c) that n_t does not vary in time. Conduction by electrons moving between trapped states is in fact important at very high trap concentrations, and assumption (a) is then invalid.

The first two assumptions means that the rate of change of n_t is determined entirely by the electron and hole capture rates, $\dot{n}_t = U_{ec} - U_{hc}$. The third assumption is included in the assumption of a steady state, defined as a state in which all concentrations are constant in time, and it leads to

$$\dot{n}_t = 0, \; U_{ec} = U_{hc} \qquad \text{(steady state).} \tag{D.7.1}$$

Using (D.6.27), (D.6.28), and replacing p_t by $N_t - n_t$, (D.7.1) is equivalent to

$$G[n(N_t - n_t) - n_1 n_t] = H[pn_t - p_1(N_t - n_t)], \tag{D.7.2}$$

an equation which can be solved for n_t, giving

$$n_t = \frac{N_t(Gn + Hp_1)}{G(n + n_1) + H(p + p_1)}. \tag{D.7.3}$$

Substituting this for n_t in either (D.6.27) or (D.6.28) gives the steady-state recombination rate via traps as

$$U_t \equiv U_{ec} = U_{hc} = \frac{N_t GH(np - n_i^2)}{G(n + n_1) + H(p + p_1)} \tag{D.7.4}$$

after using the relation $n_1 p_1 = n_i^2$. This can finally be combined with (D.6.26) to give the total steady-state recombination rate

$$U_s = U_{\text{direct}} + U_t = \left[F + \frac{N_t GH}{G(n + n_1) + H(p + p_1)}\right](np - n_i^2). \tag{D.7.5}$$

It is usual to define steady-state lifetimes τ_n, τ_p for electrons and holes respectively by

$$U_s = \frac{n - n_0}{\tau_n} = \frac{p - p_0}{\tau_p}, \tag{D.7.6}$$

or, introducing the excess electron and hole concentrations $\delta_n \equiv n - n_0$, $\delta_p \equiv p - p_0$, by

$$U_s = \frac{\delta_n}{\tau_n} = \frac{\delta_p}{\tau_p} \tag{D.7.7}$$

so that τ_n is the average time spent in the conduction band by the δ_n excess electrons, and similarly for the valence band. In general, (D.7.5) shows that these lifetimes will be functions of n and p. They are, however, approximately constant under certain conditions.

Consider first the case of a strongly p-type material, in which the steady state is sufficiently near equilibrium that

$$p_0 \gg n_0, |\delta_n|, |\delta_p| \tag{D.7.8}$$

and suppose that $B_1 \sim B_2$, $T_1^s \sim T_2^s$, $T_1 \sim T_2 \sim T_3 \sim T_4$, where the sign $\sim$ denotes 'is of the same order as' and is used in the sense that, for example, $p_0 \gg n_0$ implies $B_2 p_0 \gg B_1 n_0$. Then inspection of (D.6.29) shows that $F \approx B^s + B_2 p_0$, $G \approx T_1^s + T_2 p_0$, $H \approx T_2^s + T_4 p_0$ and F, G, H are approximately constants. Also (D.7.6) implies

$$np - n_i^2 = np - n_0 p_0 = (n_0 + \delta_n)(p_0 + \delta_p) - n_0 p_0 \approx p_0 \delta_n \tag{D.7.9}$$

so that, comparing (D.7.5) with (D.7.7),

$$\frac{1}{\tau_n} \approx \left[F + \frac{N_t GH}{G(n_0 + n_1) + H(p_0 + p_1)} \right] p_0 \tag{D.7.10a}$$

and τ_n is approximately constant. A similar argument applies to τ_p in strongly n-type material. This situation is described by saying that the lifetime of minority carriers in extrinsic material is approximately constant.

A further simplification occurs if direct band–band recombination is neglected ($F = 0$) and if $p_0 \gg n_1, p_1$. (D.7.10a) then becomes

$$\frac{1}{\tau_n} \approx N_t G \approx N_t(T_1^s + T_2 p_0), \tag{D.7.10b}$$

so that if concentration-dependent parameters τ_n^*, τ_p^* are defined by

$$\tau_n^* \equiv \frac{1}{N_t G}, \; \tau_p^* \equiv \frac{1}{N_t H}, \; U_t = \frac{np - n_i^2}{\tau_p^*(n + n_1) + \tau_n^*(p + p_1)}, \tag{D.7.11}$$

then τ_n, as defined in (D.7.7), and τ_n^* are equal in strongly p-type material if recombination via traps is dominant, and similarly τ_p and τ_p^* are equal in strongly n-type material. (D.7.11) is a natural extension of the result of Shockley and Read[11], which in the present notation is

$$\tau_{n0} \equiv \frac{1}{N_t T_1^s}, \qquad \tau_{p0} \equiv \frac{1}{N_t T_2^s}, \qquad U_t = \frac{np - n_i^2}{\tau_{p0}(n + n_1) + \tau_{n0}(p + p_1)}, \tag{D.7.12}$$

and was derived on the assumption that processes (d) and (h) of fig. D.5.1 are dominant.

A constant steady-state lifetime may also be obtained if conditions are such that the semiconductor is electrically neutral, $n + n_t - p =$ constant or $\delta_n + \delta n_t - \delta p = 0$, and if the fraction of excess electrons which is taken up by the traps is small, $|\delta n_t| \ll |\delta n|$. These conditions, known together as the no-trapping approximation, lead simply to

$$\delta n \approx \delta p, \qquad n = n_0 + \delta_n, \qquad p \approx p_0 + \delta n. \tag{D.7.13}$$

It is left to the reader to verify that if δn is sufficiently small, (D.7.5) with (D.7.13) leads to constant lifetimes for electrons and holes, while if δn is sufficiently large, (D.7.12) with (D.7.13) leads to

$$\tau_n \approx \tau_p \approx \tau_{p0} + \tau_{n0}. \tag{D.7.14}$$

Thus in the limit $\delta n \to 0$ (low excitation) all the recombination mechanisms considered lead to a constant lifetime, while in the limit $\delta n \to \infty$ (high excitation) only the Shockley–Read processes (d) and (h) do so. For all other processes the lifetime ultimately decreases with increasing excitation as $1/\delta n$ or $1/(\delta n)^2$.

D.7b Transient decay

The previous results apply only to steady-state conditions. It is also of interest to consider how excess electrons and holes decay if the forces maintaining them are removed. To see this problem in its simplest terms, consider a sample in which electron and hole concentrations are uniform in space, and in which there is no electric field, so that carriers do not leave the sample by diffusion or drift. The rates of change of carrier concentrations are then determined entirely by recombination,

$$\frac{\mathrm{d}(\delta n)}{\mathrm{d}t} = -U_{\text{direct}} - U_{ec}, \qquad \frac{\mathrm{d}(\delta p)}{\mathrm{d}t} = -U_{\text{direct}} - U_{hc}. \tag{D.7.15}$$

Since carriers do not enter or leave the sample, its total charge is constant, so that

$$n + n_t - p = \text{constant}, \quad n_t = n_{t0} + p - p_0 - (n - n_0) = n_{t0} + \delta p - \delta n \tag{D.7.16}$$

[11] W. Shockley and W. T. Read, *Phys. Rev.*, **87**, 835 (1952).

where suffix 0 denotes the final equilibrium state reached. The recombination rates (D.6.26)–(D.6.28) can now be expressed in terms of δn and δp only, and (D.7.15) becomes a pair of coupled non-linear differential equations for δn and δp, which can, in principle, be solved. However, no analytical solution is known for this general form of the problem, and special cases must be considered.

Consider first the case in which δn and δp are known to be small initially, and therefore throughout the decay. The rates U_{direct}, U_{ec}, U_{hc} can each be expanded in a Taylor series about their equilibrium values of zero. One finds, to first order in δn, δp, that

$$\frac{\mathrm{d}(\delta n)}{\mathrm{d}t} = a_{11}\delta n + a_{12}\delta p, \qquad \frac{\mathrm{d}(\delta p)}{\mathrm{d}t} = a_{21}\delta n + a_{22}\delta p, \tag{D.7.17}$$

where

$$a_{11} \equiv -\frac{\partial(U_{\text{direct}} + U_{ec})}{\partial n}, \qquad a_{12} \equiv -\frac{\partial(U_{\text{direct}} + U_{ec})}{\partial p}$$

$$a_{21} \equiv -\frac{\partial(U_{\text{direct}} + U_{hc})}{\partial n}, \qquad a_{22} \equiv -\frac{\partial(U_{\text{direct}} + U_{hc})}{\partial p}, \tag{D.7.18}$$

all derivatives being evaluated at equilibrium. (D.7.17) can be written as a matrix equation

$$\frac{\mathrm{d}}{\mathrm{d}t}\begin{pmatrix}\delta n\\ \delta p\end{pmatrix} = \begin{pmatrix}a_{11} & a_{12}\\ a_{21} & a_{22}\end{pmatrix}\begin{pmatrix}\delta n\\ \delta p\end{pmatrix} \equiv \mathbf{a}\begin{pmatrix}\delta n\\ \delta p\end{pmatrix}$$

and has solutions of the form

$$\delta n = A\,\mathrm{e}^{-t/\tau_1} + B\,\mathrm{e}^{-t/\tau_2}, \qquad \delta p = C\,\mathrm{e}^{-t/\tau_1} + D\,\mathrm{e}^{-t/\tau_2} \tag{D.7.19}$$

where $(-1/\tau_1)$ and $(-1/\tau_2)$ are eigenvalues of the matrix $\mathbf{a}$, and A, B, C, D are constants. The times τ_1 and τ_2 can be regarded as transient lifetimes.

A fuller account of this approach is given by Evans and Landsberg[12]. It has the disadvantage that most information on recombination mechanisms comes from experiments in which δn and δp are large.

An alternative approach is to look for situations in which one of δn, δp can be eliminated. Three such situations suggest themselves—(i) the pure semiconductor, (ii) the n-type semiconductor with traps near the conduction band in energy, or its converse, (iii) the no-trapping situation.

[12] D. A. Evans and P. T. Landsberg, *Proc. Roy. Soc.*, **A267**, 464 (1962); *J. Phys. Chem. Solids*, **26**, 315 (1965). See also E. A. B. Cole, *Proc. Phys. Soc.*, **85**, 135 (1965).

In case (i) only unavoidable processes occur and $\delta n = \delta p$. One finds that

$$-\frac{d(\delta n)}{dt} = U_{\text{direct}}$$

$$= [B^s + B_1 n_0 + B_2 p_0 + (B_1 + B_2)\delta n][(n_0 + p_0)\delta n + (\delta n)^2]. \tag{D.7.20}$$

An exact integration would be possible [see problem (D.7.1)] but we shall consider here only limiting cases. For small δn,

$$-\frac{d(\delta n)}{dt} = \frac{\delta n}{\tau}, \qquad \delta n = (\Delta n)\,e^{-t/\tau}, \qquad \frac{1}{\tau} \equiv (B^s + B_1 n_0 + B_2 p_0)(n_0 + p_0), \tag{D.7.21}$$

where Δn is the initial value of δn, assumed to be also small. If Δn and δn are $\gg (n_0 + p_0)$, but Auger effects are negligible, $B_1 = B_2 = 0$, then

$$-\frac{d(\delta n)}{dt} = B^s(\delta n)^2, \quad \frac{1}{\delta n} = \frac{1}{\Delta n} + B^s t, \qquad \delta n = \frac{\Delta n}{1 + (t/\tau)[\Delta n/(n_0 + p_0)]}, \tag{D.7.22}$$

where now $\tau \equiv [B^s(n_0 + p_0)]^{-1}$. It will be noted that however large Δn is, δn is reduced to less than $(n_0 + p_0)$ within the fixed time interval τ. If Auger effects are not neglected, the decay from large initial values of δn will be even faster than this. In particular, if $\delta n \gg n_0 + p_0$ and Auger effects are predominant,

$$-\frac{d(\delta n)}{dt} = (B_1 + B_2)(\delta n)^3, \qquad \delta n = \frac{\Delta n}{\sqrt{[1 + 2(\Delta n)^2(B_1 + B_2)t]}}. \tag{D.7.23}$$

In case (ii) it will be assumed that electrons have been excited from the traps into the conduction band, and that neither trap nor conduction band interacts significantly with the valence band in the time scale of interest. Thus

$$\delta p \approx 0, \qquad \delta n_t \approx -\delta n, \qquad \delta p_t \approx \delta n, \tag{D.7.24}$$

and

$$-\frac{d(\delta n)}{dt} = (T_1^s + T_1 n)(np_t - n_1 n_t)$$

$$= (T_1^s + T_1 n_0 + T_1 \delta n)[(n_0 + p_{t0})\delta n + \delta n^2]. \tag{D.7.25}$$

It will be noted that (D.7.25) has a similar form to (D.7.20). The three limiting cases are also similar. They are:

Small $\delta n, \Delta n$

$$\delta n = \Delta n\,\mathrm{e}^{-t/\tau}, \qquad \tau \equiv [(T_1^s + T_1 n_0)(n_0 + p_{t0})]^{-1}. \tag{D.7.26}$$

Large $\delta n, \Delta n$, but Auger effects unimportant

$$\delta n = \frac{\Delta n}{1 + (t/\tau)[\Delta n/(n_0 + p_{t0})]}, \qquad \tau \equiv [T_1^s(n_0 + p_{t0})]^{-1} \tag{D.7.27}$$

Large δn, Δn, Auger effects dominant

$$\delta n = \frac{\Delta n}{\sqrt{[1 + 2(\Delta n)^2 T_1 t]}}. \tag{D.7.28}$$

Again an exact integration of (D.7.25) is possible and is described in problem (D.7.1). For a more complete account, see Evans and Landsberg[12]. Experiments of this type are reported in Phelan and Love[13].

Case (iii) is applicable when the number of traps is small compared to the number of excess carriers, but recombination is still predominantly via the traps. The no-trapping approximation is used in the form

$$|\delta n_t| \ll |\delta n|, |\delta p| \quad \text{and} \quad \left|\frac{\mathrm{d}(\delta n_t)}{\mathrm{d}t}\right| \ll \left|\frac{\mathrm{d}(\delta n)}{\mathrm{d}t}\right|, \left|\frac{\mathrm{d}(\delta p)}{\mathrm{d}t}\right|, \tag{D.7.29}$$

or equivalently

$$\delta n \approx \delta p \quad \text{and} \quad -\frac{\mathrm{d}(\delta n)}{\mathrm{d}t} = U_{ec} \approx U_{hc} = -\frac{\mathrm{d}(\delta p)}{\mathrm{d}t}. \tag{D.7.30}$$

It is therefore assumed that at no stage do a significant number of recombining electrons or holes accumulate on the traps. Since this leads to $U_{ec} \approx U_{hc}$, equation (D.7.1) is true as an approximation. One can, therefore, approximately replace U_{ec} and U_{hc} by U_t, which has the same form as in the steady-state equation (D.7.4) deduced from (D.7.1). Finally, n is replaced by $n_0 + \delta n$ and p by $p_0 + \delta n$, giving

$$-\frac{\mathrm{d}(\delta n)}{\mathrm{d}t} = \frac{GH\,N_t[(n_0 + \delta n)(p_0 + \delta n) - n_0 p_0]}{G(n_0 + n_1 + \delta n) + H(p_0 + p_1 + \delta n)} \tag{D.7.31}$$

and

$$G = T_1^s + T_1 n_0 + T_2 p_0 + (T_1 + T_2)\delta n,$$
$$H = T_2^s + T_3 n_0 + T_4 p_0 + (T_3 + T_4)\delta n. \tag{D.7.32}$$

[13] R. J. Phelan and W. F. Love, *Phys. Rev.*, **133**, A1134 (1964).

(D.7.31) can be written as

$$-\frac{\mathrm{d}(\delta n)}{\mathrm{d}t} = \frac{1}{R(\delta n)}, \qquad R(x) \equiv \frac{M_2 x^2 + M_1 x + M_0}{x(x+\alpha)(x+\beta)(x+\nu)}, \qquad \text{(D.7.33)}$$

where

$$\alpha \equiv \frac{T_1^s + T_1 n_0 + T_2 p_0}{T_1 + T_2}, \qquad \beta \equiv \frac{T_2^s + T_3 n_0 + T_4 p_0}{T_3 + T_4}, \qquad \nu \equiv n_0 + p_0. \qquad \text{(D.7.34)}$$

and $M_{0,1,2}$ are constants. From this point on the same procedure can be used as in problem (D.7.1), the results being rather more complex[12]. Here it will be sufficient to note that in the high-excitation limit Δn, $\delta n \gg \alpha$, β, ν the decay has the form

$$\delta n = \frac{\Delta n}{1 + at\Delta n}, \qquad a \equiv \frac{N(T_1 + T_2)(T_3 + T_4)}{T_1 + T_2 + T_3 + T_4}. \qquad \text{(D.7.35)}$$

This decay law is similar to (D.7.22) and (D.7.27), valid for predominantly single-electron recombination, whereas here Auger processes are dominant. For smaller values of δn, the decay is exponential whenever any two of α, β, ν differ widely in magnitude and δn lies between them. The final decay with $\delta n \ll \alpha, \beta, \nu$ is also exponential, so that in theory up to three distinct exponential ranges are possible. This use of the non-trapping approximation was first introduced by Goureau[14].

Problem

(D.7.1) Show that the decay laws (D.7.20) and (D.7.25) can each be rewritten as

$$-\frac{\mathrm{d}(\delta n)}{\mathrm{d}t} = A\,\delta n(\alpha + \delta n)(\nu + \delta n)$$

and find A, α, ν for each case. Show also that if $R(x)$ is the rational function $[Ax(\alpha + x)(\nu + x)]^{-1}$, then $R(x)$ can be expressed in partial fractions as

$$R(x) = \frac{1}{A(\alpha - \nu)}\left[\frac{1}{\alpha}\left(\frac{1}{\alpha + x} - \frac{1}{x}\right) - \frac{1}{\nu}\left(\frac{1}{\nu + x} - \frac{1}{x}\right)\right].$$

Hence show that, since $t = \int_{\Delta n}^{\delta n} R(x)\,\mathrm{d}x$,

$$\left(\frac{1 + \alpha/\delta n}{1 + \alpha/\Delta n}\right)\left(\frac{1 + \nu/\delta n}{1 + \nu/\Delta n}\right)^{-\alpha/\nu} = \exp[-A\alpha(\alpha - \nu)t].$$

Check that this result reduces to (D.7.26) or to (D.7.28) under the appropriate limiting conditions.

[14] G. M. Goureau, *Soviet Phys-Solid State*, **4**, 1419 (1962).

D.8 CHARGE TRANSPORT AND POISSON'S EQUATION

In discussing the flow of electrons and holes in a semiconductor it is necessary to consider electrons and holes as being to some extent localised in space, and also to consider that the electrostatic potential may vary from point to point. The first consideration appears incompatible with the description of electrons by Bloch waves, which are essentially non-localized. However, any desired localized wavefunction can be formed, by a process analogous to Fourier synthesis, from a combination of Bloch waves with appropriate weighting factors. It will therefore be assumed that electrons and holes can be considered as particles. A rough calculation shows that if an electron's kinetic energy is uncertain by approximately its classical value, $1{\cdot}5\,k_0 T$ at 300°K, its position can be defined to within about 10^{-7} cm, which is sufficiently accurate in most cases. The variation of electrostatic potential with position poses further problems in that the potential energy of an electron no longer has lattice periodicity. It will be assumed without proof that, provided the electric field is not so high that the electrostatic potential varies significantly over a few interatomic distances, the concepts of band structure can be retained, with E_c and E_v becoming functions of position. The quasi-Fermi levels μ_e and μ_h may also vary with position, but in equilibrium the Fermi level μ_0 will be a constant throughout the material.

The energy of an electron at rest (i.e. an electron of zero wave-vector) in the conduction band bears, for a given material, a fixed relation to E_c and therefore to E_i. It follows that E_i is a measure of the electrostatic potential energy of an electron, and therefore that $-(E_i/e)$ is a measure of the electrostatic potential, and that the electric field is

$$\mathbf{E} = \frac{1}{e}\,\text{grad}\,E_i. \tag{D.8.1}$$

It is found experimentally that in fields up to a few hundred volts/cm the effect of the field is to superimpose on the random motions of electrons and holes average drift velocities $\mathbf{V}_{de}$, $\mathbf{V}_{dh}$ respectively, where

$$\mathbf{V}_{de} = -\mu_n \mathbf{E}, \qquad \mathbf{V}_{dh} = \mu_p \mathbf{E}. \tag{D.8.2}$$

The coefficients μ_n, μ_p are approximately independent of field and are called the electron and hole mobilities. (For a more detailed discussion of mobility see §A.2.)

The steady drift velocities here described result from a balance between kinetic energy gained from the field and kinetic energy lost in collision with other carriers, phonons and impurity atoms. The mobilities therefore vary with temperature (affecting the numbers of phonons) and also with impurity density. Their values at 300° K for high-purity germanium and silicon, and also for silicon containing phosphorus atoms in a concentration of 10^{16} cm^{-3}, are given in table D.8.1.

Table D.8.1 Room-temperature mobilities in Ge and Si (cm^2 V^{-1} sec^{-1}).

	μ_n	μ_p
Ge, pure	3900	1900
Si, pure	1350	480
Si, 10^{16}P/cm^{-3}	1100	250

A particle current density for electrons can be defined as a vector $\mathbf{i}_e$ parallel to the net flow of electrons whose magnitude is equal to the number of electrons crossing unit area normal to the flow in unit time. If $\mathbf{i}_h$ is similarly defined for holes, and if the current is due entirely to the drift of carriers in the electric field, then

$$\mathbf{i}_e = n\mathbf{V}_{de} = -n\mu_n\mathbf{E}, \qquad \mathbf{i}_h = p\mathbf{V}_{de} = p\mu_p\mathbf{E}. \tag{D.8.3}$$

The net transport of positive charge due to electrons is described by a current density $\mathbf{j}_e = -e\mathbf{i}_e$, and that due to holes by a current density $\mathbf{j}_h = +e\mathbf{i}_h$. One therefore has

$$\mathbf{j}_e(\text{drift}) = en\mu_n\mathbf{E}, \qquad \mathbf{j}_h(\text{drift}) = ep\mu_p\mathbf{E}, \tag{D.8.4}$$

and if $\mathbf{j} \equiv \mathbf{j}_e + \mathbf{j}_h$ is the total charge current density,

$$\mathbf{j}(\text{drift}) = e(n\mu_n + p\mu_p)E \equiv \sigma\mathbf{E}, \tag{D.8.5}$$

σ being the conductivity of the material. Since μ_n and μ_p are of the same order, it appears that extrinsic materials can have a much greater conductivity than intrinsic, the ratio being of the order (n/n_i) or (p/n_i) in n-type and p-type materials respectively.

The results (D.8.3)–(D.8.5) are valid if n and p are constant in space. If not, additional current densities arise because of diffusion, acting in such a direction as to smooth out variations in n and p. These current densities have the form

$$\mathbf{i}_e(\text{diff.}) = -D_n \operatorname{grad} n, \qquad \mathbf{i}_h(\text{diff.}) = -D_p \operatorname{grad} p, \tag{D.8.6}$$

where D_n, D_p are diffusion coefficients varying with temperature and impurity density as do the mobilities. The current densities will, in general, each be a sum of diffusion and drift terms,

$$\mathbf{i}_e = -n\mu_n\mathbf{E} - D_n \operatorname{grad} n, \qquad \mathbf{i}_h = p\mu_p\mathbf{E} - D_p \operatorname{grad} p, \tag{D.8.7}$$

and the total charge current density will be the sum

$$\mathbf{j} = \mathbf{j}_e + \mathbf{j}_h = -e\mathbf{i}_e + e\mathbf{i}_h. \tag{D.8.8}$$

The current densities can be written in a different form for non-degenerate bands by noting that

$$n = n_i \exp\left(\frac{\mu_e - E_i}{k_0 T}\right), \qquad \text{grad}\, n = \frac{n}{k_0 T} \text{grad}\,(\mu_e - E_i),$$

$$p = n_i \exp\frac{(E_i - \mu_h)}{k_0 T}, \qquad \text{grad}\, p = \frac{p}{k_0 T} \text{grad}\,(E_i - \mu_h). \tag{D.8.9}$$

When also $\mathbf{E}$ is replaced by $(1/e)$ grad E_i from (D.8.1), (D.8.7) becomes

$$\mathbf{i}_e = -\frac{n\mu_n}{e} \text{grad}\, E_i + \frac{nD_n}{k_0 T} \text{grad}\,(E_i - \mu_e),$$

$$\mathbf{i}_h = \frac{p\mu_p}{e} \text{grad}\, E_i - \frac{pD_p}{k_0 T} \text{grad}\,(E_i - \mu_h). \tag{D.8.10}$$

Now in equilibrium, when both quasi-Fermi levels become equal and constant in space, the particle currents $\mathbf{i}_e, \mathbf{i}_h$ must vanish individually. This means that the coefficients of grad E_i in (D.8.10) must vanish, leading to the Einstein relations

$$D_n = \frac{k_0 T}{e} \mu_n, \qquad D_p = \frac{k_0 T}{e} \mu_p. \tag{D.8.11}$$

Using these relations in (D.8.10) gives

$$\mathbf{i}_e = -\frac{n\mu_n}{e} \text{grad}\, \mu_e = -nD_n \,\text{grad}\, F_e,$$

$$\mathbf{i}_h = \frac{p\mu_p}{e} \text{grad}\, \mu_h = pD_p \,\text{grad}\, F_h, \tag{D.8.12}$$

so that the currents are related simply to the gradients of the quasi-Fermi levels. The total charge current density is, from (D.8.8) and (D.8.12),

$$\mathbf{j} = n\mu_n \,\text{grad}\, \mu_e + p\mu_p \,\text{grad}\, \mu_h. \tag{D.8.13}$$

Since electrons remain in the conduction band until they recombine or are captured, the electron concentration in any small element of volume will decrease at a net rate equal to the recombination and capture rates *plus* the excess of electron flow out of the volume element over the electron flow into it. This is expressed mathematically, with the corresponding result for holes, as a pair of carrier conservation equations

$$-\frac{\mathrm{d}n}{\mathrm{d}t} = U_{\text{direct}} + U_{ec} - g + \operatorname{div} \mathbf{i}_e$$

$$-\frac{\mathrm{d}p}{\mathrm{d}t} = U_{\text{direct}} + U_{hc} - g' + \operatorname{div} \mathbf{i}_h \tag{D.8.14}$$

where g and g' denote any electron and hole generation processes not covered in §D.6. In steady-state conditions, $\mathrm{d}n/\mathrm{d}t = \mathrm{d}p/\mathrm{d}t = 0$ and $U_{ec} = U_{hc} = U_t$ as in (D.7.4). Assuming that $g = g' = 0$ then gives

$$\operatorname{div} \mathbf{i}_e = \operatorname{div} \mathbf{i}_h = -U_s, \quad \operatorname{div} \mathbf{j} = e(\operatorname{div} \mathbf{i}_h - \operatorname{div} \mathbf{i}_e) = 0 \quad \text{(steady state)}, \tag{D.8.15}$$

where $U_s \equiv U_{\text{direct}} + U_t$ is the total recombination rate.

So far the field $\mathbf{E}$ has appeared as an independent variable. It is, in fact, connected to the carrier concentration by Poisson's equation

$$\operatorname{div} \mathbf{E} = \frac{1}{e} \operatorname{div} \operatorname{grad} E_i = \frac{4\pi\rho}{\epsilon}, \tag{D.8.16}$$

where ρ is the net positive charge density, which can be expressed as

$$\rho = e(p - n - N^*) \tag{D.8.17}$$

N^* being the net fixed concentration of electrons (on donors, acceptors and other centres).

In equilibrium, Poisson's equation determines the electric field if N^* is known. This is possible because the Fermi level μ is constant, so that if a dimensionless potential $\phi \equiv (E_i - \mu)/k_0 T$ is defined, then

$$\mathbf{E} = k_0 T \operatorname{grad} \phi. \tag{D.8.18}$$

Also, from (D.4.25), $n = n_i e^{-\phi}$ and $p = n_i e^{\phi}$, so that

$$\rho = en_i(\mathrm{e}^{\phi} - \mathrm{e}^{-\phi} - N^*/n_i) = 2en_i(\sinh\phi - \sinh\phi_1), \tag{D.8.19}$$

where ϕ_1 has been defined by

$$2n_i \sinh\phi_1 = N^*. \tag{D.8.20}$$

Substituting (D.8.18) and (D.8.19) in (D.8.16) gives

$$\operatorname{div} \operatorname{grad} \phi = \frac{8\pi e^2 n_i}{\epsilon k_0 T}[\sinh\phi - \sinh\phi_1]. \tag{D.8.21}$$

In the case of intrinsic material, $\phi_1 = 0$. If it is also assumed that $\phi \ll 1$, (D.8.21) reduces to

$$\operatorname{div}\operatorname{grad}\phi = \frac{2\phi}{L_{Di}^2}, \text{ where } L_{Di} \equiv \left(\frac{\epsilon k_0 T}{4\pi e^2 n_i}\right)^{\frac{1}{2}}. \tag{D.8.22}$$

Assuming a one-dimensional solution $\phi = \phi(x)$ which vanishes at $x = +\infty$, one finds that

$$\phi(x) = \phi(0)\exp(-\sqrt{2}x/L_{Di}), \tag{D.8.23}$$

so that any small deviation of ϕ from zero disappears in a distance of the order L_{Di}. L_{Di} is known as the Debye length for intrinsic material, and is about 10^{-4} cm for Ge at 300°K. Since the right hand side of (D.8.21) with $\phi_1 = 0$ increases rapidly in magnitude for $|\phi| > 1$, a large initial value of ϕ will also disappear in a total distance not much greater than L_{Di}. Thus any departure from neutrality in pure material in equilibrium is damped out within a few Debye lengths from the source.

It is left to the reader to show [problem (D.8.1)] that in a uniformly extrinsic material ($|N^*| = \text{constant} \gg n_i$) any departure from the neutral condition $\phi = \phi_1$ is damped out in a distance of order L_D, where $L_D \equiv L_{Di}(n_i/|N^*|)^{\frac{1}{2}}$ is the extrinsic Debye length. It will be assumed in Chapter VIII that this result can be extended to non-equilibrium situations, so that regions of constant donor and acceptor concentrations are electrically neutral except in the neighbourhood of a discontinuity.

A different approximation is useful when electrons and holes are less numerous in a given region than fixed charges, i.e. $n, p \ll |N^*|$. This case is likely to arise in a region of high electric field, where a given current can be maintained by relatively few carriers. It leads, with (D.8.16) and (D.8.17), to

$$e\operatorname{div}\mathbf{E} = \operatorname{div}\operatorname{grad}E_i \approx -\frac{4\pi e^2 N^*}{\epsilon}, \tag{D.8.24}$$

a result which will be applied in Chapter VIII to the analysis of p–n junctions.

From now on attention will be centred on problems of one-dimensional steady-state current flow. If currents and fields are defined as positive in the direction of increasing x, the relevant results of this section simplify to:

$$j_e = e\left(n\mu_n E + D_n\frac{dn}{dx}\right), \qquad j_h = e\left(p\mu_p E - D_p\frac{dp}{dx}\right), \tag{D.8.25}$$

$$j_e = n\mu_n\frac{d\mu_e}{dx}, \qquad j_h = p\mu_p\frac{d\mu_h}{dx}, \tag{D.8.26}$$

$$\frac{\mathrm{d}j_e}{\mathrm{d}x} = -\frac{\mathrm{d}j_h}{\mathrm{d}x} = eU_s, \qquad j = j_e + j_h = \text{constant}, \qquad \text{(D.8.27)}$$

$$e\frac{\mathrm{d}E}{\mathrm{d}x} = \frac{\mathrm{d}^2 E_i}{\mathrm{d}x^2} = \frac{4\pi e\rho}{\epsilon} = \frac{4\pi e^2}{\epsilon}(p - n - N^*). \qquad \text{(D.8.28)}$$

Problem

(D.8.1) Consider Poisson's equation in the form (D.8.21) in a region in which the net donor density N^* is constant so that ϕ_1 is constant. Show that (D.8.21), for a one-dimensional problem, can be written

$$\frac{\mathrm{d}^2\phi}{\mathrm{d}x^2} = \frac{16\pi e^2 n_i}{\epsilon k_0 T}\cosh\left(\frac{\phi + \phi_1}{2}\right)\sinh\left(\frac{\phi - \phi_1}{2}\right).$$

Assume next that $|\phi - \phi_1| \ll 1$ but that ϕ_1 is numerically greater than $1(|N^*| \gg n_i)$. Show that $\cosh\frac{(\phi + \phi_1)}{2} \approx \cosh\phi_1 \approx \frac{|N^*|}{2n_i}$, and that Poisson's equation reduces approximately to

$$\frac{\mathrm{d}^2\phi}{\mathrm{d}x^2} = \frac{4\pi e^2|N^*|}{\epsilon k_0 T}(\phi - \phi_1),$$

with solutions $\phi - \phi_1 \sim \exp(\pm x/L_D)$, where $L_D \equiv (\epsilon k_0 T/4\pi e^2|N^*|)^{\frac{1}{2}}$. Deduce that any deviation of ϕ from the value ϕ_1 will become small in a distance of order L_D. Verify that, in strongly extrinsic materials ($|N^*| \gtrsim 10^{17}$ cm^{-3} at 300°K), L_D is of order 10^{-6} cm or less.

CHAPTER VIII

Aspects of *p–n* Junction Theory

D.9 DESCRIPTION OF THE JUNCTION

A *p–n* junction consists of a semiconductor lattice which contains a *p*-type region and an *n*-type region adjacent to each other, the basic band structure being the same in both regions (junctions in which the band structure alters are known as heterojunctions and will not be considered further). The transition from *p*-type impurity concentration (an excess of acceptors) to *n*-type (an excess of donors) may be abrupt or gradual, depending on the techniques used to form the junction. Since the equilibrium Fermi level μ_0 is near the valence band in *p*-type and near the conduction band in *n*-type material, some bending of the band structure is needed for μ_0 to remain constant. This and other consequences are illustrated in fig. D.9.1 for an ideally abrupt junction in equilibrium, assuming the structure to be one-dimensional with contacts at the planes $x = -a$ and $x = c$.

Diagram (a) shows the assumed distribution of acceptors and donors in terms of the net effective donor concentration N_d^*. Diagram (b) shows that, as discussed in connexion with (D.8.19), there is effective neutrality with a resulting flat band structure except for a region near the junction. The width of this region, known as the space-charge region, is defined as $W = l_p + l_n$, and the centre of the region is taken to be at x_0, the actual junction plane being at the origin. Diagram (c) shows the equilibrium majority carrier densities p_p in the p region and n_n in the n region, and also that as the band edges bend away from the Fermi level, both n and p drop to very low values in the space-charge region. The net charge density ρ is then largely determined by the donors and acceptors, as indicated in diagram (d). Finally, Poisson's equation with effectively constant ρ gives a linear variation of electric field in the space-charge region, as illustrated in diagram (e). Donors and acceptors are assumed fully ionized, so that $N_d^* = N_d - N_a$.

The width of the region can be determined on the assumption that n and p are negligible throughout the region. Integrating (D.8.28) once with the condition of zero field at $x = -l_p$ gives the field at x as

$$E(x) = \frac{4\pi}{\epsilon} \int_{-l_p}^{x} \rho \, dx, \qquad \text{(D.9.1)}$$

and since the field is also zero at $x = l_n$,

$$0 = \frac{4\pi}{\epsilon}\int_{-l_p}^{l_n} \rho\, dx = \frac{4\pi}{\epsilon} e(l_p N_a - l_n N_d), \qquad l_p N_a = l_n N_d. \quad \text{(D.9.2)}$$

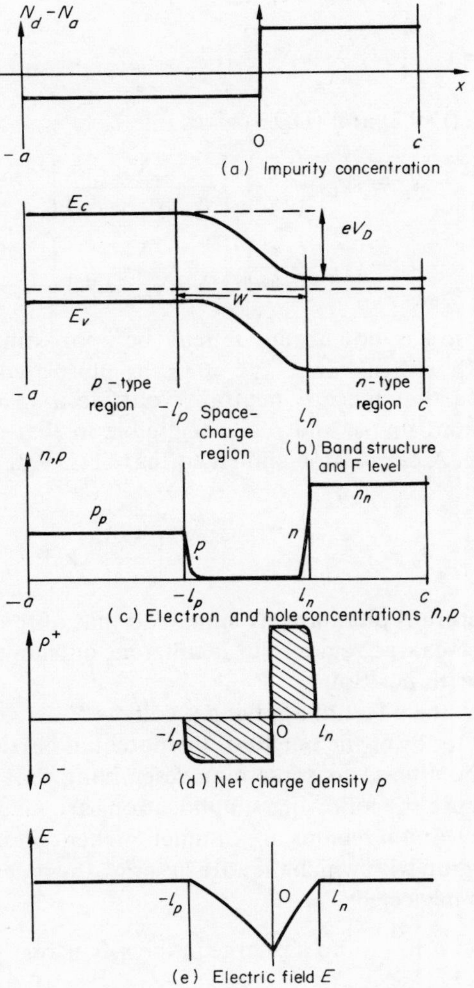

Fig. D.9.1. The one-dimensional abrupt *p–a* junction in equilibrium.

Integrating a second time gives the potential drop eV_D:

$$V_D = -\int_{-l_p}^{l_n} E(x)\,dx = -\int_{-l_p}^{l_n} |E(x)| + \int_{-l_p}^{l_n} x\frac{dE}{dx}\,dx$$

$$= 0 + \frac{4\pi}{\epsilon}\int_{-l_p}^{l_n} x\rho\,dx$$

$$= \frac{4\pi}{\epsilon}e\left[N_d\frac{l_n^2}{2} + N_a\frac{l_p^2}{2}\right], \tag{D.9.3}$$

and combining (D.9.2) and (D.9.3) gives

$$l_p = (V_D)^{\frac{1}{2}}\left(\frac{\epsilon}{2\pi e}\right)^{\frac{1}{2}}\left[\frac{N_d}{N_a(N_a + N_d)}\right]^{\frac{1}{2}},$$

$$l_n = (V_D)^{\frac{1}{2}}\left(\frac{\epsilon}{2\pi e}\right)^{\frac{1}{2}}\left[\frac{N_a}{N_d(N_a + N_d)}\right]^{\frac{1}{2}}. \tag{D.9.4}$$

If the transition is not abrupt it may be approximated by a linear variation of N_d^* with distance, $N_d^* = \lambda x$, as illustrated in fig. D.9.2. It will be assumed that there is neutrality outside a space-charge region $-l_p < x < l_n$, and that n and p are negligible in that region, so that ρ varies as shown. A calculation similar to that of (D.9.2) and (D.9.3) leads to

$$l_p = l_n = \frac{W}{2}, \qquad W = \left(\frac{3\epsilon}{\pi e\lambda}\right)^{\frac{1}{3}}(V_D)^{\frac{1}{3}}. \tag{D.9.5}$$

The band structure is qualitatively similar to that of the abrupt junction within the space-charge region, but is different outside the region as N_d^* is still varying with position.

Consider now the effect of applying a voltage V across the contacts in such a sense as to lower the height of the potential barrier V_D. The band structure will be altered to something resembling that of fig. D.9.3. In drawing this figure the following simplifications are made:

(a) Since the p and n regions are of much higher conductivity than the space-charge region with which they are in series, the potential drop across these regions is neglected;

(b) since $j_e = e\mu_n n\frac{d\mu_e}{dx}$ and n is large in the n-type region, the drop of μ_e in that region is neglected, and similarly for μ_n in the p-type region, and

(c) the drop of μ_e and μ_h in the space-charge region is neglected. The last simplification can be justified only with reference to actual numerical values; this will be left to a later section.

These simplifications mean that

$$p \approx p_p \quad \text{for} \quad x < -l_p,$$
$$n \approx n_n \quad \text{for} \quad x > l_n,$$
$$np \approx n_i^2 \exp\left(\frac{eV}{kT}\right) \quad \text{for} \quad -l_p \leqslant l \leqslant l_n. \tag{D.9.6}$$

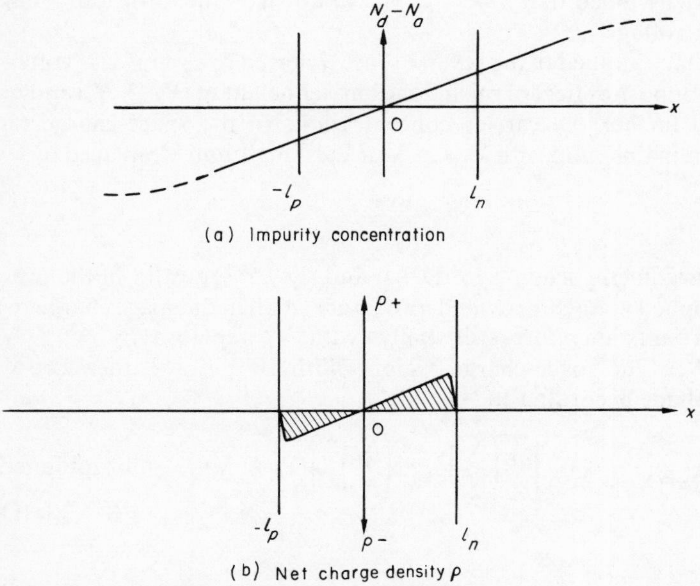

Fig. D.9.2. The one-dimensional linear junction.

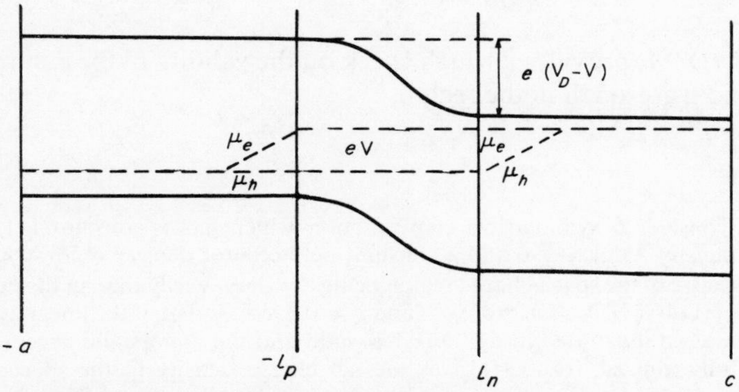

Fig. D.9.3. Energy bands and quasi-Fermi levels for a forward applied voltage $V < V_D$.

The effect of applying a voltage in this sense (referred to as a forward voltage) is thus to reduce the potential barrier height to $(V_D - V)$ and to increase carrier concentration in the space-charge region. A limit to this process is evidently reached when $V = V_D$, as the barrier would then vanish and the space-charge region become indistinguishable from the p and n regions. In practice part of the applied voltage would appear across the p and n regions and (D.9.6) would cease to hold. In what follows it is therefore assumed that $V < V_D$. V_D is known as the diffusion voltage or built-in voltage.

A voltage applied in the reverse sense (referred to as a reverse voltage V_r) has the opposite effect of raising the barrier height to $(V_D + V_r)$ and reducing still further the carrier concentrations in the space-charge region. (D.9.6) remains valid with $V = -V_r$ if V_r is small, and is replaced by

$$np \ll n_i^2 \quad \text{for} \quad -l_p \leqslant x \leqslant l_n \tag{D.9.7}$$

if V_r is large.

The arguments leading to (D.9.4) and (D.9.5) are valid in the presence of an applied voltage, provided n and p are small in the space charge region. These results therefore still apply, with V_D replaced by $(V_D - V)$ or $(V_D + V_r)$. The space-charge region width $W = l_p + l_n$ therefore varies with voltage according to:

$$W = \left[\frac{\epsilon}{2\pi e(N_a + N_d)}\right]^{\frac{1}{2}}\left[\left(\frac{N_d}{N_a}\right)^{\frac{1}{2}} + \left(\frac{N_a}{N_d}\right)^{\frac{1}{2}}\right](V_D - V)^{\frac{1}{2}} \quad \text{(abrupt junction)}, \tag{D.9.8}$$

and

$$W = \left(\frac{3\epsilon}{\pi e\lambda}\right)^{\frac{1}{3}}(V_D - V)^{\frac{1}{3}} \quad \text{(linear junction)}. \tag{D.9.9}$$

Problem (D.9.1) provides a rough check on the validity of the assumption that n and p are small in the region.

Problems

(D.9.1) Consider a symmetrical step junction which has a constant net donor density N for $x > 0$ and a constant net acceptor density N for $x < 0$, the centre of the space-charge region being at $x = 0$. Verify that, in the notation of (D.10.32), $N = n_i \exp(\eta_D/2)$ and $\delta = 0$. Deduce that, if the linear approximation illustrated in fig. D.10.1 is valid and the donors and acceptors are fully ionized, (D.10.34) gives the net charge density in the space-charge region as

$$\rho = e(p - n \pm N) = en_i[-2\exp(\eta/2)\sinh(\theta x/W) \pm \exp(\eta_D/2)],$$

the plus sign referring to $x > 0$ and the minus sign to $x < 0$. Using this result in

$$V_D - V = \frac{kT\theta}{e} = \frac{4\pi}{\epsilon}\int_{-W/2}^{W/2} x\rho \, dx,$$

show that

$$W^2\left[\left(\frac{4}{\theta^2}\sinh\frac{\theta}{2} - \frac{2}{\theta}\cosh\frac{\theta}{2}\right)\exp\frac{\eta}{2} + \frac{\exp(\eta_D/2)}{4}\right] = \frac{kT\epsilon\theta}{4\pi e^2 n_i},$$

and hence that

$$W = B\theta^{\frac{1}{2}}\left[1 + \frac{8}{\theta^2}(1 - e^{-\theta}) - \frac{4}{\theta}(1 + e^{-\theta})\right]^{-\frac{1}{2}}$$

where B is independent of θ.

Check that (D.9.8), under the same conditions, becomes $W = B\theta^{\frac{1}{2}}$, with the same value of B, so that if θ is fairly large the fractional error in W due to neglecting n and p is about $2/\theta$. Verify also that as θ varies from 20 to 4, W varies from about $5B$ to about $3B$.

(D.9.2) Consider a junction in which the net effective donor density N_d^* varies as

$$N_d^* = +N \qquad (x > c)$$

$$N_d^* = -N \qquad (x < -c)$$

$$N_d^* = \frac{Nx}{c} \qquad (-c \leqslant x \leqslant c).$$

Assuming that n and p can be neglected in the space-charge region, prove that the region is symmetrical about $x = 0$. Prove also that, with the notation of the last problem,

$$W = 2\left[\frac{c^2}{3} + \frac{\epsilon(V_D - V)}{4\pi e N}\right]^{\frac{1}{2}} \qquad \text{if } W > 2c,$$

$$W = \left[\frac{3\epsilon c}{\pi e N}(V_D - V)\right]^{\frac{1}{3}} \qquad \text{if } W < 2c,$$

and find for what applied voltage W is equal to $2c$.

D.10 THE CURRENT–VOLTAGE RELATION

D.10a General

The current density through the junction can be written as the sum of electron and hole current densities at any point,

$$j = j_e(-a) + j_h(-a) = j_e(-a) + j_h(c) + e\int_{-a}^{c} U_s \, dx, \qquad \text{(D.10.1)}$$

where (D.8.27) has been used. Now if the contact at $x = -a$ is far from the junction, n and p will be nearly constant in space and near their equilibrium values, $\mathrm{d}n/\mathrm{d}t \approx \mathrm{d}p/\mathrm{d}t \approx 0$ and $n \ll p$. According to (D.8.25), the current will then be carried almost entirely by holes, $j_e(-a) \ll j_h(-a) \approx j$. Similarly $j_h(c) \ll j$. Thus the total current density j can be expressed with small error as a recombination–generation current

$$j \approx e \int_{-a}^{c} U_s \, \mathrm{d}x \equiv j_{bp} + j_{sc} + j_{bn}, \tag{D.10.2}$$

where

$$j_{bp} \equiv e \int_{-a}^{-l_p} U_s \, \mathrm{d}x, \qquad j_{sc} \equiv e \int_{-l_p}^{l_n} U_s \, \mathrm{d}x, \qquad j_{bn} \equiv e \int_{l_n}^{c} U_s \, \mathrm{d}x. \tag{D.10.3}$$

The current is positive (i.e. charge flows from left to right) if recombination predominates, which is the case for forward voltage since, according to fig. D.9.3, $\mu_e \geqslant \mu_h$ everywhere. As will be seen later, a p–n junction will act as a good rectifier only if j_{sc} is small. We therefore consider first the so-called ideal junction, one for which j_{sc} can be neglected.

D.10b The ideal junction

Since the p and n regions are of high conductivity, the electric field in them is small. Considering the electron current in the p region, we therefore neglect the drift term, so that

$$j_e \approx eD_n \frac{\mathrm{d}n}{\mathrm{d}x}, \qquad eU_s = \frac{\mathrm{d}j_e}{\mathrm{d}x} \approx eD_n \frac{\mathrm{d}^2 n}{\mathrm{d}x^2} \tag{D.10.4}$$

Also, provided $n \ll p$, (D.9.6) means that U_s can be described by a steady-state lifetime for electrons

$$U_s = \frac{n - n_p}{\tau_n}, \tag{D.10.5}$$

where τ_n is defined by (D.7.10); $n_p \equiv n_i^2/p_p$ is the equilibrium value of n within the p region. Combining (D.10.4) and (D.10.5) gives an equation

$$\frac{\mathrm{d}^2 n}{\mathrm{d}x^2} = \frac{n - n_p}{D_n \tau_n} \equiv \frac{n - n_p}{L_n^2}, \qquad L_n \equiv (D_n \tau_n)^{\frac{1}{2}}. \tag{D.10.6}$$

L_n is called a diffusion length for electrons in the p region. It is convenient to define a new coordinate

$$z \equiv \frac{x + l_p}{L_n}, \qquad w_p \equiv \frac{a - l_p}{L_n} \equiv \frac{W_p}{L_n}, \tag{D.10.7}$$

where W_p is the width of the p region. The differential equation for n is now

$$\frac{d^2n}{dz^2} = n - n_p, \tag{D.10.8}$$

with general solution

$$n = n_p + A\cosh z + B\sinh z = n_p + C\,e^z + D\,e^{-z}. \tag{D.10.9}$$

The constants A and B, or C and D, are determined by boundary conditions at $x = -l_p$ ($z = 0$) and at the contact $x = -a$ ($z = -w_p$). From (D.9.6),

$$n \approx \frac{n_i^2}{p_p}\exp\left(\frac{eV}{k_0T}\right) = n_p\exp\left(\frac{eV}{k_0T}\right) \equiv n_p\,e^{\eta} \text{ at } x = -l_p, \tag{D.10.10}$$

where the dimensionless applied voltage

$$\eta \equiv eV/k_0T \tag{D.10.11}$$

has been introduced. The boundary condition at the contact depends on the width of the region. If $W_p \gg L_n$, the term $D\,e^{-z}$ in (D.10.9) becomes unacceptably large unless $D = 0$, and then

$$C = n_p(e^{\eta} - 1), \qquad n = n_p(e^{\eta} - 1)\,e^z. \tag{D.10.12}$$

The current density j_{bp} is obtained by integrating (D.10.5) over the region:

$$\begin{aligned} j_{bp} &= eL_n\int_{-w_p}^{0}\frac{(n - n_p)}{\tau_n}dz \\ &= \frac{en_p(e^{\eta} - 1)L_n}{\tau_n}\int_{-w_p}^{0} e^z\,dz \\ &\approx \frac{en_p(e^{\eta} - 1)L_n}{\tau_n}, \end{aligned} \tag{D.10.13}$$

the term e^{-w_p} being neglected compared to 1. An identical argument for the n region, assuming that $c - l_n \equiv W_p \gg L_p \equiv (D_p\tau_p)^{\frac{1}{2}}$, leads to the total current for an ideal junction

$$j \approx j_{bp} + j_{bn} \approx e\left(\frac{n_pL_n}{\tau_n} + \frac{p_nL_p}{\tau_p}\right)[\exp(eV/k_0T) - 1]. \tag{D.10.14}$$

For reverse voltages of more than a few tenths of a volt ($-eV \gg k_0T$) this leads to a saturated current

$$j\,(\text{reverse saturated}) = -j_{rs} = -e\left(\frac{n_pL_n}{\tau_n} + \frac{p_nL_p}{\tau_p}\right) \tag{D.10.15}$$

and the general expression for the current, (D.10.14), can be written as

$$j = j_{rs}[\exp(eV/k_0T) - 1]. \tag{D.10.16}$$

The highest voltage for which (D.10.16) can apply is determined by the requirements

$$n \ll p_p \quad \text{at} \quad x = -l_p, \qquad p \ll n_n \quad \text{at} \quad x = l_n,$$

which are equivalent to

$$eV \ll k_0 T \log_e(p_p/n_p), \qquad eV \ll k_0 T \log_e(n_n/p_n). \qquad \text{(D.10.17)}$$

Thus the current j is of the form (D.10.16) as long as it is less than $j_{rs}(p_p/n_p)$ and $j_{rs}(n_n/p_n)$, both of which may be of order $10^6 j_{rs}$ or more. The ideal diode can therefore act as a very effective rectifier, with a resistance in the reverse direction 10^6 or more times the resistance in the forward direction.

If the p and n regions are not wide compared to the diffusion length it becomes necessary to consider the contacts in more detail. The contacts often contain a large number of recombination centres resulting from localised electron states at the edge of the crystal structure. Recombination at such centres results in a recombination particle current of electrons and holes into the surface; in the same way as volume recombination is described by a lifetime, surface recombination can be described by a velocity, so that at $x = -a$, for example

$$i_e = -s(n - n_p) \qquad \text{(D.10.18)}$$

where s is the surface recombination velocity for electrons at this contact. It will be assumed that s, like τ_n, is independent of n under the conditions obtaining in the p region. Also, since such a current can be due only to diffusion, the electric field and the concentration of electrons being small, one finds

$$D_n\left(\frac{dn}{dx}\right)_{x=-a} = s[n(-a) - n_p]. \qquad \text{(D.10.19)}$$

If s is known this provides the boundary condition at $x = -a$ which, in conjunction with (D.10.10), determines the constants A and B in (D.10.9).

A special case of interest is that in which s is so large as effectively to make $n = n_p$ at the contact. This condition in (D.10.9) gives $B = A \coth w_p$, and (D.10.10) gives $A = n_p(e^\eta - 1)$. The current is obtained as in (D.10.13):

$$\begin{aligned} j_{bp} &= \frac{eL_n}{\tau_n}\int_{-w_p}^{0} (A\cosh z + B\sinh z)\,dz \\ &= \frac{eL_n}{\tau_n}[A\sinh w_p + B(1 - \cosh w_p)] \\ &= \frac{en_pL_n}{\tau_n}(e^\eta - 1)\left[\frac{1 + \cosh w_p}{\sinh w_p}\right]. \end{aligned} \qquad \text{(D.10.20)}$$

For large w_p this reduces to (D.10.13) as expected. For small w_p its application is complicated by the facts (i) that l_p (and hence w_p) varies with voltage, and (ii) that the electron current at the contact $j_e(-a)$, neglected in (D.10.2), may now be significant.

It is left to the reader to obtain the expression corresponding to (D.10.20) for a finite or zero value of s, and to verify that in all cases it reduces to (D.10.13) for large w_p, so that the result is independent of the precise nature of the contact.

D.10c Space-charge region recombination

The current j_{sc} due to recombination–generation in the space charge region is determined in principle by (D.10.3) if the shape of the band edges is known, since n and p are then determined. However, both the abrupt and linear junctions as illustrated in figs. D.9.1 and D.9.2 give rise to a variation of n and p with position of the form $n, p \propto \exp[ax + bx^2]$ (abrupt) or $\exp[ax + bx^2 + cx^3]$ (linear), which make the integral (D.10.3) very intractable. It is therefore advisable to get as much information as possible by considering special cases.

Consider first the case of a reverse voltage V_r sufficiently large to remove almost all electrons and holes from the space-charge region. The generation rate is found as minus the value of (D.7.5) with $n = p = 0$,

$$g = -U_s = n_i^2\left[B^s + \frac{N_t T_1^s T_2^s}{T_1^s n_1 + T_2^s p_1}\right] = \text{constant.} \qquad \text{(D.10.21)}$$

The current due to generation in the space-charge region is

$$j_{sc} = \int_{-l_p}^{l_n} eU_s \, dx = -eg(l_p + l_n) = -egW. \qquad \text{(D.10.22)}$$

The current $j_{bp} + j_{bn}$ approaches its reverse saturated value (D.10.15),

$$j_{bp} + j_{bn} = -e\left(\frac{n_p L_n}{\tau_n} + \frac{p_n L_p}{\tau_p}\right), \qquad \text{(D.10.23)}$$

and the ratio of the currents is

$$\frac{j_{sc}}{j_{bp} + j_{bn}} \propto W. \qquad \text{(D.10.24)}$$

From (D.9.8) and (D.9.9), therefore, both j_{sc} and the ratio $j_{sc}/(j_{bn} + j_{bp})$ increase with reverse voltage,

$$j_{sc} \propto (V_D + V_r)^{\frac{1}{2}} \text{ (abrupt junction)}, \quad j_{sc} \propto (V_D + V_r)^{\frac{1}{3}} \text{ (linear junction).} \qquad \text{(D.10.25)}$$

Thus a junction in which j_{sc} is important does not have a limiting current in the reverse direction, in contrast to the behaviour of an ideal junction. It was shown by Sah, Noyce and Shockley[15] that for typical values of diffusion lengths, lifetimes etc., and for small reverse voltage,

$$j_{sc}/(j_{bp} + j_{bn}) \approx 0{\cdot}1\,(\text{Ge}) \quad \text{or} \quad 3000\,(\text{Si}). \tag{D.10.26}$$

The result (D.10.25) agrees well with the actual behaviour of silicon junctions for moderate reverse voltage.

A second case of interest occurs when the dominant recombination processes are radiative, either band-to-band or via localized levels very close in energy to the conduction band. For the band-to-band process,

$$U_{\text{direct}} = B^s(np - n_i^2), \tag{D.10.27}$$

while if the localized levels are near (say) the conduction band, so that $T_1^s > T_2^s$ and, in most of the space-charge region, $n_1 \gg n, p, p_1$, (D.7.4) becomes

$$U_t \approx \frac{N_t T_1^s T_2^s}{T_1^s n_1}(np - n_i^2) = \frac{N_t T_2^s}{n_1}(np - n_i^2). \tag{D.10.28}$$

The total rate is then

$$U_s = \left[B^s + \frac{N_t T_2^s}{n_1}\right](np - n_1^2) = \left[B^s + \frac{T_2^s N_t}{n_1}\right]n_i^2(\mathrm{e}^\eta - 1), \tag{D.10.29}$$

using (D.9.6) and (D.10.11). Since this rate is constant in most of the space-charge region, integration leads to a current

$$j_{sc} \approx eWU_s = en_i^2\left[B^s + \frac{N_t T_2^s}{n_1}\right](\mathrm{e}^\eta - 1)W. \tag{D.10.30}$$

The variation of W with voltage is much slower than that of the exponential term except for reverse voltage [see problem (D.9.1)] so that for these processes

$$j_{sc} \propto (\mathrm{e}^\eta - 1) = [\exp(eV/k_0T) - 1] \text{ (forward voltage)} \tag{D.10.31}$$

and j_{sc} has the same voltage dependence as j_{bn} and j_{bp} for the ideal junction. The radiative processes discussed are certainly important, and possibly dominant, in GaAs and GaP junctions used as semiconductor lamps and lasers. It should be noted, however, that such devices often operate with degenerate conduction or valence bands, so that the mass-action laws on which (D.10.30) depends would not be exact.

[15] C. T. Sah, R. N. Noyce and W. Shockley, *Proc. I.R.E.*, **45**, 9 (1957).

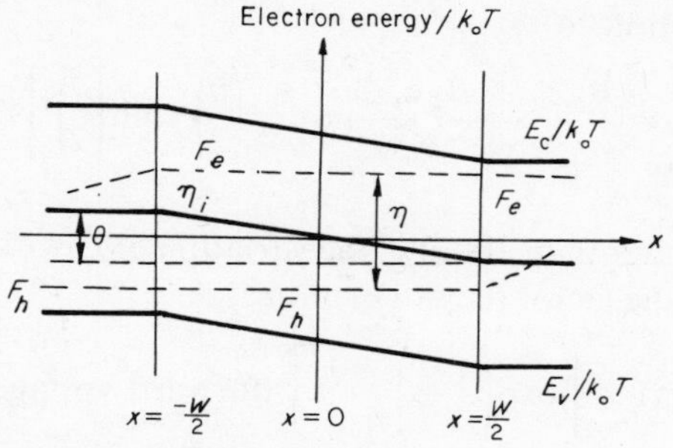

Fig. D.10.1. Simplified dimensionless energy band diagram, for the space-charge region for a forward voltage $V < V_D$ notation as in (D.10.32).

In order to gain some idea of the behaviour of j_{sc} under less restrictive conditions, band edges in the space-charge region will be assumed straight. Also the origin of x will be taken temporarily as the centre of the region, and E_i will be measured from its value at the centre of the region. The resulting picture is shown in fig. D.10.1. It is convenient to define dimensionless energies and quasi-Fermi levels

$$\eta_i \equiv E_i/k_0T, \qquad F_{e,h} \equiv \mu_{e,h}/k_0T, \qquad \eta \equiv \mathrm{eV}/k_0T,$$

$$\delta \equiv \tfrac{1}{2}(F_e + F_h) = \tfrac{1}{2}\log_e(n_n/p_p), \qquad \theta \equiv \mathrm{e}(\mathrm{V_D} - \mathrm{V})/k_0T, \qquad \eta_D \equiv \mathrm{eV_D}/k_0T. \tag{D.10.32}$$

Then, from the figure,

$$\eta_i = -\theta x/W, \qquad F_e = \eta/2 + \delta, \qquad F_h = -\eta/2 + \delta, \tag{D.10.33}$$

and from (D.6.3)

$$n = n_i \exp(\eta/2 + \delta + \theta x/W), \qquad p = n_i \exp(\eta/2 - \delta - \theta x/W). \tag{D.10.34}$$

The identification of δ in (D.10.32) follows from $n(W/2) = n_n$ and $p(-W/2) = p_p$. The current j_{sc} can now be calculated by substituting (D.10.34) in (D.7.5) for U_s. The different types of process will be considered separately.

For the band–band processes, (D.10.34) leads to

$$U_{\text{direct}} = \left[B^s + B_1 n_i \exp\left(\frac{\eta}{2} + \delta + \frac{\theta x}{W}\right) + B_2 n_i \exp\left(\frac{\eta}{2} - \delta - \frac{\theta x}{W}\right)\right] n_i^2(\mathrm{e}^{\eta} - 1), \tag{D.10.35}$$

and after integration to

$$j_{sc}(\text{direct}) = \left[B^s + \frac{(B_1\,e^{\delta} + B_2\,e^{-\delta})}{\theta} n_i \exp\left(\frac{\eta}{2}\right) \sinh\left(\frac{\theta}{2}\right)\right] W n_i^2 (e^{\eta} - 1). \tag{D.10.36}$$

Provided θ is greater than about 4, the variation of j_{sc} with forward voltage is due mainly to the factor $(e^{\eta} - 1)$, so that

$$j_{sc}\ (\text{direct}) \propto \left[\exp\left(\frac{eV}{k_0 T}\right) - 1\right] \ (\text{forward voltage} < V_D). \tag{D.10.37}$$

The electron and hole capture processes involving one electron have been studied[15,16], the result being that for $V < V_D$

$$j_{sc} \propto \left[\exp\left(\frac{eV}{k_0 T}\right) - 1\right] \text{ for } V < V_1, \qquad j_{sc} \propto \exp\left(\frac{eV}{2k_0 T}\right) \text{ for } V > V_1, \tag{D.10.38}$$

where the voltage V_1 is defined by

$$eV_1 = 2k_0 T \log_e [(n_1 + p_1)/2n_i]. \tag{D.10.39}$$

If the trap energy is near the centre of the forbidden band V_1 is small and $j_{sc} \propto \exp(eV/2k_0 T)$ for all forward voltages up to V_D.

The electron and hole capture processes involving electron collision have been studied[16] with the assumption that the four mass-action constants $T_1 \ldots T_4$ are equal, as are T_1^s and T_2^s. It was found that

$$j_{sc} \propto \left[\exp\left(\frac{eV}{k_0 T}\right) - 1\right] \text{ for } V < V_D. \tag{D.10.40}$$

The results of this section can be summarized as follows: for forward voltage less than the built-in voltage almost all recombination mechanisms, whether occurring inside or outside the space-charge region, contribute a term proportional to $[\exp(eV/k_0 T) - 1]$ to the current. The exception is single-electron recombination in the space-charge region via centres lying near the centre of the forbidden band, which contributes a term proportional to $\exp(eV/2k_0 T)$. For reverse voltage, the current due to generation outside the space-charge region saturates, while the current due to generation in the region increases as $(V_D + V_r)^{\frac{1}{2}}$ or $(V_D + V_r)^{\frac{1}{3}}$.

Sources other than recombination in the form discussed above may contribute to or even dominate the observed current. These sources include (i) leakage along the surface of the device, (ii) electron tunnelling,

[16] D. A. Evans and P. T. Landsberg, *Solid-State Electronics*, **6**, 169 (1963).

which occurs when there is an energy overlap between the valence band in the p region and the conduction band in the n region so that an electron can cross from one band to the other at constant energy, (iii) photon-assisted tunnelling, in which the bands need not overlap, since, for example, a valence band electron can absorb a photon and thus be raised into a virtual state in the forbidden gap, from which it can cross into the conduction band at constant energy, and (iv) avalanche multiplication of current in the space-charge region. Surface leakage is usually small in useful devices. An account of tunnelling in diodes is given by Gentile[17], and avalanche multiplication is discussed in §D.12 below. For an account of current–voltage relations at voltages higher than V_D, see Jonscher[18] or Grinberg[19].

D.11 SPACE CHARGE REGION CAPACITANCE FOR REVERSE VOLTAGE

It is evident from figs. D.9.1d and D.9.2 that the space-charge region contains a layer of net positive charge to the right of $x = 0$ and a layer of net negative charge to the left. The total charge per unit area in each layer is numerically the same, since

$$\int_{-l_p}^{l_n} \rho \, \mathrm{d}x = \frac{\epsilon}{4\pi}[E(l_n) - E(-l_p)] = 0, \qquad \text{(D.11.1)}$$

the electric field being zero at the edges of the region. Denoting this charge per unit area by Q, a differential capacitance per unit area for the space-charge region can be defined by

$$C \equiv \frac{\mathrm{d}Q}{\mathrm{d}V_r}. \qquad \text{(D.11.2)}$$

The physical basis is that as V_r increases and the space-charge region extends further into, for example, the n region, electrons are displaced from that region into the external circuit. The junction therefore acts as a condenser.

To calculate C, note first that

$$Q = -\int_{-l_p}^{0} \rho \, \mathrm{d}x = +\int_{0}^{l_n} \rho \, \mathrm{d}x. \qquad \text{(D.11.3)}$$

For reverse voltage ρ is due almost entirely to fixed charges. Any change in Q must then come about through a change in the limits of integration,

[17] S. P. Gentile, *Basic Theory and Application of Tunnel Diodes*, Van Nostrand, 1962.
[18] A. K. Jonscher, *Principles of Semiconductor Device Operation*, Bell, London, 1960.
[19] A. A. Grinberg, *Fvzika Tverdogo Tela*, **4**, 99 1962.

so that differentiating (D.11.3) gives

$$C = -\rho(-l_p)\frac{\mathrm{d}l_p}{\mathrm{d}V_r} = +\rho(l_n)\frac{\mathrm{d}l_n}{\mathrm{d}V_r}. \tag{D.11.4}$$

Also, the extension of (D.9.3) to non-zero reverse voltage is

$$V_D + V_r = \frac{4\pi}{\epsilon}\int_{-l_p}^{l_n} x\rho\,\mathrm{d}x, \tag{D.11.5}$$

and differentiating this in the same way gives

$$1 = \frac{4\pi}{\epsilon}\left[-l_p\rho(-l_p)\frac{\mathrm{d}l_p}{\mathrm{d}V_r} + l_n\rho(l_n)\frac{\mathrm{d}l_n}{\mathrm{d}V_r}\right]. \tag{D.11.6}$$

Finally, substituting (D.11.4) in (D.11.6) gives

$$C = \frac{\epsilon}{4\pi(l_n + l_p)} = \frac{\epsilon}{4\pi W}, \tag{D.11.7}$$

the classical result for a parallel plate condenser if W is the distance between the plates.

The variation of capacitance with voltage now follows from (D.9.8) or (D.9.9) for a simple junction. In fact

$$C^{-2} \propto (\mathrm{V}_D + \mathrm{V}_r)\ \text{(abrupt junction)}, \quad C^{-3} \propto (\mathrm{V}_D + \mathrm{V}_r)\ \text{(linear junction)}. \tag{D.11.8}$$

A graph of C^{-2} or C^{-3} against voltage would then be a straight line for sufficiently large reverse voltage, the intercept on the voltage axis being equal to V_D. Other possibilities exist[20].

The treatment given above applies essentially to D.C. or low-frequency A.C. situations. For a discussion of A.C. capacitance in general, see Gossick[21].

D.12 AVALANCHE BREAKDOWN UNDER REVERSE VOLTAGE

The mean electric field in the space-charge region is $\bar{E} = (\mathrm{V}_D + \mathrm{V}_r)/W \propto (\mathrm{V}_D + \mathrm{V}_r)^m$, where $m = \frac{1}{2}$ for an abrupt junction and $\frac{2}{3}$ for a linear junction. The mean distance travelled by an electron or hole between collisions, l, does not vary greatly with voltage. As the reverse voltage is increased, a point may be reached at which $e\bar{E}l$ is of order E_g. At this point a significant number of electrons and holes will gain an amount of kinetic

[20] In the case of metal–semiconductor contacts, see, for instance, P. T. Landsberg, *Proc. Roy. Soc.*, **A206**, 477 (1951).

[21] B. R. Gossick, *Potential Barriers in Semiconductors*, Academic Press, 1964.

energy greater than E_g between collisions, and it is possible for such an electron or hole to collide with a valence electron, exciting it into the conduction band and producing an electron–hole pair. The newly created carriers will drift in opposite directions, and each may in turn produce an electron–hole pair by collision. This process, also known as impact ionization, can evidently lead to an increase in current above that which would be predicted by the theory of §D.10. To discuss this we define impact ionization coefficients α and β by

α = mean number of electron–hole pairs produced in 1 cm travel by an electron,
β = mean number of electron–hole pairs produced in 1 cm travel by a hole. (D.12.1)

These are of course functions of electric field and therefore of position. Consideration of the number of electron–hole pairs produced by this method in unit volume leads to

$$\text{impact-ionization generation rate} = \frac{\alpha}{e}|j_e| + \frac{\beta}{e}|j_h|, \qquad \text{(D.12.2)}$$

so that the *effective* recombination rate per unit volume becomes

$$U_{\text{effective}} = U_s - \frac{\alpha}{e}|j_e| - \frac{\beta}{e}|j_h| = U_s + \frac{\alpha}{e}j_e + \frac{\beta}{e}j_h, \qquad \text{(D.12.3)}$$

since, for reverse voltage, j_e and j_h are both negative. The current conservation equations (D.8.27) are therefore replaced by

$$\frac{\mathrm{d}j_e}{\mathrm{d}x} = \frac{-\mathrm{d}j_h}{\mathrm{d}x} = eU_s + \alpha j_e + \beta j_h. \qquad \text{(D.12.4)}$$

In the equation for j_e, j_h can be eliminated using $j_e + j_h = j$ = constant. The result is

$$\frac{\mathrm{d}j_e}{\mathrm{d}x} + (\beta - \alpha)j_e = \beta j + eU_s, \qquad \text{(D.12.5)}$$

which has an integrating factor

$$\exp\left[\int_{-l_p}^{x} (\beta - \alpha)\,\mathrm{d}x'\right] \equiv \mathrm{e}^{K(x)}, \qquad \text{(D.12.6)}$$

and the solution

$$j_e = \left[j_{e0} + \int_{-l_p}^{x} (\beta j + eU_s)\,\mathrm{e}^{K(x')}\,\mathrm{d}x'\right]\mathrm{e}^{-K(x)}. \qquad \text{(D.12.7)}$$

Here j_{eo} is the value of j_e at $x = -l_p$, and from (D.10.13)

$$j_{eo} \approx -en_pL_n/\tau_n. \tag{D.12.8}$$

A similar treatment of the equation for j_h gives

$$j_h = \left[\int_x^{l_n} (\alpha j + eU_s)\,\mathrm{e}^{K(x')}\,\mathrm{d}x' + j_{ho}\,\mathrm{e}^{K(l_n)}\right]\mathrm{e}^{-K(x)}, \tag{D.12.9}$$

where j_{ho} is the value of j_h at $x = l_n$:

$$j_{ho} \approx -ep_nL_p/\tau_p. \tag{D.12.10}$$

Adding (D.12.7) and (D.12.9) gives an equation in which $j \equiv j_e + j_h$ appears on both sides. Solving, we find

$$j = M\left[j_{eo} + \mathrm{e}^{K(l_n)}j_{ho} + \int_{-l_p}^{l_n} eU_s\,\mathrm{e}^{K(x)}\,\mathrm{d}x\right], \tag{D.12.11}$$

where

$$M \equiv \left\{\mathrm{e}^{K(x)} - \left[\int_{-l_p}^{x} \beta\,\mathrm{e}^{K(x')}\,\mathrm{d}x' + \int_x^{l_n} \alpha\,\mathrm{e}^{K(x')}\,\mathrm{d}x'\right]\right\}^{-1}. \tag{D.12.12}$$

M is called the avalanche multiplication factor. As l_n, l_p and the mean values of α and β all increase with reverse voltage, a point will be reached at which $M \to \infty$. The current then increases by many orders of magnitude for a very small increase in voltage, being limited only by the resistance of the p and n regions; this increase in current is described as an avalanche breakdown.

The interpretation of M becomes clearer under the (quite artificial) assumption that $\alpha = \beta$ throughout. (D.12.11) then becomes, since $K(x) \equiv 0$,

$$j = M\left(j_{eo} + j_{ho} + \int_{-l_p}^{l_n} eU_s\,\mathrm{d}x\right) = -M(j_{rs} + egW), \tag{D.12.13}$$

where the reverse-saturated current j_{rs} has been introduced from (D.10.15) and g is the generation rate in the absence of carriers, introduced in (D.10.22). The actual current j is therefore just M times the current predicted without considering impact ionization. If $\alpha = \beta$, M also becomes simpler:

$$M = \left[1 - \int_{-l_p}^{l_n} \alpha\,\mathrm{d}x\right]^{-1}, \tag{D.12.14}$$

and the breakdown condition is

$$\int_{-l_p}^{l_n} \alpha\,\mathrm{d}x = 1. \tag{D.12.15}$$

If the variation of α and β with electric field is known, the variation of M with voltage can be estimated, using (D.9.1) to calculate the field at any point. This is done for particularly simple conditions in problem (D.12.1). For a more general discussion and comparison with experiment, see Tyagi[22].

Problem

(D.12.1) Consider a symmetrical step junction as described in problem (D.9.1). Neglecting n and p in the space-charge region, prove that the electric field E in that region satisfies $|E| = A\{W/2 - |x|\}$ where A is independent of applied voltage. Using this result and (D.9.8), deduce that if the ionization coefficients satisfy $\alpha = \beta \propto |E|^n$, the multiplication factor M of (D.12.14) can be expressed as $M = \left[1 - \left(\dfrac{V_D + V_r}{V_D + V_{rb}}\right)^{(n+1)/2}\right]^{-1}$, where V_{rb} is the value of reverse voltage for which breakdown occurs. Verify that, in the corresponding result for a linear junction, the exponent $(n + 1)/2$ is replaced by $(2n + 1)/3$.

D.13 THE APPROXIMATION OF CONSTANT QUASI-FERMI LEVELS IN THE SPACE-CHARGE REGION

The current–voltage relations of §D.10 rest on the assumption that the quasi-Fermi levels are constant across the space-charge region. The error involved will be small, provided that the drop ΔF_e of the dimensionless quasi-Fermi level F_e across the region is much smaller than 1, and similarly for ΔF_h. An estimate for ΔF_e starts from (D.8.26), which can be written

$$j_e = enD_n \frac{\mathrm{d}F_e}{\mathrm{d}x}, \qquad \text{(D.13.1)}$$

and from

$$n = n_i \exp(F_e - \eta_i), \qquad \text{(D.13.2)}$$

giving

$$j_e = en_i D_n \exp(-\eta_i) \frac{\mathrm{d}}{\mathrm{d}x}[\exp(F_e)],$$

$$\exp(F_{e1}) - \exp(F_{e2}) = \int_{-l_p}^{l_n} \frac{j_e \exp(\eta_i)\,\mathrm{d}x}{en_i D_n}. \qquad \text{(D.13.3)}$$

Here F_{e1} and F_{e2} are the values of F_e at $x = l_n$ and $x = -l_p$ respectively.

[22] M. Singh Tyagi, *Solid-State Electronics*, **11**, 99 (1968).

Thus $\Delta F_e = F_{e1} - F_{e2}$, so that (D.13.3) is

$$\exp(\Delta F_e) - 1 = \frac{1}{en_i D_n}\int_{-l_p}^{l_n} j_e \exp(\eta_i - F_{e2})\,\mathrm{d}x. \tag{D.13.4}$$

Now $j_e(x) = j_e(-l_p) + \int_{-l_p}^{x} eU_s\,\mathrm{d}x' < j_e(-l_p) + j_{sc}$, and

$$j_e(-l_p) = \frac{en_p L_n}{\tau_n}(\mathrm{e}^{\eta} - 1) < \frac{en_p L_n}{\tau_n}\mathrm{e}^{\eta}$$

$$= \frac{eL_n}{\tau_n} n(-l_p) = \frac{eL_n n_i}{\tau_n}\exp[F_{e2} - \eta_i(-l_p)], \tag{D.13.5}$$

so that if $\lambda \equiv 1 + j_{sc}/j_e(-l_p)$,

$$j_e < \lambda \frac{eL_n n_i}{\tau_n}\exp[F_{e2} - \eta_i(-l_p)] \quad \text{for } -l_p < x < l_n. \tag{D.13.6}$$

Substituting (D.13.6) in (D.13.4) gives

$$\exp(\Delta F_e) - 1 < \frac{\lambda L_n}{D_n \tau_n}\int_{-l_p}^{l_n} \exp[\eta_i - \eta_i(-l_p)]\,\mathrm{d}x, \tag{D.13.7}$$

and since the integrand is less than or equal to 1 (η_i decreases from its value at $x = -l_p$), then

$$\exp(\Delta F_e) - 1 < \frac{\lambda L_n(l_n + l_p)}{D_n \tau_n} = \frac{\lambda W}{L_n}. \tag{D.13.8}$$

A possibly closer estimate is obtained by again assuming linear band edges, so that (see fig. D.10.1)

$$\eta_i - \eta_i(-l_p) = -\frac{\theta(x + l_p)}{W}, \tag{D.13.9}$$

leading, with (D.13.7), to

$$\exp(\Delta F_e) - 1 < \frac{\lambda}{L_n}\int_0^W \exp\left(-\frac{\theta z}{W}\right)\mathrm{d}z$$

$$\approx \frac{\lambda W}{\theta L_n}. \tag{D.13.10}$$

Since the actual band edges are far from linear, (D.13.10) cannot be considered an exact inequality. It does, however, suggest (since θ is often of order 10 or 20) that some inequality more powerful than (D.13.8) may be valid.

Typical values for Ge diodes are $W \sim 10^{-4}$ cm, $L_n \sim 5 \times 10^{-2}$ cm and $\lambda \sim 1$ giving, from (D.13.8), $\Delta F_e < 0{\cdot}002$. The position is less clear for Si diodes. For the diode considered in Sah, Noyce and Shockley[15], $W \sim 10^{-5}$ cm, $L_n \sim 10^{-2}$ cm, $L_p \sim 3 \times 10^{-4}$ cm and $\lambda \approx 600$ for small forward voltages. Thus (D.13.8) gives $\Delta F_e < 0{\cdot}6$, but the corresponding result for ΔF_h gives only $\exp(\Delta F_h) - 1 < 20$. The work referred to, however[15], contains a discussion of this point in which the variation of j_e with position is taken into account, and it is concluded that the approximation is valid. Clearly its validity cannot be taken for granted in considering any new type of junction.

D.14 SEMICONDUCTOR LAMPS AND LASERS

D.14a General

Any *p–n* junction in which a significant fraction of recombination is radiative will emit some electromagnetic radiation when a forward voltage is applied. Interest is currently centred on materials of large energy gap [gallium arsenide ($E_{g0} = 1{\cdot}52$ eV), gallium phosphide ($E_{g0} = 2{\cdot}32$ eV) and others] in which a photon emitted with an energy near E_g falls in the visible or near infra-red region of the spectrum, and a junction used in this way is described as a semiconductor lamp. An external quantum efficiency θ_e is defined as the mean number of photons leaving the device per electron (in conduction or valence band) crossing the device, so that if I is the rate of flow of electrons across the device (the total current being $J \equiv eI$) then

$$\text{number of photons emitted per second} = \theta_e I. \qquad \text{(D.14.1)}$$

One can also define an internal quantum efficiency θ_i as the ratio of radiative to total recombination current density,

$$\theta_i = \{j_{bp}\,(\text{radiative}) + j_{bn}\,(\text{radiative}) + j_{sc}\,(\text{radiative})\}/(j_{bp} + j_{bn} + j_{sc}), \qquad \text{(D.14.2)}$$

which can, in principle, be calculated from the results of §D.10.

In general $\theta_e < \theta_i$ because of losses due to internal absorption and reflection. If $h\bar{\nu}$ is the average energy of emitted photons and V the applied voltage, the power conversion efficiency of the lamp is defined by

$$\eta_e = \frac{\text{light energy output}}{\text{electrical energy input}} = \frac{\theta_e I h\bar{\nu}}{J\text{V}} = \frac{\theta_e h\bar{\nu}}{e\text{V}}. \qquad \text{(D.14.3)}$$

If a semiconductor lamp has two parallel surfaces, normal to the plane of the junction, treated to act as partial reflectors, and is operated at a sufficiently high forward voltage, a condition can be reached in which a photon emitted in a direction normal to the reflecting surfaces has a high probability of stimulating the emission of a second photon in the same direction. Provided the probability that the photon leaves the emitting region, either through the reflecting surfaces or by diffraction into the p and n regions, is small, a chain reaction can be initiated in which an intense parallel beam of nearly monochromatic light is built up and emitted through the reflecting surfaces. This phenomenon is known as light amplification and the lamp is said to be acting as a *laser*. A lower limit for the required voltage can be derived as follows.

A photon of energy hv which encounters an electron in state E_1 may be absorbed by the electron, which is excited into a previously empty state at energy $E_2 = E_1 + hv$. If, on the other hand, the photon encounters an electron in state E_2, and state E_1 is empty, it may stimulate the electron to fall back into state E_1, and emit a second photon also of energy hv. The quantum mechanical probabilities of the absorption and stimulated emission processes are the same, so that for this particular pair of states, we have:

$$\begin{aligned} \text{absorption rate} &= A(E_1, E_2)N(hv)f_e(E_1)[1 - f_e(E_2)], \\ \text{stimulated emission rate} &= A(E_1, E_2)N(hv)[1 - f_e(E_1)]f_e(E_2), \end{aligned} \tag{D.14.4}$$

where $N(hv)$ is the number of photons per electromagnetic mode with energy hv and $A(E_1, E_2)$ is a quantum mechanical transition probability. Evidently this pair of states will contribute to a net increase in the number of photons only if the stimulated emission rate is greater than the absorption rate, and with (D.14.4) this condition reduces to

$$[1 - f_e(E_1)]f_e(E_2) > f_e(E_1)[1 - f_e(E_2)]. \tag{D.14.5}$$

If it is now assumed that E_1 is a valence band state and E_2 a conduction band state, and that the occupation probabilities are describable by quasi-Fermi levels as in (D.5.1) and (D.5.2), (D.14.5) is equivalent to

$$\mu_e - \mu_h > E_2 - E_1 = hv \tag{D.14.6}$$

and is therefore independent of the particular conduction or valence band states considered, for a given photon energy. This argument has been generalized[23] to give the condition

$$\mu_2 - \mu_1 > hv, \tag{D.14.7}$$

where the transition can be between any free or localized electron states for which a quasi-Fermi level can be defined: the result is also valid if some energy is taken up by non-radiative means so that $hv \neq E_2 - E_1$.

[23] M. G. A. Bernard and B. Duraffourg, *Phys. Stat. Solidi*, **1**, 699 (1961); P. T. Landsberg, *Phys. Stat. Solidi*, **19**, 777 (1967).

In the space-charge region of a *p–n* junction, $\mu_e - \mu_h \approx e\text{V}$ for applied voltages $\text{V} < \text{V}_D$. For larger applied voltages, the potential drop in the *p* and *n* regions means that $\mu_e - \mu_h < e\text{V}$. A necessary condition for light amplification in the space-charge region is therefore

$$e\text{V} > h\nu \qquad \text{(D.14.8)}$$

if the reasonable assumption is made that the quasi-Fermi levels for any localized states will be between μ_e and μ_h. (D.14.8) is evidently not a sufficient condition for lasing action, both because of potential drops in the *p* and *n* regions and because of the inevitable loss of photons by diffraction and through the reflecting surfaces. (D.14.8) means that in any given junction a minimum current *J*, called the threshold current, is required for lasing action. It is evidently desirable to reduce the threshold current as far as possible.

D.14b Conditions for maximum efficiency (as a lamp) and minimum threshold current (as a laser)

These conditions will evidently be similar in that any non-radiative recombination is undesirable from both points of view. One should therefore choose a material with a large energy gap to minimize band-to-band Auger effects, and try to minimize the number of recombination centres with energies near the centre of the gap, as these are more likely to capture electrons and holes non-radiatively. Recombination centres with energies near the band edges can contribute to radiative recombination; the number of these should be increased. Finally, a material with a direct band gap should be chosen to increase the probability of band-to-band radiative recombination. Since the current due to single-electron processes via traps with energies near the middle of the band, (D.10.38), increases more slowly with voltage than the radiative currents, quantum efficiency would be expected to rise with increasing forward voltage up to V_D. The behaviour of the power conversion efficiency (D.14.3) is uncertain, as it is proportional to $(\text{voltage})^{-1}$.

The minimum threshold current condition for lasers differs in that even radiative recombination outside the space-charge region (defined as the region in which the separation of quasi-Fermi levels is largest) is undesirable. One therefore wishes to minimize j_{bn} and j_{bp} as far as possible. Now, from (D.10.13), $j_{bp} \propto n_p \propto (p_p)^{-1}$, so that the majority carrier densities in the *p* and *n* regions should be made large. This will have the further desirable effect of reducing the voltage drop across these regions. For an example of the theoretical calculation of threshold current, see Adams[24].

[24] M. J. Adams, *Solid State Electrons* (1969).

D.14c Thermodynamic limits on the efficiency of a lamp

The form of (D.14.3) suggests that if a high quantum efficiency can be achieved with an applied voltage less than $h\bar{\nu}/e$, a power conversion efficiency greater than unity is possible. Since junctions have been observed[25] for which $h\bar{\nu} > e\mathrm{V}$, it is of interest to consider if a power conversion efficiency greater than unity is thermodynamically possible.

We consider a region of space, known as the box, which contains all of the semiconductor junction and the contacts, so that all recombination and light emission occurs within the box. The box is connected electrically to an external circuit and is supposed to be in thermal contact with a heat reservoir, both the reservoir and the surroundings being at temperature T. We define energies and entropies transferred in unit time by

E_L, S_L = energy and entropy of photons emitted in one second.
E_e, S_e = energy and entropy given up to the box by the electrical circuit in one second.
Q = heat energy given up to the box by the heat reservoir in one second.

The conservation of energy in steady state is expressed by

$$E_L = E_e + Q, \tag{D.14.9}$$

indicating that the box is neither a source nor a sink of energy. The second law of thermodynamics can be put in the form 'no finite region which is in a steady state can be an entropy sink'. This leads us to define a non-negative quantity, the entropy generated in the box in one second, S_i, and to conclude that

$$S_L = S_e + Q/T + S_i. \tag{D.14.10}$$

It is convenient to define an effective temperature T_L for photons by

$$E_L = T_L S_L, \tag{D.14.11}$$

and to introduce the Helmholtz free energy given up by the electrical circuit,

$$F_e \equiv E_e - TS_e. \tag{D.14.12}$$

Note that (D.14.11) is equivalent to saying that photons have zero free energy. Substituting for Q from (D.14.9) in (D.14.10) and using (D.14.11) leads to

$$\frac{E_L}{T_L} = \frac{E_L - F_e}{T} + S_i, \tag{D.14.13}$$

[25] G. C. Dousmanis, C. W. Mueller, H. Nelson and G. C. Petzinger, *Phys. Rev.*, **113**, A316 (1964).

or, defining the power conversion efficiency in conformity with (D.14.3) as

$$\eta_e \equiv \frac{E_L}{E_e}, \tag{D.14.14}$$

$$\eta_e = \left[1 - \frac{T}{T_L}\right]^{-1}\left[\frac{F_e - TS_i}{E_e}\right] \tag{D.14.15}$$

and

$$\eta_e \leqslant \left[1 - \frac{T}{T_L}\right]^{-1}\frac{F_e}{E_e}. \tag{D.14.16}$$

To calculate F_e/E_e, consider first that the Fermi level at the contact on the p side is lower than that at the contact on the n side by eV, V being the applied voltage. Since electrons flow from the n side to the p side, every electron crossing the box undergoes a drop eV in its Fermi level. But the Gibbs free energy of a gas of N electrons is

$$G \equiv E - TS + pv = F + pv = \mu N, \tag{D.14.17}$$

so that the transfer of a single electron from one side of the box to the other lowers the Gibbs free energy of the electrons outside the box by eV. Assuming that the pv term does not change, we conclude that the Helmholtz free energy F lost by the external circuit is eV for each electron crossing, so that

$$F_e = Ie\mathrm{V} = J\mathrm{V} = E_e, \tag{D.14.18}$$

where I is the particle current and J the charge current as in (D.14.1) and (D.14.3). (D.14.16) now becomes simply

$$\eta_e \leqslant \left[1 - \frac{T}{T_L}\right]^{-1} \tag{D.14.19}$$

and this is the thermodynamic limit for the efficiency of a lamp.

The calculation of the effective temperature T_L has been discussed by Weinstein[26]. For isotropic radiation from a flat surface with a Gaussian spectral distribution of bandwidth $\Delta\nu$, the result is

$$T_L = \frac{h\bar{\nu}}{k_0}[\tfrac{3}{2} - \log_e \rho_L(\bar{\nu})]^{-1}, \tag{D.14.20}$$

where $\rho_L(\bar{\nu})$ is the mean number of photons per electromagnetic mode at the peak frequency $\bar{\nu}$,

$$\rho_L(\bar{\nu}) = c^2\phi_L/2\pi^{\frac{3}{2}}h(\bar{\nu})^3\Delta\nu. \tag{D.14.21}$$

[26] M. A. Weinstein, *J. Opt. Soc. Amer.*, **50**, 597 (1960); *Phys. Rev.*, **119**, 499 (1960).

ϕ_L is the energy flux density of photons and, in our notation, if A is the emitting area,

$$\phi_L = E_L/A = \theta_e Jh\bar{\nu}/Ae. \tag{D.14.22}$$

Radiation temperatures have been calculated[27] for a particular diode, and range from 744°K to 1786°K at an ambient temperature of 300°K. It is also shown that if the quantum efficiency of this diode were raised to 100% at 1mA current, other parameters being unchanged, the power conversion efficiency would become 120%, and that this would not violate the thermodynamic limit (D.14.19). The diode would then absorb heat from its surroundings and act as a refrigerator. Since external quantum efficiencies of 40% have been measured[28], the thermodynamic limit may be approached in the future. The limit is likely to be approached first at low temperatures, as (D.14.19) is then nearer unity and measured efficiencies are higher than at room temperature.

Further reading

General

J. P. McKelvey, *Solid State and Semiconductor Physics*, Harper & Row, New York, 1966.

Recombination statistics

V. L. Bonch-Bruevich and E. G. Landsberg, *Phys. Stat. Solidi*, **29**, 9 (1968).

P. T. Landsberg in Festhörperprobleme **VI**, 174 (1967).

p–n junctions

A. K. Jonscher, *Principles of Semiconductor Device Operation*, Bell, London, 1960.

Semiconductor lasers

P. R. Thornton, *Physics of Electroluminescent Devices*, Spon, London, 1967.

M. H. Pilkuhn, The Injection Laser, *Phys. Stat. Solidi*, **29**, 9–62 (1968).

M. J. Adams and P. T. Landsberg in *Gallium Arsenide Lasers*, Wiley, London, 1969.

[27] P. T. Landsberg and D. A. Evans, *Phys. Rev.*, **166**, 242 (1968).

[28] W. N. Carr, *IEEE Trans. Electron. Devices*, **12**, 531 (1965).

PART E

THEORY OF LATTICE VIBRATIONS WITH SOME APPLICATIONS

J. E. Parrott

CHAPTER IX

The Principles of Lattice Dynamics

E.1 INTRODUCTION

Lattice dynamics is concerned with the motion of the nuclei (and more closely bound electrons) in a solid. It falls into three parts. The first, dealt with in this chapter, covers the most general part of the theory based on very general assumptions about interatomic forces. This establishes the necessary framework for tackling the dynamics of specific lattices and materials, usually in model form. This is the second part of the theory dealt with in the next chapter. Finally we have the applications of lattice dynamics to the solution of physical problems and the elucidation of phenomena. Only a few examples of this will be given.

Underlying the whole treatment to be presented here is what is called the Born–Oppenheimer or adiabatic principle. This asserts that if one considers the problem of solving the Schrödinger equation for the collection of nuclei and electrons, which are the raw material from which a solid is constructed, we can separate the motion of the electrons from that of the nuclei. If we are mainly interested in the electrons the nuclear coordinates are then merely parameters. If, on the other hand, the nuclear motion is our main concern, then the effect of the electronic motion can be described in terms of suitable cohesive forces. The justification for the adiabatic approximation is that compared to the electrons the nuclei move slowly and have energy level spacings much smaller than those of the electronic states. Probably the best empirical support for this comes from optical absorption, where we have infra-red absorption due to the lattice vibrations and visible and ultra-violet absorption due to electronic interactions with the radiation. There are, however, some exceptions to this, and indeed the adiabatic approximation itself is not free from weaknesses, particularly in the case of metals. A brief discussion of this question is given in Peierls' book[1] on the theory of solids.

The adiabatic approximation, leading as it does to a valid concept of interatomic forces, justifies the classical approach to lattice dynamics which will be pursued here. In this the system of ions or atoms in the crystal is treated as one in classical dynamics, the problem being to find

[1] R. F. Peierls, *Quantum Theory of Solids*, Oxford University Press, 1955.

the normal modes, their frequencies and eigenvectors. Quantum mechanics only enters after this stage, specifying the allowed states of the equivalent harmonic oscillators. This is essentially the approach of Born and von Karman dating from 1913.

E.2 SOME PRELIMINARIES

Much of the material dealt with in this section is covered more thoroughly elsewhere, in particular in Part B, so most of the results presented here are in summary form.

The mean positions of atoms in a crystal are arranged in a regular array or lattice. This means that the crystal can be divided into identical small volumes, the smallest such volume being the unit cell. If there is only one atom in the unit cell we speak of a primitive structure or Bravais lattice. In general, of course, we have more than one atom per unit cell giving a structure with a basis. For simplicity, most of the results derived here will be for Bravais lattices.

The lattice is specified by primitive basic lattice translation vectors $\mathbf{a}_1$, $\mathbf{a}_2, \mathbf{a}_3$. Then a translation through

$$\mathbf{R}_n = n_1\mathbf{a}_1 + n_2\mathbf{a}_2 + n_3\mathbf{a}_3, \tag{E.2.1}$$

where n_1, n_2, n_3 are integers, takes one to a point geometrically equivalent to the starting point. (This is only strictly true in an infinite crystal, but for the moment we shall neglect the effects of finite size.) If the origin is at an atomic site, then the position vectors of the other sites will be of the form $\mathbf{R}_n$.

One type of unit cell will be a parallelepiped whose edges are given by $\mathbf{a}_1, \mathbf{a}_2, \mathbf{a}_3$, and whose volume is

$$\Omega = \mathbf{a}_1 \cdot (\mathbf{a}_2 \times \mathbf{a}_3). \tag{E.2.2}$$

Another form of unit cell is the Wigner–Seitz cell which has an atomic site at its centre and boundaries made up of planes bisecting the lattice vectors joining the atom to its neighbours.

The symmetry property of the infinite lattice is of the greatest importance. In particular, many functions of position will have the same symmetry. For example, the single electron potential in a stationary lattice will have the property

$$\phi(\mathbf{r}) = \phi(\mathbf{r} + \mathbf{R}_n).$$

Such functions can be expressed in terms of three dimensional Fourier series, for which purpose the concept of the reciprocal lattice proves to be of great value.

The basic vectors of the reciprocal lattice $\mathbf{b}_1, \mathbf{b}_2, \mathbf{b}_3$ are related to the direct lattice basic vectors by equations of the form

$$\mathbf{b}_1 = \frac{2\pi \mathbf{a}_2 \times \mathbf{a}_3}{\Omega}, \tag{E.2.3}$$

so that $\mathbf{a}_i \cdot \mathbf{b}_j = 2\pi\delta_{ij}$. The reader is warned that the reciprocal lattice vectors are sometimes defined without the factor 2π.

The lattice points in the reciprocal lattice are given by

$$\mathbf{K}_m = m_1\mathbf{b}_1 + m_2\mathbf{b}_2 + m_3\mathbf{b}_3 \tag{E.2.4}$$

where the m_i are integers. The direction of $\mathbf{K}_m$ is normal to the plane defined by (m_1, m_2, m_3) in the direct lattice. In crystallographic literature (h, k, l) is often used instead of (m_1, m_2, m_3). The magnitude of $\mathbf{K}_m$ is given by 2π divided by the spacing of the corresponding lattice planes.

The unit cell in the reciprocal lattice is defined in the same way as for the direct lattice and its volume is given by

$$\Omega_r = \mathbf{b}_1 \cdot (\mathbf{b}_2 \times \mathbf{b}_3) = \frac{8\pi^3}{\Omega}. \tag{E.2.5}$$

The most important type of reciprocal unit cell is that constructed in the same way as the Wigner–Seitz cell. This is called the first Brillouin zone.

Now let us suppose that $F(\mathbf{r})$ has the periodicity of the direct lattice, i.e. $F(\mathbf{r} + \mathbf{R}_n) = F(\mathbf{r})$ for all $\mathbf{R}_n$. Then using equation (B.7.2) we can represent $F(\mathbf{r})$ by

$$F(\mathbf{r}) = \sum_m F_m \mathrm{e}^{\mathrm{i}\mathbf{K}_m \cdot \mathbf{r}}, \tag{E.2.6}$$

where the Fourier coefficients F_m are given by

$$F_m = \frac{1}{\Omega}\int_\Omega F(\mathbf{r})\, e^{-\mathrm{i}\mathbf{K}_m \cdot \mathbf{r}}\, \mathrm{d}^3 r, \tag{E.2.7}$$

and the integration is over a single unit cell.

Problem

(E.2.1) Prove the result given in equation (E.2.7).

E.3 THE CLASSICAL APPROXIMATION TO THE LATTICE ENERGY

The classical model of the vibrating crystal lattice consists of a number of atoms, essentially point masses, coupled to one another by forces such that the total potential energy of a finite crystal is itself finite. Furthermore the displacements of the atoms from the lattice points are small, so that in a crystal free from static lattice defects the atoms can be labelled by the sites. Thus $\mathbf{u}(n)$ is the displacement of the atom (n) from the lattice point $\mathbf{R}_n$.

The kinetic energy of a perfect crystal can then be expressed immediately as

$$T = \tfrac{1}{2}M \sum_{n,i} \dot{u}_i^2(n) = \frac{1}{2M} \sum_{n,i} p_i^2(n). \tag{E.3.1}$$

Here M is the atomic mass, the same for all atoms in a perfect mono-isotopic Bravais crystal, $\mathbf{p}$ is the momentum $M\dot{\mathbf{u}}$ and i labels Cartesian coordinates.

The potential energy is assumed to be expressible as a power series in the displacements

$$U = U_0 + \sum_{\substack{m \\ i}} B_i^m u_i(m) + \frac{1}{2} \sum_{\substack{mn \\ ij}} B_{ij}^{mn} u_i(m) u_j(n)$$

$$+ \frac{1}{3!} \sum_{\substack{mno \\ ijk}} B_{ijk}^{mno} u_i(m) u_j(n) u_k(o) + \cdots. \tag{E.3.2}$$

U_0 is simply a constant representing the potential of the lattice when the atoms are all at their rest positions and plays no further role. The quantities B_i^m, B_{ij}^{mn}, B_{ijk}^{mno} etc. are called coupling constants of the first order, second order, etc. The first order coupling constants can be seen to vanish because of the requirement that the lattice must be in equilibrium when all the $\mathbf{u}(n)$ are zero.

The most important quantities in lattice dynamics are the second order coupling constants B_{ij}^{mn}; clearly

$$B_{ij}^{mn} = \left[\frac{\partial^2 U}{\partial u_i(m)\, \partial u_j(n)} \right]_0 \tag{E.3.3}$$

evaluated at equilibrium. Similar expressions hold for the higher order constants. The importance of the second order terms arises because so many of the properties of the lattice vibrations can be satisfactorily discussed when all terms higher than the second order are neglected. This is called the *Harmonic Approximation.* With this assumed the significant part of the lattice energy can be written as a Hamiltonian:

$$H = \frac{1}{2M} \sum_{n,i} p_i^2(n) + \frac{1}{2} \sum_{\substack{mn \\ ij}} B_{ij}^{mn} u_i(m) u_j(n). \tag{E.3.4}$$

The equations of motion are

$$M\ddot{u}_i(m) + \sum_{n,j} B_{ij}^{mn} u_j(n) = 0. \tag{E.3.5}$$

For a crystal consisting of N atoms there will be $3N$ such equations. Our main problem is to solve these. In this we are assisted very greatly by the symmetry properties of the coupling constants.

The symmetry and other invariance properties of the B_{ij}^{mn} are discussed in detail by Maradudin, Montroll and Weiss[2] and by Leibfried and Ludwig[3]. The latter is particularly useful for the higher order coupling constants. The principal consequence of all this is to greatly reduce the number of independent coupling constants. We shall only touch on this question here, giving a few results of particular importance.

(i) Because B_{ij}^{mn} is a second differential (E.3.3),

$$B_{ij}^{mn} = B_{ji}^{nm}. \tag{E.3.6}$$

(ii) Translational symmetry means that B_{ij}^{mn} depends only on $(\mathbf{R}_m - \mathbf{R}_n)$ and not on $\mathbf{R}_m$, $\mathbf{R}_n$ separately. Hence

$$B_{ij}^{mn} = B_{ij}^{\mathrm{O}m-n}, \tag{E.3.7}$$

where O refers to an origin of coordinates.

(iii) The potential energy and the force on a given atom should be invariant under rigid body displacement of the whole crystal. This requires that

$$\sum_n B_{ij}^{\mathrm{O}n} = 0. \tag{E.3.8}$$

(iv) The point symmetry of the lattice gives rise to further restrictions. In particular where we have inversion symmetry (as in all Bravais lattices):

$$B_{ij}^{\mathrm{O}n} = B_{ij}^{\mathrm{O}-n}. \tag{E.3.9}$$

All of these results will be used subsequently.

Problem

(E.3.1) Prove the result given in equation (E.3.8).

E.4 THE SOLUTION OF THE EQUATIONS OF MOTION

Before attempting to solve the equations of motion it is necessary to consider the question of boundary conditions. The advantages of translational symmetry can only be fully obtained in an infinite crystal. On the other hand in order that, for instance, the energy shall be finite, a finite

[2] A. A. Maradudin, E. W. Montroll and G. H. Weiss, *Theory of Lattice Dynamics in the Harmonic Approximation*. Academic Press, 1963.

[3] G. Leibfried and W. Ludwig, *Solid State Phys.*, **12**, 276 (1961).

crystal is desirable. Also it would be convenient if in some way the group of translational symmetry operations could be made finite, since the theory of finite groups is simpler than that of infinite groups. All these ends can be achieved by the device of periodic boundary conditions.

We suppose that we divide an infinite crystal into an (infinite) number of parallelepipeds whose edges are defined by $G\mathbf{a}_1$, $G\mathbf{a}_2$ and $G\mathbf{a}_3$. Thus the parallelepiped, or periodic volume, contains $N = G^3$ unit cells and is of volume $G^3\Omega$. We now impose the conditions that the motion in each volume is the same, so that, for example,

$$\mathbf{u}(n) = \mathbf{u}(n + G). \tag{E.4.1}$$

These are the periodic or cyclic boundary conditions. For one or two dimensions they can be given a simple realization. In one dimension we take a finite lattice and join the ends producing a closed chain, whilst in two dimensions we can do the same thing to produce a torus.

It is highly desirable that periodic boundary conditions produce results similar to that which would be found for a real finite crystal. This point will be returned to subsequently (§E.12).

The solution of the equations of motion is equivalent to finding a coordinate transformation which will reduce T, U and H to a sum of squares without cross terms. We know from matrix algebra that if the quadratic forms for T and U are positive definite there is a transformation which will simultaneously diagonalize T and U. The new coordinates are called normal coordinates.

Let us first consider a function

$$F_m(n) = \mathrm{e}^{\mathrm{i}\mathbf{K}_m \cdot \mathbf{R}_n/G}. \tag{E.4.2}$$

Using (E.2.1) and (E.2.4) we can write

$$F_m(n) = \exp\left(\frac{2\pi i}{G}\sum_i m_i n_i\right).$$

Thus $F_m(n_1, n_2, n_3) = F_m(n_1 + G, n_2, n_3)$ and satisfies the periodic boundary conditions. We now define a new vector [see equation (B.6.4)]

$$\mathbf{q} = \sum_i \frac{m_i \mathbf{b}_i}{G}. \tag{E.4.3}$$

If we consider the values of $\mathbf{q}$ corresponding to the range $0 < m_i/G \leqslant 1$ we will see that they all lie within, or on half of the surface of, the reciprocal unit cell or Brillouin zone. Furthermore there will be $G^3 = N$ distinct values of $\mathbf{q}$ so that because the zone volume is $8\pi^3/\Omega$ and $\Omega = V/N$, where V is the periodicity volume to which the boundary conditions apply, the density of allowed values of $\mathbf{q}$ in reciprocal space is $(8\pi^3)^{-1}$ per unit volume in direct space.

We can now write

$$F_m(n) = F_q(n) = e^{i\mathbf{q}\cdot\mathbf{R}_n}. \tag{E.4.4}$$

Let us sum $F_q(n)$ over n. Then [see equation (B.6.20)]

$$\sum_n F_q(n) = \sum_n e^{i\mathbf{q}\cdot\mathbf{R}_n}$$

$$= \prod_{i=1,2,3}\left[\frac{1 - e^{2\pi i m_i}}{1 - e^{2\pi i m_i/G}}\right] = N\sum_l \delta_{\mathbf{q},\mathbf{K}_l}, \tag{E.4.5}$$

where $\mathbf{K}_l$ is a reciprocal lattice translation vector which may be zero. If $\mathbf{q}$ is restricted within the first Brillouin zone all non-zero values of $\mathbf{K}_l$ are excluded.

We now define a reduced displacement

$$\mathbf{w}(n) = M^{\frac{1}{2}}\mathbf{u}(n) \tag{E.4.6}$$

and rewrite the Hamiltonian as

$$H = \frac{1}{2}\left[\sum_{n,i} \dot{w}_i^2(n) + \sum_{\substack{mn\\ ij}} C_{ij}^{mn} w_i(m) w_j(n)\right] \tag{E.4.7}$$

where $C_{ij}^{mn} = B_{ij}^{mn}/M$. We next introduce the Fourier transform of $\mathbf{w}(n)$, $\mathbf{v}(q)$

$$w_i(n) = \frac{1}{\sqrt{N}}\sum_q v_i(q)\, e^{i\mathbf{q}\cdot\mathbf{R}_n}, \tag{E.4.8a}$$

where

$$v_i(q) = \frac{1}{\sqrt{N}}\sum_n w_i(n)\, e^{-i\mathbf{q}\cdot\mathbf{R}_n}. \tag{E.4.8b}$$

Because $\mathbf{w}(n)$ is real, $v_i(q) = v_i^*(-q)$.

Then:

$$U = \frac{1}{2N}\sum_{\substack{ij\\ qq'}} v_i(q)v_j(q')\sum_m e^{i(\mathbf{q}+\mathbf{q}')\cdot\mathbf{R}_m}\sum_n C_{ij}^{mn}\, e^{i\mathbf{q}'\cdot(\mathbf{R}_n-\mathbf{R}_m)}. \tag{E.4.9}$$

Translation symmetry (E.3.7) enables the last sum to be written as

$$D_{ij}(q) = \sum_n C_{ij}^{On}\, e^{i\mathbf{q}\cdot\mathbf{R}_n} = \frac{1}{M}\sum_n B_{ij}^{On}\, e^{i\mathbf{q}\cdot\mathbf{R}_n}, \tag{E.4.10}$$

where D is the dynamical matrix; it is Hermitian $[D_{ij}(q) = D_{ji}^{*}(q)]$ because of (E.3.9). Then, using (E.4.5) and (E.4.9),

$$U = \frac{1}{2}\sum_{\substack{q\\ij}} D_{ij}(q)v_i^{*}(q)v_j(q). \tag{E.4.11}$$

The secular equation

$$|D_{ij}(q) - \omega^2\delta_{ij}| = 0 \tag{E.4.12}$$

has roots $\omega_s^2(q)$ which are the eigenvalues of $D(q)$. Clearly there are three such roots ($s = 1, 2, 3$) and because $D(q)$ is Hermitian these roots are real. Sometimes two or more of these roots are equal or 'degenerate', as, for example, at $\mathbf{q} = 0$ [see problem (E.4.1)]. In order that they shall be positive the principal minors of $D(q)$ must be positive. We shall see later that this corresponds to a stability condition. $D(q)$ will also have three eigenvectors $\mathbf{e}^s(q)$ so that

$$\sum_j D_{ij}(q)\mathrm{e}_j^s(q) = \omega_s^2(q)e_i^s(q). \tag{E.4.13}$$

These eigenvectors are real for a Bravais lattice and obey the orthogonality and closure conditions

$$\sum_i e_i^s(q)e_i^{s'}(q) = \delta_{ss'} \tag{E.4.14a}$$

and

$$\sum_s e_i^s(q)e_j^s(q) = \delta_{ij}. \tag{E.4.14b}$$

We are now able to define normal coordinates $Q_s(q)$ by means of the relations

$$v_i(q) = \sum_s e_i^s(q)Q_s(q). \tag{E.4.15a}$$

The inverse relation is

$$Q_s(q) = \sum_i e_i^s(q)\, v_i(q). \tag{E.4.15b}$$

Because the displacements must be real there are restrictions on $\mathbf{e}^s(q)$ and $Q_s(q)$. There is a choice to be made here; we select

$$\mathbf{e}^s(q) = -\mathbf{e}^s(-q) \tag{E.4.16a}$$

and

$$Q_s(q) = -Q_s^{*}(-q). \tag{E.4.16b}$$

With this choice the eigenvectors change sign as the vector $\mathbf{q}$ passes through the origin.

The normal coordinates defined in (E.4.15a) and (E.4.15b) are complex numbers, and we can express the kinetic and potential energies in terms of them. However in order to derive the equations of motion rigorously from either a Lagrangian or Hamiltonian it is necessary to have real normal coordinates. Conventionally there are two kinds of these:

(1) *Real normal coordinates of the first kind,* $X_\lambda^s(q)$. These are defined by

$$Q_s(\pm q) = \pm \frac{1}{\sqrt{2}}[X_1^s(q) \pm \mathrm{i}X_2^s(q)] \tag{E.4.17}$$

and are required only for $q > 0$. Substitution in the expressions for T and U gives the Hamiltonian:

$$H = \frac{1}{2} \sum_{\substack{q>0 \\ s}} \sum_{\lambda=1,2} \{[\dot{X}_\lambda^s(q)]^2 + \omega_s^2(q)[X_\lambda^s(q)]^2\}. \tag{E.4.18}$$

The momentum conjugate to $X_\lambda^s(q)$ is $\dot{X}_\lambda^s(q)$, and the equations of motion are

$$\ddot{X}_\lambda^s(q) + \omega_s^2(q)X_\lambda^s(q) = 0. \tag{E.4.19}$$

We have now succeeded in reducing the equations of motion (E.3.5) into a simple form by analysing the motion of the lattice into the superposition of simple harmonic oscillators. These are known as the normal modes; they are labelled by (q, s) and as may be seen from (E.4.19) the eigenvalues of (E.4.12) are the squares of the frequencies characterizing these modes. Reverting to complex normal coordinates we now have

$$\ddot{Q}_s(q) + \omega_s^2(q)Q_s(q) = 0. \tag{E.4.20}$$

Solving this gives

$$Q_s(q, t) = Q_s^- \mathrm{e}^{\mathrm{i}\omega t} + Q_s^+ \mathrm{e}^{-\mathrm{i}\omega t}, \tag{E.4.21}$$

where Q_s^- and Q^+ are determined by the initial conditions. Using (E.4.6), (E.4.8) and (E.4.15), we can now write the displacements as

$$u_i(n) = \frac{1}{\sqrt{(NM)}} \sum e_i^s(q)[Q_s^-(q)\,\mathrm{e}^{\mathrm{i}(\mathbf{q}\cdot\mathbf{R}_n+\omega t)} + Q_s^+\,\mathrm{e}^{\mathrm{i}(\mathbf{q}\cdot\mathbf{R}_n-\omega t)}], \tag{E.4.22}$$

i.e. in the form of a sum of complex travelling waves. This establishes that **q** is the propagation vector of the lattice waves and that the eigenvectors $\mathbf{e}^s(q)$ are, in fact, polarization vectors.

(2) *Real normal coordinates of the second kind,* $Z^s(q)$. These are obtained by a canonical transformation which mixes $Z^s(q)$ and $\dot{Z}^s(q)$; this is:

$$Q_s(q) = \frac{1}{2}[Z^s(q) - Z^s(-q)] - \frac{i}{2\omega_s(q)}[\dot{Z}^s(q) + \dot{Z}^s(-q)], \tag{E.4.23}$$

where $\omega_s(q)$ must be a positive root of $\omega_s^2(q)$. The Hamiltonian can be shown to be:

$$H = \frac{1}{2}\sum_{q,s} \{[\dot{Z}^s(q)]^2 + \omega_s^2(q)[Z^s(q)]^2\}, \tag{E.4.24}$$

giving equations of motion similar to (E.4.20).

One particularly simple state of affairs arises when all B_{ij}^{mn} vanish except those for which $m = n$. This uncouples the motion of the different atoms and if we use a suitable set of axes we can make B_{ij}^{mn} vanish unless $i = j$. The equation of motion is then

$$M\ddot{u}_i(m) + B_{ii}^{\mathrm{OO}}u_i(m) = 0. \tag{E.4.25}$$

The motion is then simply an independent vibration of each atom about its lattice point with frequency

$$\omega^2 = B_{ii}^{\mathrm{OO}}/M. \tag{E.4.26}$$

This will be independent of i for a cubic lattice, so that in that case each atom has three degenerate modes of vibration with the same frequency ω_E. This is the Einstein model of a solid as used in his theory of specific heats.

It should be noticed that this model does not satisfy the invariance condition (E.3.8). Unless the atomic vibrations have random phase, the model appears to prevent the possibility of vibration of the whole crystal, i.e. rigid body displacement.

Problems

(E.4.1) Prove that for the Bravais lattices discussed above $\omega_s^2(0) = 0$ for all s.

(E.4.2) Prove that for Bravais lattices the eigenvectors are real.

(E.4.3) Prove that the normal coordinates $X_\lambda^s(q)$ provide a real standing wave representation of the lattice vibrations.

(E.4.4) Prove that the normal coordinates $Z^s(q)$ afford a real travelling wave representation.

(E.4.5) Prove that the frequency function $\omega_s(q)$ is periodic in reciprocal space. What does this suggest about the slope of $\omega_s(q)$ at the zone boundary?

(E.4.6) Show that using $Q_s(q)$ the energy may be written:

$$H = \frac{1}{2}\sum_{q,s} [\dot{Q}_s^*(q)\dot{Q}_s(q) + \omega_s^2(q)Q_s^*(q)Q_s(q)].$$

E.5 THE MONATOMIC LINEAR CHAIN

We will now consider, using the formalism developed in §E.3 and §E.4, the familiar 'ball and spring' illustration of a one dimensional lattice.

The balls have a mass M and are connected by means of springs of force constant α whose equilibrium length (lattice constant) is a. The periodic conditions correspond to joining the ends of our chain of N 'atoms' so as to form a ring. We label the atoms as shown in fig. E.5.1. The boundary condition limits q to values:

$$q_i = \frac{2\pi l_i}{Na} \qquad (l_i \text{ integral}). \qquad \text{(E.5.1)}$$

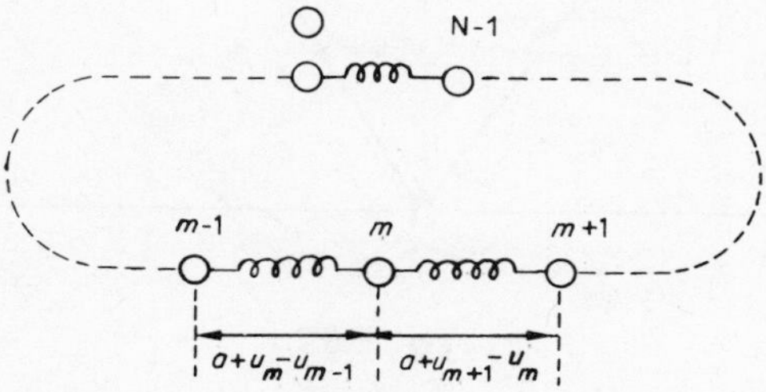

Fig. E.5.1.

The only distinct normal modes will be those for which

$$-\frac{N}{2} + 1 \leqslant l_i \leqslant \frac{N}{2}, \qquad \text{(E.5.2)}$$

N being chosen to be an even number.

The equation of motion of 'atom' m is

$$\begin{aligned} M\ddot{u}_m &= \alpha(u_{m+1} - u_m) + \alpha(u_{m-1} - u_m) \\ &= -2\alpha u_m + \alpha u_{m-1} + \alpha u_{m+1}. \end{aligned} \qquad \text{(E.5.3)}$$

Thus, using (E.3.5),

$$B^{OO} = 2\alpha; \quad B^{01} = B^{0-1} = -\alpha$$

and all other B^{Om} are zero. We note that ΣB^{On} is also zero, in accordance with (E.3.8).

The dynamical matrix in this case becomes a scalar (E.4.10),

$$D(q) = \frac{1}{M}\sum_n B^{On}\,\mathrm{e}^{iqna} = \frac{4\alpha}{M}\sin^2\left(\frac{qa}{2}\right), \qquad \text{(E.5.4)}$$

so that the frequencies are

$$\omega(q) = 2\sqrt{\left(\frac{\alpha}{M}\right)}\left|\sin\left(\frac{qa}{2}\right)\right| \qquad \text{(E.5.5)}$$

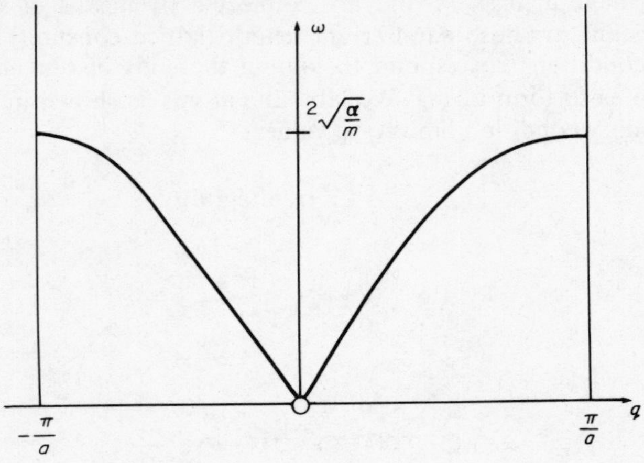

Fig. E.5.2.

This dispersion curve, as it is called, is illustrated in fig. E.5.2. It shows two typical features common to lattices in more than one dimension. The first is that at long wavelengths ($\lambda = 2\pi/q$) ω becomes proportional to q, whilst the second is that the slope of $\omega(q)$ decreases as q increases.

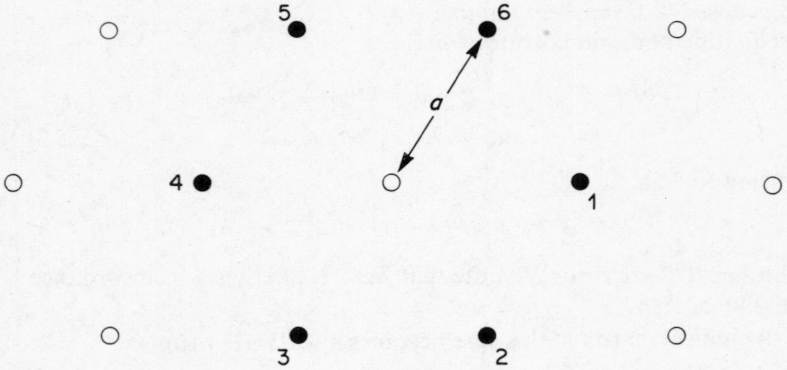

Fig. E.5.3.

Problem

(E.5.1) Consider a two-dimensional close packed lattice (fig. E.5.3) with interatomic distance a, the atoms of mass M being joined to nearest neighbours by central forces of strength α. Find the coupling constant matrices

B^{On} $(n = 0, 1, \ldots, 6)$, and the dynamical matrix, showing that the secular equation is

$$\begin{vmatrix} \frac{\alpha}{M}\left(3 - 2\cos q_1 a - \cos\frac{q_1 a}{2}\cos\frac{\sqrt{3}q_2 a}{2}\right) - \omega^2 & -\frac{\sqrt{3}\alpha}{M}\sin\frac{q_1 a}{2}\sin\frac{\sqrt{3}q_2 a}{2} \\ -\frac{\sqrt{3}\alpha}{M}\sin\frac{q_1 a}{2}\sin\frac{\sqrt{3}q_2 a}{2} & \frac{3\alpha}{M}\left(1 - \cos\frac{q_1 a}{2}\cos\frac{\sqrt{3}q_2 a}{2}\right) - \omega^2 \end{vmatrix} = 0.$$

The Brillouin zone for such a lattice is shown in fig. E.5.4, where the labelling is derived from the group theoretical labelling of the three dimensional hexagonal cell. Find the frequencies at Γ, M, K and along T, T′ and Σ. Give sketches of the dispersion curves.

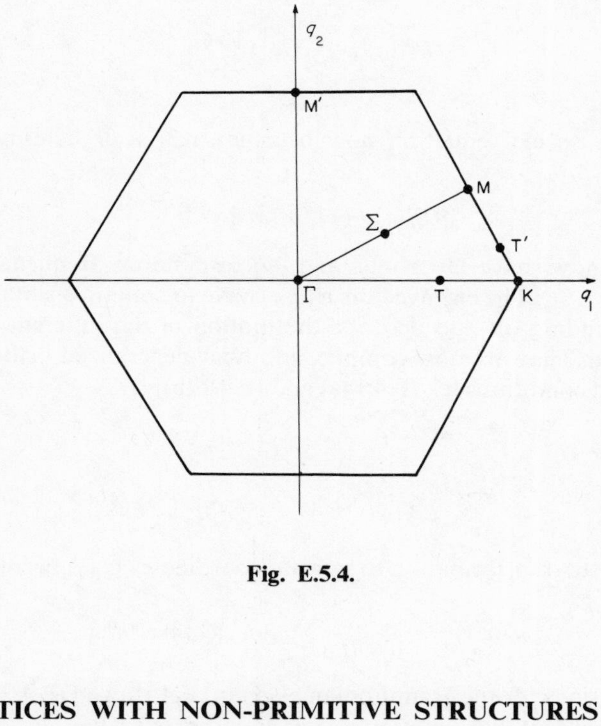

Fig. E.5.4.

E.6 LATTICES WITH NON-PRIMITIVE STRUCTURES

(By this is meant what, in §B.2, have been called 'lattices with a basis'.) We suppose that we have r atoms in the unit cell located at points $\mathbf{R}^\nu$ referred to some convenient origin in the cell; $\nu = 0, 1, \ldots r - 1$. Thus the position of the νth atom in the nth unit cell is

$$\mathbf{R}_n^\nu = \mathbf{R}_n + \mathbf{R}^\nu. \tag{E.6.1}$$

The displacements must be written $\mathbf{u}(n, \nu)$ and we have to distinguish the masses of different atoms, M_ν. The coupling constants must also be generalized thus [cf. (E.3.3)]:

$$B_{ij}^{mn}{}^{\mu\nu} = \left[\frac{\partial^2 U}{\partial u_i(m, \mu)\, \partial u_j(m, \nu)}\right]_0 \tag{E.6.2}$$

and

$$C_{ij}^{mn}{}^{\mu\nu} = (M_\mu M_\nu)^{-\frac{1}{2}} B_{ij}^{mn}{}^{\mu\nu}. \tag{E.6.3}$$

The dynamical matrix becomes [cf. (E.4.10)]:

$$D_{ij}^{\mu\nu}(q) = \sum_n C_{ij}^{On}{}^{\mu\nu}\, \mathrm{e}^{\mathrm{i}\mathbf{q}\cdot\mathbf{R}_n}, \tag{E.6.4}$$

so that the secular equation now becomes a $3r \times 3r$ determinant [cf. (E.4.12)]:

$$|D_{ij}^{\mu\nu}(q) - \omega^2\, \delta_{ij}\, \delta_{\mu\nu}| = 0. \tag{E.6.5}$$

Thus we now have $3r$ roots and corresponding frequencies $\omega_s(q)$ ($s = 1, 2, \ldots, 3r$). The eigenvectors $\mathbf{e}^s(q, \nu)$ have to contain a suitable phase factor depending on ν to describe the motion of the different atoms in the unit cell. They are now complex and obey generalized orthogonality and closure conditions [cf. (E.4.14a) and (E.4.14b)]:

$$\sum_{i,\nu} e_i^{s*}(q, \nu) e_i^{s'}(q, \nu) = \delta_{ss'}, \tag{E.6.6a}$$

$$\sum_s e_i^{s*}(q, \nu) e_j^s(q, \mu) = \delta_{ij}\, \delta_{\mu\nu}. \tag{E.6.6b}$$

The complete transformation to normal coordinates $Q_s(q)$ becomes

$$u_i(n, \nu) = \frac{1}{\sqrt{(NM_\nu)}} \sum_n e_i^s(q, \nu) Q_s(q)\, \mathrm{e}^{\mathrm{i}\mathbf{q}\cdot\mathbf{R}_n}. \tag{E.6.7}$$

The expressions for the Hamiltonian given in (E.4.18) and (E.4.24) remain unchanged except that s now runs over an increased range.

Thus the most important change is that in $\omega_s(q)$. In particular it can be proved that for only three of the $3r$ branches (s) does $\omega_s(0) = 0$. We will now illustrate this in one dimension.

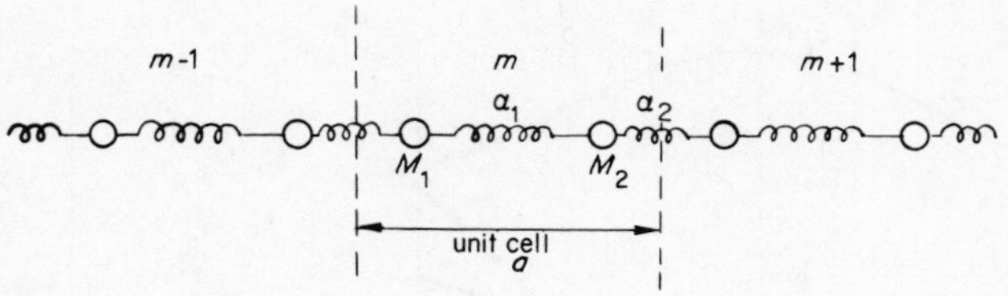

Fig. E.6.1.

The one dimensional diatomic lattice is shown in fig. E.6.1. We assume similar boundary conditions as for the monatomic case (§E.5). Both the masses of the atoms and the strength of the springs differ one from another. The equations of motion are

$$M_1\ddot{u}_1(m) = \alpha_1[u_2(m) - u_1(m)] + \alpha_2[u_2(m-1) - u_1(m)], \quad \text{(E.6.8a)}$$

$$M_2\ddot{u}_2(m) = \alpha_2[u_1(m+1) - u_2(m)] + \alpha_1[u_1(m) - u_2(m)]. \quad \text{(E.6.8b)}$$

Thus:

$$B_{11}^{OO} = B_{22}^{OO} = \alpha_1 + \alpha_2$$
$$B_{12}^{OO} = B_{21}^{OO} = -\alpha_1$$
$$B_{12}^{O1} = B_{21}^{O1} = -\alpha_2,$$

all other $B_{\mu\nu}^{On}$ being zero. The dynamical matrix is now 2 × 2:

$$D(q) = \begin{bmatrix} \dfrac{\alpha_1 + \alpha_2}{M_1} & -\dfrac{\alpha_1 + \alpha_2\,e^{-i\mathbf{q}\cdot\mathbf{a}}}{\sqrt{(M_1M_2)}} \\ \dfrac{-\alpha_1 + \alpha_2\,e^{i\mathbf{q}\cdot\mathbf{a}}}{\sqrt{(M_1M_2)}} & \dfrac{\alpha_1+\alpha_2}{M_2} \end{bmatrix}, \quad \text{(E.6.9)}$$

so that the frequencies are given by

$$\omega^2 = \frac{(\alpha_1+\alpha_2)(M_1+M_2)}{2M_1M_2}$$
$$\left\{1 \pm \sqrt{\left[1 - \frac{8\alpha_1\alpha_2M_1M_2}{(\alpha_1\alpha_2)^2(M_1+M_2)^2}(1-\cos qa)\right]}\right\}. \quad \text{(E.6.10)}$$

This is illustrated in fig. E.6.2. Thus the frequency spectrum is divided into two parts separated by a gap. The lower branch is called 'acoustical' for reasons made clearer in the next section, and the upper 'optical' because in real solids the frequencies are often in the near infrared.

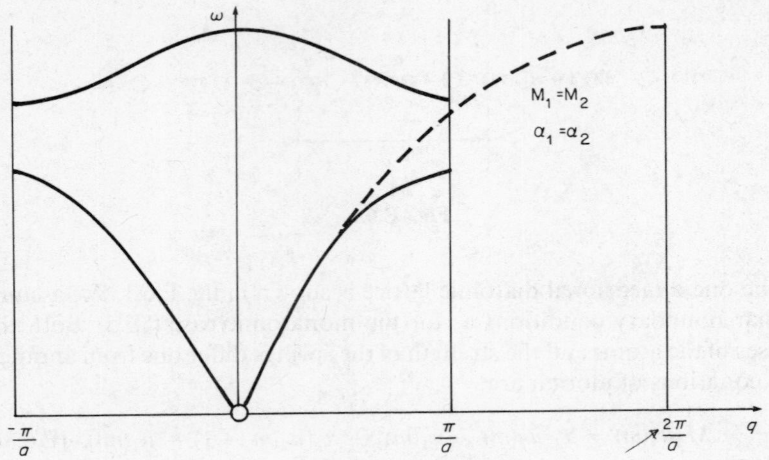

Fig. E.6.2.

If $\alpha_1 \to \alpha_2$ and $M_1 \to M_2$ the gap at $q = \pi/a$ vanishes and the 'zone' doubles in size corresponding to a halving of the unit cell. Inspection of fig. E.6.1 shows that this is exactly what is to be expected.

Problem

(E.6.1) Show that for $q = 0$ the eigenvectors of the diatomic linear chain are:

$$e_1^a = \frac{1}{\sqrt{\left(1 + \dfrac{M_2}{M_1}\right)}}, \qquad e_1^o = \frac{1}{\sqrt{\left(1 + \dfrac{M_1}{M_2}\right)}},$$

$$e_2^a = \frac{1}{\sqrt{\left(1 + \dfrac{M_1}{M_2}\right)}}, \qquad e_2^o = -\frac{1}{\sqrt{\left(1 + \dfrac{M_2}{M_1}\right)}}.$$

Note that for acoustic modes the atoms in the unit cell are in phase, but in the optical modes they move in antiphase.

E.7 WAVE MOTION IN AN ELASTIC CONTINUUM

Let us begin by examining the behaviour of the dynamical matrix at long wavelengths. Referring to (E.4.10) and expanding the exponential we find

$$D_{ij}(q) = \frac{1}{M} \sum B_{ij}^{On}(1 + i\mathbf{q} \, . \, \mathbf{R}_n - \tfrac{1}{2}(\mathbf{q} \, . \, \mathbf{R}_n)^2 + \cdots). \qquad \text{(E.7.1)}$$

Clearly the first term in the series vanishes because of (E.3.8). The second term also vanishes because of symmetry; in the case of a Bravais lattice this is given in a straightforward way by (E.3.9). Thus the first non-vanishing term appears to contain q^2. If we consider the secular equation (E.4.12) it is clear that at long wavelengths the frequency is proportional to q, implying a constant phase velocity of the waves.

This is precisely what is observed for sound waves in solid bodies up to quite high frequencies. In fact it is clear that for these branches, where $\omega \to 0$ as $q \to 0$, the lattice vibrations are the same as sound waves, hence the name 'acoustic modes' for those having that property. It is thus clearly of interest to examine the vibrations of elastic solids in the present context.

In this discussion of continuum mechanics[4] we limit ourselves to the isotropic case. The first step is the determination of the Lagrangian density function, $\mathscr{L}$. The kinetic energy density is given by

$$\mathscr{T} = \tfrac{1}{2}\rho \ddot{u}_i^2, \tag{E.7.2}$$

where ρ is the density of the medium and u_i is a component of the displacement field $\mathbf{u}(r)$. (In this section the tensor summation convention is used.) The potential energy is in fact the elastic strain energy. For small strains we can write this in terms of the displacement gradients,

$$\mathscr{U} = \frac{1}{4}\mu\left(\frac{\partial u_i}{\partial r_j} + \frac{\partial u_j}{\partial r_i}\right)^2 + \frac{\lambda}{2}\left(\frac{\partial u_k}{\partial r_k}\right)^2. \tag{E.7.3}$$

Here μ and λ are the Lamé constants of isotropic elasticity; μ is the ordinary rigidity modulus; the bulk modulus is given by $\frac{1}{3}(3\lambda + 2\mu)$.

Using (E.7.2) and (E.7.3) we can easily obtain the Lagrangian density $\mathscr{L} = \mathscr{T} - \mathscr{U}$. The extremal condition of classical mechanics may then be applied to the total Lagrangian $L = \int \mathscr{L}\, \mathrm{d}^3 r$, where the integration is over a volume V, later to be identified with the periodic volume. This condition gives the equations of motion

$$\rho \ddot{u}_i + \frac{\partial}{\partial r_j}\left(\frac{\partial \mathscr{L}}{\partial(\partial u_i/\partial r_j)}\right) = 0,$$

i.e.

$$\rho \ddot{u}_i = (\lambda + \mu)\frac{\partial^2 u_j}{\partial r_i\, \partial r_j} + \mu \frac{\partial^2 u_i}{\partial r_k^2}. \tag{E.7.4}$$

[4] R. N. Thurston, *Wave Propagation in Fluids and Normal Solids*, in W. P. Mason (ed.), Physical Acoustics, Vol. 1A, Academic Press, 1964.

We now assume a solution of the form

$$u_i = e_i\, e^{i(\mathbf{q}\cdot\mathbf{r}-\omega t)}, \tag{E.7.5}$$

which, substituted into (E.7.4), gives

$$\rho\omega^2 e_i = (\lambda + \mu)q_i q_j e_j + \mu q^2 e_i. \tag{E.7.6}$$

If these equations for e_i are to have non-trivial solutions we must have

$$\begin{vmatrix} (\lambda+\mu)q_1^2 + \mu q^2 - \rho\omega^2 & (\lambda+\mu)q_1 q_2 & (\lambda+\mu)q_1 q_3 \\ (\lambda+\mu)q_2 q_1 & (\lambda+\mu)q_2^2 + \mu q^2 - \rho\omega^2 & (\lambda+\mu)q_2 q_3 \\ (\lambda+\mu)q_3 q_1 & (\lambda+\mu)q_3 q_2 & (\lambda+\mu)q_3^2 + \mu^2 q^2 - \rho\omega^2 \end{vmatrix} = 0, \tag{E.7.7}$$

which is a secular equation for ω^2, having roots

$$\omega^2 = \omega_l^2 = \frac{\lambda + 2\mu}{\rho} q^2 = v_l^2 q^2, \tag{E.7.8a}$$

$$= \omega_t^2 = \frac{\mu}{\rho} q^2 = v_t^2 q^2 \text{ (twice)}. \tag{E.7.8b}$$

v_l and v_t are the corresponding sound velocities. The eigenvectors $\mathbf{e}^l$ and $\mathbf{e}^t$ must satisfy the conditions derived from (E.7.6) and (E.7.8a, b):

$$(q_i q_j - q^2\,\delta_{ij})e_j^l = 0, \tag{E.7.9a}$$

$${}_iq_j e_j^t = 0. \tag{E.7.9b}$$

Multiplying the former by e_i^l and summing over the repeated suffices leads to

$$(\mathbf{q}\cdot\mathbf{e}^l)^2 = q^2 e^{l2},$$

whence $\mathbf{e}^l$ is parallel to $\mathbf{q}$ and the vibrations provided by $\mathbf{e}^l$ are longitudinal. The latter condition can only be satisfied if $\mathbf{e}^t$ is perpendicular to $\mathbf{q}$ so that these modes are transverse (and also doubly degenerate). The eigenvectors may be made to satisfy the orthonormality condition

$$\mathbf{e}^s(q)\cdot\mathbf{e}^{s'}(q) = \delta_{ss'}. \tag{E.7.10}$$

We can now write down the Hamiltonian for the medium contained in the volume V:

$$H = \int_V \left[\frac{p_i^2}{2\rho} + \frac{\mu}{4}\left(\frac{\partial u_i}{\partial r_j} + \frac{\partial u_j}{\partial r_i}\right)^2 + \frac{\lambda}{2}\left(\frac{\partial u_k}{\partial r_k}\right)^2\right] d^3r, \tag{E.7.11}$$

where $p_i = \rho\dot{u}_i$ is the momentum conjugate to u_i.

Up to this point $\mathbf{q}$ has been regarded as a continuous variable. However this is not possible for a finite solid, so we impose periodic boundary conditions on a cube whose edge is of length $V^{\frac{1}{3}}$. This limits the allowed values of **q** to

$$q_i = \frac{2\pi}{V^{\frac{1}{3}}} n_i, \tag{E.7.12}$$

where n_i is an integer. This equation is analogous to (E.4.3) in the lattice case.

We can now introduce normal coordinates in the form of Fourier coefficients of **u** and **p**. We write

$$\mathbf{u} = \frac{1}{\sqrt{(\rho V)}} \sum_{q,s} \mathbf{e}^s(q) Q_s^q(t)\, \mathrm{e}^{i\mathbf{q}\cdot\mathbf{r}} \tag{E.7.13a}$$

and

$$\mathbf{p} = \sqrt{\left(\frac{\rho}{V}\right)} \sum_{q,s} \mathbf{e}^s(q) P_s^q(t)\, \mathrm{e}^{i\mathbf{q}\cdot\mathbf{r}}. \tag{E.7.13b}$$

The inverse transformations are

$$Q_s^q = \sqrt{(\rho V)}\mathbf{e}^s(q) . \int_V \mathbf{u}\, \mathrm{e}^{-i\mathbf{q}\cdot\mathbf{r}}\, \mathrm{d}^3 r \tag{E.7.14a}$$

and

$$P_s^q = \sqrt{\left(\frac{V}{\rho}\right)} \mathbf{e}^s(q) . \int_V \mathbf{p}\, \mathrm{e}^{-i\mathbf{q}\cdot\mathbf{r}}\, \mathrm{d}^3 r, \tag{E.7.14b}$$

which may be proved with the assistance of the relation [equation (B.6.24)]

$$\int_V \mathrm{e}^{\mathrm{i}(\mathbf{q}+\mathbf{q}')\cdot\mathbf{r}}\, \mathrm{d}^3 r = V\delta_{q',-q}. \tag{E.7.15}$$

To prove the normal coordinate character of Q_s^q we need the expression for the displacement gradient, which can be obtained from (E.7.13a). It is:

$$\frac{\partial u_i}{\partial r_j} = \frac{\mathrm{i}}{\sqrt{(\rho V)}} \sum e_i^s(q) q_j Q_s^q(t)\, \mathrm{e}^{i\mathbf{q}\cdot\mathbf{r}}. \tag{E.7.16}$$

The fact that the displacements are real leads to relations involving $\mathbf{e}^s$ and Q_s^q similar to (E.4.16a, b).

We can now write the Hamiltonian as

$$H = \frac{1}{2} \sum_{q,s} \left\{ P_s^q P_s^{q*} + \frac{1}{\rho}[\mu q^2 + (\lambda + \mu)(\mathbf{e}^s \cdot \mathbf{q})^2] Q_s^q Q_s^{q*} \right\}, \tag{E.7.17}$$

which, if we recall (E.7.8a, b), is of the same form as the Hamiltonian given in problem (E.4.6) with frequencies

$$\omega_s = v_s q. \tag{E.7.18}$$

Thus the elastic continuum model behaves in a similar way to the lattice model taken to the long wavelength limit (E.7.1). It is to be noted however that it is impossible to have a genuinely isotropic lattice model.

One of the most important differences between the lattice and the continuum model is the fact that the latter has no unit cell and hence no Brillouin zone. There is no periodicity in reciprocal space and all distinct values of **q**, no matter how large, correspond to different possible oscillations. We are concerned with the difference between (E.7.15) and

$$\sum_{\substack{\text{all } n \\ \text{in } V}} e^{i(\mathbf{q}+\mathbf{q}')\cdot\mathbf{R}_n} = N \sum_l \delta_{\mathbf{q}+\mathbf{q}',\mathbf{K}_l}, \tag{E.7.19}$$

which can be derived from (E.4.5), there being N unit cells in the volume V. In both (E.7.15) and (E.7.19) the possible values of **q** are limited to these allowed by the periodic boundary conditions, but in the latter equation $\mathbf{q}, \mathbf{q}'$ describe the same motion, $\mathbf{K}_l$ being a reciprocal lattice translation. Thus the continuum model has an infinite number of normal modes unless some *ad hoc* method of limiting the number is adopted. This is of importance in the theory of specific heats.

Problem

(E.7.1) In using the elastic continuum model of a crystal lattice it is necessary to cut off the range of **q** in order to give the correct number of normal modes. One method of doing these is to describe a sphere in reciprocal space which will contain just the right number of allowed values of **q**, which is N per branch, N being the number of unit cells in the volume V. Show that the radius of this sphere is

$$q_D = \left(\frac{6\pi^2 N}{V}\right)^{\frac{1}{3}}.$$

E.8 OPTICAL MODES IN IONIC CRYSTALS

In the previous section we showed how the long wavelength acoustic modes of vibration were related to the macroscopic elastic properties of the solid. We will now see whether the frequencies of the optical modes can be determined from the macroscopic properties for the particular case of ionic crystals. We again restrict ourselves to very long wavelengths so that the only important motion is relative motion within the unit cell, this motion being almost identical in each cell. This point was touched on in problem (E.6.1).

In the case of ionic crystals the optical modes result in the ions of opposite charge moving in opposite directions. This is equivalent to a local varying dipole moment, which when transverse will be shown later to interact strongly with electromagnetic radiation. The macroscopic polarization of the crystal, **P**, can be written as:

$$\mathbf{P} = N_0(e^*\mathbf{s} + \alpha\mathbf{E}_l). \tag{E.8.1}$$

We have two terms here, the first of which is due to the relative displacement of the positive and negative ions. e^* is the effective charge on these ions and $\mathbf{s} = \mathbf{u}_+ - \mathbf{u}_-$ is the relative displacement. The second is due to the electronic polarisation of the ions in the local electric field $\mathbf{E}_l$, which is not the same as the real field **E**. The magnitude of this effect is specified by the polarisability, α; N_0 is the number of unit cells in unit volume. The relation of $\mathbf{E}_l$ to **E** is given by the Lorentz formula

$$\mathbf{E}_l = \mathbf{E} + \mathbf{P}/3\epsilon_0. \tag{E.8.2}$$

In this equation we use M.K.S. units and ϵ_0 is the free space permittivity. Combining (E.8.1) and (E.8.2) gives us **P** in terms of **E** and **s**:

$$\mathbf{P} = N_0 \frac{(e^*\mathbf{s} + \alpha\mathbf{E})}{\left(1 - \dfrac{N_0\alpha}{3\epsilon_0}\right)}. \tag{E.8.3}$$

Now at very high frequencies the ions will be unable to follow the applied field and $\mathbf{s} \to 0$. In this case we can write

$$\mathbf{P} = (\epsilon_\infty - 1)\epsilon_0\mathbf{E}, \tag{E.8.4}$$

where ϵ_∞ is the high frequency dielectric constant. Because $\mathbf{s} \to 0$, we can now obtain from (E.8.3) and (E.8.4) the polarizability in terms of ϵ_∞, and then eliminate it in (E.8.3) giving

$$\mathbf{P} = \frac{N_0 e^*(2 + \epsilon_\infty)}{3}\mathbf{s} + (\epsilon_\infty - 1)\epsilon_0\mathbf{E}. \tag{E.8.5}$$

The next step is to consider the equations of motion of the ions. These will be of the form

$$M_+ \frac{d^2\mathbf{u}_+}{dt^2} = -K(\mathbf{u}_+ - \mathbf{u}_-) + e^*\mathbf{E}_l, \tag{E.8.6a}$$

$$M_- \frac{d^2\mathbf{u}_-}{dt^2} = K(\mathbf{u}_+ - \mathbf{u}_-) - e^*\mathbf{E}_l, \tag{E.8.6b}$$

where M_+, M_- are the masses of the ions and K simply defines the strength of the restoring forces. These two equations can be combined using $\mathbf{s}$ and a reduced mass $M_r = M_+ M_- /(M_+ + M_-)$ to give

$$M_r \frac{\mathrm{d}^2\mathbf{s}}{\mathrm{d}t^2} = -K\mathbf{s} + e^*\mathbf{E}_l, \tag{E.8.7}$$

so substituting for $\mathbf{E}_l$ we get

$$M_r \frac{\mathrm{d}^2\mathbf{s}}{\mathrm{d}t^2} = -M_r\omega_0^2\mathbf{s} + \frac{e^*(2+\epsilon_\infty)}{3}\mathbf{E}, \tag{E.8.8}$$

where

$$\omega_0^2 = \frac{K}{M_r} - \frac{N_0 e^{*2}(2+\epsilon_\infty)}{9M_r\epsilon_0}. \tag{E.8.9}$$

We now make another change of variables:

$$\mathbf{w} = \sqrt{(N_0 M_r)}\mathbf{s}, \tag{E.8.10}$$

to give the following symmetrical pair of basic equations:

$$\frac{\mathrm{d}^2\mathbf{w}}{\mathrm{d}t^2} = -\omega_0^2\mathbf{w} + \sqrt{\left(\frac{N_0}{M_r}\right)}\frac{e^*(2+\epsilon_\infty)}{3}\mathbf{E}, \tag{E.8.11a}$$

$$\mathbf{P} = \sqrt{\left(\frac{N_0}{M_r}\right)}\frac{e^*(2+\epsilon_\infty)}{3}\mathbf{w} + (\epsilon_\infty - 1)\epsilon_\infty\mathbf{E}. \tag{E.8.11b}$$

When the applied field is static, $\mathrm{d}^2\mathbf{w}/\mathrm{d}t^2 = 0$. We can then find $\mathbf{w}$ in terms of $\mathbf{E}$ and thence write $\mathbf{P}$ in terms of $\mathbf{E}$. By definition,

$$\mathbf{P} = (\epsilon_s - 1)\epsilon_0\mathbf{E} \tag{E.8.12}$$

in this case, so

$$\epsilon_s - \epsilon_\infty = \frac{N_0 e^{*2}(2+\epsilon_\infty)^2}{9M_r\omega_0^2\epsilon_0}, \tag{E.8.13}$$

where ϵ_s is the static dielectric constant. Thus we can write our basic pair in terms of the macroscopic parameters ϵ_s and ϵ_∞, and ω_0:

$$\frac{\mathrm{d}^2\mathbf{w}}{\mathrm{d}t^2} = -\omega_0^2\mathbf{w} + \omega_0\sqrt{[\epsilon_0(\epsilon_s - \epsilon_\infty)]}\mathbf{E}, \tag{E.8.14a}$$

$$\mathbf{P} = \omega_0\sqrt{[\epsilon_0(\epsilon_s - \epsilon_\infty)]}\mathbf{w} + (\epsilon_\infty - 1)\epsilon_0\mathbf{E}. \tag{E.8.14b}$$

We now wish to find what type of motion $\mathbf{w}$ is able to undergo in the absence of any applied field. To do this we first of all divide $\mathbf{w}$ into a solenoidal and an irrotational part:

$$\mathbf{w} = \mathbf{w}_t + \mathbf{w}_l, \tag{E.8.15}$$

where div $\mathbf{w}_t = 0$ and curl $\mathbf{w}_l = 0$. Such a decomposition is known to be unique.

In the absence of space charge the Laplace equation for **D** is

$$\text{div}\,\mathbf{D} = 0. \tag{E.8.16}$$

Now $\mathbf{D} = \epsilon_0\mathbf{E} + \mathbf{P}$ and **P** can be expressed in terms of **w** and **E** from (E.8.14b). Thus

$$\text{div}\,\mathbf{E} + \frac{\omega_0}{\epsilon_\infty}\sqrt{\left(\frac{\epsilon_s - \epsilon_\infty}{\epsilon_0}\right)}\,\text{div}\,\mathbf{w} = 0. \tag{E.8.17}$$

A solution of this is clearly

$$\mathbf{E} = -\frac{\omega_0}{\epsilon_\infty}\sqrt{\left(\frac{\epsilon_s - \epsilon_\infty}{\epsilon_0}\right)}\,\mathbf{w}_l. \tag{E.8.18}$$

Thus (E.8.14a) can be written

$$\frac{d^2}{dt^2}(\mathbf{w}_t + \mathbf{w}_l) = -\omega_0^2\mathbf{w}_t - \omega_0^2\frac{\epsilon_s}{\epsilon_\infty}\mathbf{w}_l.$$

Separating the solenoidal and irrotational parts we have:

$$\frac{d^2\mathbf{w}_t}{dt^2} = -\omega_0^2\mathbf{w}_t, \tag{E.8.19a}$$

$$\frac{d^2\mathbf{w}_l}{dt^2} = -\frac{\epsilon_s}{\epsilon_\infty}\omega_0^2\mathbf{w}_l. \tag{E.8.19b}$$

If we write $\mathbf{w}_t$ and $\mathbf{w}_l$ in the form of a wave, $e^{i\mathbf{q}\cdot\mathbf{r}}$, the conditions which require them to be solenoidal and irrotational will demand that $\mathbf{w}_t$ is normal to **q** and $\mathbf{w}_l$ parallel to **q**. Thus $\mathbf{w}_t$ corresponds to transverse vibrations with frequency $\omega_{t0}^2 = \omega_0^2$ and $\mathbf{w}_l$ to longitudinal vibrations with a higher frequency $\omega_{l0}^2 = (\epsilon_s/\epsilon_\infty)\omega_0^2$. The equation

$$\frac{\omega_{l0}^2}{\omega_{t0}^2} = \frac{\epsilon_s}{\epsilon_\infty} \tag{E.8.20}$$

is the Lyddane–Sachs–Teller relation. Since ω_{t0} is much more easily found it enables ω_{l0} to be determined. For covalent crystals, where $\epsilon_s = \epsilon_\infty$, it predicts that the longitudinal and transverse optical modes are degenerate at $\mathbf{q} = 0$ as is indeed the case.

Szigeti showed that the restoring force constant K is related to the elastic properties of the crystal. He predicted that the bulk modulus should be given by

$$B = \frac{N_0 M_r a^2(\epsilon_s + 2)\omega_{t0}^2}{3(\epsilon_\infty + 2)}, \tag{E.8.21}$$

a result confirmed experimentally to within 10% for the alkali halides.

Problem

(E.8.1) Prove the result quoted above, that a solenoidal vector field may only support transverse vibrations, and an irrotational field only longitudinal vibrations.

E.9 THE QUANTIZATION OF THE LATTICE VIBRATIONS

Before proceeding to the main business of this section it is helpful to consider a fourth type of normal coordinate, denoted by $\alpha_s(q)$ and its complex conjugate $\alpha_s^*(q)$. These are conveniently defined in terms of the real quantities $Z^s(q)$ thus:

$$Z^s(q) = \sqrt{\left[\frac{\hbar}{2\omega_s(q)}\right]}[\alpha_s(q) + \alpha_s^*(q)] \tag{E.9.1a}$$

and

$$\dot{Z}^s(q) = \mathrm{i}\sqrt{\left[\frac{\hbar\omega_s(q)}{2}\right]}[\alpha_s(q) - \alpha_s^*(q)]. \tag{E.9.1b}$$

The reverse transformation is readily obtained. At present we will regard $\hbar$ as simply an undefined constant. If we substitute (E.9.1a) and (E.9.1b) into (E.4.24) the Hamiltonian takes on a particularly simple form:

$$H = \hbar \sum \omega_s(q)\alpha_s^*(q)\alpha_s(q). \tag{E.9.2}$$

The quantities $\alpha_s(q)$ also have extremely simple equations of motion:

$$\dot{\alpha}_s(q) = \mathrm{i}\omega_s(q)\alpha_s(q) \tag{E.9.3a}$$

and

$$\dot{\alpha}_s^*(q) = -\mathrm{i}\omega_s(q)\alpha_s^*(q). \tag{E.9.3b}$$

One relation of great importance is that which relates the displacement to $\alpha_s(q)$, $\alpha_s^*(q)$. This can be obtained by combining (E.4.6), (E.4.8a), (E.4.15a) and (E.4.23) with (E.9.1a, b), giving

$$u_i(n) = \sum_{q,s}\sqrt{\left[\frac{\hbar}{2NM\omega_s(q)}\right]}\, e_i^s(q)\, \mathrm{e}^{\mathrm{i}\mathbf{q}\cdot\mathbf{R}_n}[\alpha_s(q) - \alpha_s^*(-q)]. \tag{E.9.4}$$

This enables us to express any function of the displacements in terms of $\alpha_s(q)$, $\alpha_s^*(q)$.

Equations (E.4.18) or (E.4.24) expressed the Hamiltonian as a sum of simple harmonic oscillator Hamiltonians for the individual modes. Furthermore the coordinates and momenta were real variables. This means that the problem of quantizing the lattice vibrations is simply that for an harmonic oscillator, a problem dealt with in all elementary courses of quantum mechanics. We will therefore deal with it in fairly summary fashion.

Using the conventional notation of p for momentum and q for coordinate, the Hamiltonian of the harmonic oscillator is

$$H = \tfrac{1}{2}(p^2 + \omega^2 q^2). \tag{E.9.5}$$

Employing the Schrödinger representation with the usual transformation

$$p \to \frac{\hbar}{\mathrm{i}} \frac{\partial}{\partial q} \quad \text{and} \quad q \to q,$$

the Schrödinger equation is

$$\frac{1}{2}\left\{-\hbar^2 \frac{\partial^2}{\partial q^2} + \omega^2 q^2\right\}\psi = \epsilon\psi \tag{E.9.6}$$

where ϵ is the energy. It is shown in standard texts on quantum mechanics that ϵ is restricted to certain values only:

$$\epsilon = (n + \tfrac{1}{2})\hbar\omega \tag{E.9.7}$$

where the quantum number n is a positive integer or zero. The wave functions are

$$\psi_n = \frac{\mathrm{e}^{-\omega q^2/2\hbar} H_n\left[\left(\frac{\omega}{\hbar}\right)^{\frac{1}{2}} q\right]}{\pi^{\frac{1}{4}} 2^{n/2} \sqrt{n!}} \tag{E.9.8}$$

where H_n is an Hermite polynomial.

The operators corresponding to the variables $\alpha_s(q)$ and $\alpha_s^*(q)$ will be, dropping the (s, q),

$$\alpha = \sqrt{\left(\frac{\omega}{2\hbar}\right)} q + \sqrt{\left(\frac{\hbar}{2\omega}\right)} \frac{\partial}{\partial q} \tag{E.9.9a}$$

and

$$\alpha^\dagger = \sqrt{\left(\frac{\omega}{2\hbar}\right)} q - \sqrt{\left(\frac{\hbar}{2\omega}\right)} \frac{\partial}{\partial q}. \tag{E.9.9b}$$

This result will follow directly from (E.9.1a,b). These operators are non-Hermitian. Using the recurrence properties of the Hermite polynomials it can be shown that

$$\alpha\psi_n = \sqrt{n}\psi_{n-1} \tag{E.9.10a}$$

$$\alpha^\dagger\psi_n = \sqrt{(n+1)}\psi_{n+1}. \tag{E.9.10b}$$

Thus α is an annihilation operator which decreases the quantum number of normal mode by one and $\alpha^\dagger$ is a creation operator which increases it by one. The lattice vibrational quanta which are annihilated or created by these operators are called phonons, by analogy with the photons of the electromagnetic field. Using vector notation where $|0\rangle$ describes the ground state where there are no phonons and $|n\rangle$ the state where there are n phonons one obtains the interesting result

$$|n\rangle = (\alpha^\dagger)^n|0\rangle/\sqrt{(n!)}. \tag{E.9.11}$$

The operators α and $\alpha^\dagger$ have a number of other interesting properties. Using (E.9.5) and the fundamental quantum condition $(pq - qp) = \hbar/\mathrm{i}$, it can be shown that

$$\alpha\alpha^\dagger = \frac{H}{\hbar\omega} + \frac{1}{2}. \tag{E.9.12a}$$

Similarly

$$\alpha^\dagger\alpha = \frac{H}{\hbar\omega} - \frac{1}{2}, \tag{E.9.12b}$$

so that the commutator of α and $\alpha^\dagger$ is

$$[\alpha, \alpha^\dagger] = (\alpha\alpha^\dagger - \alpha^\dagger\alpha) = 1. \tag{E.9.13}$$

The classical Hamiltonian in terms of α, α^* was given by (E.9.2). In quantum mechanics this must be symmetrised to give

$$H = \frac{\hbar\omega}{2}(\alpha^\dagger\alpha + \alpha\alpha^\dagger) \tag{E.9.14a}$$

$$= \hbar\omega(\alpha^\dagger\alpha + \tfrac{1}{2}) \tag{E.9.14b}$$

using (E.9.13). Comparison with (E.9.7) shows that $\alpha^\dagger\alpha$ is the phonon number operator, i.e.

$$\alpha^\dagger\alpha\psi_n = n\psi_n. \tag{E.9.15}$$

Let us now return to the complete set of normal modes describing the lattice vibrations. The total energy is just the sum of the individual oscillator energies (E.9.7):

$$\mathscr{E} = \hbar \sum (N_s^q + \tfrac{1}{2})\omega_s(q) \qquad \text{(E.9.16)}$$

where N_s^q is the quantum number of the mode (q, s), or put another way there are N_s^q phonons of type (q, s). The wavefunctions of the whole lattice is the product of the individual oscillator wavefunctions:

$$\Psi\{N_s^q\} = \prod_{q,s} \psi(N_s^q). \qquad \text{(E.9.17)}$$

The action of an individual $\alpha_s(q)$, $\alpha_s^\dagger(q)$ is restricted to the particular mode having the same label, the others being unaffected. Thus

$$\alpha_s(q)\Psi\{N_{s'}^{q'}\} = \sqrt{(N_s^q)}\Psi\{\text{all same except } (N_s^q - 1)\}, \qquad \text{(E.9.18a)}$$

$$\alpha_s^\dagger(q)\Psi\{N_{s'}^{q'}\} = \sqrt{(N_s^q + 1)}\Psi\{\text{all same except } (N_s^q + 1)\}. \qquad \text{(E.9.18b)}$$

Similarly the expression for the commutator (E.9.13) needs generalizing to give

$$[\alpha_s(q), \alpha_{s'}^\dagger(q')] = \delta_{ss'}\delta_{qq'}. \qquad \text{(E.9.19)}$$

This completes the basic quantum mechanical machinery necessary to cope with the vibrations of a crystal lattice, in the harmonic approximation, where the states defined by the N_s^q are permanent and unchanging.

If one regards the lattice vibrational phonon from the viewpoint of many-body theory it provides a particularly simple example of a quasiparticle. Let us consider the Einstein model, described in the final part of §E.4., which decouples the motion of different atoms. If we now quantize the vibrations of the individual atoms we produce what may be called Einstein phonons. The terms in the potential energy neglected in the Einstein model give rise to a strong interaction between the Einstein phonons. This 'clothes' the bare Einstein phonon producing a quasiparticle which is in fact the phonon corresponding to the correct harmonic oscillations of the lattice. The treatment of this question using diagrammatic techniques is given by Mattuck[5].

Problem

(E.9.1) Prove that the constant $\hbar$ in (E.9.1a, b) is in fact the fundamental quantum constant of, for example, (E.9.7).

[5] R. D. Mattuck *A Guide to Feynman Diagrams in the Many-Body Problem*, McGraw–Hill, London, 1967.

E.10 PHONON STATISTICS AND PHONON TRANSITIONS

In the applications of lattice dynamics it is frequently necessary to know the thermodynamic average values of the occupation numbers, N_s^q, are. This question can be approached in two ways. In the first we emphasize the role of N_s^q as a quantum number belonging to an harmonic oscillator, whilst in the second we direct attention to the phonons as particles having certain properties.

(i) The first method commences with the expression for the probability W_n of finding an oscillator in state n with energy $(n + \frac{1}{2})\,\hbar\omega$. From the basic equation of statistical mechanics,

$$W_n = \frac{e^{-(n+\frac{1}{2})\hbar\omega/k_0T}}{\sum_n e^{-(n+\frac{1}{2})\hbar\omega/k_0T}}. \tag{E.10.1}$$

Here k_0 is Boltzmann's constant and T the thermodynamic temperature. We now cancel out the zero point energy, $\frac{1}{2}\hbar\omega$, and consider the denominator of (E.10.1):

$$\sum_n e^{-n\hbar\omega/k_0T} = 1 + e^{-\hbar\omega/k_0T} + e^{-2\hbar\omega_0/k_0T} + \cdots$$
$$= (1 - e^{-\hbar\omega/k_0T})^{-1},$$

so that

$$W_n = e^{-n\hbar\omega/k_0T}(1 - e^{-\hbar\omega/k_0T}). \tag{E.10.2}$$

The average value of n found in thermal equilibrium will then be:

$$\tilde{n} = \sum nW_n = \sum n\, e^{-n\hbar\omega/k_0T}(1 - e^{-\hbar\omega/k_0T})$$
$$= e^{-\hbar\omega/k_0T}(1 - e^{-\hbar\omega/k_0T})(1 + 2\,e^{-\hbar\omega/k_0T} + 3e^{-2\hbar\omega/k_0T} + \cdots).$$

Now $(1 + 2x + 3x^2 + \cdots)(1 - x) = 1 + x + x^2 + x^3 + \cdots = (1 - x)^{-1}$, so

$$\tilde{n} = \frac{e^{-\hbar\omega/k_0T}}{1 - e^{-\hbar\omega/k_0T}} = [e^{\hbar\omega/k_0T} - 1]^{-1},$$

or

$$\tilde{N}_s^q = [e^{\hbar\omega_s(q)/k_0T} - 1]^{-1}. \tag{E.10.3}$$

This is the Planck function originally deduced for photons in the theory of black body radiation.

(ii) Regarded as particles the phonons have the following properties:

(A) all phonons belonging to the same mode are indistinguishable;

(B) phonons have creation and annihilation operators obeying the commutation rule (E.9.13) or (E.9.19). They are therefore bosons, and the number in any one mode is quite unlimited.

(C) Phonons are not conserved; their total number $\sum N_s^q$ is not fixed. This means that they have no chemical potential.

From these three premises the standard methods of statistical mechanics will give the same form for $\bar{N}_s^q$ as (E.10.3).

We have just shown what the phonon occupation numbers are in thermal equilibrium. However no mechanism has yet been mentioned by which an initial arbitrary set of N_s^q can change to that characteristic of thermal equilibrium. In the harmonic approximation the state specified by a set $\{N_s^q\}$ is a permanent eigenstate because the quantum mechanical problem was solved exactly. Changes in $\{N_s^q\}$ can only arise from perturbations outside the harmonic approximation. These might be of two kinds: firstly there are those which are intrinsic to the system such as the anharmonic terms in the lattice potential energy, and secondly there are effects imposed from outside the system such as electromagnetic radiation. We can describe both types of possible effect by an extra term in the Hamiltonian of the system, H', which forms the starting point for a calculation based on time dependent perturbation theory.

We define a transition probability Q_i^f such that $Q_i^f\,\mathrm{d}t$ is the probability that the system will undergo a transition from an initial state (i) to a final state (f) in a time $\mathrm{d}t$. If it refers to a situation where the system is known to be definitely in state (i) initially, whilst state (f) is definitely empty, we use an intrinsic transition probability. Time-dependent perturbation theory proves that

$$Q_i^f = \frac{2\pi}{\hbar}|\langle f|H'|i\rangle|^2\delta(\epsilon_f - \epsilon_i). \tag{E.10.4}$$

The δ-function ensures energy conservation; this means that the above expression is only approximate, since if the perturbation is so strong that the time during which the system remains in a state is less than $\hbar/(\epsilon_f - \epsilon_i)$, the uncertainty principle tolerates some degree of non-conservation of energy because the energies are not really well-defined.

The quantity $\langle f|H'|i\rangle$ is called the matrix element of H' with respect to the initial and final states of the system. In wave mechanics

$$\langle f|H'|i\rangle = \int \psi_f^* H' \psi_i \,\mathrm{d}^3 r \tag{E.10.5}$$

where the integration is carried out over the whole space for which the wavefunctions are defined.

In the case of lattice vibrations H' will certainly contain the atomic displacements $u_i(n)$. These can be written in terms of $\alpha_s(q)$ and $\alpha_s^\dagger(q)$ using (E.9.4). This suggests that the matrix elements of $\alpha_s(q)$ and $\alpha_s^\dagger(q)$ may be of particular importance. Let us consider this question for a single harmonic oscillator. Now, because of (E.9.10a),

$$\langle n'|\alpha|n\rangle = \sqrt{n}\int \psi_{n'}^*\psi_{n-1}\,\mathrm{d}^3r = \sqrt{n}\,\delta_{n',n-1} \tag{E.10.6}$$

since the functions ψ_n are orthonormal. Thus the only non-vanishing matrix element of α is

$$\langle n-1|\alpha|n\rangle = \sqrt{n} \tag{E.10.7a}$$

since all other final states give zero. Similarly using (E.9.10b) the only non vanishing matrix element of $\alpha^\dagger$ is

$$\langle n+1|\alpha^\dagger|n\rangle = \sqrt{(n+1)}. \tag{E.10.7b}$$

Thus, under the influence of a perturbation linear in α and $\alpha^\dagger$, the state of a normal mode can only be changed by one phonon, either up or down. Also, in the case of an assembly of modes, only one mode can be affected by a given $\alpha_s(q)$ or $\alpha_s^\dagger(q)$.

Let us consider as an example the matrix elements of the third order term in the potential energy. From (E.3.2) this is

$$U_3 = \frac{1}{3!}\sum_{\substack{mno\\ijk}} B_{ijk}^{mno}u_i(m)u_j(n)u_k(o). \tag{E.10.8}$$

Substitution of the displacements using (E.9.4) will yield products like

$$\alpha_s(q)\alpha_{s'}(q')\alpha_{s''}^\dagger(q'').$$

It will be seen that this term will produce transitions in which two phonons (q, s) and (q', s') disappear and one phonon (q'', s'') is created. These three phonon processes are very important in the theory of lattice thermal conductivity.

Problem

(E.10.1) Consider a crystal, perfect except that the atom at the lattice point $\mathbf{R}_n$ has been replaced by one whose mass is $M + \Delta M$, but identical in other respects. Prove that this gives rise to a transition probability that a phonon (q, s) is scattered into being a phonon (q', s') given by

$$Q_{qs}^{q's'} = \frac{(\Delta M)^2}{2N^2M^2}\omega_s(q)\omega_{s'}(q')[\mathbf{e}^s(q)\cdot\mathbf{e}^{s'}(q')]^2N_s^q(N_{s'}^{q'}+1)\,\delta[\omega_s(q)-\omega_{s'}(q')].$$

Notice that this result does not contain the quantum constant $\hbar$.

CHAPTER X

The Applications of Lattice Dynamics

E.11 INTRODUCTION

The applications to be discussed in this chapter are of two kinds. Firstly we use the basic theory of Chapter IX to find out more about the lattice vibrations of real substances. From the point of view of theory this means the study of models which are hoped to resemble, in important particulars, the real solid of interest. The essential piece of information for most purposes in the spectrum or density of states over frequency, knowledge of which will enable us to make headway with the second type of application. In this we are directly concerned with actual experimental phenomena such as specific heat or absorption of electromagnetic radiation. It is also possible to feed back from these kinds of experiment data which will enable us to construct the spectrum empirically.

E.12 THE DENSITY OF STATES FUNCTION[2]

Let us suppose that we wish to determine the value of a quantity of the form

$$F = \sum_{s,\mathbf{q}} f(\omega). \qquad \text{(E.12.1)}$$

As an example one might consider any of the thermodynamic properties of the lattice, such as the internal energy. This is the energy of the lattice vibrations in thermodynamic equilibrium, and will be given by

$$\bar{\mathscr{E}} = \hbar \sum_{s,\mathbf{q}} (\tilde{N}_s^q + \tfrac{1}{2})\omega_s(q) \qquad \text{(E.12.2)}$$

which, if we remember that $\tilde{N}_s^q$ is given by (E.10.3), is of the general form of (E.12.1).

Because there are a very large number of allowed values of $\mathbf{q}$ having a density of $(8\pi^3)^{-1}$ per unit volume of direct space, we can change (E.12.1) from a sum to an integral over $\mathbf{q}$:

$$F = \frac{V}{8\pi^3} \sum_s \int_{\text{B.Z.}} f(\omega)\, \mathrm{d}^3 q \qquad \text{(E.12.3)}$$

where the integration is over the whole of the first Brillouin Zone. It would clearly be of assistance if the integral could be changed to one over frequency. Let us restrict ourselves to a particular value of s and consider the constant frequency surfaces for ω and $\omega + \mathrm{d}\omega$. If $\mathrm{d}S$ is an element of area of the constant frequency surface ω, and the distance between that surface and that corresponding to $\omega + \mathrm{d}\omega$ is $\mathrm{d}q_n$ (n for normal) then

$$\mathrm{d}^3 q = \mathrm{d}q_n \, \mathrm{d}S.$$

Now, by the definition of a gradient,

$$\mathrm{d}\omega = |\nabla_q \omega| \, \mathrm{d}q_n$$

(because $\nabla_q \omega$ is normal to the constant frequency surface) so

$$\mathrm{d}^3 q = \frac{\mathrm{d}\omega \, \mathrm{d}S}{|\nabla_q \omega|}. \tag{E.12.4}$$

Using this important result we can now write our quantity F as

$$F = V \int f(\omega) g(\omega) \, \mathrm{d}\omega, \tag{E.12.5}$$

where

$$g(\omega) = \frac{1}{8\pi^3} \sum_s \int \frac{\mathrm{d}S}{|\nabla_q \omega|} \tag{E.12.6}$$

is the density of states (function) over frequency or 'spectrum'. In this case the surface integral is over the three surfaces having the same frequency ω; sometimes it is convenient to divide up $g(\omega)$ into separate $g_s(\omega)$. We know that $\int g(\omega) \, \mathrm{d}\omega = 3N/V$, the total number of normal modes.

It is important that the density of states function be not significantly affected by the choice of boundary conditions. According to Ledermann's Theorem[2] in a given frequency range, only G^{-1} of the mode frequencies in that range can be shifted out of it by alteration of the boundary conditions where there are G^3 unit cells in the crystal. This result can be upset, however, if the interatomic forces are of very long range.

The crucial quantity $\nabla_q \omega$ has another important significance. It can be shown as a standard result of wave theory that it is the group velocity $\mathbf{V}_s(q)$, the phase velocity being $\omega_s(q)/|q|$. In the long wavelength limit these have the same magnitude, but it must be remembered that it is the group velocity which determines the rate of energy flow in the lattice.

For the monatomic linear chain discussed in §E.5 the sound velocity is easily calculated from (E.5.5):

$$V(q) = \frac{\mathrm{d}\omega}{\mathrm{d}q} = a\left(\frac{\alpha}{M}\right)^{\frac{1}{2}} \cos qa = \frac{a}{2}\left(\frac{4\alpha}{M} - \omega^2\right)^{\frac{1}{2}}. \tag{E.12.7}$$

For long wavelengths this is just $a(\alpha/M)^{\frac{1}{2}}$, which is the sound velocity in a string whose mass per unit length is M/a and whose tension is αa. Thus we have made contact with the continuum model for one dimension.

The density of states is then

$$g(\omega) = \frac{1}{\pi}\frac{dq}{d\omega} = \frac{2}{\pi a}\left(\frac{4\alpha}{M} - \omega^2\right)^{-\frac{1}{2}}, \tag{E.12.8}$$

where a factor of two has been introduced to allow for the two possible directions of q corresponding to the same frequency. Equation (E.3.1) is sketched in fig. E.12.1, from which it is clear that $g(\omega)$ has a singularity at $\omega = 2\sqrt{(\alpha/M)}$. Such a singularity is obviously integrable, since we must have

$$\int g(\omega)\, d\omega = \frac{1}{a}. \tag{E.12.9}$$

These singularities are called 'critical points'.

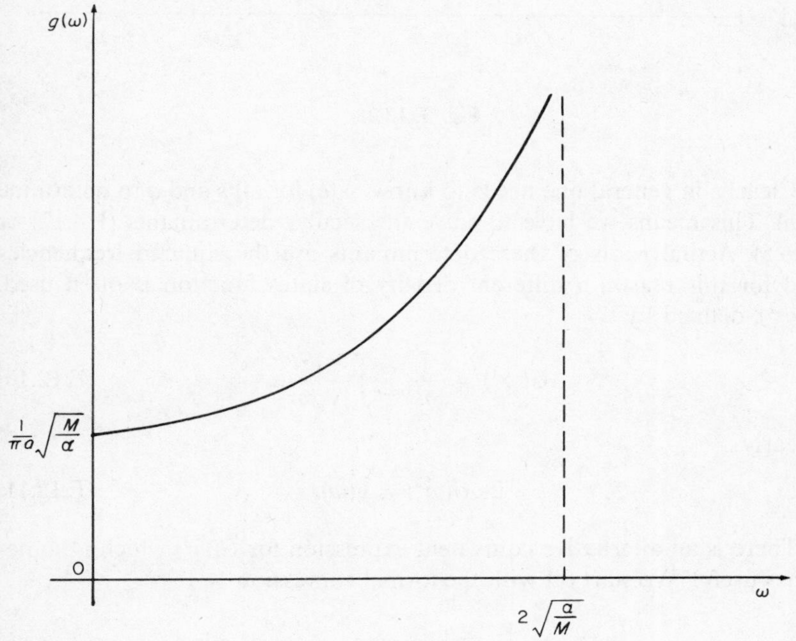

Fig. E.12.1.

A similar density of states curve is plotted in fig. E.12.2 for the diatomic linear chain discussed in §E.6. The three critical points correspond to the three points in fig. E.6.2 when the $\omega_s(q)$ is flat; one of them belongs to $q = 0$, the other two to $q = \pi/a$. We shall see later that the location of the critical points is often at zone corners and the zone centre.

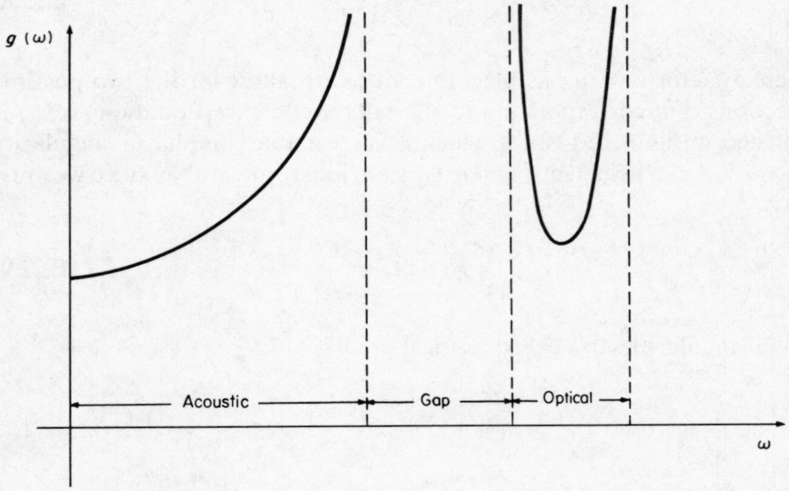

Fig. E.12.2.

Clearly, in general one needs to know $\omega_s(q)$ for all s and q to determine $g(\omega)$. This means we have to solve the secular determinants (E.4.12) or (E.6.5). Actual roots of these determinants are the squared frequencies and for this reason a different density of states function is often used, $G(\omega^2)$, defined by

$$G(\omega^2) = \frac{1}{8\pi^3}\sum_s \int \frac{dS}{|\nabla_q \omega^2|}. \tag{E.12.10}$$

Clearly

$$2\omega G(\omega^2) = g(\omega). \tag{E.12.11}$$

There is an alternative equivalent expression for $G(\omega^2)$ which is sometimes useful. We start off with the formal expression

$$G(\omega^2) = \frac{1}{8\pi^3}\sum_s \int_{\text{B.Z.}} \delta[\omega^2 - \omega_s^2(q)]\, d^3q. \tag{E.12.12}$$

Now, from the theory of Fourier transforms,

$$\delta(x) = \frac{1}{2\pi}\int_{-\infty}^{\infty} \mathrm{e}^{-\mathrm{i}xy}\,\mathrm{d}y,$$

so that

$$G(\omega^2) = \frac{1}{16\pi^4}\sum_s \int\int \mathrm{e}^{-\mathrm{i}[\omega^2 - \omega_s^2(q)]y}\,\mathrm{d}y\,\mathrm{d}^3q$$

$$= \frac{1}{(2\pi)^{\frac{1}{2}}}\int \mathrm{e}^{-\mathrm{i}\omega^2 y} f(y)\,\mathrm{d}y, \qquad \text{(E.12.13)}$$

where

$$f(y) = \frac{1}{(2\pi)^{\frac{7}{2}}}\sum_s \int_{\mathrm{B.Z.}} \mathrm{e}^{\mathrm{i}\omega_s^2(q)y}\,\mathrm{d}^3q \qquad \text{(E.12.14)}$$

is the Fourier transform of $G(\omega^2)$. We are defining the Fourier transform relationship by

$$F(y) = \frac{1}{(2\pi)^{\frac{1}{2}}}\int \mathrm{e}^{\mathrm{i}xy} f(x)\,\mathrm{d}x \qquad \text{(E.12.15a)}$$

and

$$f(x) = \frac{1}{(2\pi)^{\frac{1}{2}}}\int \mathrm{e}^{-\mathrm{i}xy} F(y)\,\mathrm{d}y. \qquad \text{(E.12.15b)}$$

We shall make use of this later.

It will be seen that if we know a sufficiently large number of the coupling constants we can find the frequencies to any required degree of accuracy. Alternatively we might hope to use measured values of $\omega_s(q)$ to give us information about the coupling constants. Theoretically the $B_{ij}^{\mathrm{O}n}$ should be obtainable from the interatomic force law; a number of different possibilities have been investigated. However we shall show how perturbation theory, group theory and topology enable us to make estimates of $G(\omega^2)$ without having obtained all the roots of the secular equation.

Problem

(E.12.1) Sketch $G(\omega^2)$ for the monatomic linear chain.

E.13 A TWO-DIMENSIONAL LATTICE MODEL

Models of this kind have been investigated by Blackman, Montroll and others[6]; one based on a hexagonal lattice was given as an example in

[6] A useful description of work with simple models is given in J. de Launay, *Solid State Physics*, **2**, 220 (1956).

§E.5. In this section we will deal with a square lattice whose equilibrium spacing is a and where the atoms are of mass M. The atoms will be assumed to interact with nearest and next nearest neighbours only by a central force.

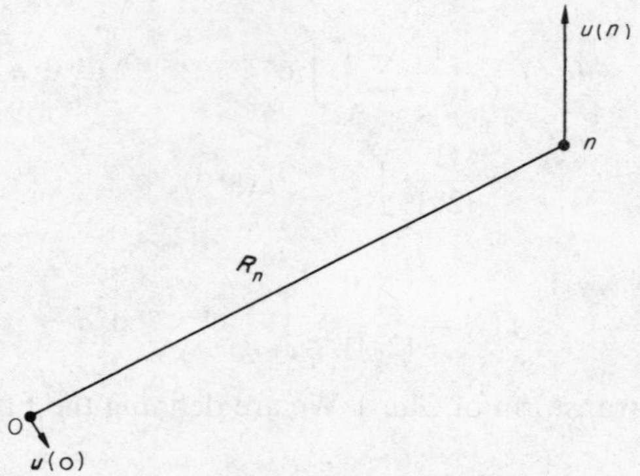

Fig. E.13.1.

Let us consider two atoms whose lattice points are separated by $\mathbf{R}_n$ (see fig. E.13.1); let $\boldsymbol{\epsilon}_n$ be a unit vector parallel to $\mathbf{R}_n$. Then if the displacement of atom (O) is $\mathbf{u}(\mathrm{O})$ and that of atom (n) is $\mathbf{u}(n)$ and the force constant is α_n, the force due to these displacements is

$$\mathbf{F}_n = -\alpha_n\{\boldsymbol{\epsilon}_n \cdot [\mathbf{u}(\mathrm{O}) - \mathbf{u}(n)]\}\boldsymbol{\epsilon}_n \qquad \text{(E.13.1)}$$

and the total force due to interaction with a number of atoms is

$$\mathbf{F} = -\sum_n \alpha_n\{\boldsymbol{\epsilon}_n \cdot [\mathbf{u}(\mathrm{O}) - \mathbf{u}(n)]\}\boldsymbol{\epsilon}_n. \qquad \text{(E.13.2)}$$

From equation (E.3.2) for the potential energy of the lattice we see that the force on the (O) atom is

$$F_i = -\sum B_{ij}^{\mathrm{O}n} u_j(n), \qquad \text{(E.13.3)}$$

so that

$$B_{ij}^{\mathrm{OO}} = \sum_n \alpha_n \epsilon_i(n)\epsilon_j(n) \qquad \text{(E.13.4a)}$$

and

$$B_{ij}^{\mathrm{O}n} = -\alpha_n \epsilon_i(n)\epsilon_j(n) \quad (n \neq 0). \qquad \text{(E.13.4b)}$$

Now let us consider our two-dimensional square lattice (fig. E.13.2). The nearest neighbours are numbered 1 to 4, the next nearest 5 to 8; we have

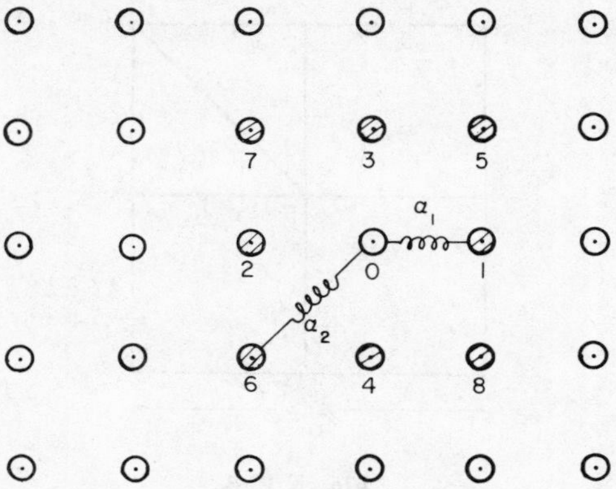

Fig. E.13.2.

two force constants, α_1 and α_2. The next step is to calculate a set of $\epsilon_i(n)$ for each atom and then the coupling constants can be determined. We find that

$$\left.\begin{aligned} &B^{OO} = 2(\alpha_1 + \alpha_2)\begin{pmatrix} 1 & 0 \\ 0 & 1 \end{pmatrix}, \\ &B^{O1} = B^{O2} = -\alpha_1\begin{pmatrix} 1 & 0 \\ 0 & 0 \end{pmatrix}, \qquad B^{O3} = B^{O4} = -\alpha_1\begin{pmatrix} 0 & 0 \\ 0 & 1 \end{pmatrix}, \\ &B^{O5} = B^{O6} = -\frac{\alpha_2}{2}\begin{pmatrix} 1 & 1 \\ 1 & 1 \end{pmatrix}, \qquad B^{O7} = B^{O8} = -\frac{\alpha_2}{2}\begin{pmatrix} -1 & 1 \\ 1 & -1 \end{pmatrix}. \end{aligned}\right\} \tag{E.13.5}$$

From this we can calculate the dynamical matrix (E.4.10) and the secular equation (E.4.12). This is:

$$\begin{vmatrix} \dfrac{2}{M}[\alpha_1(1-\cos q_1 a) + \alpha_2(1-\cos q_1 a \cos q_2 a)] - \omega^2 & \dfrac{2\alpha_2}{M}\sin q_1 a \sin q_2 a \\ \dfrac{2\alpha_2}{M}\sin q_1 a \sin q_2 a & \dfrac{2}{M}[\alpha_1(1-\cos q_2 a) + \alpha_2(1-\cos q_1 a \cos q_2 a)] - \omega^2 \end{vmatrix} = 0 \tag{E.13.6}$$

To discuss the results contained in (E.13.6) we need the Brillouin Zone for the square lattice. This is shown in fig. E.13.3, the points and line of high symmetry being labelled as in Kittel's *Quantum Theory of Solids*[7].

[7] C. Kittel, *Quantum Theory of Solids*, Wiley, New York, 1965.

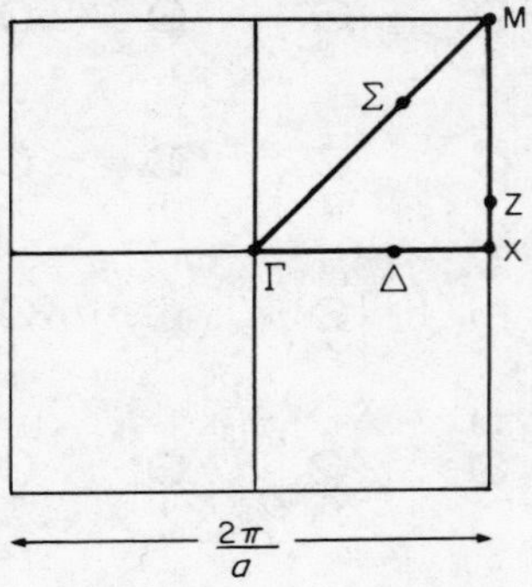

Fig. E.13.3.

If we solve the secular equation along the high symmetry directions Σ, Δ and Z we shall also find the frequencies at the symmetry points Γ, M and X. We will find these in turn.

1. *The Σ-line.* Here we have $q_1 = q_2 = q/\sqrt{2}$. The secular equation factorizes easily giving two branches, of which we will label the upper L and the lower T.

$$\omega^2_{\Sigma L} = \frac{2}{M}\left[\alpha_1\left(1 - \cos\frac{qa}{\sqrt{2}}\right) + \alpha_2(1 - \cos\sqrt{2}qa)\right], \quad \text{(E.13.7a)}$$

$$\omega^2_{\Sigma T} = \frac{2\alpha_1}{M}\left(1 - \cos\frac{qa}{\sqrt{2}}\right). \quad \text{(E.13.7b)}$$

The eigenvectors may be found using (E.4.13). These are

$$\mathbf{e}^{L}_{\Sigma} = \left(\frac{1}{\sqrt{2}}, \frac{1}{\sqrt{2}}\right) \quad \text{and} \quad \mathbf{e}^{T}_{\Sigma} = \left(-\frac{1}{\sqrt{2}}, \frac{1}{\sqrt{2}}\right). \quad \text{(E.13.8)}$$

The results confirm, as expected, that the higher frequency branch is longitudinal and the lower transverse. At both Γ and M the two branches are degenerate, being zero at Γ and $4\alpha_1/M$ at M. Although the eigenvectors along Σ are given by (E.13.8), at the two end points their directions become indeterminate, so that any pair of orthogonal unit vectors will suffice. The results in (E.13.7a, b) are shown in fig. E.13.4. It will be seen that $\omega^2_{\Sigma L}$ shows a maximum if $\alpha_1 < 4\alpha_2$ but not otherwise.

2. *The Δ-line.* Here we have $q_1 = q$ and $q_2 = 0$ and the eigenvalues (see fig. E.13.5) are

$$\omega^2_{\Delta L} = \frac{2}{M}(\alpha_1 + \alpha_2)(1 - \cos qa), \quad \text{(E.13.9a)}$$

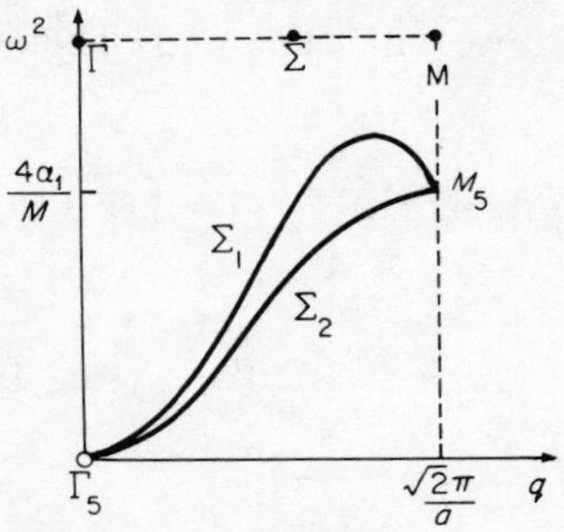

Fig. E.13.4.

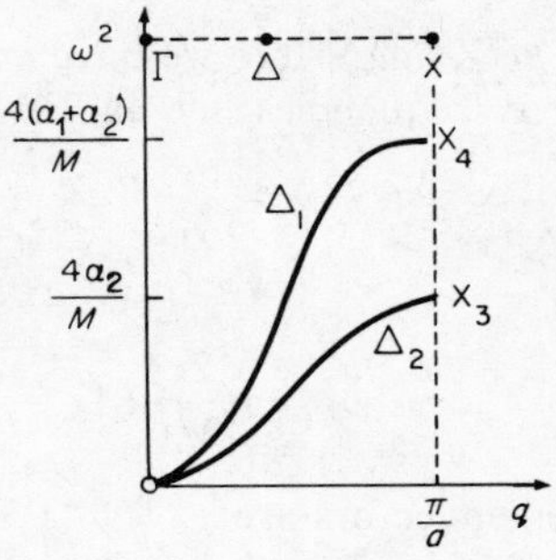

Fig. E.13.5.

$$\omega_{\Delta T}^2 = \frac{2\alpha_2}{M}(1 - \cos qa). \tag{E.13.9b}$$

The eigenvectors are

$$\mathbf{e}_\Delta^L = (1, 0) \quad \text{and} \quad \mathbf{e}_\Delta^T = (0, 1). \tag{E.13.10}$$

The frequencies are non-degenerate now at the point X with $\omega^2 = 4(\alpha_1 + \alpha_2)/M$ for the upper branch and $4\alpha_2/M$ for the lower. Furthermore the assignment to longitudinal and transverse is again clear and as expected.

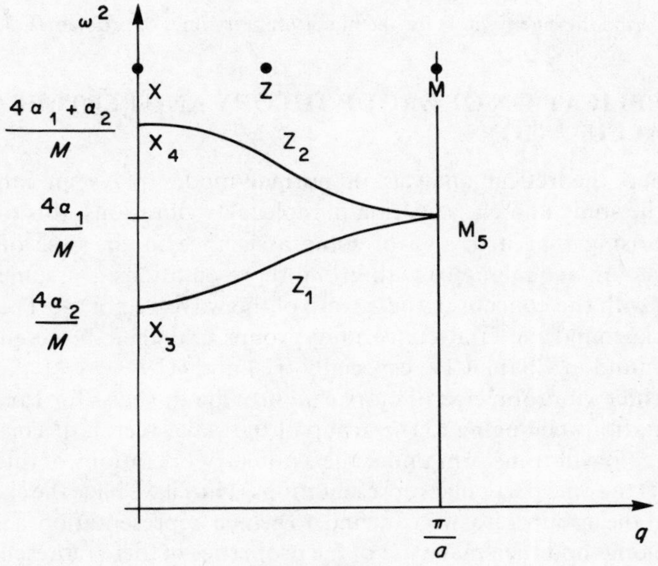

Fig. E.13.6.

3. *The Z line.* Here we have $q_1 = \pi/a$ and we will set $q_2 = q'$. Then the squared frequencies (see fig. E.13.6) are

$$\omega_{ZL}^2 = \frac{2}{M}[2\alpha_1 + \alpha_2(1 + \cos q'a)] \qquad \text{(E.13.11a)}$$

and

$$\omega_{ZT}^2 = \frac{2}{M}[\alpha_1(1 - \cos q'a) + \alpha_2(1 + \cos q'a)]. \qquad \text{(E.13.11b)}$$

The eigenvectors are

$$\mathbf{e}_Z^L = (1, 0) \quad \text{and} \quad \mathbf{e}_Z^T = (0, 1). \qquad \text{(E.13.12)}$$

These are neither parallel nor perpendicular to q, so the title longitudinal or transverse is purely formal in this case.

In this particular case there would be no special difficulty in finding the frequency at any point in the zone, but in more realistic examples, whilst it is relatively easy to find the values at high symmetry points, the general point requires a large quantity of computation. However by combining group theory and perturbation theory the form of ω^2 in the neighbourhood of high symmetry points can frequently be discovered without too much difficulty.

Problem

(E.13.1) Find the eigenvectors for the high symmetry lines of problem (E.5.1).

E.14 APPLICATION OF GROUP THEORY AND PERTURBATION CALCULATIONS

The group theoretical analysis of normal modes plays an important part in the study and classification of molecular vibrations. It is therefore not surprising that it is also of some assistance in the case of lattice vibrations. An acquaintance with group representations is assumed here, together with the concept of the 'group of the wave-vector, **q**'. The necessary background material concerning groups and their representations is to be found in Chapter IV, especially §C.3 and §C.9.

The lattice vibration eigenvectors $\mathbf{e}^s(q)$ provide the basis for irreducible representations belonging to the group of the wave-vector, **q**. This means that the $\mathbf{e}^s(q)$ will transform under the symmetry operations of this group according to some particular representations. Thus if we have the character table for the group of **q** we can find to which representation a normal mode belongs and then make use of the properties of these representations in subsequent work.

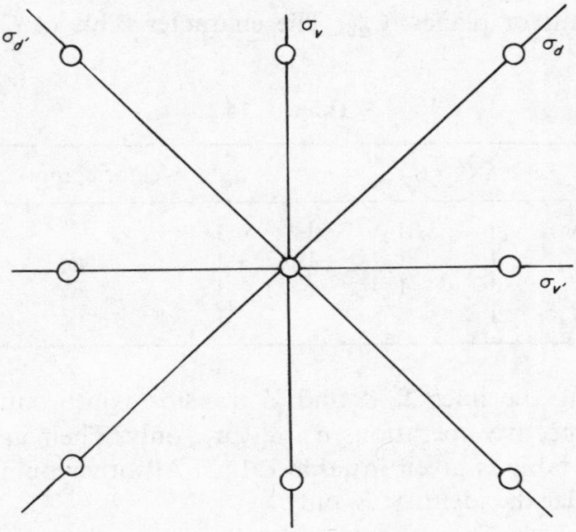

Fig. E.14.1. 4-fold axis vertical to plane of diagram.

The point group of the two-dimensional square lattice is C_{4v} (4 mm) consisting of a four-fold axis and two pairs of mirror planes (fig. E.14.1). The character table of C_{4v} is given in table E.14.1, using the notation of Kittel[7]. See §C.5 for the meaning and use of character tables, and also problems (C.5.5), (C.6.1) and (C.6.2).

Table E.14.1

C_{4v}	E	$2C_4$	C_2	$2\sigma_v$	$2\sigma_d$	Vector components
Γ_1, M_1	1	1	1	1	1	
Γ_2, M_2	1	1	1	−1	−1	R_z
Γ_3, M_3	1	−1	1	1	−1	
Γ_4, M_4	1	−1	1	−1	1	
Γ_5, M_5	2	0	−2	0	0	R_x, R_y

Also shown in the table are the representations corresponding to the transformation of vector components.

If we now refer to the Brillouin Zone in fig. E.13.3 we see immediately that the Γ point ($\mathbf{q} = 0$) has the complete symmetry of C_{4v}. Furthermore the point M is moved to equivalent points by all the operations of C_{4v} so it also has the complete point group symmetry. If, however, we consider the point X we will find that C_{4v} and $\sigma_{d'}$ move it to a non-equivalent point. Thus X has the symmetry of the group consisting of a two-fold axis and one pair of mirror planes, C_{2v}. The character table of C_{2v} is shown in table E.14.2.

Table E.14.2

C_{2v}	E	C_2	σ_v	$\sigma_{v'}$	Vector components
X_1	1	1	1	1	
X_2	1	1	-1	-1	R_z
X_3	1	-1	1	-1	R_y
X_4	1	-1	-1	1	R_x

Points along the lines Σ, Δ and Z transform into equivalent points under the respective operations σ_d, $\sigma_{v'}$, σ_v, only. Their group is C_s and the character table is given in table E.14.3. All other points in the zone transform under the identity, *E*, only.

Table E.14.3

C_s	E	$\sigma_{v'}, \sigma_d, \sigma_v$	Vector components
Δ_1, Σ_1, Z_1	1	1	Component of R in mirror plane
Δ_2, Σ_2, Z_2	1	-1	Component of R normal to mirror plane

We can now assign the normal modes to their appropriate representations, remembering that as the eigenvectors are also vectors in direct space only these representations which transform vectors into themselves are going to take part.

(1). At points Γ and M, vectors in the plane transform according to Γ_5, M_5. Both of these are 2-fold degenerate, so without any of the calculations in §E.13 we would know that the frequencies at Γ and M must also be two-fold degenerate.

(2). The eigenvectors at X are given by (E.13.10) and illustrated in fig. E.14.2a. Because $\mathbf{e}^{\mathrm{T}}$ is odd under C_2, even under σ_v and odd under $\sigma_{v'}$, $\mathbf{e}^T$ belongs to X_3; similarly $\mathbf{e}^2$ belongs to X_4. This merely confirms the vector assignments in table E.14.2.

(3). The eigenvectors at Δ, Σ and Z are given in (E.13.8), (E.13.10) and (E.13.12) and illustrated in fig. E.14.2b, c, d.

(a) Δ; $\mathbf{e}^T$ odd with respect to $\sigma_{v'}$, $\mathbf{e}^{\mathrm{L}}$ even, so

$$\mathrm{T} \sim \Delta_2, \qquad \mathrm{L} \sim \Delta_1 .$$

Similarly

(b) Σ; $\mathrm{T} \sim \Sigma_{12}$, $\quad \mathrm{L} \sim \Sigma_1$

(c) Z; $\mathrm{T} \sim Z_1$, $\quad \mathrm{L} \sim Z_2$.

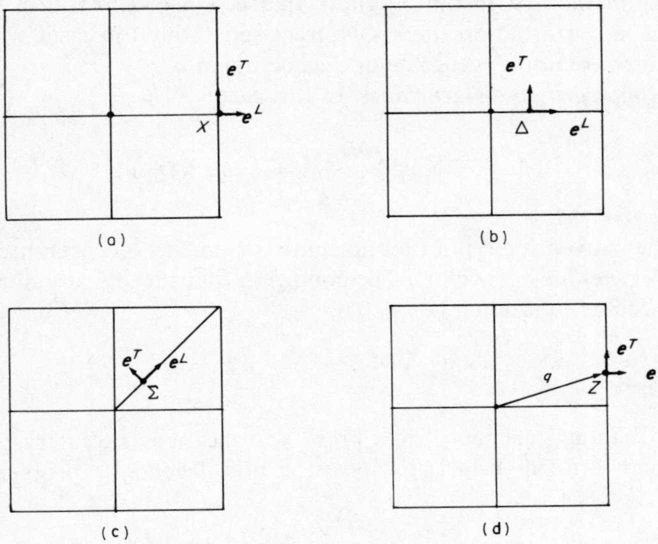

Fig. E.14.2.

Some of these assignments could have been obtained from those at Γ, M and X by the use of compatibility tables, as discussed in §C.10.

Let us now suppose that we are able to determine the eigenvalues and eigenvectors for all the branches at a particular point in the zone, $\mathbf{q}_0$. We can now write down $D_{ij}(\mathbf{q})$ as a Taylor expansion about the point $\mathbf{q}_0$:

$$D_{ij}(\mathbf{q}) = D_{ij}(\mathbf{q}_0) + q_l' \left.\frac{\partial D_{ij}}{\partial q_l'}\right|_{\mathbf{q}=\mathbf{q}_0} + \frac{1}{2} q_l' q_m' \left.\frac{\partial^2 D_{ij}}{\partial q_l' \, \partial q_m'}\right|_{\mathbf{q}=\mathbf{q}_0} + \cdots, \tag{E.14.1}$$

where $\mathbf{q}' = \mathbf{q} - \mathbf{q}_0$. Furthermore, because all the eivenvectors are ordinary vectors in direct space, the eigenvectors $\mathbf{e}^s(\mathbf{q}_0)$ form a complete set in terms of which the eigenvectors $\mathbf{e}^s(\mathbf{q})$ can be expressed thus:

$$\mathbf{e}^s(\mathbf{q}) = \sum_{s'} A_{qq_0}^{ss'} \mathbf{e}^{s'}(\mathbf{q}_0). \tag{E.14.2}$$

The $A_{qq_0}^{ss'}$ are coefficients which may be determined subsequently. Finally we will write

$$\omega_s^2(\mathbf{q}) = \Delta\omega^2 + \omega_s^2(\mathbf{q}_0). \tag{E.14.3}$$

We have now laid the foundations for a perturbation treatment of $\Delta\omega^2$.

Let us, in the first instance, ignore the second order term in (E.14.1). As usual in perturbation theory we treat separately the cases where the eigenvalues are non-degenerate or degenerate at $\mathbf{q}_0$.

(1). *Non-degenerate eigenvalues.* In this case:

$$\Delta\omega^2 = \sum_{\substack{i,j \\ l}} e_i^s q_l' \frac{\partial D_{ij}}{\partial q_l'} e_j^s = \langle s|\mathbf{q}' \cdot \nabla D|s\rangle, \tag{E.14.4}$$

where the conventional notation for matrix elements has been used.

(2). *Degenerate eigenvalues.* The condition that the $A_{qq_0}^{ss'}$ are non-trivial gives a secular equation

$$|\langle s'|\mathbf{q}' \cdot \nabla D|s\rangle - \Delta\omega^2\, \delta^{ss'}| = 0 \tag{E.14.5}$$

for $\Delta\omega^2$.

If the diagonal element $\langle s|\mathbf{q}' \cdot \nabla D|s\rangle$ vanishes it is necessary to go to the second term in (E.14.1) to solve the non-degenerate problem. This gives

$$\Delta\omega^2 = \langle s|\mathbf{q}\mathbf{q}' : \nabla\nabla D|s\rangle + \sum_{\substack{s' \\ \text{except} \\ s'=s}} \frac{\langle s|\mathbf{q}' \cdot \nabla D|s'\rangle\langle s'|\mathbf{q}' \cdot \nabla D|s\rangle}{\omega_s^2(q_0) - \omega_{s'}^2(q_0)}. \tag{E.14.6}$$

The second term in (E.14.6) is in second order perturbation theory. If all the first order matrix elements vanish the secular equation for the degenerate case becomes

$$|\langle s'|\mathbf{q}\mathbf{q}' : \nabla\nabla D|s\rangle - \nabla\omega^2\, \delta^{ss'}| = 0. \tag{E.14.7}$$

This completes the necessary equipment for making a perturbation theoretical investigation of the behaviour of $\omega_s^2(\mathbf{q})$ in the neighbourhood of $\mathbf{q}_0$.

Group theory can be used to find the points where $\langle s'|\mathbf{q}' \cdot \nabla D|s\rangle$ vanishes. The condition for the vanishing of such a matrix element is that if the eigenvector $\mathbf{e}^s(\mathbf{q}_0)$ transforms according to the irreducible representation Γ_s, $\mathbf{e}^{s'}(\mathbf{q}_0')$ according to $\Gamma_{s'}$ and the perturbation according to Γ_t, then the reduction[8] of the product representation $\Gamma_{s'} \times \Gamma_t \times \Gamma_s$ must not contain the identity representation Γ_1. This is equivalent (for real representations) to saying that the reduction of $\Gamma_t \times \Gamma_s$ must contain $\Gamma_{s'}$. In our case the perturbation transforms like a vector.

It will be seen that at such points $\nabla_q\omega^2$ will vanish so we are dealing with a particular class of critical points, the 'ordinary critical points'. The determinations of the density of states function will present special

[8] V. Heine, *Group Theory in Quantum Mechanics*, Pergamon, 1960, Chapter III.

difficulties near these points. The set of critical points which can be found by the above group theoretical method is called the symmetry set, S, but there may be other ordinary critical points which cannot be found in this way.

Let us now enumerate S for the two-dimensional square lattice, considering all the points having greater symmetry than the general point.

(1). At Γ and M the relevant representations for the normal modes were Γ_5 and M_5. Similarly the vector representation is Γ_5 and M_5. Now using standard methods:

$$\Gamma_5 \times \Gamma_5 = \Gamma_1 + \Gamma_2 + \Gamma_3 + \Gamma_4 \tag{E.14.8a}$$

which does not include Γ_5. Hence there are critical points at Γ and M.

(2). At X, Γ_s and $\Gamma_{s'}$ may be X_3 or X_4 and the vector representations are the same. We therefore need:

$$X_3 \times X_3 = X_1 \tag{E.14.8b}$$

$$X_3 \times X_4 = X_4 \times X_3 = X_2 \tag{E.14.8c}$$

$$X_4 \times X_4 = X_1, \tag{E.14.8d}$$

showing that X is a critical point.

(3). Let us consider the Δ-line. Then we need:

$$\Delta_2 \times \Delta_1 = \Delta_2 \tag{E.14.8e}$$

$$\Delta_2 \times \Delta_2 = \Delta_1 \tag{E.14.8f}$$

$$\Delta_1 \times \Delta_1 = \Delta_1, \tag{E.14.8g}$$

so there is no symmetry critical point along the Δ-line. The same applies to Σ and Z. The results (E.14.8e) and (E.14.8f) do show, however, that the component of $\nabla_q \omega^2$ normal to the symmetry line vanishes because Δ_2 is the representation belonging to vectors normal to the line.

Thus, in the two-dimensional square lattice,

$$S = \{\Gamma, \mathrm{M}, \mathrm{X}\} \tag{E.14.9}$$

is the set of symmetry critical points.

We will now apply the perturbation equations (E.14.6) and (E.14.7) to find the frequencies in the neighbourhood of Γ, M and X for the two dimensional square lattice. Since the modes are non-degenerate at X we will consider that point first.

(1). X *point* L *branch*. Since $\mathbf{e}_X^L = (1, 0)$ the only element of D with which we are concerned is D_{11}. Direct calculation shows that at X:

$$\frac{\partial^2 D_{11}}{\partial q_1^2} = -\frac{2a^2}{M}(\alpha_1 + \alpha_2); \quad \frac{\partial^2 D_{11}}{\partial q_1\, \partial q_2} = 0; \quad \frac{\partial^2 D_{11}}{\partial q_2^2} = -\frac{2a^2\alpha_2}{M},$$

so

$$\omega_L^2(q) = \frac{4(\alpha_1 + \alpha_2)}{M} - \frac{a^2}{M}[(\alpha_1 + \alpha_2)q_1'^2 + \alpha_2 q_2'^2], \qquad \text{(E.14.10)}$$

where

$$q_1' = q_1 - \frac{\pi}{a}, \quad q_2' = q_2.$$

Thus we have an analytic maximum in the L branch at X. A frequency contour is shown in fig. E.14.3a.

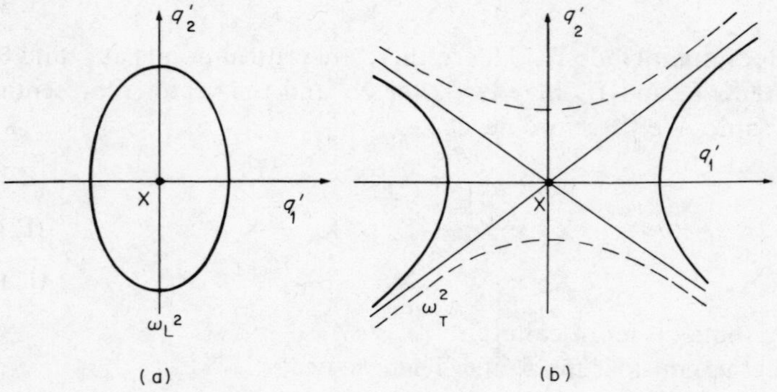

Fig. E.14.3.

(2). X *point*, T *branch*. Since $\mathbf{e}_L^T = (0, 1)$ we are only concerned with D_{22}. Making direct calculations of the necessary second derivatives leads to the result

$$\omega_T^2(q) = \frac{4\alpha_2}{M} - \frac{a^2}{M}[\alpha_2 q_1'^2 - (\alpha_1 - \alpha_2)q_2'^2]. \qquad \text{(E.14.11)}$$

Now for a physically reasonable model, $\alpha_1 > \alpha_2$ because α_1 refers to nearest and α_2 to next-nearest neighbours. In this event there is a saddle point in the T branch at X, as shown in fig. E.14.3b.

(3). M *point*. Here we must apply degenerate perturbation theory. All the components $\partial^2 D_{ij}/\partial q_l \, \partial q_m$ must be evaluated, leading to the secular equation

$$\begin{vmatrix} \frac{a^2}{M}[(\alpha_2 - \alpha_1)q_1'^2 + \alpha_2 q_2'^2] - \Delta\omega^2 & \frac{2a^2\alpha_2}{M} q_1' q_2' \\ \frac{2a^2\alpha_2 q_1' q_2'}{M} & \frac{a^2}{M}[\alpha_2 q_1'^2 + (\alpha_2 - \alpha_1)q_2'^2] - \Delta\omega^2 \end{vmatrix} = 0 \qquad \text{(E.14.12)}$$

whose roots are

$$\omega^2 = \omega_M^2 = \frac{q'^2 a^2}{2M}\{(2\alpha_2 - \alpha_1) \pm [\alpha_1^2 - (\alpha_1^2 - 4\alpha_2^2)\sin^2 2\theta]^{\frac{1}{2}}\}, \tag{E.14.13}$$

where $q'_1 = q' \cos\theta$, $q'_2 = q' \sin\theta$. This result is illustrated in fig. E.14.4 for the case where $\alpha_1 > 4\alpha_2$. The surface is 'fluted' for both ω_T^2 and ω_L^2 giving a maximum for the former and a kind of generalized saddle point for the latter.

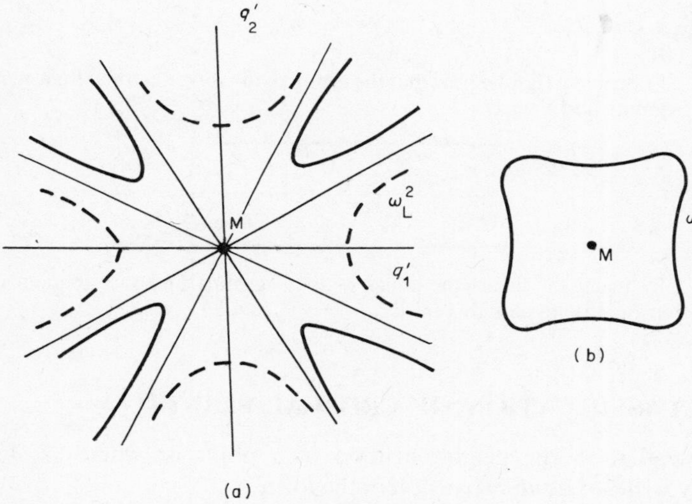

Fig. E.14.4.

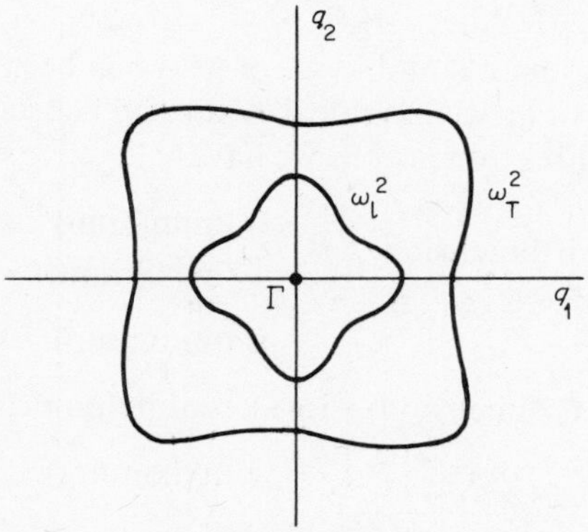

Fig. E.14.5.

(4). Γ *point.* The squared frequencies in the neighbourhood of Γ are given by

$$\omega^2 - \frac{q^2a^2}{M}\{(\alpha_1 + 2\alpha_2) \pm [\alpha_1^2 - (\alpha_1^2 - 4\alpha_2^2)\sin^2 2\theta]^{\frac{1}{2}}\}. \quad \text{(E.14.14)}$$

This gives two fluted minima as shown in fig. E.14.5.

In this lengthy section we have shown how group theory and perturbation theory can help us to find ω^2 in the neighbourhood of some of the critical points.

Problem

(E.14.1) In problem (E.5.1) show that the group of the wave vector at high symmetry points and lines is:

Γ	K	M	T, T', Σ
C_{6v}	C_{3v}	C_{2v}	C_s

By using the character tables of these groups find the symmetry set of critical points for this lattice.

E.15 CLASSIFICATION OF CRITICAL POINTS[9]

Suppose that in the neighbourhood of a point $\mathbf{q}_0$ where $|\nabla_q\omega^2| = 0$, we may write ω^2 as an expansion of the form

$$\omega^2 = \omega_0^2 - \sum_{i=1}^{j} b_i q_1'^2 + \sum_{i=j+1}^{3} b_i q_i'^2 \qquad (b_i > 0), \qquad \text{(E.15.1)}$$

where $\mathbf{q}' = \mathbf{q} - \mathbf{q}_0$ and a suitable set of axes has been selected. Then the point $\mathbf{q}_0$ is called an analytic critical point (c.p.) specified by its index, j. For one, two and three dimensions we have

$$\text{1 dimension} \left\{ j \begin{array}{l} = 0 \text{ (minimum)} \\ = 1 \text{ (maximum)} \end{array} \right.$$

$$\text{2 dimensions} \left\{ j \begin{array}{l} = 0 \text{ (minimum)} \\ = 1 \text{ (saddle point)} \\ = 2 \text{ (maximum)} \end{array} \right.$$

[9] J. C. Phillips, *Phys. Rev.*, **104**, 1263 (1956).

$$3 \text{ dimensions} \begin{cases} & = 0 \text{ (minimum)} \\ j & = 1 \text{ (saddle point)} \\ & = 2 \text{ (saddle point)} \\ & = 3 \text{ (maximum).} \end{cases}$$

This type of expansion (E.15.1) around ω_0^2 will not be possible if ω^2 is degenerate at the c.p. Here we shall find the 'fluted' c.p. giving an expansion in the neighbourhood of $\mathbf{q}_0$ of the form

$$\omega^2 = \omega_0^2 + |\mathbf{q} - \mathbf{q}_0|^2 \chi\left(\frac{\mathbf{q} - \mathbf{q}_0}{|\mathbf{q} - \mathbf{q}_0|}\right). \tag{E.15.2}$$

Such were the expressions obtained for the two dimensional square lattice near Γ and M.

To assist in classifying these, Phillips introduced the concept of positive and negative sectors. A sector is an angle or solid angle with the c.p. at its apex where $\omega^2 - \omega_0^2$ has the same sign; > 0 for positive < 0 for negative sectors. We can also supply this idea to analytic c.p.'s. The sector numbers are written as (P, N) so for analytic c.p.'s in two dimensions we have for maxima (0, 1) for minima (1, 0) and for saddle points (2, 2). For fluted maxima and minima the sector numbers are the same as for analytic, but fluted c.p.'s of the kind illustrated in fig. E.14.3a have sector numbers (4, 4).

The fluted and analytic c.p.'s together make up the class of ordinary c.p.'s defined in §E.4. There are also singular c.p.'s. These arise where branches cross either at a point (2 dimensions) or a line (3 dimensions). If we use a labelling convention such that if the ith branch at $\mathbf{q}$ is the one of frequency $\omega_i^2(\mathbf{q})$, and $i < j$ implies $\omega_i^2 \leqslant \omega_j^2$, then the singular c.p. is one at which one or more components of $\nabla_q \omega^2$ change discontinuously and the remainder vanish. In two dimensions these can only be generalized minima and maxima.

An important question is whether there is any way of finding all the c.p.'s. We have already shown in §E.4 how group theory can be used to find the symmetry set, S, of c.p.'s. Another method of finding c.p.'s is by making use of a topological argument. The principle of this is very simple and can be seen most easily in a two-dimensional illustration (fig. E.15.1). Since ω^2 is periodic in reciprocal space it must have at least one maximum and one minimum in each zone. Referring to fig. E.15.1, if A and B are maxima in adjoining zones, then any curve joining A and B must have at least one minimum in it. These minima define a curve joining two minima C and D; this curve must have at least one maximum denoted by E.

It will be seen that E is a saddle point, and by similar arguments applied to curves joining A and F the existence of a second saddle point may be deduced. Thus in a two dimensional lattice we must have at least one maximum, one minimum and two saddle points.

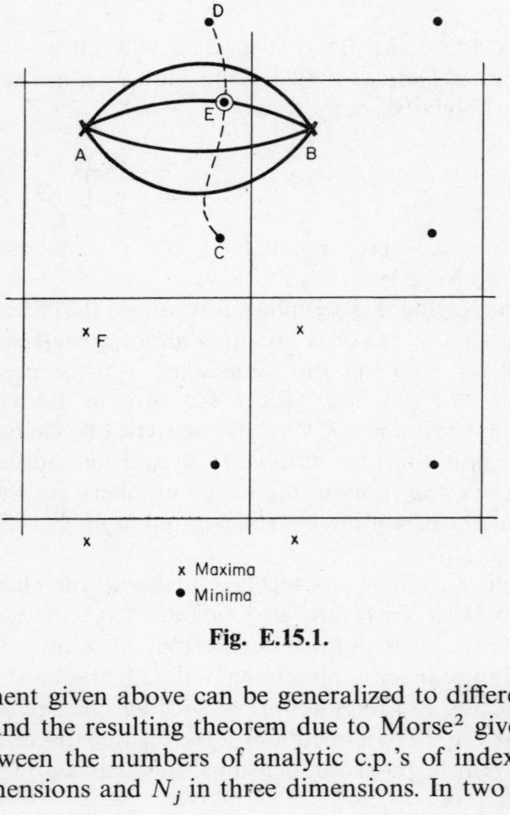

Fig. E.15.1.

The argument given above can be generalized to different numbers of dimensions and the resulting theorem due to Morse[2] gives a number of relations between the numbers of analytic c.p.'s of index j, denoted by n_j in two dimensions and N_j in three dimensions. In two dimensions we have

$$n_0 \geqslant 1; \qquad n_1 - n_0 \geqslant 1; \qquad n_2 - n_1 + n_0 = 0, \tag{E.15.3}$$

and in three dimensions

$$N_0 \geqslant 1; \qquad N_i - N_0 \geqslant 2; \qquad N_2 - N_1 + N_0 \geqslant 1;$$
$$N_3 - N_2 + N_1 - N_0 = 0. \tag{E.15.4}$$

These results can be extended to cover fluted and singular c.p.'s. Fluted maxima and minima count the same as analytic but otherwise a special weighting factor, q, must be applied, depending on the sector numbers

(P, N). This is shown in table E.15.1. It will be seen that the (4, 4) fluted point in two dimensions count as three saddle points in the Morse relations, (E.15.3). In labelling c.p.'s this is an f_1 point, the others are p_j $(j = 0, 1, 2)$.

Table E.15.1

(P, N)	q	j
(0, 1)	1	l (= number of dimensions)
(1, 0)	1	0
(n, n)	$n - 1$	1 in two dimensions
If $P > 1$ only	$P - 1$	2 (in three dimensions)
If $N > 1$ only	$N - 1$	1 (in three dimensions)
Otherwise both	$N - 1$	2 (in three dimensions)
	$P - 1$	2 (in three dimensions)

The set of c.p.'s deduced by the combination of group theoretical and topological arguments is called the minimal set, $\dot{M}$. However the complete set may be larger than M, and there appears to be no general way of finding the complete set except by direct calculation of the lattice frequencies.

Let us now apply the Morse relations and table E.15.1 to our example of the two-dimensional square lattice.

(1). *Transverse branch* (T). Here we have

$$\begin{array}{lll} \Gamma & p_0 & (1,0) \\ \mathrm{M} & p_2 & (0,1) \\ \mathrm{X} & p_1 & (2,2) \end{array}$$

Since there are two X points in the zone

$$n_0 = 1, \qquad n_1 = 2, \qquad n_2 = 1,$$

which satisfies equation (E.15.3).

(2). *Longitudinal branch* (L). The situation for this branch differs according as to whether α_1 is less or greater than $4\alpha_2$ [see (E.13.7a)].

(a) $\alpha_1 > 4\alpha_2$:

$$\begin{array}{lll} \Gamma & p_0 & (1,0) \\ \mathrm{M} & f_1 & (4,4) \\ \mathrm{X} & p_2 & (0,1). \end{array}$$

Referring to table E.15.1, we see that f_1 has a weight of three so we have:

$$n_0 = 1, \qquad n_1 = 3, \qquad n_2 = 2,$$

which again satisfies (E.15.3).

(b) $\alpha_1 < 4\alpha_2$:

$$\begin{array}{lll} \Gamma & p_0 & (1,0) \\ \mathrm{M} & p_0 & (1,0) \\ \mathrm{X} & p_2 & (0,1). \end{array}$$

This gives

$$n_0 = 2, \qquad n_1 = 0, \qquad n_2 = 2,$$

which does not satisfy (E.15.3). In fact we need to make up n_1 to 4; this is the saddle point along the Σ-line shown in fig. E.13.4.

$$\Sigma\, p_1 \quad (2, 2).$$

It is incidentally a typical result that an increase in the relative strength of long range forces leads to the appearance of extra critical points.

Problem

(E.15.1) Locate and classify the c.p.'s for problem E.5.1, checking that they satisfy the Morse relations. Note that one c.p. is singular.

E.16 CONTRIBUTION OF CRITICAL POINTS TO DENSITY OF STATES

We suppose now that the critical points have been identified and that the form of ω^2 near the c.p. has been found. A return can now be made to the problem, raised in §E.2, of determining the density of states function, $G(\omega^2)$. Let us assume we have analytic c.p. in two dimensions. By suitable choice of axes and scale we can write:

$$\omega^2(q) = \omega_0^2 + a(\epsilon_1 q_1^2 + \epsilon_2 q_2^2) \tag{E.16.1}$$

for the frequencies near a c.p. at ω_0^2. Here ϵ_1, ϵ_2 may be ± 1. In two dimensions we have, from (E.12.13) and (E.12.14),

$$G(\omega^2) = \frac{1}{(2\pi)^{\frac{1}{2}}} \int_{-\infty}^{\infty} e^{-i\omega^2 y} f(y)\, dy \tag{E.16.2}$$

and

$$f(y) = \frac{1}{(2\pi)^{\frac{5}{2}}} \int_{\mathrm{B.Z.}} e^{i\omega^2(q)y}\, d^2 q. \tag{E.16.3}$$

Then, using (E.16.1),

$$f(y) = \frac{e^{i\omega_0^2 y}}{(2\pi)^{\frac{5}{2}}} \int_{-\infty}^{\infty} e^{ia\epsilon_1 q_1^2 y} \, dq_1 \int_{-\infty}^{\infty} e^{ia\epsilon_2 q_2^2 y} \, dq_2, \tag{E.16.4}$$

where we have extended the limits of integration to from $-\infty$ to $+\infty$. To integrate (E.16.4) consider first the case where both ϵ_1 and ϵ_2 are positive. Make a change to polar coordinates q, θ where $q_1 = q \cos\theta$, $q_2 = q \sin\theta$. Then we have

$$\int_0^{2\pi} d\theta \int_0^{\infty} e^{iaq^2 y} q \, dq = \frac{i\pi}{ay}.$$

For both ϵ_1 and ϵ_2 negative we have $-i\pi/ay$ and thus for one of the individual integrals $(i\epsilon\pi/ay)^{\frac{1}{2}}$. Then

$$f(y) = \frac{e^{i\omega_0^2 y}}{(2\pi)^{\frac{5}{2}}} \frac{i\pi}{ay} (\epsilon_1)^{\frac{1}{2}} (\epsilon_2)^{\frac{1}{2}}, \tag{E.16.5}$$

care being necessary to separate the square roots of ϵ_1 and ϵ_2. Thus for the different c.p.'s we have for $f(y)$:

$$j = 0 \qquad \text{(minimum)} \qquad \frac{\pi i \, e^{-i\omega_0^2 y}}{(2\pi)^{\frac{5}{2}} ay} \tag{E.16.6a}$$

$$j = 1 \qquad \text{(saddle point)} \qquad \frac{\pi \, e^{-i\omega_0^2 y}}{(2\pi)^{\frac{5}{2}} a|y|} \tag{E.16.6b}$$

$$j = 2 \qquad \text{(maximum)} \qquad \frac{\pi i \, e^{-i\omega_0^2 y}}{(2\pi)^{\frac{5}{2}} ay}. \tag{E.16.6c}$$

Our next step requires the Fourier transforms[10] of y^{-1} and $|y|^{-1}$. To obtain this we use two theorems. The first is the Fourier inversion theorem that if $G(y)$ is the Fourier transform (F.T.) of $g(x)$ then $g(x)$ is the F.T. of $G(-y)$. The second is that if $G(y)$ is the F.T. of $g(x)$ then $iyG(y)$ is the F.T. of dg/dx.

The F.T. of y^{-1} is $-i(2\pi)^{\frac{1}{2}} \operatorname{sgn} x$ where $\operatorname{sgn} x = 1$ when $x > 0$ and $= -1$ when $x < 0$. To prove this we use the inversion theorem:

$$G = \int_{-\infty}^{\infty} \operatorname{sgn} y \, e^{-ixy} \, dy.$$

Now

$$\frac{d}{dy}(\operatorname{sgn} y) = 2\,\delta(y),$$

[10] For a more rigorous treatment than that offered here see M. J. Lighthill, *Fourier Analysis and Generalized Functions*, Cambridge University Press, 1958.

but the F.T. of $2\delta(y)$ is $(2/\pi)^{\frac{1}{2}}$, so because of the second theorem the F.T. of d(sgn y)/dy is:

$$\frac{i\alpha}{(2\pi)^{\frac{1}{2}}} G(x),$$

which proves what is required.

The F.T. of $|y|^{-1}$ is $-(2/\pi)^{\frac{1}{2}} \log_e |x|$; to prove this put

$$G(x) = \int_{-\infty}^{\infty} \log_e |y| \, e^{-ixy} \, dy.$$

Now

$$\frac{d}{dx}(\log_e |y|) = \frac{1}{y},$$

so

$$ixG(x) = \int \frac{e^{-ixy}}{y} \, dy = -\pi i \operatorname{sgn} x,$$

i.e.

$$G(x) = -\frac{\pi \operatorname{sgn} x}{x} = -\frac{\pi}{|x|},$$

so the F.T. of $\log_e |x|$ is $-(2/\pi)^{\frac{1}{2}}|x|^{-1}$, which by the inversion theorem proves what is required.

Then we have the following contributions to $G(\omega^2)$:

$$\text{Minimum} \qquad \operatorname{sgn}(\omega^2 - \omega_0^2)/8\pi a \tag{E.16.7a}$$

$$\text{Saddle point} \quad -\log_e |\omega^2 - \omega_0^2|/4\pi^2 a \tag{E.16.7b}$$

$$\text{Maximum} \qquad -\operatorname{sgn}(\omega^2 - \omega_0^2)/8\pi a. \tag{E.16.7c}$$

Thus the maxima and minima introduce positive and negative steps of $(4\pi a)^{-1}$ whilst saddle points give logarithmic infinities.

Similar calculations may be carried out for one and three dimensions. In one dimension we find that $G(\omega^2)$ is:

$$\text{Minimum} \qquad [1 - \operatorname{sgn}(\omega^2 - \omega_0^2)]/|\omega^2 - \omega_0^2|^{\frac{1}{2}} 4\pi^2 a \tag{E.16.8a}$$

$$\text{Maximum} \qquad [1 - \operatorname{sgn}(\omega^2 - \omega_0^2)]/|\omega^2 - \omega_0^2|^{\frac{1}{2}} 4\pi^2 a. \tag{E.16.8b}$$

and in three dimensions:

$$\text{Minimum} \qquad |\omega^2 - \omega_0^2|^{\frac{1}{2}}[1 + \operatorname{sgn}(\omega^2 - \omega_0^2)]/6\pi a \tag{E.16.9a}$$

Saddle point ($j = 1$) $\quad |\omega^2 - \omega_0^2|^{\frac{1}{2}}[-1 + \mathrm{sgn}(\omega^2 - \omega_0^2)]/6\pi a \qquad$ (E.16.9b)

Saddle point ($j = 2$) $\quad -|\omega^2 - \omega_0^2|^{\frac{1}{2}}[1 + \mathrm{sgn}(\omega^2 - \omega_0^2)]/6\pi a \qquad$ (E.16.9c)

Maximum $\quad |\omega^2 - \omega_0^2|^{\frac{1}{2}}[1 - \mathrm{sgn}(\omega^2 - \omega_0^2)]/6\pi a. \qquad$ (E.16.9d)

Notice that in three dimensions there are no discontinuities or infinities in $G(\omega^2)$. The form of $G(\omega^2)$ is shown in fig. E.16.1.

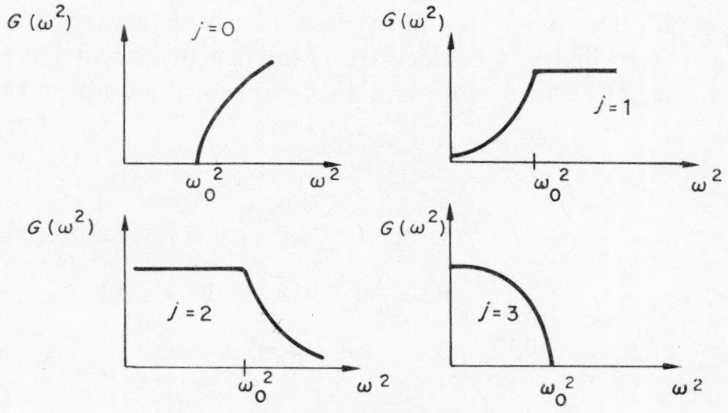

Fig. E.16.1.

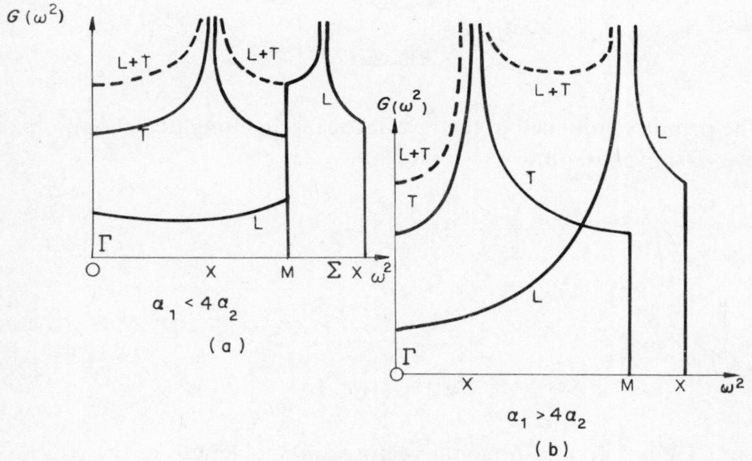

Fig. E.16.2.

We are now in a position to sketch $G(\omega^2)$ for the two-dimensional square lattice which we have studied earlier. This will take different forms for $\alpha_1 < 4\alpha_2$ and $\alpha_1 > 4\alpha_2$, and is shown in fig. E.16.2a, b. It is of interest to notice that in the former case the critical points at M cancel out and in the latter case the critical point at M for the transverse modes is obscured.

E.17 LATTICE VIBRATIONS IN THE DIAMOND STRUCTURE

Materials with the diamond structure (e.g. germanium) have been the subject of many investigations, experimental and theoretical, designed to discover the lattice vibrational spectrum. In this section we will attempt to calculate the lattice frequencies using a very simple model[6].

The diamond structure consists of two interpenetrating face-centred cubic lattices displaced with respect to each other by one quarter of the body diagonal. The most significant feature of this arrangement is that each atom is tetrahedrally bonded to its nearest neighbours as shown in fig. E.17.1. (See §C.7 for a complete discussion of the diamond lattice.)

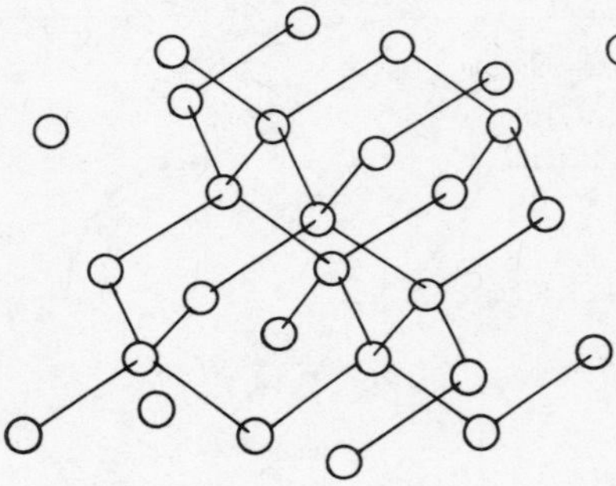

Fig. E.17.1.

The primitive unit cell of the f.c.c. lattice is an elongated rhombohedron whose basic vectors are

$$\mathbf{a}_1 = \frac{a}{2}(\mathbf{j} + \mathbf{k})$$

$$\mathbf{a}_2 = \frac{a}{2}(\mathbf{i} + \mathbf{k}) \qquad \text{(E.17.1)}$$

$$\mathbf{a}_3 = \frac{a}{2}(\mathbf{i} + \mathbf{j})$$

where **i**, **j**, **k** are unit orthogonal vectors and a is length of the edge of the cubic cell. The diamond structure has the basis

$$\mathbf{R}^0 = 0, \qquad \mathbf{R}^1 = \frac{3a}{4}(\mathbf{i} + \mathbf{j} + \mathbf{k}). \qquad \text{(E.17.2)}$$

The Brillouin zone is shown in fig. E.17.2 with the high symmetry points and lines labelled in the conventional way.

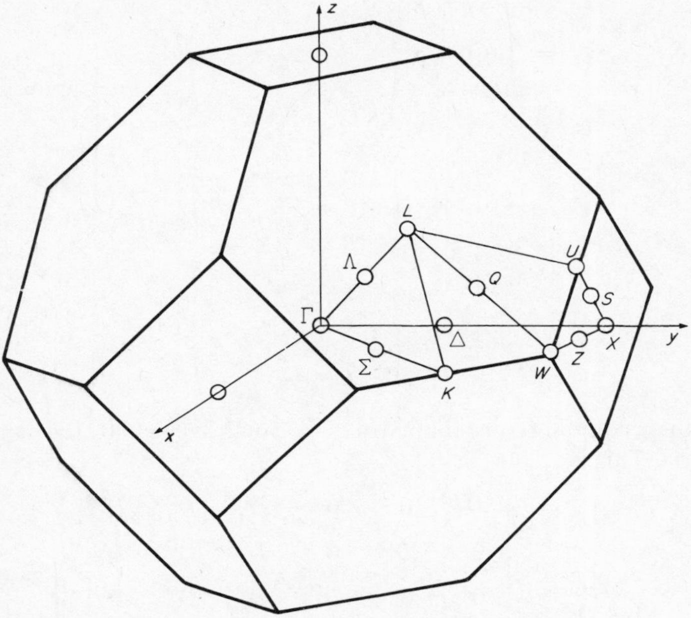

Fig. E.17.2.

Since there is no usable and valid theory of the cohesive forces in covalent materials such as germanium, it is preferable to consider a general set of coupling constants $\overset{mn}{\underset{ij}{B_{\mu\nu}}}$, limited of course by symmetry and invariance conditions. The number of independent force constants will then depend on the range assigned to the interatomic forces. If we limit ourselves to nearest neighbour forces only then we have just two parameters, α and β. There will be five distinct coupling constant matrices:

$$B_{00}^{00} = \begin{pmatrix} 4\alpha & 0 & 0 \\ 0 & 4\alpha & 0 \\ 0 & 0 & 4\alpha \end{pmatrix} \tag{E.17.3a}$$

$$B_{01}^{0\bar{1}\bar{1}\bar{1}} = \begin{pmatrix} -\alpha & -\beta & -\beta \\ -\beta & -\alpha & -\beta \\ -\beta & -\beta & -\alpha \end{pmatrix} \tag{E.17.3b}$$

$$B_{O1}^{O\bar{1}O\bar{1}} = \begin{pmatrix} -\alpha & \beta & -\beta \\ \beta & -\alpha & \beta \\ -\beta & \beta & -\alpha \end{pmatrix} \tag{E.17.3c}$$

$$B_{O1}^{OO\bar{1}\bar{1}} = \begin{pmatrix} -\alpha & \beta & \beta \\ \beta & -\alpha & -\beta \\ \beta & -\beta & -\alpha \end{pmatrix} \tag{E.17.3d}$$

$$B_{O1}^{O\bar{1}\bar{1}O} = \begin{pmatrix} -\alpha & -\beta & \beta \\ -\beta & -\alpha & \beta \\ \beta & \beta & -\alpha \end{pmatrix} \tag{E.17.3e}$$

Using these coupling constant matrices we can then calculate the dynamical matrix. This is:

$$D = D_{ij}^{\mu\nu} = \left[\begin{array}{cccccc} D_{11}^{OO} & 0 & 0 & D_{11}^{O1} & D_{12}^{O1} & D_{13}^{O1} \\ 0 & D_{11}^{OO} & 0 & D_{12}^{O1} & D_{11}^{O1} & D_{23}^{O1} \\ 0 & 0 & D_{11}^{OO} & D_{13}^{O1} & D_{23}^{O1} & D_{11}^{O1} \\ \hline \multicolumn{3}{l}{\text{Complex conjugate of opposite sub-matrix}} & \multicolumn{3}{l}{\text{Same as opposite sub-matrix}} \end{array}\right] \tag{E.17.4}$$

where

$$D_{11}^{OO} = \frac{4\alpha}{M}, \; D_{11}^{O1} = -\frac{4\alpha}{M}(c_1c_2c_3 + is_1s_2s_3),$$

$$D_{12}^{O1} = \frac{4\beta}{M}(s_1s_2c_3 + ic_1c_2s_3), \text{ etc.}$$

and

$$c_i = \cos\frac{q_i a}{4}, \qquad s_i = \sin\frac{q_i a}{4}.$$

To find the eigenfrequencies we have to solve the corresponding secular equation

$$|D_{ij}^{\mu\nu}(q) - \omega^2\delta_{ij}\delta_{\mu\nu}| = 0. \tag{E.17.5}$$

This can be reduced by multiplying the first three columns by $D_{11}^{OO} - \omega^2$ and dividing the last three rows by $D_{11}^{OO} - \omega^2$. This leaves the lower right hand quarter of the determinant as just three diagonal elements each unity. We can now reduce the upper right hand quarter to zero leaving

$$\begin{bmatrix} b_{11} - (D_{11}^{OO} - \omega^2)^2 & b_{12} & b_{13} \\ b_{12}^* & b_{22} - (D_{11}^{OO} - \omega^2)^2 & b_{23} \\ b_{13}^* & b_{23}^* & (D_{11}^{OO} - \omega^2)^2 \end{bmatrix} = 0, \quad \text{(E.17.6)}$$

where

$$M^2 b_{11} = 16\alpha^2(c_1^2c_2^2c_3^2 + s_1^2s_2^2s_3^2) + 16\beta^2(c_2^2s_3^2 + s_2^2c_3^2)$$

$$M^2 b_{22} = 16\alpha^2(c_1^2c_2^2c_3^2 + s_1^2s_2^2s_3^2) + 16\beta^2(c_1^2s_3^2 + s_1^2c_3^2)$$

$$M^2 b_{33} = 16\alpha^2(c_1^2c_2^2c_3^2 + s_1^2s_2^2s_3^2) + 16\beta^2(c_1^2s_2^2 + s_1^2c_2^2)$$

$$M^2 b_{12} = -16\beta(2\alpha - \beta)s_1c_1s_2c_2 + \mathrm{i}\,16\beta^2(c_1^2s_2^2 - s_1^2c_2^2)s_3c_3$$

$$M^2 b_{13} = -16\beta(2\alpha - \beta)s_1c_1s_3c_3 + \mathrm{i}\,16\beta^2(c_1^2s_3^2 - s_1^2c_3^2)s_2c_2$$

$$M^2 b_{23} = -16\beta(2\alpha - \beta)s_2c_2s_3c_3 + \mathrm{i}\,16\beta^2(c_2^2s_3^2 - s_2^2c_3^2)s_1c_1 .$$

Clearly the equation (E.17.6) can be solved without too much difficulty for high symmetry points and lines. Before doing this, however, it is worthwhile locating the critical points for the diamond structure. This was done by J. C. Phillips[8] who showed that the symmetry set S of c.p.'s consisted of the Γ, L, X and W points for all branches. In order to determine the nature of the c.p.'s one needs to know the form of the frequency surfaces in their neighbourhood. This can be found either from experimental evidence or by calculation, i.e. by solving an equation such as (E.17.6) near the c.p.

Let us then return to the solution of (E.17.6) for a particularly simple case where $\mathbf{q} = (q, 0, 0)$. This is (see fig. E.17.2) the Δ line joining Γ and X. The non-diagonal parts of (E.17.6) vanish in this case because $s_2 = s_3 = 0$, $c_2 = c_3 = 1$, so writing $s_1 = s$ and $c_1 = c$ we find the simple solutions

$$\omega_{\mathrm{L}}^2 = \frac{4\alpha}{M}(1 \pm c) \qquad \text{(E.17.7a)}$$

and a doubly degenerate pair of solutions

$$\omega_{\mathrm{T}}^2 = \frac{4\alpha}{M} \pm \left(\frac{16\alpha^2c^2}{M^2} + \frac{16\beta^2s^2}{M^2}\right)^{\frac{1}{2}}, \qquad \text{(E.17.7b)}$$

where the subscripts L and T for longitudinal and transverse take us beyond what we have yet proved. In both (E.17.7a) and (E.17.7b) the upper sign gives optical mode frequencies, the lower, acoustic. The results are sketched in fig. E.17.3. Notice that at X we have two degenerate longitudinal frequencies.

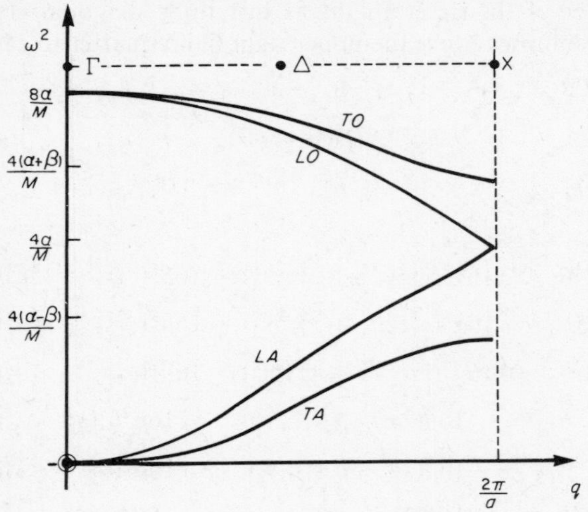

Fig. E.17.3.

To find the eigenvectors we use (E.14.2) slightly extended to cover the non-Bravais case. There are six equations for the eigenvectors $\mathbf{e}_i^\mu$:

$$\left(\frac{4\alpha}{M} - \omega^2\right)\mathbf{e}_1^{\circ} - \frac{4\alpha c}{M}\mathbf{e}_1^1 = 0 \tag{E.17.8a}$$

$$\left(\frac{4\alpha}{M} - \omega^2\right)\mathbf{e}_2^{\circ} - \frac{4\alpha c}{M}\mathbf{e}_2^1 + \frac{4\mathrm{i}\,\beta s}{M}\mathbf{e}_3^1 = 0 \tag{E.17.8b}$$

$$\left(\frac{4\alpha}{M} - \omega^2\right)\mathbf{e}_3^{\circ} + \frac{4\mathrm{i}\,\beta s}{M}\mathbf{e}_2^1 - \frac{4\alpha c}{M}\mathbf{e}_3^1 = 0 \tag{E.17.8c}$$

$$-\frac{4\alpha c}{M}\mathbf{e}_1^{\circ} + \left(\frac{4\alpha}{M} - \omega^2\right)\mathbf{e}_1^1 = 0 \tag{E.17.8d}$$

$$-\frac{4\alpha c}{M}\mathbf{e}_2^{\circ} - \frac{4\mathrm{i}\,\beta s}{M}\mathbf{e}_3^{\circ} + \left(\frac{4\alpha}{M} - \omega^2\right)\mathbf{e}_2^1 = 0 \tag{E.17.8e}$$

$$-\frac{4\mathrm{i}\,\beta s}{M}\mathbf{e}_2^{\circ} - \frac{4\alpha c}{M}\mathbf{e}_3^{\circ} + \left(\frac{4\alpha}{M} - \omega^2\right)\mathbf{e}_3^1 = 0. \tag{E.17.8f}$$

Substitution of (E.17.7a) into the above equations gives longitudinal eigenvectors of two types. In the case where the lower sign is used, i.e. for the acoustic modes, the two atoms in the unit cell are in phase, but for the upper sign, i.e. for the optic branch, the two atoms are in antiphase.

Thus

$$\mathbf{e}_{\mathrm{LA}}^{\mathrm{O}} = \frac{1}{\sqrt{2}}(1, 0, 0),\ \mathbf{e}_{\mathrm{LA}}^{1} = \frac{1}{\sqrt{2}}(1, 0, 0)$$

$$\mathbf{e}_{\mathrm{LO}}^{\mathrm{O}} = \frac{1}{\sqrt{2}}(1, 0, 0),\ \mathbf{e}_{\mathrm{LO}}^{1} = \frac{1}{\sqrt{2}}(-1, 0, 0).$$

If we substitute (E.17.7b) into the set of equations (E.17.8) we find that we have transverse eigenvectors. For one of the two transverse branches we find:

$$\mathbf{e}_{\mathrm{TA}}^{\mathrm{O}} = \frac{1}{\sqrt{2}}(0, 1, 0) \quad \mathbf{e}_{\mathrm{TA}}^{1} = \frac{1}{\sqrt{2}}\left[0, \frac{1}{\sqrt{(1+\lambda^2)}}, \frac{\mathrm{i}\lambda}{\sqrt{(1+\lambda^2)}}\right]$$

$$\mathbf{e}_{\mathrm{TO}}^{\mathrm{O}} = \frac{1}{\sqrt{2}}(0, -1, 0)\ e_{\mathrm{TO}}^{1} = \frac{1}{\sqrt{2}}\left[0, \frac{1}{\sqrt{(1+\lambda^2)}}, \frac{\mathrm{i}\lambda}{\sqrt{(1+\lambda^2)}}\right]$$

where $\lambda = \beta s/\alpha c$ so λ varies from zero at the Γ point to ∞ at the X point. Inspection shows that at the Γ point the acoustic modes have atoms (0) and (1) in phase whilst the optic modes have them exactly out of phase. At X on the other hand we have the two atoms vibrating in directions at right angles and in quadrature with each other as far as phase is concerned.

Incidentally we have now demonstrated the appropriateness of the L and T labelling in (E.17.7a) and (E.17.7b). In propagation directions (**q**) of lower symmetry in the Brillouin zone it will be found that the polarizations are not generally either exactly longitudinal or exactly transverse.

By a perturbation expansion about X the nature of the critical points there can be established. For the TA branches we find that one branch has a P_1 and the other an F_2 type c.p. The nature of the c.p.'s at Γ, X, L and W is shown in table E.17.1. To find the c.p.'s which lie on symmetry lines

Table E.17.1

Branch	Γ	X	L	W	Σ	Q	S_I
*TO*1	P_3	P_0	P_2	$P_1(2)$	$P_2, P_1(1)$	—	—
*TO*2	P_3	P_0	P_2	$P_3(2)$	$P_2(1), P_1$	P_1	P_2
LO	$\{P_3$	$P_2(1)$	P_2	$P_0(2)$	P_1	—	—
	P_3	$P_1(1)$	P_2	$P_0(2)$	—	—	—
LA	$P_0(3)$	$P_3(1)$	P_1	$P_2(2)$	—	—	—
*TA*1	$P_0(3)$	P_1	P_1	$P_1(2)$	P_3	P_2	—
*TA*2	$P_0(3)$	F_2	P_1	$P_3(2)$	—	—	—
Weight	1	3	4	6	12	24	3

The numbers in brackets (n) means that there are n directions in which the first derivative is discontinuous.

topological arguments are necessary. It appears that all phonon branches have zero first derivative normal to symmetry lines except for certain Z and Λ line branches where doubly degenerate bands split giving equal and opposite first derivatives. The critical points on Σ and Q lines are specified in table E.17.1. Finally there may be c.p.'s at general points on symmetry planes such as $S_I(q_i = 0)$ or $S_{II}(q_i = q_j)$. The c.p.'s on S_I are given in table E.17.1. Notice that for the LO branch it is not possible to be definite about the nature of the c.p.'s without more information. Both possibilities satisfy the Morse relations (E.15.4).

Another quantity which can be calculated from (E.17.6) are the sound velocities which can be found by determining frequencies in the neighbourhood of $\mathbf{q} = 0$ for the acoustic modes. In this way the elastic constants can be determined in terms of the two force constants α and β:

$$c_{11} = \alpha/a, \qquad c_{12} = (2\beta - \alpha)/a, \qquad c_{44} = (\alpha^2 - \beta^2)/\alpha a. \tag{E.17.9}$$

This leads to the Born relation between the elastic constants

$$\frac{4c_{11}(c_{11} - c_{44})}{(c_{11} + c_{12})^2} = 1. \tag{E.17.10}$$

This is satisfied quite well in the case of germanium and silicon but not by diamond.

Despite this, when the theoretically predicted frequencies of the two force-constant model are compared with experimental frequencies determined by inelastic neutron scattering, the agreement is not good, the theoretical frequencies being much too high, as shown in fig. E.17.4. In fact it was found by Herman that it was necessary to consider forces between fifth nearest neighbours to obtain agreement with experiment. As this calculation contained fifteen adjustable parameters, the good fit with experiment was probably not significant. In view of the covalent binding in germanium such long range forces seem rather implausible.

The situation was relieved by the use of the 'shell model' of Dick and Overhauser[11]. This model was designed for use with alkali halides. The ions are divided into two parts, a core consisting of the nucleus and the inner electrons and a shell comprised of the outer electrons. Each of these is coupled to the core and shell of the nearest neighbour. This type of theory proved successful initially for the case of sodium iodide and when used for germanium gave results in agreement with experiment with only two adjustable parameters. The calculated dispersion curves are shown in fig. E.17.4.

[11] An elementary account of the shell model is given by W. Cochran, *Rept. Progr. Phys.*, **25**, 1 (1963).

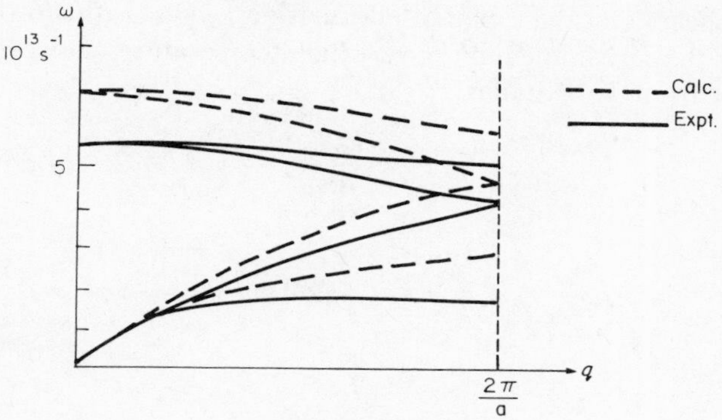

Fig. E.17.4.

E.18 THE LATTICE SPECIFIC HEAT

We now turn to the application of lattice dynamics to the solution of various important physical problems. The oldest of these is that of the specific heat. If we suppose (as is exactly true in the harmonic approximation) that there is no change of volume with temperature, then the heat capacity at constant volume, c_v, is the temperature derivative of the internal energy $\bar{\mathscr{E}}$

$$c_v = \frac{\partial \bar{\mathscr{E}}}{\partial T}. \tag{E.18.1}$$

Thus the problem of calculating the specific heat is really the same as that of calculating $\bar{\mathscr{E}}$*. This is given by (E.12.2), and making use of the results in §E.12, in particular (E.12.5), we see that

$$\bar{\mathscr{E}} = V\hbar \int \omega g(\omega)\tilde{N}(\omega)\,\mathrm{d}\omega \tag{E.18.2}$$

and

$$c_v = \frac{V\hbar^2}{k_0 T^2} \int \omega g(\omega)\tilde{N}(\tilde{N} + 1)\,\mathrm{d}\omega, \tag{E.18.3}$$

where use has been made of the result

$$\frac{\partial \tilde{N}}{\partial T} = \frac{\hbar\omega}{k_0 T^2}\tilde{N}(\tilde{N} + 1). \tag{E.18.4}$$

* In a metal where there are free electrons there will be an electronic specific heat, negligible compared to the lattice specific heat, except at low temperatures. It may be deducted from the measured specific heat without too much difficulty.

Thus it appears that the essential information needed is the form of $g(\omega)$. This is not the case in the limit of very high temperatures, however. In this case we see from (E.10.3) that

$$\tilde{N} \approx \frac{k_0 T}{\hbar\omega},$$

so

$$\bar{\mathscr{E}} = V k_0 T \int g(\omega)\,\mathrm{d}\omega$$

$$= 3Nk_0 T \tag{E.18.5a}$$

and

$$c_v = 3Nk_0. \tag{E.18.5b}$$

This is just the classical value which is obtained from the principle of equipartition of energy for an assembly of $3N$ harmonic oscillators. The classical nature of the result is emphasised by the disappearance of $\hbar$ from the equations (E.18.5a, b).

Except in the high temperature limit, the form of $g(\omega)$ is of very great importance in that the differences between one substance and another are entirely contained in $g(\omega)$. Certain simplified forms of $g(\omega)$ have been of great significance.

(1). The first of these arises from the Einstein model described at the end of §E.4. He assumed all the atoms vibrated independently at the same frequency ω_E. This means that

$$g(\omega) = 3N\delta(\omega - \omega_E), \tag{E.18.6}$$

so using (E.18.2) and (E.18.3):

$$\bar{\mathscr{E}} = 3N\hbar\omega_E \tilde{N}(\omega_E) \tag{E.18.7a}$$

and

$$c_v = \frac{3N\hbar^2\omega_E^2}{k_0 T^2}\tilde{N}(\omega_E)[\tilde{N}(\omega_E) + 1]. \tag{E.18.7b}$$

At high temperatures (E.18.7a) and (E.18.7b) lead to the classical results in (E.18.5a, b) but when $\hbar\omega_E \gg k_0 T$ both $\bar{\mathscr{E}}$ and c_v are proportional to $\exp(-\hbar\omega_E/k_0 T)$.

The Einstein theory gives qualitatively the right kind of temperature dependence but the specific heat is too low at really low temperatures. Clearly this is not surprising in view of the extremely crude assumptions underlying the theory.

(2). The next model to be applied to the calculation of specific heat was that of the elastic continuum, whose properties were discussed in §E.7. Let us calculate the density of states for this case. The group velocity $\nabla_q \omega$ has the same magnitude as the velocity of sound, so from (E.12.6)

$$g(\omega) = \frac{1}{8\pi^3} \sum_s \frac{1}{v_s} \int dS$$

$$= \frac{\omega^2}{2\pi^2} \sum_s \frac{1}{v_s^3}. \tag{E.18.8a}$$

Now Debye, in his original development of the theory, assumed that all the velocities had the same value, v. Then

$$g(\omega) = \frac{3\omega^2}{2\pi^2 v^3}. \tag{E.18.8b}$$

In the argument thus far, equations (E.18.8a, b) have the same form no matter how high the frequency. It is necessary, however, that the total number of modes be correct, so let us make $g(\omega) = 0$ above a cut-off frequency, ω_D. This means that:

$$\omega_D = v(6\pi^2 N/V)^{\frac{1}{3}}. \tag{E.18.9}$$

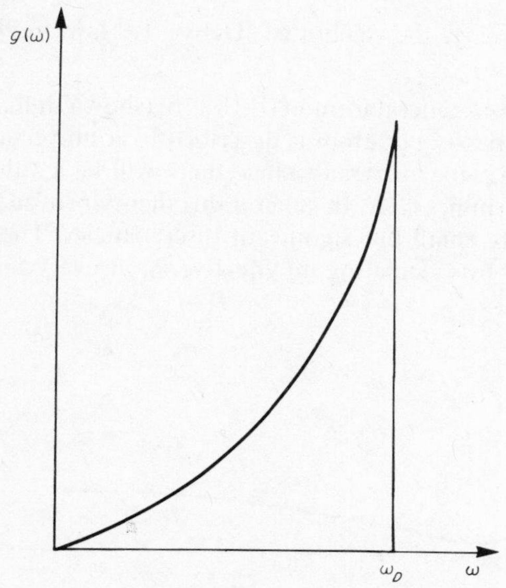

Fig. E.18.1.

The form of $g(\omega)$ is shown in fig. E.18.1. It is customary to define a Debye temperature θ_D by the equation

$$\theta_D = \hbar\omega_D/k_0. \tag{E.18.10}$$

Using the Debye form of the density of states the internal energy is

$$\bar{\mathscr{E}} = 9Nk_0T\left(\frac{T}{\theta_D}\right)\left(\frac{T}{\theta_D}\right)^3 \int_0^{\theta_D/T} z^3\tilde{N}\,\mathrm{d}z \tag{E.18.11a}$$

where $z = \hbar\omega/k_0T$, and the heat capacity is

$$c_\mathrm{v} = 9Nk_0\left(\frac{T}{\theta_D}\right)\left(\frac{T}{\theta}\right)^3 \int_0^{\theta_D/T} z^4\tilde{N}(\tilde{N}+1)\,\mathrm{d}z. \tag{E.18.11b}$$

At high temperatures these equations lead to the usual classical results, but at low temperatures important differences appear. When $T \ll \theta_D$ the upper limits in (E.18.11a, b) can be replaced by infinity with only trivial error. Now:

$$\int_0^\infty \frac{z^3\,\mathrm{d}z}{e^z - 1} = 6\xi(4) = \frac{\pi^4}{15},$$

where $\xi(x)$ is the Riemann Zeta function. Then the heat capacity is

$$c_\mathrm{v} = \frac{12\pi^4 Nk_0}{5}\left(\frac{T}{\theta_D}\right)^3. \tag{E.18.12}$$

This result expresses the celebrated 'Debye T^3' law, in close agreement with experiment.

Returning to the general form in (E.18.11b) (shown in fig. E.18.2) we see that the heat capacity per atom is described by a universal law with one parameter, θ_D, i.e. for every substance there will be a value of θ_D which completely determines c_v/N. In general this theory proved highly successful, but there are small but significant discrepancies. These are conveniently exhibited by calculating an effective θ_D at every temperature and

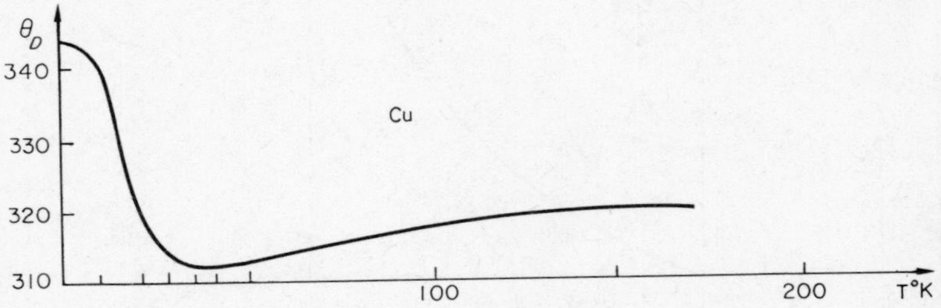

Fig. E.18.2.

plotting θ_D as a function of T. Such a curve for copper is shown in fig. E.18.2. These discrepancies may be supposed to arise from the over-simplified form of $g(\omega)$ and might be expected to be removed if a more accurate form of $g(\omega)$ could be obtained.

The most critical test might therefore be expected if an experimentally determined density of states function is used. This has been reported by Johnson and Cochran[12] for the case of germanium where dispersion curves obtained by neutrón scattering have been used to calculate $g(\omega)$ and hence c_v. The result is given in the form of a θ_D against T curve in fig. E.18.3, and the agreement, while not perfect, is quite gratifying. It certainly suggests that the formulation underlying (E.18.3) is not at fault, and that the problem of calculating c_v accurately is that of calculating a good density of states function.

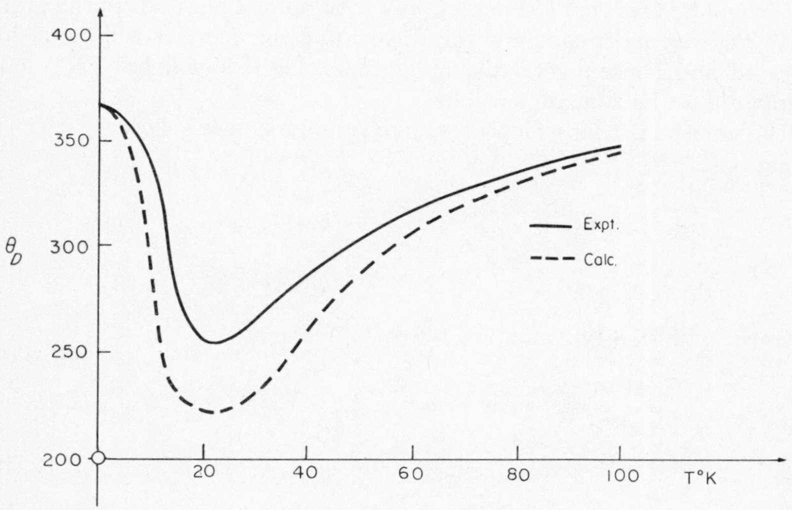

Fig. E.18.3.

Problems

(E.18.1) Carry through the Debye calculation for the one dimensional case (i.e. for a string) and show that the low temperature specific heat is proportional to T.

(E.18.2) Optical modes show very little dispersion and therefore might be added to the Debye model in the form of Einstein frequencies. Discuss the likely results in a qualitative manner.

[12] F. A. Johnson and W. Cochran, *Proc. Exeter Conf. on Semiconductors*, p. 498 (1962).

E.19 INTERACTION OF OPTICAL MODES WITH THE ELECTROMAGNETIC FIELD

We now turn our attention to the interaction of lattice vibrations with external influences. In this section we extend the results of §E.8 to cover the classical theory of interaction between optical modes and the electromagnetic field[7].

In §E.8 we derived two equations relating polarization **P**, electric field **E** and reduced ionic displacement **w** and then combined these with div **D** = 0 which relates **E** and **P**. The other Maxwell equations were neglected; this omission will be repaired in the present section. The dispersion of the optical modes was also neglected in §E.8, and will still be neglected. The justification for this is that in free space the electromagnetic waves will be of a frequency given by $\omega = ck$, where k is the wave number. These will be expected to interact only with optical phonons of the same wave number and frequency; this means that the energies will be in the infrared and consequently the magnitude of $\mathbf{q} = \mathbf{k}$ will be very small compared to the zone dimensions.

The set of equations which we wish to simultaneously solve is Maxwell's equations:

$$\operatorname{curl} \mathbf{H} = \epsilon_0 \frac{\partial \mathbf{E}}{\partial t} + \frac{\partial \mathbf{P}}{\partial t}, \qquad \operatorname{curl} \mathbf{E} = -\mu_0 \frac{\partial \mathbf{H}}{\partial t},$$
$$\operatorname{div}(\epsilon_0 \mathbf{E} + \mathbf{P}) = 0, \qquad \operatorname{div} \mathbf{H} = 0, \tag{E.19.1}$$

together with (E.8.14a) and (E.8.14b):

$$\frac{\mathrm{d}^2\mathbf{w}}{\mathrm{d}t^2} = -\omega_0^2 \mathbf{w} + \omega_0 \sqrt{[\epsilon_0(\epsilon_s - \epsilon_\infty)]}\mathbf{E}, \tag{E.19.2a}$$

$$\mathbf{P} = \omega_0 \sqrt{[\epsilon_0(\epsilon_s - \epsilon_\infty)]}\mathbf{w} + (\epsilon_\infty - 1)\epsilon_0 \mathbf{E}. \tag{E.19.2b}$$

We now assume that we have wave solutions for each of the variables, all of the form $e^{i(\mathbf{q}\cdot\mathbf{r}-\omega t)}$. We then find

$$\left.\begin{aligned} &\mathbf{q} \times \mathbf{H} = -\omega(\epsilon_0 \mathbf{E} + \mathbf{P}) \quad \mathbf{q}\cdot(\epsilon_0 \mathbf{E} + \mathbf{P}) = 0 \\ &\mathbf{q} \times \mathbf{E} = \omega \mu_0 \mathbf{H} \qquad\qquad\qquad \mathbf{q}\cdot\mathbf{H} = 0 \\ &(\omega^2 - \omega_0^2)\mathbf{w} + \omega_0\sqrt{[\epsilon_0(\epsilon_s - \epsilon_\infty)]}\mathbf{E} = 0 \end{aligned}\right\}. \tag{E.19.3}$$

This relation between **w** and **E** enables us to relate **P** and **E** by eliminating **w**:

$$\mathbf{P} = \left(\frac{\epsilon_\infty \omega^2 - \epsilon_s \omega_0^2}{\omega^2 - \omega_0^2} - 1\right)\epsilon_0 \mathbf{E}. \tag{E.19.4}$$

We can now also eliminate **P** from the Maxwell equations giving

$$\mathbf{q} \times \mathbf{H} = -\omega\epsilon_0 \frac{\epsilon_\infty \omega^2 - \epsilon_s \omega_0^2}{\omega^2 - \omega_0^2} \mathbf{E} \tag{E.19.5a}$$

and

$$(\epsilon_\infty \omega^2 - \epsilon_s \omega_0^2)\mathbf{q} \cdot \mathbf{E} = 0. \tag{E.19.5b}$$

From this last equation we can draw one of two conclusions. Either

(A) $\omega^2 = \frac{\epsilon_s}{\epsilon_\infty}\omega_0^2$

or

(B) **q** is perpendicular to **E**.

(A). We also have $\mathbf{q} \times \mathbf{H} = 0$ which, since $\mathbf{q} \cdot \mathbf{H} = 0$ also, means that $\mathbf{H} = 0$ if $\mathbf{q} \neq 0$. Furthermore we now have $\mathbf{q} \times \mathbf{E} = 0$ so **E** must be parallel to **q**, i.e. the vibrations are *longitudinal*. Thus

$$\frac{\epsilon_s}{\epsilon_\infty}\omega_0^2 = \omega_l^2 \tag{E.19.6}$$

and

$$\mathbf{E} = E_q \mathbf{q}\, \mathrm{e}^{\mathrm{i}(\mathbf{q}\cdot\mathbf{r} - \omega_l t)}.$$

Notice that $\mathbf{P} = -\epsilon_0 \mathbf{E}$, so $\mathbf{D} = 0$ for these modes.

(B). It quickly follows that **E**, **H** and **q** are mutually perpendicular and we are concerned with *transverse* vibrations. Now from (E.19.5a):

$$\mathbf{E} = -\frac{\omega^2 - \omega_0^2}{\omega^2 \epsilon_0 \mu_0 (\epsilon_\infty \omega^2 - \epsilon_s \omega_0^2)} \mathbf{q} \times (\mathbf{q} \times \mathbf{E})$$

$$= \frac{(\omega^2 - \omega_0^2)c^2 q^2}{\omega^2(\epsilon_\infty \omega^2 - \epsilon_s \omega_0^2)} \mathbf{E}$$

where c is the free space velocity of light $(\epsilon_0\mu_0)^{-\frac{1}{2}}$. We now obtain an equation for ω:

$$\omega^4 - \omega^2\left(\omega_l^2 - \frac{c^2q^2}{\epsilon_\infty}\right) + \omega_0^2 \frac{c^2q^2}{\epsilon_\infty} = 0$$

with solutions

$$\omega^2 = \frac{1}{2}\left\{\omega_l^2 + \frac{c^2q^2}{\epsilon_\infty} \pm \left[\omega_l^2 + \frac{c^4q^4}{\epsilon_\infty^2} + \frac{2c^2q^2}{\epsilon_\infty}(\omega_l^2 - 2\omega_0^2)\right]^{\frac{1}{2}}\right\}. \tag{E.19.7}$$

For very small q,

$$\omega \approx \omega_l \quad \text{or} \quad [cq/(\epsilon_s)^{\frac{1}{2}}];$$

for very large q,

$$\omega \approx [cq/(\epsilon_\infty)^{\frac{1}{2}}] \quad \text{or} \quad \omega_t,$$

where we have written $\omega_0 = \omega_t$.

To summarize (fig. E.19.1), we have a set of longitudinal normal modes of frequency ω_l and two sets of transverse normal modes. One set behaves at long wavelengths as normal electromagnetic waves in a medium of refractive index $(\epsilon_s)^{\frac{1}{2}}$, but at short wavelengths takes on a phonon-like behaviour with frequency ω_t. The other set behaves like phonons of frequency ω_l at long wavelengths and like radiation in a medium of refractive index $(\epsilon_\infty)^{\frac{1}{2}}$ at short wavelengths. Thus these transverse modes are neither photons nor phonons but a mixture of the two.

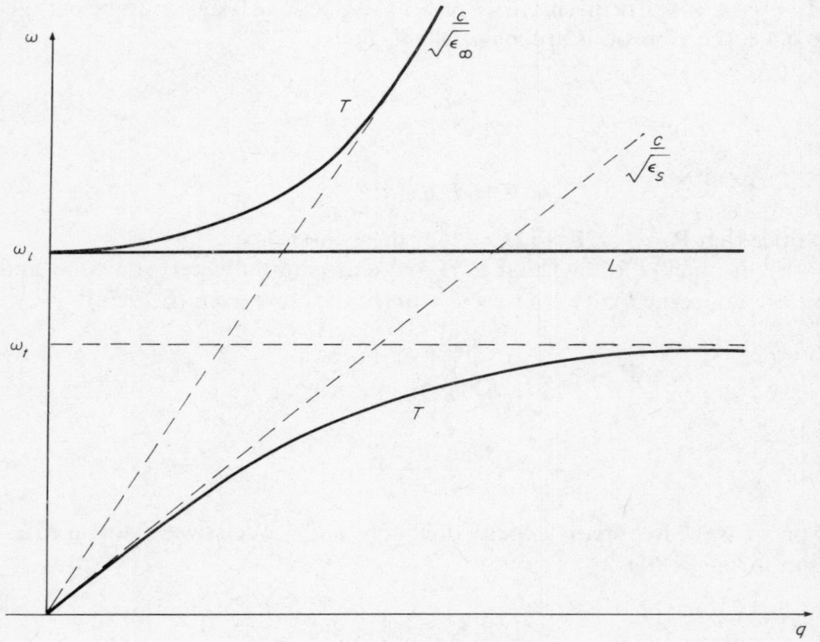

Fig. E.19.1.

It is interesting to consider the relationship of these modes to the electromagnetic modes outside the crystal. These will be transverse and of all frequencies $\omega = ck$, where we use $\mathbf{k}$ for the wave-vector in free space.

These will be able to couple to modes inside the crystal except in the frequency range $\omega_t < \omega < \omega_l$, where there are no propagating modes inside. In this range external electromagnetic waves will be reflected. The longitudinal modes inside the crystal will not be coupled to any external modes and since they give $\mathbf{D} = 0$ this is consistent with the condition that the normal (i.e. longitudinal) component of $\mathbf{D}$ must be continuous across any boundary. The meaning of the 'energy gap' between ω_t and ω_l can be seen by calculating the dielectric constant for the system, using

$$\mathbf{P} = [\epsilon(\omega) - 1]\epsilon_0\mathbf{E}. \tag{E.19.8}$$

Then

$$\epsilon(\omega) = \left(\frac{\omega^2 - \omega_l^2}{\omega^2 - \omega_t^2}\right)\epsilon_\infty, \tag{E.19.9}$$

which has a pole at ω_t and a zero at ω_l (fig. E.19.2). Therefore $\epsilon(\omega) < 0$ when $\omega_t < \omega < \omega_l$. We may formally write

$$q = \frac{\omega}{c}[\epsilon(\omega)]^{\frac{1}{2}} \tag{E.19.10}$$

so a negative ϵ gives an imaginary q and a decaying solution, i.e. reflexion of incident electromagnetic waves, as described in the previous paragraph.

Absorption by dissipative processes has been neglected in the above treatment.

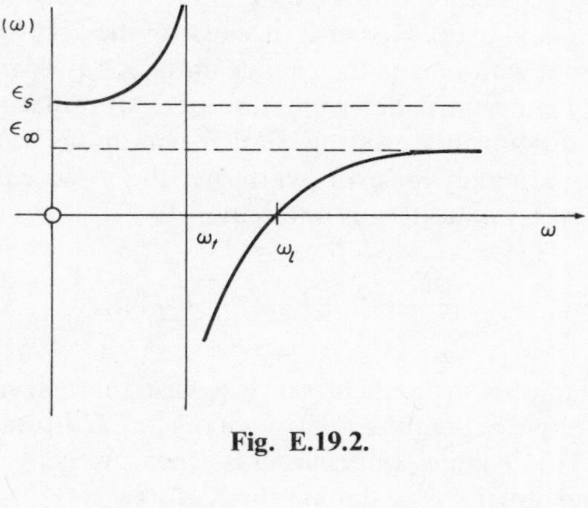

Fig. E.19.2.

Problems

(E.19.1) Calculate the effect on $\epsilon(\omega)$ of introducing a small damping term $\gamma d\mathbf{W}/dt$ into (E.19.2a). What is the physical significance of this?

E.20 PHONON ENERGIES FROM X-RAY AND NEUTRON SCATTERING

The theory of the method by which phonon energies may be derived from X-ray and neutron scattering data will not be presented in any sort of detail. Some general idea of the principles of the method is all that will be given[2].

Let us consider first the case of X-ray scattering. Suppose that $\mathbf{k}_0$ and $\mathbf{k}$ are the wave-vectors of an incident and scattered X-ray beam. Then, in the case of the zero order or Bragg scattering used in crystallography,

$$\mathbf{k} - \mathbf{k}_0 = \mathbf{K}_m, \tag{E.20.1}$$

so that the X-ray beam has been scattered by an amount equal to a reciprocal lattice vector. If the atoms are moving due to lattice vibrations we have, in addition, a first order scattering where

$$\mathbf{k} - \mathbf{k}_0 = \mathbf{K}_m - \mathbf{q}, \tag{E.20.2}$$

$\mathbf{q}$ being, as usual, a phonon wave-vector. As well as (E.20.2) we must have an energy conservation condition

$$\hbar c(k - k_0) = \pm \hbar\omega_s(\mathbf{q}). \tag{E.20.3}$$

The upper sign corresponds to the disappearance of a (q, s) phonon with increase of X-ray energy, and the lower sign to the reverse process. It will be seen that a measurement of the magnitude and direction of $\mathbf{k}_0$ and $\mathbf{k}$ will determine $\mathbf{q}$ and $\omega_s(\mathbf{q})$. However, because of the very large magnitude of $\hbar c\mathbf{k}_0$ compared with $\hbar\omega_s(\mathbf{q})$, the change in the X-ray energy is relatively small, making the accurate determination of $\omega_s(\mathbf{q})$ rather difficult.

In the case of neutron scattering, similar conditions must be satisfied. For first order coherent inelastic scattering the same equation (E.20.2) holds but the energy condition is different:

$$\frac{\hbar^2}{2M_n}(\mathbf{k}^2 - \mathbf{k}_0^2) = \pm \hbar\omega_s(\mathbf{q}), \tag{E.20.4}$$

where M_n is the mass of the neutron. If we use thermal neutrons whose wavelength is a few Å then the relative change in neutron energy is quite considerable. This enables quite accurate measurements of $\omega_s(\mathbf{q})$ to be made with much greater ease than in the X-ray case.

Groups at Harwell, Chalk River and elsewhere have used the neutron scattering method to determine the phonon dispersion curves for a number of different ionic solids, semiconductors and metals.

E.21 OPTICAL ABSORPTION DUE TO LATTICE VIBRATIONS

One approach to the question of the interaction of light and lattice vibrations was outlined in §E.19. The discussion there was (a) classical and (b) concerned only with non-dissipative effects. In this section we shall consider the quantum theory of optical absorption. It will be seen, however, that the approach is in many ways less fundamental than that of §E.19.

We will expect that the fluctuating electric polarization in a crystal will be able to interact with electromagnetic radiation and give rise to a certain absorption spectrum. Such absorption is observed in both ionic and covalent solids, generally in the infrared. The analysis of these absorption bands is a valuable source of information about lattice vibrational frequencies at certain points in the Brillouin zone. The theory presented here is a straightforward application of the semi-classical perturbation theory of the interaction of radiation with matter. It follows closely a paper by Lax and Burstein[13].

If **P** is the polarization of a dielectric, the interaction energy with an electric field **E** is

$$U = -\mathbf{P} \cdot \mathbf{E}. \tag{E.21.1}$$

Since **E** is the local field inside the dielectric it will not be exactly the same as the applied field. If this field is due to an incident light beam we may write:

$$\mathbf{E} = E_0 \mathbf{f}\, \mathrm{e}^{\mathrm{i}(\mathbf{k}\cdot\mathbf{r} - \nu t)} \tag{E.21.2}$$

where **f** is the polarization vector of the electric part of the electromagnetic wave, **k** is the wave-vector and ν the angular frequency. Let us write the polarization as a power series in the atomic displacements, thus

$$\mathbf{P} = \mathbf{P}_0 + \mathbf{P}_1 + \mathbf{P}_2 + \cdots. \tag{E.21.3}$$

$\mathbf{P}_0$ is any permanent static dipole moment which may exist. This will be zero except for ferroelectric substances. The next term is

$$\mathbf{P}_1 = \sum_{\nu,n} \mathbf{e}_n^{\nu} \,.\, \mathbf{u}_n^{\nu} = \sum_{\nu,n} \mathbf{e}^{\nu} \,.\, \mathbf{u}_n^{\nu} \tag{E.21.4}$$

where $\mathbf{e}_n^{\nu}$ is a second order tensor, the effective charge on ion ν in the nth unit cell. Now translational symmetry means that $\mathbf{e}_n^{\nu}$ cannot depend on n, which leads to the second form of (E.21.4). For simplicity we shall also assume that the effective charge is a scalar, e^{ν}. The third term in (E.21.3) will have the form:

$$\mathbf{P}_2 = \sum \mathbf{e}_{m-n}^{\mu\nu} : \mathbf{u}_m^{\mu}\mathbf{u}_n^{\nu}. \tag{E.21.5}$$

[13] M. Lax and E. Burstein, *Phys. Rev.*, **97**, 39 (1955).

Here $\mathbf{e}_{m-n}^{\mu\nu} \cdot \mathbf{u}_m^{\mu}$ is the charge induced on the νth atom in the nth cell by a displacement of the μth atom in the mth cell. $\mathbf{e}_{m-n}^{\mu\nu}$ is a third order tensor dependent only on the relative positions of m and n.

We can therefore write

$$U = (U_1 + U_2 + \cdots)E_0 \, e^{i\nu t} \tag{E.21.6}$$

where, for example,

$$U_1 = \mathbf{f} \cdot \mathbf{P}_1 \, e^{i\mathbf{k}\cdot\mathbf{r}} = \sum_{n,\nu} e^{\nu}\mathbf{f} \cdot \mathbf{u}_n^{\nu} \, e^{i\mathbf{k}\cdot\mathbf{R}_n^{\nu}}, \tag{E.21.7}$$

where, in order to set $\mathbf{r} = \mathbf{R}_n^{\nu}$, we have assumed that the displacements $\mathbf{u}_n^{\nu}$ are small compared to the interatomic spacings.

The absorption coefficient can be written as

$$\mu(\nu) = \left[\frac{n}{\epsilon}\left(\frac{E_l}{E}\right)^2\right]\frac{4\pi^2\nu}{c} I(\nu), \tag{E.21.8}$$

where n is the refractive index, ϵ the dielectric constant, (E_l/E) is the local field correction and

$$\begin{aligned} I(\nu) &= |\langle f|\mathbf{f} \cdot \mathbf{P} \, e^{i\mathbf{k}\cdot\mathbf{r}}|i\rangle|^2 \delta(\mathscr{E}_f - \mathscr{E}_i - \hbar\nu) \\ &= I_1 + I_2 + \cdots . \end{aligned} \tag{E.21.9}$$

For the first term we need:

$$\begin{aligned} \langle f|\mathbf{P}_1 \, e^{i\mathbf{k}\cdot\mathbf{r}}|i\rangle &= \sum_{n,\nu} \langle f|e^{\nu}\mathbf{u}_n^{\nu} \, e^{i\mathbf{k}\cdot\mathbf{R}_n^{\nu}}|i\rangle \\ &= \sqrt{\left(\frac{\hbar}{2N}\right)} \sum_{s,q} \frac{\langle f|\alpha_s(\mathbf{q}) - \alpha_s^{\dagger}(-\mathbf{q})|i\rangle}{\sqrt{[\omega_s(\mathbf{q})]}} \sum_{\nu} \frac{e^{\nu}\mathbf{e}^{s}(\mathbf{q},\nu)}{\sqrt{(M_{\nu})}} \sum_{n} e^{i(\mathbf{q}+\mathbf{k})\cdot\mathbf{R}_n^{\nu}} \\ &= \sqrt{\left(\frac{N\hbar}{2}\right)} \sum_{s} \frac{\langle f|\alpha_s(-\mathbf{k}) - \alpha^{\dagger}(\mathbf{k})|i\rangle}{\sqrt{[\omega_s(\mathbf{k})]}} \sum_{\nu} \frac{e^{\nu}\mathbf{e}^{s}(\mathbf{k},\nu)}{\sqrt{(M_{\nu})}}. \end{aligned} \tag{E.21.10}$$

Here we have substituted for $\mathbf{u}_n^{\nu}$ using (E.21.9) and then employed (E.4.5) and the fact that $\omega_s(\mathbf{q})$ is even in $\mathbf{q}$. The requirement of energy conservation in (E.21.9) will eliminate the process involving $\alpha_s(-k)$ because we cannot allow the simultaneous disappearance of both a photon and a phonon. We therefore have, from (E.10.7b),

$$\langle f|\mathbf{P}_1 \, e^{i\mathbf{k}\cdot\mathbf{r}}|i\rangle = \sqrt{\left(\frac{N\hbar}{2}\right)} \sum_{s} \sqrt{\left[\frac{\tilde{N}_s^k + 1}{\omega_s(\mathbf{k})}\right]} \sum_{\nu} \frac{e^{\nu}\mathbf{e}^{s*}(k,\nu)}{\sqrt{(M_{\nu})}}. \tag{E.21.11}$$

Now the conditions of $\mathbf{k} = -\mathbf{q}$ and energy conservation are illustrated in fig. E.21.1, where allowed processes are those where the phonon frequencies and the line $\nu = ck$ intersect. This means that from the phonon point of view **k** is almost zero, i.e. $\omega_s(k) = \omega_s(0)$, and that only optical modes can be involved. Finally only transverse modes can interact because $\mathbf{f}\cdot\mathbf{e}^s$ vanishes for longitudinal vibrations. Then

$$I_1(\nu) = N\frac{\tilde{N}^{\circ}_{\mathrm{TO}} + 1}{\omega_{\mathrm{TO}}(0)}\left|\sum_{\nu}\frac{e^{\nu}}{\sqrt{(M_{\nu})}}\right|^2 \delta(\omega_{\mathrm{TO}}(0) - \nu). \qquad \text{(E.21.12)}$$

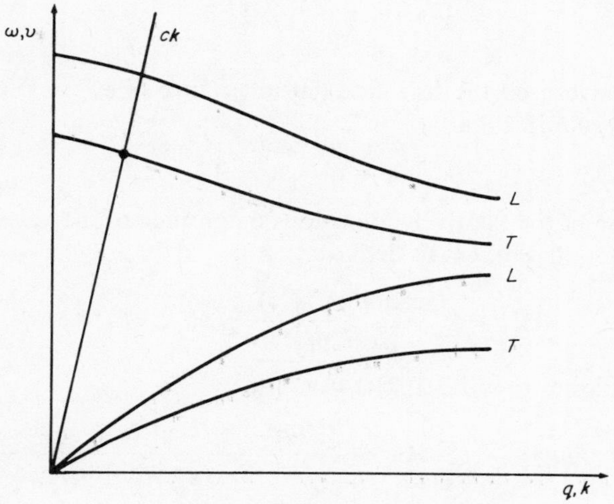

Fig. E.21.1.

A factor of two has been brought in because there are two transverse branches. I_1 therefore consists of a single narrow absorption line at the frequency of the transverse optical phonon of zero wave-vector. This is the same as the crystal Raman frequency, though we shall not prove this. In practice this single phonon line will be broadened by anharmonic effects.

In covalent crystals the atoms possess no net charge, so e^{ν} vanishes and so should the single phonon absorption. If impurities or other crystal imperfections are present, however, these may induce charges on the atoms of the lattice and then the single phonon line will appear. However two phonon processes due to $I_2(\nu)$ give the strongest absorption in covalent crystals and are, of course, observed in ionic crystals also, though here they are much weaker than the single phonon line.

The interaction term U_2 is given by

$$U_2 = \mathbf{f}\,.\,\mathbf{P}_2 = \sum_{\substack{\mu,\nu \\ m,n}} \mathbf{f}\,.\,\mathbf{e}^{\mu\nu}_{m-n}:\mathbf{u}^{\mu}_m\mathbf{u}^{\nu}_n\,\mathrm{e}^{\mathrm{i}\mathbf{k}\,.\,\mathbf{R}_n}; \qquad \text{(E.21.13)}$$

to determine I_2 we need the matrix element

$$\langle f|\mathbf{P}_2\,e^{i\mathbf{k}\cdot\mathbf{r}}|i\rangle = \frac{\hbar}{2N}\sum_{\substack{q,s\\q',s'}} \frac{\langle f|[\alpha_s(q)-\alpha^\dagger_{s*}(-q)][\alpha_s(q')-\alpha^\dagger_{s'}(-q')]|i\rangle}{\sqrt{[\omega_s(q)\omega_{s'}(q')]}}$$

$$\times \sum_{\mu,\nu,m} \mathbf{e}^{\mu\nu}_{0-m} : \mathbf{e}^s(q,\mu)\,\mathbf{e}^{s'}(q',\nu)\frac{e^{i\mathbf{q}\cdot\mathbf{R}_m}}{\sqrt{(M_\mu M_\nu)}}$$

$$\times \sum_n e^{i(\mathbf{q}+\mathbf{q}'+\mathbf{k})\cdot\mathbf{R}_n}, \tag{E.21.14}$$

where we have used the translational invariance of $\mathbf{e}^{\mu\nu}_{m-n}$. The final sum in (E.21.14) requires that

$$\mathbf{q}+\mathbf{q}'+\mathbf{k}=0, \tag{E.21.15}$$

and because of the relatively enormous magnitude of the velocity of light $k \ll q$ so $|\mathbf{q}+\mathbf{k}| \approx q$. Let us define a vector

$$\mathbf{H}^q_{ss'} = \sum_{\mu,\nu} \frac{\mathbf{e}^s(q,\mu)\mathbf{e}^{s'*}(q,\nu)}{\sqrt{(M_\mu M_\nu)}} : \sum_m \mathbf{e}^{\mu\nu}_{0-m}\,e^{i\mathbf{q}\cdot\mathbf{R}_m}, \tag{E.21.16}$$

which enables us to write (E.21.14) as

$$\langle f|\mathbf{P}_2 e^{i\mathbf{k}\cdot\mathbf{r}}|i\rangle = -\frac{\hbar}{2}\sum_{\substack{q\\ss'}} \mathbf{H}^q_{ss'} \frac{\langle f|[\alpha_s(q)-\alpha^\dagger_s(-q)][\alpha_{s'}(-q)-\alpha^\dagger_{s'}(q)]|i\rangle}{\sqrt{[\omega_s(q)\omega_{s'}(q)]}}. \tag{E.21.17}$$

It will be seen that a theoretical calculation of $\mathbf{H}^q_{ss'}$ is likely to be extremely difficult.

The matrix element will contain the various products of creation and annihilation operators appearing in (E.21.17).

(a) $\alpha^\dagger_s(-q)\alpha^\dagger_{s'}(q)$. This gives the simultaneous emission of two phonons of almost exactly opposite wave-vector. The matrix element is

$$\frac{\hbar}{2}\sum_{\substack{q\\ss'}} \sqrt{\left[\frac{(\tilde{N}^q_s+1)(\tilde{N}^q_{s'}+1)}{\omega_s(q)\omega_{s'}(q)}\right]}\mathbf{H}^q_{ss'} \tag{E.21.18}$$

and it will be seen that this 'summation' process occurs quite strongly even at low temperatures where $\tilde{N}^q_s \to 0$.

(b) $\alpha_s(q)\alpha^\dagger_s(q)$. This gives the simultaneous emission and absorption of two phonons of almost the same wave vector. Since the matrix element for this process will contain a factor $\sqrt{(N^q_s)}$ the 'difference bands' will be very weak at low temperatures.

(c) $\alpha_s(q)\alpha_{s'}(-q)$. Energy conservation prohibits these processes.

We see that at low temperatures the absorption spectrum is determined by (E.21.18) and:

$$I_2(\nu) = \frac{\hbar}{32\pi^3} \sum_{ss'} \int \mathrm{d}^3q \, \frac{|\mathbf{f} \, . \, \mathbf{H}^q_{ss'}|^2}{\omega_s(q)\omega_{s'}(q)} \delta[\omega_s(q) + \omega_{s'}(q) - \nu]. \quad \text{(E.21.19)}$$

Two factors will affect $I_2(\nu)$.

(1). The properties of $\mathbf{H}^q_{ss'}$ which in turn derive from the symmetry of $\mathbf{e}^{\mu\nu}_{m-n}$ and, through the $\mathbf{e}^s(q)$, on $D^{\mu\nu}$ the dynamical matrix. It may be shown that inversion symmetry eliminates processes involving two phonons of the same branch, i.e.

$$\mathbf{H}^q_{ss} = 0.$$

(2). The lattice vibrational dispersion functions, $\omega_s(q)$. We can write

$$\mathrm{d}^3q = \mathrm{d}S_{ss'} \frac{\mathrm{d}(\omega_s + \omega_{s'})}{|\nabla_q\omega_s + \nabla_q\omega_{s'}|}$$

where $S_{ss'}$ is a constant 'sum of frequencies' surface for branches s and s'. Points where $|\nabla_q\omega_s + \nabla_q\omega_{s'}| = 0$ are called two phonon critical points by analogy with the single phonon c.p.'s discussed earlier. The two phonon c.p.'s occur either where there are two single phonon c.p.'s at the same point in the zone but in different branches s, s', or where $\nabla_q\omega_s$ and $\nabla_q\omega_{s'}$ are equal and opposite. We can therefore write

$$I(\nu) = \frac{\hbar}{32\pi^3} \sum_{ss'} \int \mathrm{d}S_{ss'} \frac{|\mathbf{f} \, . \, \mathbf{H}^q_{ss'}|^2}{\omega_s(\nu - \omega_s)|\nabla_q\omega_s + \nabla_q\omega_{s'}|}. \quad \text{(E.21.20)}$$

If $\mathbf{H}^q_{ss'}$ is a slowly varying function, $I(\nu)$ will resemble a two phonon density of states function with discontinuities of slope where the two phonon c.p.'s occur.

An account of the application of this analysis to diamond and zinc blende semiconductors is given by Johnson[14].

[14] F. A. Johnson, *Progr. Semiconductors*, **9**, 179 (1965).

PART F

ONE-ELECTRON GREEN'S FUNCTIONS IN SOLID STATE PHYSICS

T. Lukes

CHAPTER XI

General Theory

F.1 INTRODUCTION

The applications of Green's functions in mathematical physics already cover a vast field[1,2,3,4,5,6]. In this and the following two chapters* we shall consider only a limited area of this fascinating terrain and shall, in the first instance, restrict the choice of subject matter by confining the discussion to the properties of fermions. We shall further limit ourselves, essentially, to the one-electron properties of these. This means that in the calculation of physical properties the fermions will be treated as non-interacting except in so far as the exclusion principle is obeyed.

The importance of Green's functions in solid state physics arises from the fact that exact expressions for many physical properties of interest may be written in terms of them. As a simple example we shall take a one-electron Green's function $G(\mathbf{x}, \mathbf{x}', E)$. It can be shown that quite generally the one-electron density of states $D(E)$ may be written as

$$D(E) = -\frac{1}{2\pi i} \int \langle\langle G(\mathbf{x}, \mathbf{x}, E) \rangle\rangle \, d\mathbf{x}$$

and that the static one-electron conductivity tensor σ_{ij} is given by

$$\sigma_{ij} = \frac{8\pi e^2}{\hbar} \sum_{n,m} \int\int\int\int \langle\langle \frac{\partial G}{\partial x_i}(E_n, \mathbf{x}, \mathbf{x}') \frac{\partial G}{\partial y_j}(E_m, \mathbf{y}, \mathbf{y}') \rangle\rangle$$
$$\delta(E_n - E_m) \frac{\partial f}{\partial E_n} \delta(\mathbf{x}' - \mathbf{y}) \, \delta(\mathbf{y}' - \mathbf{x}) \, d\mathbf{x} \, d\mathbf{y} \, d\mathbf{x}' \, d\mathbf{y}'$$

where f is the fermi function and the brackets $\langle\langle \ \rangle\rangle$ denote an average in a sense to be defined later.

* In this section it is convenient to use a somewhat different notation from that adopted in the rest of the book. Vectors and matrices are printed in bold, e.g. **A**. Operators are given a hat, e.g. $\hat{\mathbf{A}}$.

[1] A. A. Abrikosov, L. A. Gorkov and I. E. Dzyaloshinski, *Methods of Quantum Field Theory in Statistical Physics*, Prentice-Hall, 1963.
[2] V. L. Bonch-Bruevich and S. V. Tyablikov, *The Green Function Method in Statistical Mechanics*, North Holland, 1962.
[3] P. M. Morse and H. Feshbach, *Methods of Theoretical Physics*, McGraw-Hill, 1964.
[4] D. Ter Haar, *On the Use of Green's Functions in Statistical Mechanics*, Scottish Universities Summer School, Oliver and Boyd, 1961.
[5] D. N. Zubarev, *Soviet. Phys. Uspekhi*, **3**, 320 (1960).
[6] W. E. Parry and R. E. Turner, *Green's Functions in Statistical Mechanics*, *Rept. Progr. Phys.*, 27, 23 (1964).

From a more fundamental point of view the Green's functions are related to the probability amplitude of a quantum state by bilinear expressions and are therefore connected with probabilities rather than with probability amplitudes. Thus they have, in many cases, more physical significance than the probability amplitude, which is not directly observable.

F.2 THE CONCEPT OF A STATE VECTOR

In order to introduce more general concepts we recall that the position of a point in a three-dimensional Cartesian space can be specified by the vector

$$\mathbf{r} = x\mathbf{i} + y\mathbf{j} + z\mathbf{k} \tag{F.2.1}$$

where **i**, **j**, **k** are unit vectors, called base vectors. The three vectors **i**, **j**, **k**, are linearly independent; $\mathbf{r} = 0$ implies that $x = 0$, $y = 0$, $z = 0$. Any vector in the space may be expressed as a linear combination of the three base vectors and the number of base vectors is equal to the dimensionality of the space. Scalar multiplication is defined by

$$\mathbf{i}\cdot\mathbf{j} = \mathbf{j}\cdot\mathbf{i} = \mathbf{j}\cdot\mathbf{k} = \mathbf{k}\cdot\mathbf{j} = \mathbf{k}\cdot\mathbf{i} = \mathbf{i}\cdot\mathbf{k} = 0, \tag{F.2.2a}$$

$$\mathbf{i}\cdot\mathbf{i} = \mathbf{j}\cdot\mathbf{j} = \mathbf{k}\cdot\mathbf{k} = 1, \tag{F.2.2b}$$

and is required to obey the distributive law. Thus

$$\mathbf{r}\cdot\mathbf{r} = (x\mathbf{i} + y\mathbf{j} + z\mathbf{k})\cdot(x\mathbf{i} + y\mathbf{j} + z\mathbf{k}) = x^2 + y^2 + z^2 = r^2 \tag{F.2.3}$$

defines the length of the vector **r**.

(F.2.2a) expresses the *orthogonality* of the vectors **i**, **j**, **k**. (F.2.2b) expresses the normalization condition. The vectors **i**, **j**, **k**, satisfying both (F.2.2a) and (F.2.2b) form an orthonormal set.

An arbitrary function of position in space **A**(**r**) can be similarly represented by its components

$$\mathbf{A} = A_x\mathbf{i} + A_y\mathbf{j} + A_z\mathbf{k}$$

which are defined by taking scalar products with three base vectors **i**, **j**, **k**. If the axes are rotated to form another orthogonal system x', y', z' then the vector **A** is specified, or represented, by new components A'_x, A'_y, A'_z and, clearly, there are an infinite number of equivalent representations of the vector **A**.

In quantum mechanics we are familiar with the concept of a wavefunction $\psi(\mathbf{q})$ and its Fourier transform, the wavefunction in the momentum representation, $\psi(\mathbf{p})$. In the one-dimensional case these are related by

$$\psi(q) = \int_{-\infty}^{\infty} \psi(p)\, e^{ipq/\hbar} \frac{dp}{(2\pi\hbar)^{\frac{1}{2}}} \tag{F.2.4a}$$

$$\psi(p) = \int_{-\infty}^{\infty} \psi(q)\, e^{-ipq/\hbar} \frac{dq}{(2\pi\hbar)^{\frac{1}{2}}}. \tag{F.2.4b}$$

Consider, for the sake of simplicity, a case in which the momentum p is quantized into a set of discrete values $p_1, p_2, \ldots, p_i$, as, for example, in the case of a particle in a one dimensional box with infinitely high sides. In this case, the values of $\psi(p_1), \psi(p_2), \ldots, \psi(p_i)$ of the function $\psi(p)$ specify the function completely. This suggests that the allowed momenta $p_1, p_2, \ldots, p_i$ be used to set up an orthonormal vector space in which the function $\psi(p)$ is represented by a vector with components $\psi(p_1)$, $\psi(p_2), \ldots, \psi(p_i)$. In the example considered above this space has an infinite number of dimensions, because the magnitude of the allowed momenta has no upper bound. This idea can clearly be generalized to cases in which the allowed momenta have continuously varying values, and can therefore be applied to the function $\psi(q)$. We have now reached the essence of the idea associated with Dirac's formulation. The state of the system, independent of any particular coordinate system chosen, is represented by a vector. In any *particular* system chosen the state is represented by the components of the vector in that particular coordinate system. The momentum or coordinate representations are, from this point of view, particular coordinate systems, rotated with respect to one another, in which the wave-vector is specified by its components $\psi(p)$ or $\psi(q)$. It is physically obvious that an infinite number of such representations exist. The fact that the functions $\psi(p)$, $\psi(q)$, or any other representatives of the wave-vector, may be complex, and the possibility of an arbitrary number of dimensions both require a generalization of the definitions applicable to 'ordinary' three-dimensional vector spaces.

Although wavefunctions have been used to introduce the concept of a state vector, primarily in order to give an example of a case in which the representatives of the state vector may be complex, it is clear that the idea is a perfectly general one. A classical system of N monatomic particles is specified by giving the N position and momentum coordinates, $\mathbf{p}_1, \ldots, \mathbf{p}_N, \mathbf{q}_1, \ldots, \mathbf{q}_N$, of the particles. It is then possible to think of the state of the system as specified by a vector, say $\mathbf{R}(t)$, whose projections on the $6N$ axes are the values of the components of $\mathbf{q}_i$ and $\mathbf{p}_i$. The space in which $\mathbf{R}(t)$ moves is just the classical phase space. For a continuous system like a vibrating string the state vector may be specified by giving the displacement and momentum at each point of the string as a function of the time. Here again the possibility of alternative representations is evident; for the one-dimensional vibrating string the system may be represented by either the values of displacement $x(t)$ and momentum $p(x, t)$ in the coordinate representation, or by the equivalent set of Fourier coefficients A_n, B_n.

F.3 LINEAR VECTOR SPACES

To effect the necessary generalization mentioned in the last section, we consider complex vectors in a finite number of dimensions, n. The transition to an infinite number of dimensions will, whenever necessary, be effected by suitable limiting processes, without great regard for mathematical rigour.

Let $\mathbf{a}_i$ $(i = 1, 2, \ldots, n)$ be a set of linearly independent vectors in an n-dimensional vector space. Any vector $\mathbf{x}$ may be written as

$$\mathbf{x} = \sum_i \mathbf{a}_i x_i \tag{F.3.1}$$

where the x_i may, in general, be complex. The set $\mathbf{a}_i$ is said to form a basis and the x_i are the components of $\mathbf{x}$ in that basis. It is convenient to define a length for the vector $\mathbf{x}$ which will (1) reduce to (F.2.3) for real three-dimensional vectors, (2) be zero if $x_i = 0$ for all i, and (3) be real. Such a definition is given by

$$x^2 = \sum_{i=1}^{n} |x_i|^2. \tag{F.3.2}$$

In order to generalize the concept of a scalar product to complex vectors it is convenient to introduce the concept of a dual space. Corresponding to each vector $\mathbf{x}$ in the a space we define another vector $\mathbf{x}^\dagger$ in the dual space by

$$\mathbf{x}^\dagger = \sum_{i=1}^{n} \bar{x}_i \mathbf{a}_i^\dagger \tag{F.3.3}$$

where $\bar{x}_i$ is the complex conjugate of x_i and

$$\mathbf{a}_i^\dagger \mathbf{a}_j = \delta_{ij},$$

i.e.

$$\begin{aligned} \mathbf{a}_i^\dagger \mathbf{a}_i &= 1 \qquad (\text{all } i), \\ \mathbf{a}_i^\dagger \mathbf{a}_j &= 0 \qquad (i \neq j). \end{aligned} \tag{F.3.4}$$

(F.3.4) is the generalization of (F.2.2) to complex vectors; if (F.3.4) holds, the $\mathbf{a}_i$ are said to form a *biorthonormal* basis. If $\mathbf{a}_i^\dagger = \mathbf{a}_i$, (F.3.4) expresses the mutual orthogonality of the vectors and the fact that they are unitary, i.e. of unit length. Defining the distributive law to hold, the square of the length is given by

$$\mathbf{x}^\dagger\mathbf{x} = \sum_{i=1}^{n}\sum_{j=1}^{n}(\bar{x}_i\mathbf{a}_i^\dagger)(x_j\mathbf{a}_j) = \sum_{i=1}^{n}\sum_{j=1}^{n}\bar{x}_i\delta_{ij}x_j = \sum_{i=1}^{n}|x_i|^2. \tag{F.3.5}$$

That the biorthonormal character of the $\mathbf{a}_i$ is a sufficient condition for linear independence may be verified as follows:

$$\mathbf{x} = 0 = \sum_{i=1}^{n}\mathbf{a}_i x_i \tag{F.3.6}$$

then

$$\mathbf{a}_j^\dagger\mathbf{x} = \sum_i \mathbf{a}_j^\dagger\mathbf{a}_i x_i = \sum_i \delta_{ji}x_i = x_j = 0.$$

Note that the general scalar product

$$\mathbf{x}^\dagger\mathbf{y} = \sum_{\substack{i=1\\j=1}}^{n}(\bar{x}_i\mathbf{a}_i^\dagger)(y_j\mathbf{a}_j) = \sum_{i=1}^{n}(\bar{x}_i y_i) \tag{F.3.7}$$

is, in general, a *complex* number, and that multiplication of vectors is *not* commutative; $\mathbf{x}^\dagger\mathbf{y}$ is, from (F.3.7), a number. On the other hand $\mathbf{y}\mathbf{x}^\dagger$ corresponds to a dyadic or open product in ordinary vector calculus, as may be seen by considering

$$(\mathbf{x}^\dagger\mathbf{y}\mathbf{y}^\dagger) = \mathbf{x}^\dagger(\mathbf{y}\mathbf{y}^\dagger) = \mathbf{x}^\dagger\sum_{i=1}^{n}|y_i|^2, \tag{F.3.8}$$

which can be seen to be a vector.

From (F.3.1) the *expansion theorem* may be obtained. Multiplying (F.3.1) by $\mathbf{a}_j^\dagger$ from the left:

$$\mathbf{a}_j^\dagger\mathbf{x} = \sum_i \mathbf{a}_j^\dagger x_i\mathbf{a}_i = x_j. \tag{F.3.9}$$

Substituting this back into (F.3.1):

$$\mathbf{x} = \sum_{i=1}^{n}\mathbf{a}_i x_i = \sum_{i=1}^{n}\mathbf{a}_i(\mathbf{a}_i^\dagger\mathbf{x}) = \sum_{i=1}^{n}(\mathbf{a}_i\mathbf{a}_i^\dagger)\mathbf{x},$$

from which

$$\sum_{i=1}^{n}\mathbf{a}_i\mathbf{a}_i^\dagger = \mathbf{1}. \tag{F.3.10}$$

Equation (F.3.1) gives the expansion of a vector $\mathbf{x}$ in terms of a biorthonormal basis $\mathbf{a}$. If the set of $\mathbf{a}$'s is such that an *arbitrary* vector may be expanded in terms of them, the set is said to be *complete*. In a three-dimensional Cartesian space the **i**, **j**, **k** vectors form a complete set; the **i**, **j**, by themselves clearly do not. Equation (F.3.10), which is satisfied by a complete set, is called the *completeness* relation.

Multiplication of a vector by a number c and the addition of two vectors are defined by

$$c\mathbf{x} = \sum_{i=1}^{n} \mathbf{a}_i(cx_i), \tag{F.3.11}$$

$$\mathbf{x} + \mathbf{y} = \sum_{i=1}^{n} \mathbf{a}_i x_i + \sum_{i=1}^{n} \mathbf{a}_i y_i = \sum_{i=1}^{n} \mathbf{a}_i(x_i + y_i). \tag{F.3.12}$$

References: (4), (6), (9), (7)*.

F.4 DIRAC NOTATION

Further manipulations in vector spaces are considerably simplified by using the notation introduced by Dirac[7]. In this notation the vectors $\mathbf{a}_i$ are simply written $|i\rangle$ and are referred to as kets. The vectors $\mathbf{a}_i^\dagger$ in the dual space are called bras and are written $\langle i|$. Since the base vectors are specified by their suffices different base vectors in the same space are sometimes denoted by primes $|i'\rangle$, $|i''\rangle$, etc. A summation convention is often employed.

In Dirac notation, (F.3.10) reads

$$|i\rangle\langle i| = \mathbf{1} \tag{F.4.1}$$

and (F.3.4) reads

$$\langle i|j\rangle = \delta_{ij} \tag{F.4.2}$$

or

$$\langle i|i'\rangle = \delta_{ii'}.$$

Multiplying (F.4.1) by $|x\rangle$ from the right,

$$|i\rangle\langle i|x\rangle = |x\rangle, \tag{F.4.3}$$

which corresponds to (F.3.1) and identifies $\langle i|x\rangle$ with x_i, in accordance with (F.3.9). The $\langle i|x\rangle$, i.e. the scalar products of $|x\rangle$ with $\langle i|$, are the representatives of $|x\rangle$ in the $|i\rangle$ basis. The set $\langle i|x\rangle$ 'represents' a vector $|x\rangle$ just as the components of an 'ordinary' vector represent it in a particular coordinate system. The representatives of the kets may be associated with column vectors and the representatives of the bras with row

* The reference numbers in parentheses at the end of each section refer to the numbers in the general reference list at the end of part F.

[7] P. A. M. Dirac, *Quantum Mechanics*, Clarendon Press, Oxford, 1958.

vectors, since the term $\langle x|y\rangle$ is a pure number whereas $|x\rangle\langle y|$ is a dyadic. Therefore the operation represented by (F.3.3) involves transposition and taking the complex conjugate of each term.

The simplicity of Dirac notation can only be appreciated by comparing it with alternative notations in more complicated cases than those hitherto considered. Ample scope for its application will be found in the following sections.

References: (4), (6), (3), (7).

F.5 CHANGE OF BASIS

Consider two biorthonormal bases $\mathbf{a}_i$, $\mathbf{b}_i$, in an n-dimensional vector space, which both form complete sets. Using (F.4.1),

$$|i\rangle = |L\rangle\langle L|i\rangle \tag{F.5.1}$$

and

$$|L\rangle = |i\rangle\langle i|L\rangle. \tag{F.5.2}$$

The function $\langle L|i\rangle$ in (F.5.1) can be represented as a square matrix $\mathbf{S}$ with elements S_{Li}; it is called the transformation matrix from the i to the L representation. Similarly $\langle i|L\rangle$ is the transformation matrix from the L to the i representation which will be denoted by $\mathbf{T}$. From their definition it follows that

$$\langle L|i\rangle = \langle i|L\rangle^*$$

$$\mathbf{S} = (\mathbf{T}^T)^* \tag{F.5.3}$$

where $\mathbf{T}^T$ stands for the transpose of $\mathbf{T}$. The matrix $\mathbf{T}$ so defined is said to be the *hermitian* conjugate of $\mathbf{S}$. The operation of transposing and taking the complex conjugate is denoted by a dagger†. Thus

$$\mathbf{S} = \mathbf{T}^\dagger,$$

$$\mathbf{T} = \mathbf{S}^\dagger. \tag{F.5.4}$$

Clearly

$$\sum_i \langle L|i\rangle\langle i|L\rangle = \langle L|L\rangle = \mathbf{I},$$

i.e.

$$\sum_i s_{Li}t_{iL} = \sum_i s_{Li}s_{iL}^* = \mathbf{SS}^\dagger = \mathbf{I}. \tag{F.5.5}$$

A matrix $\mathbf{S}$ which satisfies (F.5.5) is said to be *unitary*. Using these results it is simple to express an arbitrary vector $|x\rangle$ in any biorthonormal basis once its components in any other basis are known. Thus

$$|x\rangle = \sum_i |i\rangle \langle i|x\rangle$$

where the $\langle i|x\rangle$ are the components of $|x\rangle$ in the i basis. Let us assume these are known. Then in any other basis L

$$\langle L|x\rangle = \sum_i \langle L|i\rangle \langle i|x\rangle, \tag{F.5.6}$$

which expresses the components of $|x\rangle$ in the L basis in terms of those in the i basis.

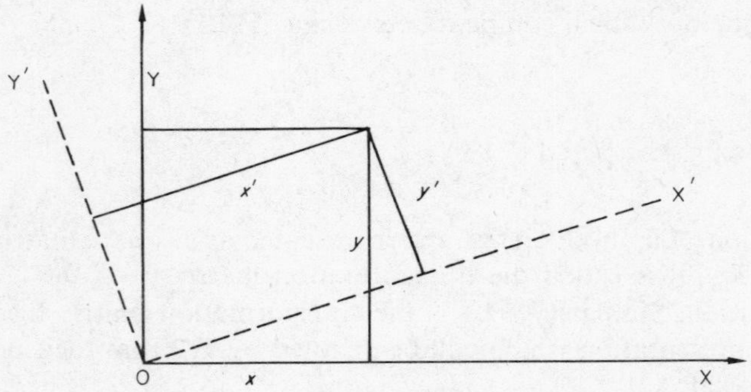

Fig. F.5.1.

Example

Consider a coordinate x, y, which by a counterclockwise rotation of angle θ is transformed into a system x', y'. In this case, the components of any vector will be real.

Writing

$$\begin{aligned} |r\rangle &= |i\rangle \langle i|r\rangle \\ &= |x\rangle \langle x|r\rangle + |y\rangle \langle y|r\rangle \\ &= |x\rangle x + |y\rangle y, \end{aligned}$$

we have

$$\begin{aligned} \langle x'|r\rangle &= \langle x'|x\rangle x + \langle x'|y\rangle y \\ \langle y'|r\rangle &= \langle y'|x\rangle x + \langle y'|y\rangle y, \end{aligned} \tag{F.5.7}$$

i.e.

$$\begin{bmatrix} \langle x'|r\rangle \\ \langle y'|r\rangle \end{bmatrix} = \begin{bmatrix} \langle x'|x\rangle & \langle x'|y\rangle \\ \langle y'|x\rangle & \langle y'|y\rangle \end{bmatrix} \begin{bmatrix} \langle x|r\rangle \\ \langle y|r\rangle \end{bmatrix}.$$

Clearly

$$x' = \cos\theta x + \sin\theta y,$$
$$y' = -\sin\theta x + \cos\theta y. \tag{F.5.8}$$

From (F.5.7) and (F.5.8),

$$\langle x|x\rangle = \cos\theta$$
$$\langle x'|y\rangle = \sin\theta$$
$$\langle y'|x\rangle = -\sin\theta$$
$$\langle y'|y\rangle = \cos\theta. \tag{F.5.9}$$

To conform with previous notation we put

$$|x\rangle, |y\rangle = |i\rangle, |i'\rangle,$$
$$|x'\rangle, |y'\rangle = |L\rangle, |L'\rangle,$$
$$\mathbf{S} = \begin{bmatrix} \cos\theta & \sin\theta \\ -\sin\theta & \cos\theta \end{bmatrix}$$

(for this particular case **S** is, of course, real and $\mathbf{S}^\dagger = \mathbf{S}^T$). It can be verified by direct multiplication that $\mathbf{SS}^\dagger = \mathbf{SS}^T = \mathbf{I}$, i.e. that the matrix **S** is unitary.

For an arbitrary vector **A** with components A_x, A_y, $A_{x'}$, $A_{y'}$ in the two systems we have

$$\langle x'|A\rangle = \langle x'|x\rangle\langle x|A\rangle + \langle x'|y\rangle\langle y|A\rangle,$$
$$\langle y'|A\rangle = \langle y'|x\rangle\langle x|A\rangle + \langle y'|y\rangle\langle y|A\rangle.$$

Using (F.5.9) these can be used to express $A_{x'}$, $A_{y'}$ in terms of A_x and A_y.
References: (4), (6).

Problem

(F.5.1) Write down the expressions for A_x and A_y in terms of $A_{x'}$, $A_{y'}$ and of the elements of **S**. (Note that **S** is unitary.)

F.6 SUMMARY OF MATRIX PROPERTIES

(a) If $\mathbf{T} = \mathbf{S}^\dagger$, i.e. $t_{ji}^* = s_{ij}$, **T** is said to be the *hermitian conjugate* of **S**. (F.6.1)

(b) If $s_{ij}^* = s_{ji}$, i.e. $\mathbf{S} = \mathbf{S}^\dagger$, **S** is hermitian or self-adjoint. (F.6.2)

(c) If $\mathbf{SS}^\dagger = \mathbf{I}$ (where **I** is the unit matrix) then **S** is unitary. (F.6.3)

(d) If $\mathbf{ST} = \mathbf{I}$, **T** is said to be the *inverse* of **S**, written $\mathbf{S}^{-1}$. A necessary and sufficient condition for the inverse of a matrix to exist is that its determinant is non-zero. If the elements of **S** are again denoted by s_{ij}, then the elements of $\mathbf{S}^{-1}$ are given by

$$(\mathbf{S}^{-1})_{ij} = \frac{S_{ji}}{|\mathbf{S}|} \tag{F.6.4}$$

where S_{ji} is the cofactor of the element s_{ji}.

(e) The trace of a matrix is defined by $\mathrm{Tr}[\mathbf{A}] = \sum_i a_{ii}$.

(f) If a function $f(x)$ has the convergent series expansion about $x = a$

$$f(x) = \sum_{n=0}^{\infty} \frac{(x-a)^n}{n!}\left(\frac{\mathrm{d}^n f}{\mathrm{d}x^n}\right)_{x=a},$$

then a function $f(\mathbf{A})$ of a matrix **A** can be defined by

$$f(\mathbf{A}) = \sum_{n=0}^{\infty} \frac{(\mathbf{A} - a\mathbf{I})^n}{n!}\left(\frac{\mathrm{d}^n f}{\mathrm{d}x^n}\right)_{x=a} \tag{F.6.5}$$

where **I** is the unit matrix.

(g) If, for a matrix **A** and a number λ there exists a column such that, in conventional notation,

$$\mathbf{A}u_i = \lambda_i u_i, \tag{F.6.6}$$

then u_i is said to be an eigenvector of **A**, λ_i is said to be an eigenvalue of **A**, and the eigenvalue λ_i and the eigenvector u_i are said to belong to each other.

(h) The eigenvalues of a hermitian matrix are real.

(i) If a matrix **A** satisfies the relation

$$\mathbf{A}\mathbf{A}^\dagger = \mathbf{A}^\dagger\mathbf{A}, \tag{F.6.7}$$

A is said to be a normal matrix. It follows that hermitian and unitary matrices are normal matrices. It can be shown that if a matrix is normal its eigenvectors form a biorthonormal set (basis)[8].

(j) A matrix function for which the expansion (F.6.6) holds can be evaluated as

$$f(\mathbf{A})u_i = \sum_{n=0}^{\infty} \left[\frac{\mathrm{d}^n f}{\mathrm{d}x^n}\right]_{x=a} \frac{(\lambda_i - a)^n u_i}{n!}, \tag{F.6.8}$$

that is

$$f(\mathbf{A})u_i = f(\lambda_i)u_i. \tag{F.6.9}$$

If the eigenvectors u_n are formed into a matrix **u**, and the eigenvalues are similarly formed into a diagonal matrix $\mathbf{\Lambda}$, (F.6.9) can be written

$$\mathbf{A}\mathbf{u} = \mathbf{u}\mathbf{\Lambda}, \qquad \mathbf{A} = \mathbf{u}\mathbf{\Lambda}\mathbf{u}^{-1} \tag{F.6.10}$$

[8] G. Goertzel and N. Tralli, *Some Mathematical Methods of Physics*, McGraw–Hill, 1960.

(F.6.10) is the *spectral representation* of a matrix in terms of its eigenvectors and eigenvalues. (F.6.9) can then be written

$$f(\mathbf{A})\mathbf{u} = \mathbf{u}f(\mathbf{\Lambda}),$$

$$f(\mathbf{A}) = \mathbf{u}f(\mathbf{\Lambda})\mathbf{u}^{-1}, \tag{F.6.11}$$

which is the spectral representation of the function $f(\mathbf{A})$.

Problems

(F.6.1) Show that, provided the matrices can be multiplied in the order indicated,

$$\mathrm{Tr}[\mathbf{AB}] = \mathrm{Tr}[\mathbf{BA}],$$

$$\mathrm{Tr}[\mathbf{ABC}] = \mathrm{Tr}[\mathbf{CAB}] = \mathrm{Tr}[\mathbf{BCA}].$$

(F.6.2) Verify that $[\mathbf{AB}]^T = \mathbf{B}^T\mathbf{A}^T$. Generalize to evaluate the expression

$$[\mathbf{ABC}\ldots]^T.$$

(F.6.3) Show that $[\mathbf{AB}]^{-1} = \mathbf{B}^{-1}\mathbf{A}^{-1}$. Generalize to evaluate $[\mathbf{ABC}\ldots]^{-1}$ in terms of the inverses $\mathbf{A}^{-1}, \mathbf{B}^{-1}, \ldots$

(F.6.4) Let

$$\epsilon = \begin{pmatrix} 0 & 0 & 0 \\ 1 & 0 & 0 \\ 0 & 1 & 0 \end{pmatrix}.$$

Evaluate ϵ^2, ϵ^3, $e^{\epsilon t}$, using the definition of a matrix function given above.

(F.6.5) Using the series expansions for e^x, $\cos x$, $\sin x$, show that

$$\exp\begin{pmatrix} 0 & 1 \\ -1 & 0 \end{pmatrix} t = \mathbf{I}\cos t + \begin{pmatrix} 0 & 1 \\ -1 & 0 \end{pmatrix} \sin t$$

(F.6.6) (a) If $\mathbf{H}$ is a hermitian matrix, show that $e^{i\mathbf{H}t}$ is unitary (t is any scalar).

(b) Show that $\dfrac{1 + i\mathbf{H}}{1 - i\mathbf{H}}$ is unitary.

(c) If $\mathbf{U}$ is unitary show that $\dfrac{i(\mathbf{U} - \mathbf{1})}{\mathbf{U} + \mathbf{1}}$ is hermitian.

(F.6.7) Find the eigenvectors and eigencolumns of the matrix

$$\begin{pmatrix} -5 & 6 \\ -4 & 5 \end{pmatrix}.$$

(F.6.8) Using equation (F.6.11), evaluate $\begin{pmatrix} 1 & 3 \\ 3 & 1 \end{pmatrix}^{15}$.

(F.6.9) Let $\mathbf{M} = \begin{pmatrix} 0 & 0 & 0 \\ 1 & 0 & 0 \\ 0 & 1 & 0 \end{pmatrix}$.

If $f(x)$ has a convergent power series expansion about $x = 0$, show that

$$f(\mathbf{M}) = f(0) + \mathbf{M}f'(0) + \frac{\mathbf{M}^2}{2!}f''(0).$$

F.7 LINEAR OPERATORS

An operator $\hat{\mathbf{L}}$ is an entity which, acting on a vector $\mathbf{x}$, converts it into a vector $\mathbf{y}$:

$$\hat{\mathbf{L}}|x\rangle = |y\rangle. \tag{F.7.1}$$

If

$$\hat{\mathbf{L}}[|x\rangle + |y\rangle] = \hat{\mathbf{L}}|x\rangle + \hat{\mathbf{L}}|y\rangle \tag{F.7.2}$$

$$\hat{\mathbf{L}}\alpha|x\rangle = \alpha\hat{\mathbf{L}}|x\rangle \tag{F.7.3}$$

(α is a scalar) then the operator is said to be linear.

A linear operator is determined if and only if it effects on every vector in a basis are known. In particular, we may write

$$\begin{aligned}\hat{\mathbf{L}}|i\rangle &= \sum_j |j\rangle\, \langle j|\, \hat{\mathbf{L}}|i\rangle \\ &= |j\rangle l_{ji},\end{aligned} \tag{F.7.4}$$

which shows that the action of the operator is represented by l_{ji}. Using (F.4.1) twice, we can write

$$\hat{\mathbf{L}} = \sum_i \sum_j |i\rangle\, \langle i|\hat{\mathbf{L}}|j\rangle\, \langle j|, \tag{F.7.5}$$

which shows that the matrix l_{ij} completely determines the operator $\hat{\mathbf{L}}$. (F.7.4) similarly shows that $\hat{\mathbf{L}}$ determines the matrix l_{ij}. From (F.7.4) we define the hermitian conjugate of an operator $\hat{\mathbf{L}}$ by

$$\begin{aligned}\hat{\mathbf{L}}^\dagger &= \sum_{i,j} |i\rangle (l_{ij})^\dagger \langle j| \\ &= |i\rangle \bar{l}_{ji} \langle j|.\end{aligned} \tag{F.7.6}$$

The identity operator $\hat{\mathbf{I}}$ is defined by

$$\hat{\mathbf{I}}|u\rangle = |u\rangle. \tag{F.7.7}$$

If the matrix elements of $\hat{\mathbf{L}}$ are known, not in the i basis but in the m basis (say), then (F.7.5) may be written [by using (F.4.1)] in terms of known matrix elements as

$$\hat{\mathbf{L}} = \sum_i \sum_j \sum_m \sum_n |i\rangle\, \langle i|m\rangle\, \langle m|\hat{\mathbf{L}}|n\rangle\, \langle n|\, j\rangle\, \langle j|. \tag{F.7.8}$$

The inverse operator is defined by the relation

$$\hat{\mathbf{L}}\hat{\mathbf{L}}^{-1} = \hat{\mathbf{I}} \tag{F.7.9}$$

An operator $\hat{\mathbf{L}}$ is said to be unitary if

$$\hat{\mathbf{L}}\hat{\mathbf{L}}^{\dagger} = \hat{\mathbf{I}} \tag{F.7.10}$$

and normal if

$$\hat{\mathbf{L}}\hat{\mathbf{L}}^{\dagger} = \hat{\mathbf{L}}^{\dagger}\hat{\mathbf{L}}. \tag{F.7.11}$$

For a function $f(x)$ with a convergent series expansion about the point $x = a$, the operator function $f(\hat{\mathbf{L}})$ may be defined by

$$f(\hat{\mathbf{L}}) = \sum_{n=0}^{\infty} \frac{(\hat{\mathbf{L}} - a\hat{\mathbf{I}})^n}{n!}\left(\frac{\mathrm{d}^n f}{\mathrm{d}x^n}\right)_{x=a}. \tag{F.7.12}$$

If an operator $\hat{\mathbf{L}}$ and a non-zero vector $|i\rangle$ are related by

$$\hat{\mathbf{L}}|i\rangle = i|i\rangle \tag{F.7.13}$$

where i is a number, $|i\rangle$ is said to be an eigenvector of $\hat{\mathbf{L}}$, i is said to be an eigenvalue of $\hat{\mathbf{L}}$, and the eigenvalue i and the eigenvector $|i\rangle$ are said to belong to each other. From the discussion of §F.6 it follows that normal operators have eigenvectors which form a biorthonormal basis. If these form a complete set one can write

$$\hat{\mathbf{L}} = \sum_i \hat{\mathbf{L}}|i\rangle\langle i| = \sum_i |i\rangle\, i\,\langle i|. \tag{F.7.14}$$

This is the *spectral representation* of an operator. From (F.7.12) and (F.7.13) it follows that

$$f(\hat{\mathbf{L}}) = \sum_i |i\rangle\, f(i)\,\langle i|, \tag{F.7.15}$$

which is the *spectral representation* of the operator $f(\hat{\mathbf{L}})$.

An illustrative example

Consider the 2 × 2 Pauli spin operator $\hat{\boldsymbol{\sigma}}_z$ which obeys the equation

$$\hat{\boldsymbol{\sigma}}_z^2 = \hat{\mathbf{I}} \tag{F.7.16}$$

and assume that $\hat{\boldsymbol{\sigma}}_z$ is hermitian. The eigenvectors of $\hat{\boldsymbol{\sigma}}_z$ satisfy

$$\hat{\boldsymbol{\sigma}}_z|l\rangle = l|l\rangle$$

and

$$\hat{\boldsymbol{\sigma}}_z^2|l\rangle = l^2|l\rangle. \tag{F.7.17}$$

From the discussion of §F.6, it follows that the $|l\rangle$ form a biorthonormal set, since $\hat{\boldsymbol{\sigma}}_z$ is a normal operator. Since each eigenvector is undefined to within a constant factor, this factor may be chosen to give $\langle l|l\rangle = 1$, i.e. normalization to unity. Multiplication of (F.7.17) by $\langle l|$ from the left gives

$$\langle l|\hat{\boldsymbol{\sigma}}_z^2|l\rangle = l^2,$$

$$l^2 = \pm 1.$$

The eigenvectors are $|+1\rangle$ and $|-1\rangle$. In this simple case we need not *assume* the Dirac expansion theorem; we are able to prove it. For an arbitrary vector $|P\rangle$ we can write

$$\begin{aligned}|P\rangle &= \tfrac{1}{2}(\hat{\mathbf{I}} + \hat{\boldsymbol{\sigma}}_z + \hat{\mathbf{I}} - \hat{\boldsymbol{\sigma}}_z)|P\rangle \\ &= \tfrac{1}{2}(\hat{\mathbf{I}} + \hat{\boldsymbol{\sigma}}_z)|P\rangle + \tfrac{1}{2}(\hat{\mathbf{I}} - \hat{\boldsymbol{\sigma}}_z)|P\rangle. \end{aligned} \tag{F.7.18}$$

From (F.7.16),

$$\hat{\boldsymbol{\sigma}}_z[\tfrac{1}{2}(\hat{\mathbf{I}} + \hat{\boldsymbol{\sigma}}_z)]|P\rangle = \tfrac{1}{2}(\hat{\mathbf{I}} + \hat{\boldsymbol{\sigma}}_z)|P\rangle$$

so that $\frac{1}{2}(\hat{\mathbf{I}} + \hat{\boldsymbol{\sigma}}_z)|P\rangle$ is an eigenket of $\hat{\boldsymbol{\sigma}}_z$ with eigenvalue $+1$. It can therefore only differ from $|+1\rangle$ by a constant and we can write

$$\tfrac{1}{2}(\hat{\mathbf{I}} + \hat{\boldsymbol{\sigma}}_z)|P\rangle = c_1| + 1\rangle. \tag{F.7.19}$$

Similarly

$$\hat{\boldsymbol{\sigma}}_z[\tfrac{1}{2}(\hat{\mathbf{I}} - \hat{\boldsymbol{\sigma}}_z)]|P\rangle = \tfrac{1}{2}(\hat{\boldsymbol{\sigma}}_z - \mathbf{1})|P\rangle = (-1)\tfrac{1}{2}(\hat{\mathbf{I}} - \hat{\boldsymbol{\sigma}}_z)|P\rangle$$

so that $\frac{1}{2}(\hat{\mathbf{I}} - \hat{\boldsymbol{\sigma}}_z)|P\rangle$ is an eigenket of $\hat{\boldsymbol{\sigma}}_z$ with eigenvalue (-1). Hence

$$\tfrac{1}{2}(\hat{\mathbf{I}} - \hat{\boldsymbol{\sigma}}_z)|P\rangle = c_2|(-1)\rangle. \tag{F.7.20}$$

From (F.7.18),

$$|P\rangle = c_1|+1\rangle + c_2|-1\rangle \tag{F.7.21}$$

which shows that an arbitrary vector $|P\rangle$ in the space may be written as a superposition of the eigenvectors $|\pm 1\rangle$. The set of eigenvectors is therefore complete—a particular case of the Dirac expansion theorem. The operator $\hat{\boldsymbol{\sigma}}_z$ has the spectral representation

$$\begin{aligned}\hat{\boldsymbol{\sigma}}_z &= |l\rangle\, l\,\langle l| \\ &= |(+1)\rangle(+1)\langle(+1)| + |(-1)\rangle(-1)\langle(-1)| \\ &= |+\rangle\langle+| - |-\rangle\langle-|, \end{aligned} \tag{F.7.22}$$

which is a particular case of (F.7.15). From (F.7.21) the representatives of the vector $|P\rangle$ in the $|i\rangle$ representation are $\langle+|P\rangle$ and $\langle-|P\rangle$, which are identified respectively with c_1 and c_2, since in the l representation, in which $\hat{\boldsymbol{\sigma}}_z$ is diagonal, (F.7.21) can be written (utilizing (F.4.3)) as

$$|P\rangle = |l\rangle\langle l|P\rangle = |+\rangle\langle +|P\rangle + |-\rangle\langle -|P\rangle. \qquad \text{(F.7.23)}$$

The representatives of P in any other representation L are given by

$$\begin{aligned}\langle L|P\rangle &= \langle L|l\rangle\langle l|P\rangle \\ &= \langle L|+\rangle\langle +|P\rangle + \langle L|-\rangle\langle -|P\rangle,\end{aligned} \qquad \text{(F.7.24)}$$

which is a particular case of (F.5.6). The matrix representing the operator $\hat{\boldsymbol{\sigma}}_z$ in the i representation has elements which are given by

$$\begin{aligned}\langle i|\hat{\boldsymbol{\sigma}}|j\rangle = \langle +1|\hat{\boldsymbol{\sigma}}_z|+1\rangle &= 1 \\ \langle -1|\hat{\boldsymbol{\sigma}}_z|-1\rangle &= -1 \\ \langle +1|\hat{\boldsymbol{\sigma}}_z|-1\rangle &= 0 \\ \langle -1|\hat{\boldsymbol{\sigma}}_z|+1\rangle &= 0,\end{aligned}$$

so that the operator may be represented by the matrix

$$\begin{bmatrix} +1 & 0 \\ 0 & -1 \end{bmatrix}.$$

In any other representation $|L\rangle$ the operator is represented by a matrix with elements

$$\begin{aligned}\langle L|\hat{\boldsymbol{\sigma}}|L'\rangle &= \langle L|l\rangle\langle l|\hat{\boldsymbol{\sigma}}|l'\rangle\langle l'|L'\rangle \\ &= \langle L|+1\rangle\langle +1|L'\rangle - \langle L|-1\rangle\langle -1|L'\rangle.\end{aligned} \qquad \text{(F.7.25)}$$

This ends our example.

Problems

(F.7.1) Show that if $\hat{\mathbf{L}}$, $\hat{\mathbf{M}}$ are two operators

$$(\hat{\mathbf{L}}\hat{\mathbf{M}})^\dagger = \hat{\mathbf{M}}^\dagger\hat{\mathbf{L}}^\dagger,$$
$$(\hat{\mathbf{L}}\hat{\mathbf{M}})^{-1} = \hat{\mathbf{M}}^{-1}\hat{\mathbf{L}}^{-1}.$$

(F.7.2) If $\hat{\mathbf{p}}$, $\hat{\mathbf{q}}$, are two operators which satisfy (in one dimension)

$$[\hat{\mathbf{q}}, \hat{\mathbf{q}}] = 0, [\hat{\mathbf{p}}, \hat{\mathbf{p}}] = 0, \qquad [\hat{\mathbf{q}}, \hat{\mathbf{p}}] = \hat{\mathbf{q}}\hat{\mathbf{p}} - \hat{\mathbf{p}}\hat{\mathbf{q}} = i\hbar,$$

show that

$$[\hat{\mathbf{q}}, \hat{\mathbf{p}}^l] = i\hbar l\hat{\mathbf{p}}^{l-1},$$

$$[\hat{\mathbf{p}}, \hat{\mathbf{q}}^l] = -i\hbar l\hat{\mathbf{q}}^{l-1}.$$

(F.7.3) If $\hat{\mathbf{H}}$ is a hermitian operator show that $e^{i\hat{\mathbf{H}}t}$ is unitary.

(F.7.4) Using Dirac notation show that

$$\mathrm{Tr}\{\hat{\mathbf{L}}\hat{\mathbf{M}}\} = \mathrm{Tr}\{\hat{\mathbf{M}}\hat{\mathbf{L}}\}.$$

Generalize to a product of operators $\hat{\mathbf{L}}$, $\hat{\mathbf{M}}$, $\hat{\mathbf{N}}$.

(F.7.5) Show that eigenvectors belonging to two different eigenvalues of a hermitian operator are orthogonal.

(F.7.6) Show that the eigenvalues of a hermitian operator are real.

(F.7.7) Using the results of problem (F.7.2) show that, if $F(\hat{\mathbf{p}})$ and $G(\hat{\mathbf{q}})$ are functions that may be expanded in a power series in $\hat{\mathbf{p}}$ and $\hat{\mathbf{q}}$ respectively,

$$[\hat{\mathbf{q}}, F(\hat{\mathbf{p}})] = i\hbar\frac{\partial F}{\partial \hat{\mathbf{p}}},$$

$$[\hat{\mathbf{p}}, G(\hat{\mathbf{q}})] = -i\hbar\frac{\partial G}{\partial \hat{\mathbf{q}}}.$$

Show further that if $F(\hat{\mathbf{p}}, \hat{\mathbf{q}})$ is a function of $\hat{\mathbf{p}}$ and $\hat{\mathbf{q}}$,

$$[\hat{\mathbf{q}}, F(\hat{\mathbf{p}}, \hat{\mathbf{q}})] = i\hbar\frac{\partial F}{\partial \hat{\mathbf{p}}},$$

$$[\hat{\mathbf{p}}, F(\hat{\mathbf{p}}, \hat{\mathbf{q}})] = -i\hbar\frac{\partial F}{\partial \hat{\mathbf{q}}}.$$

(F.7.8) If **A**, **B**, are arbitrary non-commuting matrices or operators, prove the identity

$$[\mathbf{A} - \mathbf{B}]^{-1} = [\mathbf{A}(\hat{\mathbf{I}} - \mathbf{A}^{-1}\mathbf{B})]^{-1} = [\hat{\mathbf{I}} - \mathbf{A}^{-1}\mathbf{B}]^{-1}\mathbf{A}^{-1}.$$

From this obtain the iterative expansion

$$[\mathbf{A} - \mathbf{B}]^{-1} = \mathbf{A}^{-1} + \mathbf{A}^{-1}\mathbf{B}\mathbf{A}^{-1} + \mathbf{A}^{-1}\mathbf{B}\mathbf{A}^{-1}\mathbf{B}\mathbf{A}^{-1} + \cdots.$$

F.8 REPRESENTATIONS IN QUANTUM MECHANICS

F.8a General

The term 'representation' is used in the literature in at least two distinct senses, and this article is no exception.

The most important use of the term refers to the formulation of the basic equation of motion of the state vector. In particular we distinguish between the Schrödinger representation, in which the state vector is time-dependent, and the Heisenberg representation, in which it is independent of time. A second use of the term deals essentially with the coordinate system in which the motion of the state vector is described.

The Schrödinger representation may be set up by means of the following postulates:

(a) The state of the system is determined at any time t by a state vector $\psi_s(t)$.

(b) To an observable A there corresponds an operator $\hat{\mathbf{A}}_s$ which is *independent of time*. The eigenvalues of $\hat{\mathbf{A}}_s$ are the possible results of the measurements of A, that is, denoting the eigenvalues of $\hat{\mathbf{A}}_s$ by a,

$$\hat{\mathbf{A}}_s|a\rangle = a|a\rangle \tag{F.8.1}$$

and the probability of a measurement of A yielding the value a at time t is $|\langle a|\psi_s(t)\rangle|^2$. The a's, which are the results of possible measurements, must be real. This implies that $\hat{\mathbf{A}}_s$ must be a hermitian operator.

(c) There exists a hermitian operator $\hat{\mathbf{H}}_s$ such that

$$i\hbar \frac{\partial}{\partial t}|\psi_s(t)\rangle = \hat{\mathbf{H}}|\psi_s(t)\rangle. \tag{F.8.2}$$

(d) Two classical dynamical variables a, b, which are conjugate in the Hamiltonian sense, are represented by Schrödinger operators $\hat{\mathbf{A}}$, $\hat{\mathbf{B}}$, which obey

$$\hat{\mathbf{A}}_i\hat{\mathbf{B}}_j - \hat{\mathbf{B}}_j\hat{\mathbf{A}}_i = i\hbar\,\delta_{ij}. \tag{F.8.3}$$

F.8b The position and momentum representations

Postulate (b) of the preceding section may be applied in particular to the measurement of position and momentum. We assume the existence of an operator $\hat{\mathbf{q}}$ whose eigenvalues q give the possible results of a measurement of position. From problem (F.7.7) we have the general commutation relations

$$[\hat{\mathbf{q}}, F(\hat{\mathbf{p}})] = i\hbar\frac{\partial F}{\partial \hat{\mathbf{p}}}, \tag{F.8.4}$$

$$[\hat{\mathbf{p}}, G(\hat{\mathbf{q}})] = -i\hbar\frac{\partial G}{\partial \hat{\mathbf{q}}}. \tag{F.8.5}$$

Problem (F.7.7) is based on the result of problem (F.7.2). We assume that the 'arbitrary' functions $F(\hat{\mathbf{p}})$, $F(\hat{\mathbf{q}})$ can be expanded in power series in $\hat{\mathbf{p}}$ and $\hat{\mathbf{q}}$ respectively. We then prove directly that if the result of (F.7.2) holds for the index l then it holds also for $l + 1$. The result of (F.7.7) then follows from that of (F.7.2). We postulate that since $\hat{\mathbf{p}}$ and $\hat{\mathbf{q}}$ are operators representing observables, the eigenvalues of $\hat{\mathbf{p}}$ and $\hat{\mathbf{q}}$ must be real. Hence $\hat{\mathbf{p}}$ and $\hat{\mathbf{q}}$ must be hermitian. We now introduce the basis in which $\hat{\mathbf{q}}$ is diagonal by the equation

$$\hat{\mathbf{q}}|q'\rangle = q'|q'\rangle. \tag{F.8.6}$$

Now introduce the translation operator $e^{-i\hat{\mathbf{p}}\xi/\hbar} = \hat{\mathbf{S}}$ (where ξ is a c-number and real). Its hermitian conjugate is

$$\hat{\mathbf{S}}^{\dagger} = e^{+i\hat{\mathbf{p}}^{\dagger}\xi/\hbar}$$
$$= e^{i\mathbf{p}\xi/\hbar} \tag{F.8.7}$$

since

$$\hat{\mathbf{p}}^{\dagger} = \hat{\mathbf{p}}.$$

Hence

$$\hat{\mathbf{S}}\hat{\mathbf{S}}^{\dagger} = 1, \tag{F.8.8}$$

i.e. $\hat{\mathbf{S}}$ is unitary. Now consider the application of (F.8.4) to the commutator

$$[\hat{\mathbf{q}}, \hat{\mathbf{S}}] = i\hbar\frac{\partial \hat{\mathbf{S}}}{\partial \hat{\mathbf{p}}} = i\hbar\left(\frac{-i\xi}{\hbar}\right)\hat{\mathbf{S}} = \xi\hat{\mathbf{S}}. \tag{F.8.9}$$

(F.8.8) can be written

$$\hat{\mathbf{q}}\hat{\mathbf{S}} = \hat{\mathbf{S}}\hat{\mathbf{q}} + \xi\hat{\mathbf{S}} = \hat{\mathbf{S}}(\hat{\mathbf{q}} + \xi), \tag{F.8.10}$$

$$\hat{\mathbf{q}}\hat{\mathbf{S}}|q'\rangle = \hat{\mathbf{S}}(\hat{\mathbf{q}} + \xi)|q'\rangle = \hat{\mathbf{S}}(q' + \xi)|q'\rangle. \tag{F.8.11}$$

From (F.8.10) $\hat{\mathbf{q}}$ operating on $\hat{\mathbf{S}}|q'\rangle$ produces $\hat{\mathbf{S}}|q'\rangle$ multiplied by the constant $q' + \xi$, i.e. $|\hat{\mathbf{S}}q'\rangle$ is an eigenvector of $\hat{\mathbf{q}}$ with eigenvalue $q' + \xi$. Therefore

$$\hat{\mathbf{S}}|q'\rangle = c|q' + \xi\rangle \tag{F.8.12}$$

where c is a constant which comes from the fact that eigenvectors are not determined in absolute value. We put $c = 1$. Since ξ is arbitrary any eigenvalue of $\hat{\mathbf{q}}$ can be generated; they form a continuous basis. Then

$$\hat{\mathbf{S}}(\xi)|0\rangle = |\xi\rangle,$$

$$\hat{\mathbf{S}}(\xi)\hat{\mathbf{S}}(q')|0\rangle = e^{-i\hat{\mathbf{p}}\xi/\hbar}\cdot e^{-i\hat{\mathbf{p}}q'/\hbar}|0\rangle = |q' + \xi\rangle$$
$$= \hat{\mathbf{S}}(\xi)|q'\rangle. \tag{F.8.13}$$

Hence

$$\langle q''|\hat{\mathbf{S}}(\xi)|q'\rangle = \langle q''|q' + \xi\rangle. \tag{F.8.14}$$

Since the q's are the eigenvectors of a hermitian operator they form an orthogonal basis. Thus

$$\langle q''|\hat{\mathbf{S}}(\xi)|q'\rangle = \delta(q'' - q' - \xi). \tag{F.8.15}$$

To obtain the matrix elements of p we note that (F.8.15) is true for all ξ and in particular, therefore, for small values of it. For these (F.8.14) can be expanded:

$$\langle q''|\,\mathrm{e}^{-\mathrm{i}\hat{\mathbf{p}}\xi/\hbar}|q'\rangle = \langle q''|\left[1 - \frac{\mathrm{i}\mathbf{p}\xi}{\hbar} + 0(\xi^2)\right]|q'\rangle,$$

so

$$\langle q''|\hat{\mathbf{p}}|q'\rangle = -\frac{\hbar}{\mathrm{i}}\,\frac{\delta(q'' - q' - \xi) - \delta(q'' - q')}{\xi}$$

$$= \frac{\hbar}{\mathrm{i}}\,\delta'(q'' - q'). \qquad \text{(F.8.16)}$$

In order to evaluate $\langle q''|\hat{\mathbf{p}}^n|q'\rangle$ we note that, taking $n = 2$,

$$\langle q''|\hat{\mathbf{p}}^2|q'\rangle = \langle q''|\hat{\mathbf{p}}|q'''\rangle\langle q'''|\hat{\mathbf{p}}|q'\rangle$$

$$= \int \left(\frac{\hbar}{\mathrm{i}}\right)^2 \delta'(q'' - q''')\,\delta'(q''' - q')\,\mathrm{d}q''' \qquad \text{(F.8.17)}$$

We now use the property of the δ-function:

$$\int \mathrm{d}q'\delta'(q - q')f(q') = f'(q), \qquad \text{(F.8.18)}$$

so that

$$\langle q''|\hat{\mathbf{p}}^2|q'\rangle = \left(\frac{\hbar}{\mathrm{i}}\right)^2 \frac{\partial^2}{\partial q''^2}\,\delta(q'' - q'). \qquad \text{(F.8.19)}$$

Clearly

$$\langle q''|\hat{\mathbf{p}}^n|q'\rangle = \left(\frac{\hbar}{\mathrm{i}}\right)^n \frac{\partial^n}{\partial q''}\,\delta(q'' - q').$$

For any operator function $F(\hat{\mathbf{p}})$ expressible as a power series in $\hat{\mathbf{p}}$:

$$\langle q''|F(\hat{\mathbf{p}})|q'\rangle = F\left(\frac{\hbar}{\mathrm{i}}\frac{\partial}{\partial q'}\right)\delta(q'' - q'), \qquad \text{(F.8.20)}$$

with obvious extension to three dimensions.

Since any operator $F(\hat{\mathbf{q}})$ is diagonal in the $|q'\rangle$ representation,

$$\langle q''|F(\hat{\mathbf{q}})|q'\rangle = F(q')\,\delta(q'' - q'). \qquad \text{(F.8.21)}$$

It may similarly be shown by using the function $\hat{\mathbf{S}}^\dagger = \mathrm{e}^{\mathrm{i}\hat{\mathbf{p}}'\xi/\hbar}$ that the representation $|p\rangle$ in which $\hat{\mathbf{p}}$ is diagonal also has a continuous spectrum extending from $-\infty$ to ∞. The two representations are connected by the transformation function

$$\langle q|p\rangle$$

whose value may be calculated as follows. Let

$$\hat{\mathbf{p}}|p'\rangle = p'|p'\rangle \tag{F.8.22}$$

be the defining equation for the p basis. This can be multiplied by $\langle q'|$ from the left:

$$\begin{aligned} p'\langle q'|p'\rangle = \langle q'|\hat{\mathbf{p}}|p'\rangle &= \int \langle q'|\hat{\mathbf{p}}|q''\rangle \langle q''|p'\rangle \, \mathrm{d}q'' \\ &= \int \frac{\hbar}{\mathrm{i}} \, \delta'(q'' - q')\langle q''|p'\rangle \, \mathrm{d}q'' \\ &= \frac{\hbar}{\mathrm{i}} \frac{\partial}{\partial q'} \langle q'|p'\rangle. \end{aligned} \tag{F.8.23}$$

This is a differential equation for $\langle q'|p'\rangle$ whose solution is

$$\langle q'|p'\rangle = a\, \mathrm{e}^{\mathrm{i}p'q'/\hbar}. \tag{F.8.24}$$

The value of the constant may be obtained as follows. Using the orthogonality of the $|p\rangle$'s (which follows from the fact that they are the eigenvectors of a hermitian operator):

$$\langle p''|q'\rangle\langle q'|p'\rangle = |a|^2 \int \mathrm{e}^{\mathrm{i}(p'-p'')q'/\hbar} \, \mathrm{d}q' = \langle p''|p'\rangle = \delta(p'' - p'). \tag{F.8.25}$$

The integral will, in a later chapter, be shown to be equal to $2\pi\hbar\, \delta(p'' - p')$ and therefore

$$|a|^2 2\pi\hbar = 1. \tag{F.8.26}$$

This determines a, apart from an arbitrary phase factor. Choosing this to be unity we have

$$\langle q|p\rangle = \frac{1}{(2\pi\hbar)^{\frac{1}{2}}} \mathrm{e}^{\mathrm{i}pq/\hbar}. \tag{F.8.27}$$

(F.8.27) is frequently written, putting $\hbar k = p$, in the form

$$\langle q|k\rangle = \frac{1}{(2\pi)^{\frac{1}{2}}} \mathrm{e}^{\mathrm{i}kq} \tag{F.8.28}$$

and for three dimensions

$$\langle \mathbf{q}|\mathbf{k}\rangle = \frac{1}{(2\pi)^{\frac{3}{2}}} e^{i\mathbf{k}\cdot\mathbf{q}}. \tag{F.8.29}$$

An alternative way of expressing the completeness relation (F.4.1) is obtained if matrix elements are taken in the $|l\rangle$ representation:

$$\sum_i \langle l|i\rangle\, \langle i|l'\rangle = \langle l|l'\rangle. \tag{F.8.30}$$

We shall assume that the set of $|l\rangle$'s is biorthonormal. In this case the R.H.S. is just $\delta_{ll'}$ or $\delta(l - l')$ depending on whether the $|l\rangle$ basis is discrete or continuous. F.8.30 is then an alternative way of expressing the completeness relation.

We shall be mainly concerned with the consequences of postulate (b) of §F.8a when the general operator $\hat{\mathbf{A}}$ represents the time-independent Hamiltonian $\hat{\mathbf{H}}$. Postulate (b) then tells us that the allowed energy eigenvalues of the system, denoted by $|n\rangle$, will be given by

$$\hat{\mathbf{H}}|n\rangle = n|n\rangle. \tag{F.8.31}$$

This equation can be solved in abstract vector space in particular cases —for example, for the simple harmonic oscillator. We can convert it into its more usual form by taking matrix elements in the position representation and making use of equation (F.4.1). In this way we obtain

$$\langle \mathbf{q}|\hat{\mathbf{H}}|\mathbf{q}'\rangle\, \langle \mathbf{q}'|n\rangle = n\, \langle \mathbf{q}|n\rangle. \tag{F.8.32}$$

The Hamiltonian $\hat{\mathbf{H}}$ will, in all the cases we shall encounter, be written in the form

$$\hat{\mathbf{H}} = \frac{\hat{\mathbf{p}}^2}{2m} + \hat{\mathbf{H}}_I \tag{F.8.33}$$

where the first part is diagonal in the momentum representation and the second in the position representation. Making use of (F.8.19) we have

$$\begin{aligned} \left\langle \mathbf{q}\left|\frac{\hat{\mathbf{p}}^2}{2m}\right|\mathbf{q}'\right\rangle \langle \mathbf{q}'|n\rangle &= -\frac{\hbar^2}{2m}\int \delta(\mathbf{q}-\mathbf{q}')\nabla^2_{\mathbf{q}'}\, \langle \mathbf{q}'|n\rangle \mathrm{d}\mathbf{q}' \\ &= -\frac{\hbar^2}{2m}\nabla^2_{\mathbf{q}}\langle \mathbf{q}|n\rangle \end{aligned}$$

The second term gives

$$\begin{aligned} \langle \mathbf{q}|\hat{\mathbf{H}}_I|\mathbf{q}'\rangle\langle \mathbf{q}'|n\rangle &= \int H_I(\mathbf{q}')\, \delta(\mathbf{q}-\mathbf{q}')\, \langle \mathbf{q}'|n\rangle\, \mathrm{d}\mathbf{q}' \\ &= H_I(\mathbf{q})\, \langle \mathbf{q}|n\rangle. \end{aligned}$$

In this way (F.8.31) becomes

$$\left[-\frac{\hbar^2}{2m}\nabla^2 + H_I(\mathbf{q})\right]\langle\mathbf{q}|n\rangle = n\,\langle\mathbf{q}|n\rangle. \tag{F.8.34}$$

F.8c The angular momentum representation

Postulate (b) of §F.8a can be applied to the angular momentum. In this way it is established that the z component of angular momentum is quantized with quantum number m, and that the square of the total component is quantized with quantum number $l(l+1)$, where $m = 0, \pm 1, \cdots \pm l$ and $l = 0, 1, 2, \cdots$. In Dirac notation the results may be written

$$\hat{\mathbf{M}}_z|l, m\rangle = \hbar m|l, m\rangle \tag{F.8.35}$$

$$\hat{\mathbf{M}}^2|l, m\rangle = \hbar^2 l(l+1)|l, m\rangle \tag{F.8.36}$$

If θ, ψ, denote the coordinates of $\mathbf{r}$ on a unit sphere then $\langle\theta, \psi|l, m\rangle$ is just the spherical harmonic $Y_l^m(\theta, \psi)$, which may be written $\langle\theta, \psi|l, m\rangle$. Since the angular momentum operator is hermitian the states $|l, m\rangle$ form a complete set. Hence they may be used to set up a biorthonormal basis. The biorthonormality condition implies

$$\begin{aligned}\langle l, m_z|l', m'\rangle &= \int Y_{l'}^{m'}(\theta, \psi)Y_l^{m*}(\theta, \psi)\sin\theta\,\mathrm{d}\theta\,\mathrm{d}\psi \\ &= \delta_{ll'}\,\delta_{mm'}\end{aligned} \tag{F.8.37}$$

and the completeness condition implies

$$\sum_l\sum_m Y_l^m(\theta, \psi)Y_l^{m*}(\theta', \psi') = \delta(\theta - \theta')/\sin\theta\,\delta(\psi - \psi') \tag{F.8.38}$$

F.8d The plane-wave representation

Equation (F.8.34) for a free particle reads

$$\left[-\frac{\hbar^2}{2m}\nabla^2 - E_n\right]\langle\mathbf{q}|n\rangle = 0. \tag{F.8.39}$$

Eigenfunctions of the form $\mathrm{e}^{\mathrm{i}\mathbf{k}\cdot\mathbf{q}}$ where $|\mathbf{k}|^2 = 2mE/\hbar^2$ exist and are well behaved for $E > 0$. The degeneracy may be removed by defining a solution of (F.8.39) as a function of complex energy E which is well-behaved in the upper half of the complex plane. The solution of (F.8.39) for real E is then defined as the limit of the solution for complex E, approached from the upper half of the complex plane. These considerations lead to the normalized solutions

$$\langle\mathbf{q}|\mathbf{k}\rangle = \frac{1}{(2\pi)^{\frac{3}{2}}}\mathrm{e}^{\mathrm{i}\mathbf{k}\cdot\mathbf{q}} \tag{F.8.40}$$

which will later be shown to satisfy the orthogonality condition

$$\langle \mathbf{k}'|\mathbf{k}\rangle = \frac{1}{(2\pi)^3}\int e^{i\mathbf{q}\cdot(\mathbf{k}-\mathbf{k}')}\,d\mathbf{q} = \delta(\mathbf{k}-\mathbf{k}') \tag{F.8.41}$$

and the completeness relation [cf. (B.7.7)]

$$\frac{1}{(2\pi)^3}\int e^{i\mathbf{k}.(\mathbf{q}'-\mathbf{q})}\,d\mathbf{k} = \delta(\mathbf{q}'-\mathbf{q}). \tag{F.8.42}$$

F.8e The Bloch representation

The solutions of (F.8.34) for a periodic potential are known to be the Bloch functions, $\psi_{n\mathbf{k}}(\mathbf{q})$. They satisfy the orthonormality condition

$$\langle n, \mathbf{k}|n', \mathbf{k}'\rangle = \delta_{nn'}\,\delta(\mathbf{k}-\mathbf{k}'). \tag{F.8.43}$$

They form a complete set in terms of which an arbitrary function $|f\rangle$ may be expanded:

$$|f\rangle = \sum_{n}\sum_{\substack{\mathbf{k}\\(\mathrm{B.Z.})}} |n, \mathbf{k}\rangle\,\langle n, \mathbf{k}|f\rangle \tag{F.8.44}$$

or

$$\langle \mathbf{q}|f\rangle = \sum_{n}\sum_{\substack{\mathbf{k}\\\mathrm{B.Z.}}} \langle \mathbf{q}|n, \mathbf{k}\rangle\,\langle n, \mathbf{k}|f\rangle. \tag{F.8.45}$$

Such a representation is known as the crystal momentum representation of the function $\langle \mathbf{q}|f\rangle$.

F.8f The Wannier representation

It has been shown in connexion with §B.12 that a Bloch function $\psi_{n\mathbf{k}}(\mathbf{q})$ may be expressed as a sum over direct lattice vectors $\mathbf{R}_\alpha$

$$\psi_{n\mathbf{k}}(\mathbf{q}) = \frac{1}{N^{\frac{1}{2}}}\sum_{\mathbf{R}_\alpha} e^{i\mathbf{k}\,.\,\mathbf{R}_\alpha}\,a_n(\mathbf{q}-\mathbf{R}_\alpha) \tag{F.8.46}$$

The functions $a_n(\mathbf{r}-\mathbf{R}_\alpha)$ may be defined in terms of Bloch functions $\psi_{n\mathbf{k}}(\mathbf{r})$

$$a_n(\mathbf{q}-\mathbf{R}_\alpha) = \frac{1}{N^{\frac{1}{2}}}\sum_{\mathbf{k}} e^{-i\mathbf{k}\,.\,\mathbf{R}_\alpha}\psi_{n\mathbf{k}}(\mathbf{q}). \tag{F.8.47}$$

The function $a_n(\mathbf{r} - \mathbf{R}_\alpha)$ may be shown to be both orthogonal and complete:

$$\int a_n(\mathbf{q} - \mathbf{R}_\alpha)a_n^*(\mathbf{q} - \mathbf{R}_\beta)\mathrm{d}^3\mathbf{q} = \delta_{nn'}\,\delta(\mathbf{R}_\alpha - \mathbf{R}_\beta), \tag{F.8.48}$$

$$\sum_{n,\alpha} a_n(\mathbf{q}' - \mathbf{R}_\alpha)a_n^*(\mathbf{q} - \mathbf{R}_\alpha) = \delta(\mathbf{q} - \mathbf{q}'). \tag{F.8.49}$$

The preceding results, transformed into Dirac notation, read

$$\langle n, \alpha|n', \alpha'\rangle = \delta_{nn'}\,\delta_{\alpha\alpha'}, \tag{F.8.50}$$

$$\sum_{n,\alpha} \langle \mathbf{q}|n, \alpha\rangle \langle n, \alpha|\mathbf{q}'\rangle = \delta(\mathbf{q} - \mathbf{q}'). \tag{F.8.51}$$

Equations F.8.46 and F.8.47 similarly take the form

$$\langle \mathbf{q}|n, \mathbf{k}\rangle = \langle \mathbf{q}|n, \alpha\rangle \langle n, \alpha|n, \mathbf{k}\rangle, \tag{F.8.52}$$

$$\langle \mathbf{q}|n, \alpha\rangle = \langle \mathbf{q}|n, \mathbf{k}\rangle \langle n, \mathbf{k}|n, \alpha\rangle. \tag{F.8.53}$$

Comparison of F.8.52 and F.8.46 leads to the identification

$$\langle n, \alpha|n, \mathbf{k}\rangle = \mathrm{e}^{\mathrm{i}\mathbf{k}\,.\,\mathbf{R}_\alpha}. \tag{F.8.54}$$

The expansion of an arbitrary function $f(\mathbf{q})$ in the Wannier representation is then

$$\langle \mathbf{q}|f\rangle = \sum_{n,\alpha} \langle \mathbf{q}|n, \alpha\rangle \langle n, \alpha|f\rangle. \tag{F.8.55}$$

Problem

(F.8.1) Show that in the momentum representation (F.8.34) reduces to the following integral equation for the wave amplitude $\langle \mathbf{p}|n\rangle$ in momentum space:

$$\left(E - \frac{p^2}{2m}\right)\langle \mathbf{p}|n\rangle = \int \langle \mathbf{p}|H_I|\mathbf{p}'\rangle \langle \mathbf{p}'|n\rangle\,\mathrm{d}\mathbf{p}'$$

F.9 TIME-INDEPENDENT GREEN'S FUNCTIONS

F.9a Definition of Green's functions

In §F.7 an operator $\hat{\mathbf{L}}$ was defined by its action on a state vector $|x\rangle$:

$$\hat{\mathbf{L}}|x\rangle = |y\rangle. \tag{F.7.1}$$

From equation (F.7.1) a formal solution for the vector $|x\rangle$ can be written

$$|x\rangle = (\hat{\mathbf{L}})^{-1}|y\rangle \tag{F.9.1}$$

where the inverse of the operator $\hat{\mathbf{L}}$ is defined by equation (F.7.9). The operator $(\hat{\mathbf{L}})^{-1}$ is the Green's operator corresponding to the operator $\hat{\mathbf{L}}$ and is, when defined in this way, independent of choice of basis.

Examples: (a) For the time-dependent Schrödinger equation

$$\left[\mathrm{i}\hbar \frac{\partial}{\partial t} - \hat{\mathbf{H}} \right] |\psi\rangle = 0 \tag{F.9.2}$$

the Green's operator corresponding to $\mathrm{i}\hbar(\partial/\partial t) - \hat{\mathbf{H}}$ is $[\mathrm{i}\hbar(\partial/\partial t) - \hat{\mathbf{H}}]^{-1}$.

(b) For the time-independent Schrödinger equation

$$(E - \hat{\mathbf{H}})|\psi\rangle = 0 \tag{F.9.3}$$

the Green's operator corresponding to $E - \hat{\mathbf{H}}$ is $(E - \hat{\mathbf{H}})^{-1}$.

In this account, for expository reasons, attention will be confined to (F.9.3) or equations related to it. The relation between the above definition and the Green's function of the classical theories of ordinary and partial differential equations can be obtained from equation (F.9.1) by taking representatives in a continuous representation. The position and momentum representations were established to be such representations in the last chapter. (By a continuous representation here is meant one with a continuous spectrum of eigenvalues.) Taking $|l\rangle$ to be such a continuous basis we obtain from equation (F.9.1)

$$\begin{aligned} \langle l|x\rangle &= \langle l|(\hat{\mathbf{L}})^{-1}|y\rangle \\ &= \langle l|(\hat{\mathbf{L}})^{-1}|l'\rangle \langle l'|y\rangle \\ &= \int G(l, l')y(l')\, \mathrm{d}l'. \end{aligned} \tag{F.9.4}$$

Here $G(l, l')$ is the matrix element of the operator $(\hat{\mathbf{L}})^{-1}$ in the continuous representation l; it is the Green's *function* of the operator $\hat{\mathbf{L}}$ in the $|l\rangle$ representation. $\langle l|x\rangle$, $\langle l|y\rangle$, similarly become continuous functions of l, $x(l)$ and $y(l)$, respectively. The matrix elements $G(l, l')$ completely represent the operators $(\hat{\mathbf{L}})^{-1}$; one could write

$$\begin{aligned} (\hat{\mathbf{L}})^{-1} &= |l\rangle \langle l|(\hat{\mathbf{L}})^{-1}|l'\rangle \langle l'| \\ &= \sum_{l',l} |l\rangle G(l, l') \langle l'| \end{aligned} \tag{F.9.5}$$

to support this statement.

Equation (F.9.4) shows that whatever boundary conditions are imposed on the function $x(l)$ must be reflected in the Green's function $G(l, l')$; the boundary conditions of the solution are embodied in the Green's function.

The particular form of the Green's function, of course, depends on the representation chosen. Taking (F.9.3) as an example, we have, using the notation $\hat{\mathbf{G}} = (\hat{\mathbf{L}})^{-1}$,

$$(E - \hat{\mathbf{H}})\hat{\mathbf{G}} = \hat{\mathbf{I}}. \tag{F.9.6}$$

Putting $\hat{\mathbf{H}} = \hat{\mathbf{p}}^2/2m + \hat{\mathbf{V}}$, one obtains in the position representation, using equations (F.8.20) and (F.8.21),

$$\int d\mathbf{q}'' \langle \mathbf{q}|E - \hat{\mathbf{H}}|\mathbf{q}''\rangle \langle \mathbf{q}''|\hat{\mathbf{G}}|\mathbf{q}'\rangle = \langle \mathbf{q}|\mathbf{q}'\rangle = \delta(\mathbf{q} - \mathbf{q}'),$$

$$\int \left[E + \frac{\hbar^2}{2m}\nabla^2_{\mathbf{q}''} - V(\mathbf{q}'') \right] \delta(\mathbf{q} - \mathbf{q}'')G(\mathbf{q}'', \mathbf{q}')\, d\mathbf{q}'' = \delta(\mathbf{q} - \mathbf{q}').$$

This yields the differential equation obeyed by the Green's function:

$$\left[E + \frac{\hbar^2}{2m}\nabla^2_{\mathbf{q}} - V(\mathbf{q}) \right] G(\mathbf{q}, \mathbf{q}') = \delta(\mathbf{q} - \mathbf{q}'). \tag{F.9.7}$$

Equation (F.9.7) is the starting point in more classical approaches to Green's functions.

Problem

(F.9.1) Obtain the equation of motion of the Green's function of equation (F.9.3) in the momentum representation.

F.10 THE SPECTRAL REPRESENTATION OF THE GREEN OPERATOR

Applying the general equation (F.7.15) to the operator $\hat{\mathbf{G}} = (\hat{\mathbf{L}})^{-1}$ we obtain

$$\langle l|G|l'\rangle = \sum_m \langle l|m\rangle \frac{1}{m} \langle m|l'\rangle. \tag{F.10.1}$$

For a continuous representation l this is an expression for the Green's function $G(l, l')$. Inspection of (F.10.1) suggests that the Green's function is undefined if the operator $\hat{\mathbf{G}}$ has zero as an eigenvalue. The condition that the operator $\hat{\mathbf{L}}$ must not have zero as an eigenvalue can be shown to be both necessary and sufficient for the existence of a Green's function.

For the time-independent Schrödinger operator $E - \hat{\mathbf{H}}$ we can expand in terms of eigenfunctions of $\hat{\mathbf{H}}$ provided that zero is not an eigenvalue of $\hat{\mathbf{H}}$. Then

$$\hat{\mathbf{G}} = (E - \hat{\mathbf{H}})^{-1}|n\rangle \langle n| = \sum_n \frac{|n\rangle \langle n|}{E - E_n} \tag{F.10.2}$$

where the $|n\rangle$ satisfy

$$(E - \hat{\mathbf{H}})|n\rangle = 0. \tag{F.10.3}$$

The matrix elements of G in an arbitrary representation are now given by

$$G(l, l') = \sum_n \frac{\langle l|n\rangle \langle n|l'\rangle}{E - E_n} \tag{F.10.4}$$

and, in particular, in the coordinate representation,

$$\langle \mathbf{q}|G|\mathbf{q}'\rangle = G(\mathbf{q}, \mathbf{q}', E) = \sum_n \frac{\langle \mathbf{q}|n\rangle \langle n|\mathbf{q}'\rangle}{E - E_n}. \tag{F.10.5}$$

If the energy E in equation (F.10.2) is considered to be a complex function then (F.10.5) shows this to have poles at the single particle energies E_n. The boundary conditions are embodied in (F.10.5) in the eigenfunctions $\langle \mathbf{q}|n\rangle$; these imply solution of Schrödinger's equation plus satisfaction of whatever boundary conditions are imposed.

We shall now illustrate the application of (F.10.5) to some typical cases. The general analytic computation of Green's functions is a difficult problem; it involves at least as much as the solution of Schrödinger's equation since all the eigenvalues and eigenfunctions are involved in (F.10.5). As we shall see, however, it is possible to write down an integral equation for the Green's function which can form the starting point for *numerical* calculations for reasonably behaved potentials $V(\mathbf{q})$. Series expansions of the Green's function can be used to obtain approximate analytic expressions.

Problems

(F.10.1) Green's function for a particle in a one dimensional box of side a.

For this case the functions $\langle q|n\rangle$ obey the equation

$$-\frac{\hbar^2}{2m}\frac{d^2}{dq^2}[\langle q|n\rangle] = E_n[\langle q|n\rangle] \tag{F.10.6}$$

subject to the boundary conditions

$$\langle q|n\rangle = 0, \qquad q = 0;$$

$$\langle q|n\rangle = 0, \qquad q = a.$$

The normalized eigenfunctions are $(2/a)^{\frac{1}{2}} \sin(n\pi q/a)$ and the eigenvalues are given by

$$\frac{2mE_n}{\hbar^2} = \left(\frac{n\pi}{a}\right)^2 \qquad (n = 1, 2, 3, \dots), \tag{F.10.7}$$

i.e. $E_n = n^2 b$ where b is a constant.

Substituting into equation (F.10.5):

$$G(q, q', E) = \sum_{n=1}^{\infty} \frac{2/a \sin(n\pi q/a) \sin(n\pi q'/a)}{E - n^2 b}. \tag{F.10.8}$$

It will be shown later that this expression for the Green's function can be written in closed form [equation (F.11.16)].

(F.10.2) The completeness condition for the basis $|n\rangle$ may be written

$$|n\rangle\langle n| = \mathbf{1}, \text{ or}$$

$$\langle q|n\rangle\langle n|q'\rangle = \delta(q - q').$$

Verify that the eigenfunctions of (F.10.8) form a complete set.

(F.10.3) One dimensional Green's function for the free particle in a circular domain of length L (periodic boundary conditions).

The equation for the eigenfunction $\langle q|n\rangle$ is the same as before with the new boundary condition

$$\langle q|n\rangle = \langle q + L|n\rangle, \tag{F.10.9}$$

$$\frac{d}{dq}\langle q|n\rangle_{q=q} = \frac{d}{dq}\langle q|n\rangle_{q=q+L} \tag{F.10.10}$$

A complete set satisfying these conditions is

$$\langle q|n\rangle = \frac{1}{L^{\frac{1}{2}}} e^{ik_0 q} \tag{F.10.11}$$

where

$$\frac{2mE_n}{\hbar^2} = \left(\frac{2\pi n}{L}\right)^2 = k_0^2 \quad (n = \cdots -2, -1, 0, 1, 2 \ldots) \tag{F.10.12}$$

Thus

$$G(q, q', E) = \sum_{n=-\infty}^{\infty} \frac{1}{L} \frac{e^{(2\pi i n/L)(q-q')}}{E - \hbar^2/2m(2\pi n/L)^2} \tag{F.10.13}$$

(F.10.4) Prove that the eigenfunctions of equation (F.10.11) form a complete set.

(F.10.5) If the eigenstates of the Schrödinger equation have both a discrete and continuous spectrum then the expansion in equation (F.10.2) includes both a sum over the discrete states and an integral over the continuous states. Construct the Green's function in this case.

(F.10.6) Show that in the crystal momentum representation the matrix elements of the Green's function of the periodic potential are

$$\langle n, \mathbf{k}|G|n', \mathbf{k}'\rangle = \frac{\delta(\mathbf{k} - \mathbf{k}')\,\delta_{nn'}}{E - E(n, \mathbf{k})}.$$

(F.10.7) Show that in the Wannier representation the matrix elements of the Green's function of the periodic potential are given by

$$\langle n, \alpha|G|n', \alpha'\rangle = \sum_{\mathbf{k}} \frac{e^{i\mathbf{k}\cdot(\mathbf{R}_\alpha - \mathbf{R}_{\alpha'})}}{E - E(n, \mathbf{k})} \delta_{nn'}.$$

Solution: for the periodic lattice Hamiltonian $\hat{\mathbf{H}}$ we may express the Green's operator as

$$\hat{\mathbf{G}} = \frac{1}{E - \hat{\mathbf{H}}} = \sum_{n'',\mathbf{k}''} \frac{1}{E - E(n'', \mathbf{k}'')} |n'', \mathbf{k}''\rangle \langle n'', \mathbf{k}''|$$

We now take matrix elements in the Wannier representation and use equation (F.8.54).

$$\begin{aligned}\langle n, \alpha|G|n', \alpha'\rangle &= \sum_{n'',\mathbf{k}''} \frac{\langle n, \alpha|n'', \mathbf{k}''\rangle \langle n'', \mathbf{k}''|n', \alpha'\rangle}{E - E(n'', \mathbf{k}'')} \\ &= \sum_{\mathbf{k}''} \frac{\langle n, \alpha|n, \mathbf{k}''\rangle \langle n, \mathbf{k}''|n, \alpha'\rangle}{E - E(n, \mathbf{k}'')} \\ &= \sum_{\mathbf{k}''} \frac{e^{i\mathbf{k}''\cdot(\mathbf{R}_\alpha - \mathbf{R}_{\alpha'})}}{E - E(n, \mathbf{k}'')} \delta_{nn'}.\end{aligned}$$

(The sums over **k** are, of course, over the first Brillouin zone.)

F.11 THE DIRECT METHOD FOR THE GREEN'S FUNCTION

We shall now consider an alternative technique for obtaining Green's functions which will enable us to obtain some one-dimensional Green's functions in closed form and which is valuable for the insight it affords into the properties of Green's functions.

Consider the one-dimensional Schrödinger equation

$$\left[-\frac{\hbar^2}{2m}\frac{d^2}{dq^2} + V(q)\right]\langle q|n\rangle = E_n\langle q|n\rangle. \tag{F.11.1}$$

Putting

$$\frac{2m}{\hbar^2} V(q) = v(q) \tag{F.11.2}$$

and

$$\frac{2m}{\hbar^2} E_n = e_n, \tag{F.11.3}$$

we have

$$\left[\frac{d^2}{dq^2} - v(q) + e_n\right]\langle q|n\rangle = 0, \tag{F.11.4}$$

with corresponding Green's function

$$\left[\frac{d^2}{dq^2} - v(q) + e_n\right]G(q, q', E) = \delta(q - q'). \tag{F.11.5}$$

By definition of the Dirac delta function the r.h.s is zero unless $q = q'$. Hence for $q \neq q'$, the Green's function must satisfy the homogeneous equation (F.11.4). [We shall consider only the case of *unmixed* boundary conditions, that is, boundary conditions such that each boundary condition involves just one point of the boundary and not both. A condition such as $y(0) = 0$ is unmixed; a condition such as $y(0) = y(1)$ is mixed.] We may therefore put

$$G(q, q', E) = f(q)g(q'), \qquad (q < q') \tag{F.11.6a}$$

$$= h(q)i(q'), \qquad (q > q') \tag{F.11.6b}$$

where $f(q)$, $h(q)$ are solutions of the *homogeneous* equation (F.11.4) and $g(q')$, $i(q')$, are arbitrary functions.

We now choose $h(q)$ to satisfy *one* boundary condition for $q > q'$ and $f(q)$ to satisfy the other boundary condition for $q < q'$. We introduce the function

$$\begin{aligned} \eta(q - q') &= 1 \qquad (q > q') \\ &= 0 \qquad (q < q') \end{aligned} \tag{F.11.7}$$

so that

$$G(q, q', E) = f(q)g(q')\eta(q' - q) + h(q)i(q')\eta(q - q'). \tag{F.11.8}$$

From equation (F.11.7), $\eta'(q - q') = \delta(q - q')$. Hence

$$\begin{aligned} \frac{\mathrm{d}G(q, q', E)}{\mathrm{d}q} &= f'(q)g(q')\eta(q' - q) - f(q)g(q')\,\delta(q - q') \\ &\quad + h'(q)i(q')\eta(q - q') + h(q)i(q')\,\delta(q - q'). \end{aligned} \tag{F.11.9}$$

The terms in $\delta(q - q')$ must vanish, since otherwise $(\mathrm{d}^2G/\mathrm{d}q^2)$ would contain a term proportional to $\delta'(q - q')$, which does not appear in equation (F.11.5). Thus

$$\delta(q - q')[h(q)i(q') - f(q)g(q')] = 0,$$

i.e.

$$h(q)i(q) - f(q)g(q) = 0 \tag{F.11.10}$$

and

$$\begin{aligned} \frac{\mathrm{d}^2G(q, q', E)}{\mathrm{d}q^2} &= f''(q)g(q')\eta(q' - q) - f'(q)g(q')\,\delta(q' - q) \\ &\quad + h''(q)i(q')\eta(q - q') + h'(q)i(q')\,\delta(q - q'). \end{aligned} \tag{F.11.11}$$

Comparing equation (F.11.11) with equation (F.11.5)

$$\delta(q - q')[h'(q)i(q') - f'(q)g(q')] = 1,$$

i.e

$$h'(q)i(q) - f'(q)g(q) = 1. \tag{F.11.12}$$

From equations (F.11.12) and (F.11.10),

$$g(q) = \frac{h(q)}{h'(q)f(q) - f'(q)h(q)}, \tag{F.11.13}$$

$$i(q) = \frac{f(q)}{h'(q)f(q) - f'(q)h(q)}. \tag{F.11.14}$$

The denominator of equation (F.11.14) and equation (F.11.13) will be recognized as the Wronskian of the two solutions $h(q)$ and $f(q)$. Substituting equation (F.11.13) and equation (F.11.14) into equation (F.11.8):

$$G(q, q', E) = \frac{f(q)h(q')\eta(q' - q) + h(q)f(q')\eta(q - q')}{h'(q)f(q) - h(q)f'(q)} \tag{F.11.15}$$

[Note that the Wronskian of any two solutions of the Schrödinger equation belonging to the same energy level is independent of q, and also that $f(q)$ and $h(q)$ are not *eigenfunctions* of the Schrödinger equation; they do not satisfy *both* boundary conditions.]

From equation (F.11.15) a number of general properties of the Green's function for one dimension may be seen. If the two solutions $f(q)$, $h(q)$ are independent their Wronskian is non-vanishing. This is the case when both boundary conditions are not satisfied simultaneously. The functions $f(q)$ and $h(q)$ are, in this case, different functions and, in fact, linearly independent. At eigenvalues of the energy both boundary conditions are satisfied and $f(q)$ and $h(q)$ become the same function. Therefore, they are no longer linearly independent and the Wronskian vanishes. Thus, again, considered as a function of E the Green's function has poles at the eigenvalues of the energy.

As an example we shall consider the case of a particle in a box of side a (one dimension).

The functions $h(q)$ and $f(q)$ satisfying one fixed end point boundary condition at the left-hand and right-hand side respectively are given by

$$f(q) = C_1 \sin k_0 q$$

$$h(q) = C_2 \sin k_0(q - a),$$

and we have

$$f'(q) = k_0 C_1 \cos k_0 q$$

$$h'(q) = k_0 C_2 \cos k_0(q - a)$$

(here we put $e = k_0^2$),

$$W = h'(q)f(q) - f'(q)h(q) = k_0 C_1 C_2 \sin k_0 a.$$

Hence

$$G(q, q', E) = \frac{1}{k_0 \sin k_0 a}[\sin(k_0 q) \sin k_0(q' - a)\eta(q' - q) + \sin k_0(q - a)\sin(k_0 q')\eta(q - q')]. \quad \text{(F.11.16)}$$

Note that the poles (given by $\sin k_0 a = 0$) give the eigenvalues and that $\sin k_0 q$ and $\sin k_0(q - a)$ are different functions of q except at the eigenvalues, where they become identical. Note also that $G(q, q', E)$ satisfies the same boundary conditions as the eigenfunctions, i.e.

$$G(0, q', E) = 0,$$
$$G(a, q', E) = 0.$$

Problems

(F.11.1) Show that equation (F.11.16) is identical with equation (F.10.8).

(F.11.2) Consider the differential equation

$$\hat{\mathrm{L}} f(q) = \frac{\mathrm{d}^2}{\mathrm{d}q^2} f(q) = s(q) \quad (0 \leqslant q \leqslant 1).$$

(a) Write down the solution for $f(q)$ in terms of $s(q)$ and the Green's function for the operator $\hat{\mathrm{L}}$.
(b) Show by the direct method that this Green's function for the associated boundary conditions $f(0) = 0$, $f(1) = 0$, is

$$G(q, q') = (1 - q')q \quad (q \leqslant q'),$$
$$= (1 - q)q' \quad (q > q').$$

(F.11.3) By considering the eigenfunction expansion for the Green's function in the preceding example, prove that

$$\frac{2}{\pi^2} \sum_{n=1}^{\infty} \frac{\sin n\pi q \sin n\pi q'}{n^2} = (1 - q')q \quad (q \leqslant q'),$$
$$= (1 - q)q' \quad (q > q').$$

(F.11.4) If, in problem (F.11.2), the boundary conditions are changed to

$$f(0) = 0, \quad \left(\frac{\mathrm{d}f}{\mathrm{d}x}\right)_{q=1} = 0,$$

show that the Green's function, for the same operator and the new boundary conditions, is

$$G(q, q') = q \quad (q \leqslant q'),$$
$$= q' \quad (q > q').$$

(F.11.5) By considering the eigenfunction expansion for the Green's function of problem (F.11.4) show that

$$\frac{2}{\pi^2}\sum_{n=0}^{\infty}\frac{\sin(n+\frac{1}{2})\pi q\,\sin(n+\frac{1}{2})\pi q'}{(n+\frac{1}{2})^2} = q \qquad (q \leqslant q'),$$

$$= q' \qquad (q > q').$$

F.12 THE SOMMERFELD RADIATION CONDITION

In §F.1 it was mentioned that certain Green's functions could be used to evaluate the density of states and electrical conductivity of a system. These Green's functions arise naturally in scattering problems and entail special consideration of boundary conditions. They will be introduced in this section.

It will clarify the problem at issue if a particular example is considered. Consider the operator

$$\hat{\mathbf{L}} = 1 - \alpha^2 \frac{d^2}{dq^2} \qquad \text{(F.12.1)}$$

(where α is a constant $\neq 0$) defined in the infinite domain $-\infty < q < +\infty$, and suppose that we wish to find solutions of

$$\hat{\mathbf{L}} f(q) = g(q) \qquad \text{(F.12.2)}$$

subject to

$$f(\pm\infty) = g(\pm\infty) = 0. \qquad \text{(F.12.3)}$$

The solution of (F.12.2), written in the usual way as

$$f(q) = \hat{\mathbf{L}}^{-1} g(q), \qquad \text{(F.12.4)}$$

requires the evaluation of the Green's function subject to the boundary conditions (F.12.3). In the infinite domain the normalized eigenfunctions are $[1/(2\pi)^{\frac{1}{2}}]\,e^{ipq}$ (the normalization factor will be explained later), which form a complete set with eigenvalues $1 + p^2\alpha^2$; the values of p form a continuous spectrum. To conform with the notation of equation (F.10.1) we have $|q\rangle = |l\rangle$, $m = 1 + p^2\alpha^2$. Since the operator $\mathbf{L}$ does not have zero as an eigenvalue, the Green's function exists. From equation (F.10.1) it is given by [noting the correspondence $\langle l|m\rangle = [1/(2\pi)^{\frac{1}{2}}]\,e^{ipq}$]

$$G(q, q', \alpha) = \frac{1}{2\pi}\int_{-\infty}^{\infty}\frac{e^{ip(q-q')}}{1 + p^2\alpha^2}\,dp. \qquad \text{(F.12.5)}$$

In this equation put $p\alpha = z$. Then:

$$G(q, q', \alpha) = \frac{1}{2\pi\alpha}\int_{-\infty}^{\infty}\frac{e^{(iz/\alpha)(q-q')}}{(z+i)(z-i)}\,dz. \qquad \text{(F.12.6)}$$

The contour used in evaluating equation (F.12.6) must now be decided. This is, in the present instance, determined by the boundary conditions. For $q > q'$ the integrand of equation (F.12.6) will only remain finite in the upper half of the complex plane as $q \to \infty$. Hence the contour shown gives (since the semicircle gives no contribution as $|z| \to \infty$):

$$G(q, q', \alpha) = \frac{1}{2\pi} \cdot 2\pi i \cdot \frac{1}{2i} \cdot \frac{1}{\alpha} e^{-(1/\alpha)(q-q')} = \frac{1}{2\alpha} e^{-(1/\alpha)(q-q')} \quad \text{(F.12.7)}$$

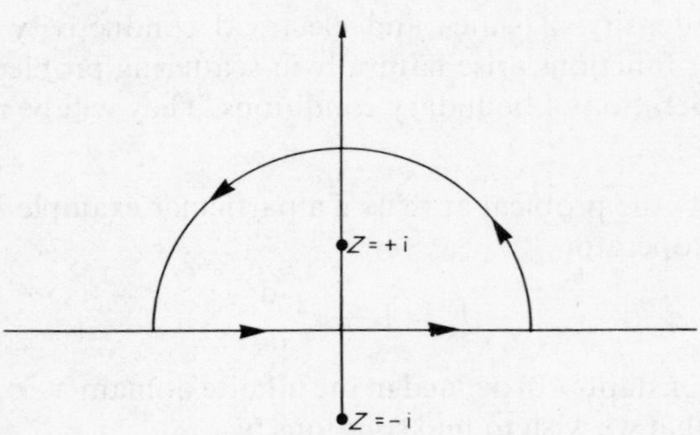

Fig. F.12.1.

For $q < q'$ the boundary conditions similarly determine the contour to be enclosed by a semicircle in the lower half of the complex plane, which gives

$$\begin{aligned} G(q, q', \alpha) &= \frac{1}{2\pi} 2\pi i \left(-\frac{1}{2i} \right) e^{(1/\alpha)(q-q')} \left(-\frac{1}{\alpha} \right) \\ &= \frac{1}{2\alpha} e^{-(1/\alpha)(q'-q)}. \end{aligned} \quad \text{(F.12.8)}$$

Equation (F.12.7) and (F.12.8) may be summarized by writing

$$G(q, q', \alpha) = \frac{1}{2\alpha} e^{-(1/\alpha)|q-q'|}, \quad \text{(F.12.9)}$$

in which it is once again seen that the boundary conditions have been satisfied. In evaluating (F.12.7) and (F.12.8) we used Cauchy's theorem

$$f(a) = \frac{1}{2\pi i} \oint \frac{f(z)}{z-a} dz$$

(where the arrow shows the direction in which the contour is traversed in a positive sense). The last minus sign in equation (F.12.8) comes from the fact that the contour used in the lower half of the complex plane was traversed in a clockwise direction.

Fortified by consideration of this example, we now wish to consider the operator

$$\hat{\mathbf{L}} = -\frac{d^2}{dq^2} - k_0^2, \tag{F.12.10}$$

which is obtained from Schrödinger's equation for a free particle by putting $2mE/\hbar^2 = k_0^2$. We wish to consider $\hat{\mathbf{L}}$ defined over an infinite domain and again limit ourselves to solutions finite at infinity. In this case a complete set of eigenfunctions is again given by the set

$$\langle q|k\rangle = \frac{1}{(2\pi)^{\frac{1}{2}}} e^{ik_0 q} \tag{F.12.11}$$

with eigenvalues $k^2 - k_0^2$ where the k's have a continuous spectrum. However, since the equation

$$\hat{\mathbf{L}}|n\rangle = 0|n\rangle$$

corresponding to $k = k_0$ *does* have solutions, the operator $\hat{\mathbf{L}}$ has zero as an eigenvalue and its Green's function does not exist. Another way of looking at the problem is to write down the spectral representation, equation (F.10.1), i.e. the equation corresponding to equation (F.12.5) in the present problem. It is

$$G(q, q', k_0) = \frac{1}{2\pi}\int_{-\infty}^{\infty} \frac{e^{ik(q-q')}}{k_0^2 - k^2} dk. \tag{F.12.12}$$

The integrand of equation (F.12.12) has two poles on the real axis and the boundary conditions on $G(q, q', E)$ are now no longer sufficient to enable a contour to be defined. In fact, the Green's function of equation (F.12.12) does not have a unique value and cannot be calculated unless additional restrictions are imposed; this is connected with the two-fold degeneracy of the eigenfunctions of the corresponding Schrödinger equation.

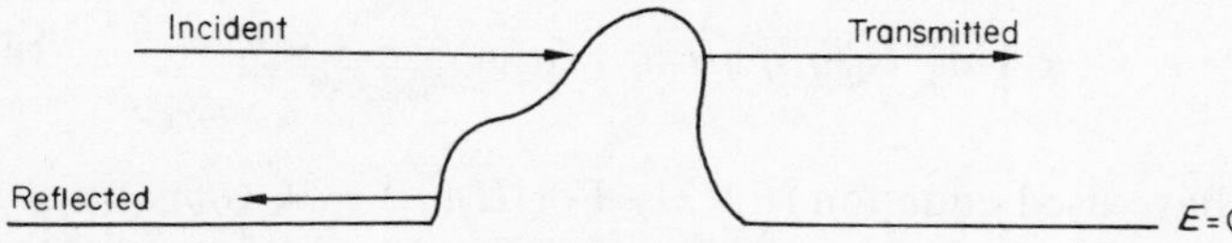

Fig. F.12.2.

While still confining ourselves to the one-dimensional case, we now try to elucidate the problem of evaluating (F.12.12) by considering a scattering problem of the conventional kind in which an incident wave is scattered by a potential $H_I(q)$ and gives rise to a reflected wave and a transmitted wave. If the origin of energy is taken as shown, and we consider a potential for which $H_I \to 0$ as $|q| \to \infty$, then the spectrum of energy is continuous and positive from the usual Sturm–Liouville theory; moreover, it is two-fold degenerate. Boundary conditions must be imposed if the eigenfunctions of the continuous spectrum are to be determined. The simplest one in the one-dimensional case is derived from the argument that if the incident wave is travelling to the right and incident as shown, then the transmitted wave on the right-hand side of the potential should have the form of a wave travelling to the right and the reflected wave the form of a wave travelling to the left. These boundary conditions suffice to uniquely determine the constants in the solution, apart from normalizing factors. These boundary conditions may now be connected with the Green's function of equation (F.12.12). By equation (F.8.34) the observed values of the energy are given by

$$(\hat{\mathbf{H}} - E)|n\rangle = 0. \qquad \text{(F.12.13)}$$

We now put $\hat{\mathbf{H}} = \hat{\mathbf{H}}_0 + \hat{\mathbf{H}}_I$ and write (F.12.13) in the form

$$\mathbf{H}_I|n\rangle = (E - \hat{\mathbf{H}}_0)|n\rangle$$

i.e.

$$|n\rangle = \frac{1}{E - \hat{\mathbf{H}}_0} \hat{\mathbf{H}}_I|n\rangle \qquad \text{(F.12.14)}$$

Equation (F.12.14) is true in any number of dimensions and independent of representation; we shall return to it later. For one dimension and in the coordinate representation,

$$\begin{aligned}\langle q|n\rangle &= \langle q|\frac{1}{E - \hat{\mathbf{H}}_0}|q'''\rangle \langle q'''|\hat{\mathbf{H}}_I|q''\rangle \langle q''|n\rangle \\ &= \int\int \mathrm{d}q''\,\mathrm{d}q'''\,G_0(q, q''')H_I(q'')\,\delta(q'' - q''')\,\langle q''|n\rangle, \\ &= \int \mathrm{d}q''\,G_0(q, q'')H_I(q'')\,\langle q''|n\rangle. \end{aligned} \qquad \text{(F.12.15)}$$

Here we have used equation (F.8.21). For $H_I(q'') = 0$, $\langle q|n\rangle$ must reduce to the incident wave. Putting this equal to $\langle q|n\rangle_0$, we obtain for the complete solution of equation (F.12.14)

$$\langle q|n\rangle = \langle q|n\rangle_0 + \int \mathrm{d}q'\,G_0(q, q')H_I(q')\,\langle q'|n\rangle. \qquad \text{(F.12.16)}$$

We now connect equation (F.12.16) with our original problem of evaluating equation (F.12.12) by requiring $G_0(q, q')$ to be such that it only gives rise to waves conforming to the boundary conditions specified for the scattering problem. To this end we consider the effect of including only *one* of the poles in equation (F.12.12) in the integration. Formally, this can be effected by displacing the poles off the axis by a small amount (see fig. F.12.3). Again, this is equivalent to formally putting $E = E + i\epsilon$

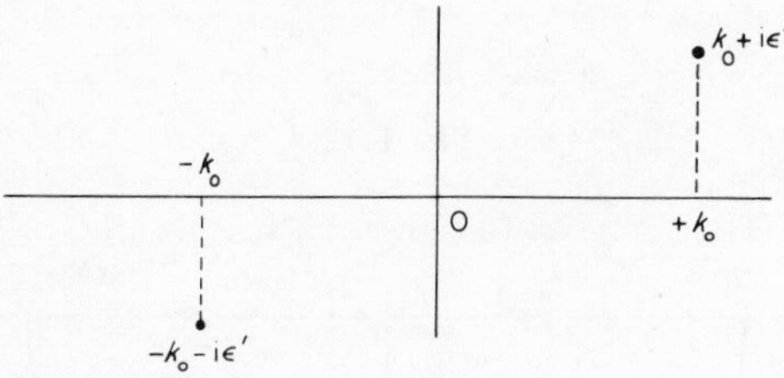

Fig. F.12.3.

where $\epsilon \to 0$ at the end of the calculation. This is equivalent in turn to putting $k_0 = k_0 + i\epsilon'$. Equation (F.12.12) now becomes

$$G(q, q', E) = \frac{1}{2\pi} \lim_{\epsilon' \to 0} \int_{-\infty}^{\infty} \frac{e^{ik(q-q')}}{(k_0 + i\epsilon' - k)(k_0 + i\epsilon' + k)} dk \tag{F.12.17}$$

where the poles are now at $k = k_0 + i\epsilon'$ and $k = -(k_0 + i\epsilon')$. Thus including the imaginary parts has had the effect of displacing the poles off the axis as in the figure. It is now possible to fix the contour of integration as in the example at the beginning of this section. If $q > q'$ we take the contour in the upper half of the complex plane, because as $|k| \to \infty$ by Jordan's lemma the integrand remains finite. Thus we only include the pole at $k_0 + i\epsilon$ (see fig. F.12.4). This gives

$$-2\pi i \cdot \frac{1}{2\pi} \cdot \frac{1}{2k_0} \cdot e^{ik_0(q-q')} = -\frac{i}{2k_0} e^{ik_0(q-q')}. \tag{F.12.18}$$

On the other hand, for $q < q'$, we use a contour in the lower half of the complex plane (fig. F.12.5). Noting that the contour is anticlockwise, we get

$$-2\pi i \cdot \frac{1}{2\pi} \cdot \left(-\frac{1}{2k_0}\right) \cdot (-1)\, e^{-ik_0(q-q')} = -\frac{i}{2k_0} e^{-ik_0(q-q')}.$$

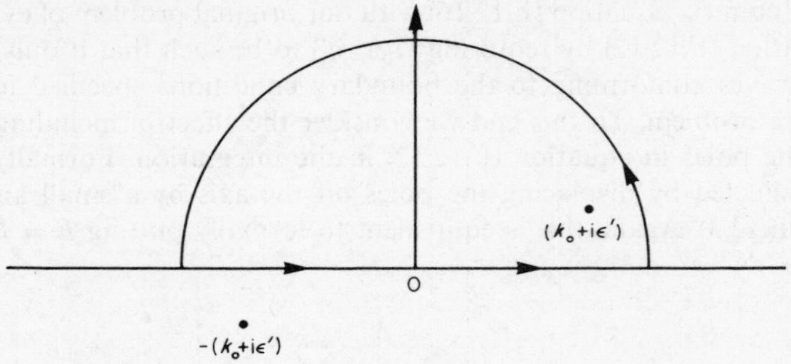

Fig. F.12.4.

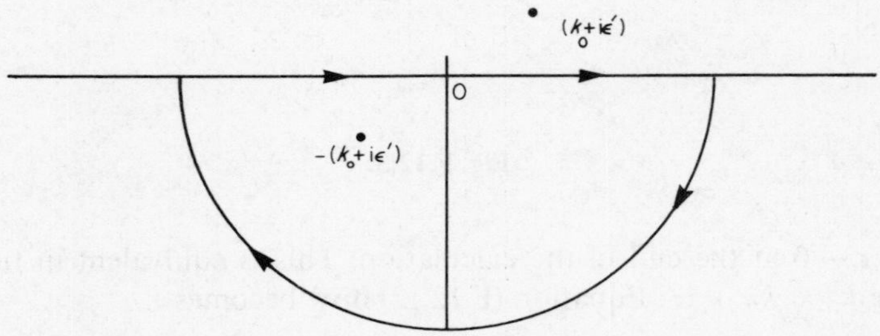

Fig. F.12.5.

Both results can now be incorporated by writing

$$G_0^+(q, q', k_0) = -\frac{i}{2k_0} e^{ik_0|q-q'|} \tag{F.12.19}$$

where the + labels, for the moment, the direction in which the poles have been displaced; $G^+ = G(k_0 + i\epsilon)$.

We now verify that G^+ is the Green's function we want, i.e. that it satisfies the boundary conditions of the problem. To this end we substitute equation (F.12.19) into equation (F.12.16) and get

$$\langle q|n\rangle = \langle q|n\rangle_0 - \frac{i}{2k_0}\int_{R_1}^{R_2} dq'\, e^{ik_0|q-q'|}\langle q'|n\rangle\, H_I(q') \tag{F.12.20}$$

where we assume the potential to be finite within the limits R_1, R_2, and zero outside these. For $q > R_2$ all the contributions are from points to the left of it, and we may write

$$\langle q|n\rangle = \langle q|n\rangle_0 - \frac{i}{2k_0} e^{ik_0 q}\int_{R_1}^{R_2} dq'\, e^{-ik_0 q'}\langle q'|n\rangle\, H_I(q'). \tag{F.12.21}$$

Since $\langle q|n\rangle_0$ is assumed to have the form e^{ikq} this expression may be written $t\,e^{ik_0q}$ and the boundary indications are satisfied to the right of the potential. For $q < R_1$ we have similarly

$$\langle q|n\rangle = \langle q|n\rangle_0 + \frac{i}{2k_0} e^{-ik_0q} \int_{R_1}^{R_2} dq'\, e^{ik_0q'} \langle q'n\rangle H_I(q') \quad \text{(F.12.22)}$$

where the first term represents the incident wave and the second the reflected wave $r\,e^{-ik_0q}$. Therefore G^+ satisfies the boundary conditions of the problem. The condition that the solution of the Schrödinger equation should correspond to outgoing waves at infinity is the *Sommerfeld radiation condition.* The addition of a small imaginary part to the energy is a convenient mathematical device for ensuring that this condition is satisfied.

Problems

(F.12.1) Show that if we let $E \to E - i\epsilon'$ where ϵ' is real and positive a Green's function for the free particle for positive energies is obtained with the form

$$G(q, q', E) = \frac{+i}{2k_0} e^{-ik_0|q-q'|}.$$

(F.12.2) Show that for negative energies the Green's function of the previous problem which satisfies the condition being finite at infinity is given by

$$G(q, q', k_0) = \frac{-1}{2k_0} e^{-k_0|q-q'|}.$$

(F.12.3) Obtain the result of problem (F.12.2) by the direct method.

(F.12.4) Show that if we take $E \to E \pm i\epsilon$, the corresponding Schrödinger operator no longer has zero as an eigenvalue and that the corresponding Green's functions $G^{\pm}$ therefore exist.

These considerations may be extended to define Green's operators $\hat{G}\pm$ by the equation

$$\hat{G}^+(E) = \frac{1}{E - \hat{H} + i\epsilon}, \quad \text{(F.12.23)}$$

$$\hat{G}^-(E) = \frac{1}{E - \hat{H} - i\epsilon}, \quad \text{(F.12.24)}$$

which are defined for the total Hamiltonian $\hat{H}$ and for an arbitrary number of dimensions. An example of the evaluation of these for various potentials will be given later.

We end this section by evaluating the expression (F.12.23) for $\hat{H} = \hat{H}_0$ and three dimensions.

From equation (F.10.5) once again,

$$\langle q|G^+|q'\rangle = G^+(q, q', E + i\epsilon) = \left\langle q\left|\frac{1}{E - H_0 + i\epsilon}\right|q'\right\rangle$$

$$= \sum_n \frac{\langle q|n\rangle\langle n|q'\rangle}{E - E_n + i\epsilon}. \quad \text{(F.12.25)}$$

Hence the eigenfunctions of the Hamiltonian can again be treated in two ways. Firstly, we can define them over the infinite domain and argue that normalized eigenfunctions are given by

$$\langle q|n\rangle = \frac{1}{(2\pi)^{\frac{3}{2}}} e^{i\mathbf{q}\cdot\mathbf{k}}, \quad \text{(F.12.26)}$$

with **k** assuming a continuous range of values. In this case

$$E_n = \frac{\hbar^2}{2m}(\mathbf{k}\cdot\mathbf{k}), \quad \text{(F.12.27)}$$

and putting

$$E = \frac{\hbar^2 k_0^2}{2m}, \quad \text{(F.12.28)}$$

equation (F.12.25) becomes

$$G^+(q, q', E) = G^+(q, q', k_0) = \frac{2m}{\hbar^2}\int \frac{d\mathbf{k}}{(2\pi)^3} \frac{e^{i\mathbf{k}\cdot(\mathbf{q}-\mathbf{q}')}}{\mathbf{k}_0^2 - \mathbf{k}^2 + i\epsilon}, \quad \text{(F.12.29)}$$

Conceptually, a more satisfactory way perhaps of looking at the eigenfunctions is to consider the $\langle q|n\rangle$ as obeying periodic boundary conditions in a box of side L, so that

$$k_x = \frac{2\pi n_x}{L} \qquad (n_x = 0, \pm 1, \pm 2, \ldots). \quad \text{(F.12.30)}$$

and

$$\langle \mathbf{q}|n\rangle = \frac{1}{L^{\frac{3}{2}}} e^{i\mathbf{k}\cdot\mathbf{q}}. \quad \text{(F.12.31)}$$

We then use the fact that $\Delta k_x = 2\pi/L$ (separation of successive values of k_x) and, in the limit $L \to \infty$, once again we arrive at equation (F.12.30).

Returning to equation (F.12.29), we put

$$\boldsymbol{\rho} = \mathbf{q} - \mathbf{q}' \quad \text{(F.12.32)}$$

and take $\boldsymbol{\rho}$ as the direction of the z axis. Then

$$\mathbf{k}\cdot\boldsymbol{\rho} = |\mathbf{k}||\boldsymbol{\rho}|\cos\theta; \qquad \mathrm{d}\mathbf{k} = k^2 \sin\theta\, \mathrm{d}\theta\, \mathrm{d}\phi\, \mathrm{d}k.$$

Then equation (F.12.29) becomes

$$\begin{aligned} G(\mathbf{q}, \mathbf{q}', k_0) &= \frac{2m}{\hbar^2}\frac{1}{(2\pi)^3}\int_0^\infty k^2\,\mathrm{d}k \int_0^\pi \sin\theta\,\mathrm{d}\theta \int_0^{2\pi} \frac{\mathrm{e}^{\mathrm{i}\mathbf{k}\cdot\boldsymbol{\rho}}\,\mathrm{d}\phi}{\mathbf{k}_0^2 - \mathbf{k}^2 + \mathrm{i}\epsilon} \\ &= \frac{2m}{\hbar^2}\frac{1}{(2\pi)^2}\int_0^\infty \frac{k^2\,\mathrm{d}k}{k_0^2 - k^2 + \mathrm{i}\epsilon}\int_{-1}^{+1} \mathrm{e}^{\mathrm{i}kx\rho}(-\mathrm{d}x) \\ &= \frac{2m}{\hbar^2}\frac{1}{(2\pi)^2}\int_0^\infty \frac{k^2\,\mathrm{d}k}{k_0^2 - k^2 + \mathrm{i}\epsilon}\left[\frac{\mathrm{e}^{\mathrm{i}k\rho} - \mathrm{e}^{-\mathrm{i}k\rho}}{\mathrm{i}k\rho}\right] \\ &= \frac{2m}{\hbar^2}\frac{1}{(2\pi)^2}\frac{1}{2\mathrm{i}\rho}\int_{-\infty}^\infty \frac{k\,\mathrm{d}k}{k_0^2 - k^2 + \mathrm{i}\epsilon}[\mathrm{e}^{\mathrm{i}k\rho} - \mathrm{e}^{-\mathrm{i}k\rho}]. \end{aligned}$$

We now proceed as in the one-dimensional case, i.e., we put

$$k_0^2 - k^2 + \mathrm{i}\epsilon = [(k_0 + k) + \mathrm{i}\epsilon'][(k_0 - k) + \mathrm{i}\epsilon']$$

and use Cauchy's theorem. The two terms in the square bracket have to be considered separately. If ρ is positive, the first term can be evaluated by the contour shown.

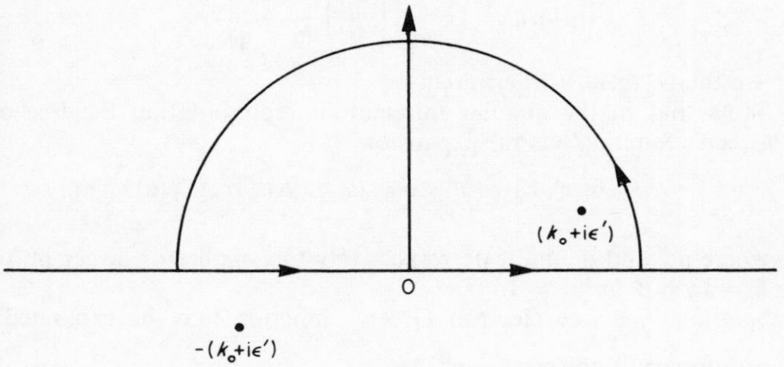

Fig. F.12.6.

This gives

$$\int_{-\infty}^{\infty} \frac{\mathrm{e}^{\mathrm{i}k\rho}k\,\mathrm{d}k}{k_0^2 - k^2 + \mathrm{i}\epsilon} = \frac{(2\pi\mathrm{i})}{2k_0}(-1)(k_0)\,\mathrm{e}^{\mathrm{i}k_0\rho}$$

The second term may, for positive ρ, be evaluated by a similar contour closed in the lower half of the complex plane and gives exactly the same result.

(F.12.5) Prove this result.

For $\rho > 0$
$$G(\mathbf{q}, \mathbf{q}', k_0) = \frac{2m}{\hbar^2}\left(-\frac{1}{4\pi}\right)\frac{e^{ik_0|\boldsymbol{\rho}|}}{\mathbf{1\rho1}}.$$

By considering similarly the case of $\rho < 0$, one obtains

$$G^+(\mathbf{q}, \mathbf{q}', k_0) = \frac{2m}{\hbar^2}\left(-\frac{1}{4\pi}\right)\frac{e^{ik_0|\boldsymbol{\rho}|}}{|\boldsymbol{\rho}|} = -\frac{2m}{\hbar^2}\left(\frac{1}{4\pi}\right)\frac{e^{ik_0|\mathbf{q}-\mathbf{q}'|}}{|\mathbf{q}-\mathbf{q}'|}. \qquad \text{(F.12.33)}$$

(F.12.6) Prove this result by explicitly considering the calculation of the integral for $\boldsymbol{\rho} < 0$.

(F.12.7) Show that in the momentum representation the matrix elements of $G_0^{\pm}$ are given by

$$G_0^{\pm}(\mathbf{k}, \mathbf{k}', k_0) = \frac{\delta(\mathbf{k}-\mathbf{k}')}{E - \dfrac{\hbar^2 k^2}{2m} \pm i\epsilon}.$$

The dependence of $G_0^+(\mathbf{q}, \mathbf{q}', k_0)$ on $\mathbf{q} - \mathbf{q}'$ only is a direct consequence of the absence of a potential and the resultant invariance of the system to a simultaneous translation of $\mathbf{q}'$ and $\mathbf{q}$. Similarly, for a potential independent of time, the Green's function of the time dependent Schrödinger equation is independent of the origin of the time axis, i.e. a function of $t - t'$ only

(F.12.8) Show that $G_0^-(\mathbf{q}, \mathbf{q}', k_0)$ is given by

$$G_0^-(\mathbf{q}, \mathbf{q}', k_0) = \frac{2m}{\hbar^2}\left(\frac{-1}{4\pi}\right)\frac{e^{-k_0|\mathbf{q}-\mathbf{q}'|}}{|\mathbf{q}-\mathbf{q}'|}.$$

(F.12.9) Are the $G_0^+(\mathbf{q}, \mathbf{q}', k_0)$ hermitian?

(F.12.10) Show that in the angular momentum representation the free electron Green's function has the expansion

$$G_0^+(\mathbf{q}, \mathbf{q}', E) = \sum_{l,m} -ik_0 j_l(k_0 q_<) h_l(k_0 q_>) Y_l^m(\mathbf{q}) Y_l^{m*}(\mathbf{q}')$$

where $q_<$ and $q_>$ indicate respectively the smaller or larger of q and q', $E = k_0^2$ and $2m/\hbar^2 = 1$.

Solution. The free electron Green's function may be expanded in the orthonormalized set $\dfrac{1}{(2\pi)^{\frac{3}{2}}} e^{i\mathbf{k}\cdot\mathbf{q}}$:

$$\langle \mathbf{q}|G_0^+|\mathbf{q}'\rangle = \sum_l \frac{\langle \mathbf{q}|\mathbf{k}\rangle\langle \mathbf{k}|\mathbf{q}'\rangle}{k_0^2 - k^2}$$

$$= \frac{1}{(2\pi)^3}\int_{-\infty}^{\infty} d\mathbf{k}\, \frac{e^{i\mathbf{k}\cdot(\mathbf{q}-\mathbf{q}')}}{k_0^2 - k^2}.$$

In this expression the exponentials may be expanded by Bauer's equation:

$$e^{i\mathbf{k}\cdot\mathbf{q}} = 4\pi \sum_{m=-l}^{+l}\sum_{l=0}^{\infty} i^l j_l(kq) Y_l^{m*}(\mathbf{k}) Y_l^m(\mathbf{q}).$$

Thus

$$G(\mathbf{q}, \mathbf{q}', E) = \frac{1}{(2\pi)^3} \int_{-\infty}^{\infty} k^2 \,\mathrm{d}\, k \,\mathrm{d}\boldsymbol{\Omega}_k \sum_{l,m} 4\pi i^l j_l(kq)\, Y_l^{m*}(\mathbf{k}) Y_l^m(\mathbf{q})$$

$$\frac{1}{k_0^2 - k^2} \times \sum_{l',m'} 4\pi i^{l'} j_{l'}(kq) Y_{l'}^{m'}(\mathbf{k}) Y_{l'}^{m'*}(\mathbf{q}').$$

Integrating over $\mathrm{d}\boldsymbol{\Omega}_k$, we make use of the orthonormality relation for spherical harmonics:

$$\int d\Omega_k Y_l^m(\mathbf{k}) Y_{l'}^{m'*}(\mathbf{k}) = \delta_{ll'}\, \delta_{mm'}$$

Thus

$$G(\mathbf{q}, \mathbf{q}', E) = \sum_{l,m} G_l(q, q', E) Y_l^m(\mathbf{q}) Y_l^{m*}(\mathbf{q}'),$$

where

$$G_l(q, q', E) = \frac{2}{\pi} \int_0^{\infty} \frac{j_l(kq) j_l(kq')}{k_0^2 - k^2} k^2 \,\mathrm{d}k.$$

Since this is even in k, it may be written

$$G_l(q, q', E) = \frac{1}{\pi} \int_{-\infty}^{\infty} \frac{j_l(kq) j_l(kq') k^2}{k_0^2 - k^2 + \mathrm{i}\epsilon} \mathrm{d}k.$$

This integral may be evaluated by contour integration. In general,

$$j_l(kq) = \tfrac{1}{2}[h_l(kq) + h_l^*(kq)]$$

where, asymptotically,

$$h_l(kq) \to i^{-l-1} \frac{\mathrm{e}^{ikq}}{kq}.$$

Consider now the case $q < q'$. Then we expand $j_l(kq')$ in terms of Hankel functions where, asymptotically,

$$\lim_{kq \to \infty} [j_l(kq)] \to \frac{1}{kq} \sin\left| kq - \frac{l\pi}{2} \right|.$$

Therefore in a product like $j_l(kq) h_l(kq')$ the $h_l(kq')$ term dominates the $j_l(kq)$ terms for $q' > q$.

In particular, for $q' > q$, we may evaluate the term $h_l(kq') j_l(kq)$ by a contour in the upper half of the complex plane, since the product behaves like e^{ikR} where $R > 0$. In this way we get the contribution

$$-\frac{ik_0}{2} h_l(k_0 q') j_l(k_0 q).$$

The term $h_l^*(kq') j_l(kq)$ similarly behaves like e^{-ikR} for large values of the argument and can be evaluated by a contour in the lower half plane, giving a contribution

$$+\frac{ik_0}{2} h_l^*(-k_0 q') j_l(k_0 q).$$

Here $j_l(-x) = (-1)^l j_l(x)$ and $h_l^*(-x) = (-1)^l h_l(x)$. Hence

$$\frac{ik_0}{2} h_l^*(-k_0 q') j_l(-k_0 q) = \frac{ik_0}{2} h_l(k_0 q') j_l(k_0 q).$$

For $q' > q$ we get the total contribution $-ik_0 h_l(k_0 q') j_l(k_0 q)$. For $q > q'$ we similarly expand the term $j_l(k_0 q)$ in Hankel functions and obtain $G_l(q, q', E) = -ik_0 h_l(k_0 q) j_l(k_0 q')$. Hence

$$G_0^+(\mathbf{q}, \mathbf{q}', E) = \sum_{l,m} -ik_0 h_l(k_0 q_>) j_l(k_0 q_<) Y_l^m(\mathbf{q}) Y_l^{m*}(\mathbf{q}').$$

(F.12.11) Obtain the function $G(\mathbf{q}, \mathbf{q}', E)$ in this example by the direct method. *Note.* The fact that the function $G(\mathbf{q}, \mathbf{q}', E)$ is diagonal in the angular momentum indices is a direct consequence of the invariance of free space under rotation.

(F.12.12) Show that any function of the variables $f(\mathbf{q}, \mathbf{q}')$ may be expanded in the form

$$f(\mathbf{q}, \mathbf{q}') = \sum f_{ll'}^{mm'}(q, q') Y_l^m(\mathbf{q}) Y_{l'}^{m'*}(\mathbf{q}').$$

[*Hint.* Expand $f(\mathbf{q}, \mathbf{q}')$ as a double Fourier integral

$$f(\mathbf{q}, \mathbf{q}') = \frac{1}{(2\pi)^6} \int\int \mathrm{e}^{-i\mathbf{k}.\mathbf{q}}\, \mathrm{e}^{-i\mathbf{k}'.\mathbf{q}'} f(\mathbf{k}, \mathbf{k}')\, d\mathbf{k}\, d\mathbf{k}',$$

use Bauer's equation,

$$\mathrm{e}^{i\mathbf{k}.\mathbf{r}} = 4\pi \sum_{l=0}^{\infty} \sum_{m=-l}^{m=+l} i^l j_l(kr) Y_l^{m*}(\mathbf{k}) Y_l^m(\mathbf{q})$$

and the orthonormality of the spherical harmonics[9].]

F.13 THE REPRESENTATION OF δ-FUNCTIONS IN THE COMPLEX PLANE

It will be shown in this section that the presence of terms $i\epsilon$ in the representation of Green's functions may also be associated with the representation of δ-functions in the complex plane.

We define

$$\delta(t) = \operatorname*{Lt}_{\epsilon \to 0} \delta(t, \epsilon) = \operatorname*{Lt}_{\epsilon \to 0} \frac{1}{2\pi} \int_{-\infty}^{\infty} \mathrm{e}^{-\epsilon|\omega|}\, \mathrm{e}^{i\omega t}\, d\omega. \qquad \text{(F.13.1)}$$

It is convenient to break equation (F.13.1) up into

$$\delta_+(t, \epsilon) = \frac{1}{2\pi} \int_0^{\infty} \mathrm{e}^{-\epsilon|\omega|}\, \mathrm{e}^{i\omega t}\, d\omega \qquad \text{(F.13.2)}$$

and

$$\delta_-(t, \epsilon) = \frac{1}{2\pi} \int_{-\infty}^{0} \mathrm{e}^{-\epsilon|\omega|}\, \mathrm{e}^{i\omega t}\, d\omega. \qquad \text{(F.13.3)}$$

[9] T. Lukes and M. Roberts, *Phys. Lett.*, **25A**, 508 (1967).

Carrying out the integration gives

$$\delta_+(t, \epsilon) = \frac{\mathrm{i}}{2\pi} \frac{1}{t + \mathrm{i}\epsilon}, \tag{F.13.4}$$

$$\delta_-(t, \epsilon) = \frac{-\mathrm{i}}{2\pi} \frac{1}{t - \mathrm{i}\epsilon}. \tag{F.13.5}$$

For an arbitrary function $f(t)$ the action of the δ's is defined by

$$\delta_{(\pm)} f(t) = \lim_{\epsilon \to 0} \int_{-\infty}^{\infty} f(t)\, \delta_{(\pm)}(t)\, \mathrm{d}t. \tag{F.13.6}$$

From equations (F.13.1), (F.13.4) and (F.13.5)

$$\delta(t, \epsilon) = \frac{1}{\pi} \frac{\epsilon}{t^2 + \epsilon^2}. \tag{F.13.7}$$

This may be represented as in the diagram.

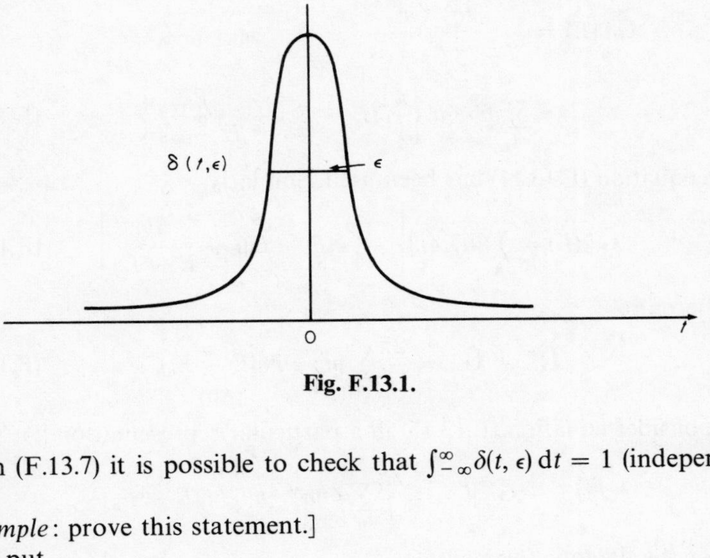

Fig. F.13.1.

From (F.13.7) it is possible to check that $\int_{-\infty}^{\infty} \delta(t, \epsilon)\, \mathrm{d}t = 1$ (independent of ϵ).
[*Example*: prove this statement.]
Now put

$$P(t, \epsilon) = \frac{t}{t^2 + \epsilon^2}. \tag{F.13.8}$$

Then

$$\delta_+(t, \epsilon) = \frac{i}{2\pi}\left[\frac{t - i\epsilon}{t^2 + \epsilon^2}\right] = \frac{1}{2}\delta(t, \epsilon) + \frac{i}{2\pi}P(t, \epsilon), \qquad \text{(F.13.9)}$$

$$\delta_-(t, \epsilon) = -\frac{i}{2\pi}\left[\frac{t + i\epsilon}{t^2 + \epsilon^2}\right] = \frac{1}{2}\delta(t, \epsilon) - \frac{i}{2\pi}P(t, \epsilon). \qquad \text{(F.13.10)}$$

It follows from equation (F.13.8) that

$$P(t, \epsilon)f(t) = \operatorname*{Lt}_{\epsilon \to 0} \int_{-\infty}^{\infty} \frac{f(t)t}{t^2 + \epsilon^2} dt = P\int_{-\infty}^{\infty} \frac{f(t)}{t} dt$$

where P denotes taking the principal value. Thus one can write symbolically

$$\delta_+(t) = \frac{1}{2}\delta(t) + \frac{i}{2\pi}P, \qquad \text{(F.13.11)}$$

$$\delta_-(t) = \frac{1}{2}\delta(t) - \frac{i}{2\pi}P. \qquad \text{(F.13.12)}$$

These results may now be applied to the Green's functions G^+ and G^-. From equation (F.12.25):

$$\begin{aligned}\hat{\mathbf{G}}^+(E) &= \sum_n \frac{|n\rangle\langle n|}{E - E_n + i\epsilon} \\ &= \sum_n |n\rangle\langle n| \left\{\frac{\pi}{i}\delta(E - E_n) + \frac{P}{E - E_n}\right\} \end{aligned} \qquad \text{(F.13.13)}$$

where equation (F.13.11) has been used. Similarly

$$\hat{\mathbf{G}}^-(E) = \sum_n |n\rangle\langle n| \left[-\frac{\pi}{i}\delta(E - E_n) + \frac{P}{E - E_n}\right] \qquad \text{(F.13.14)}$$

and therefore

$$\hat{\mathbf{G}}^+ - \hat{\mathbf{G}}^- = \frac{2\pi}{i}\sum_n |n\rangle\langle n|\delta(E - E_n). \qquad \text{(F.13.15)}$$

Now consider equation (F.13.15) in a particular representation $|l\rangle$. Then

$$\langle l|\hat{\mathbf{G}}^+ - \hat{\mathbf{G}}^-|l'\rangle = \frac{2\pi}{i}\sum_n \langle l|n\rangle\langle n|l'\rangle\,\delta(E - E_n)$$

and, for the diagonal elements,

$$\langle l|\hat{\mathbf{G}}^+ - \hat{\mathbf{G}}^-|l\rangle = \frac{2\pi}{i}\sum_n \langle n|l\rangle\langle l|n\rangle\,\delta(E - E_n).$$

Now sum over all the states l. Interchanging the order of the summations,

$$\sum_l \langle l|\hat{\mathbf{G}}^+|l\rangle - \langle l|\hat{\mathbf{G}}^-|l\rangle = \frac{2\pi}{\mathrm{i}}\sum_n\sum_l \langle n|l\rangle\langle l|n\rangle\,\delta(E - E_n)$$

$$= \frac{2\pi}{\mathrm{i}}\sum_n \delta(E - E_n). \qquad \text{(F.13.16)}$$

The r.h.s. of equation (F.13.16) is a summation over all the states in the neighbourhood of the energy E. For a system with discrete energy levels this function, considered as a function of E, is highly discontinuous. However, for a system with a continuous spectrum, equation (F.13.16) is to be interpreted as a sum over a large number of energy levels in the neighbourhood of E. The number of such levels per unit energy interval $\mathrm{d}E$ is there defined to be the density of states $D(E)$. It is better to visualize the δ-function in equation (F.13.16) as defined by equation (F.13.7) as of finite width ϵ, so that for a quasi-continuous spectrum ϵ contains a large number of such levels. From eqn. (F.13.16) then, for an arbitrary representation l,

$$\frac{\mathrm{i}}{2\pi}\sum_l \langle l|\hat{\mathbf{G}}^+|l\rangle - \langle l|\hat{\mathbf{G}}^-|l\rangle = D(E). \qquad \text{(F.13.17)}$$

In particular we shall consider the coordinate representation $|l\rangle = |q\rangle$, for which the density of states can be expressed by

$$\frac{\mathrm{i}}{2\pi}\int_{-\infty}^{\infty} [\hat{\mathbf{G}}^+(\mathbf{q},\mathbf{q},E) - \hat{\mathbf{G}}^-(\mathbf{q},\mathbf{q},E)]\,\mathrm{d}\mathbf{q} = D(E) \qquad \text{(F.13.18)}$$

and for the momentum representation $|l\rangle = |k\rangle$

$$\frac{\mathrm{i}}{2\pi}\int_{-\infty}^{\infty} [\hat{\mathbf{G}}^+(\mathbf{k},\mathbf{k},E) - \hat{\mathbf{G}}^-(\mathbf{k},\mathbf{k},E)]\frac{\mathrm{d}\mathbf{k}}{(2\pi)^3} = D(E) \qquad \text{(F.13.19)}$$

Here the factor of $(2\pi)^{-3}$ comes from the density of states in momentum space which enters through the change from summation to integration in equation (F.13.19) (we consider unit volume of coordinate space)

$$\sum_k \to \int \frac{\mathrm{d}\mathbf{k}}{(2\pi)^3}.$$

(see equation B.7.17). From equation (F.13.13) and equation (F.13.14) the two functions $\hat{\mathbf{G}}^+$ and $\hat{\mathbf{G}}^-$ are complex conjugates so that

$$\hat{\mathbf{G}}^+ - \hat{\mathbf{G}}^- = 2\mathrm{i}\mathscr{I}(\hat{\mathbf{G}}^+). \qquad \text{(F.13.20)}$$

Finally,

$$D(E) = -\frac{1}{\pi}\sum_{\mathbf{l}} \mathscr{I}\,[(G^+(\mathbf{l},\mathbf{l})]. \qquad \text{(F.13.21)}$$

[*Note.* The relations (F.13.18) and (F.13.19) are sometimes written

$$D(E) = \int_{-\infty}^{\infty} \rho(\mathbf{q}, E)\, \mathrm{d}\mathbf{q},$$

$$D(E) = \int_{-\infty}^{\infty} \rho(\mathbf{k}, E) \frac{\mathrm{d}\mathbf{k}}{(2\pi)^3}.$$

where the terms $\rho(\mathbf{q}, E)$ and $\rho(\mathbf{k}, E)$ may be interpreted as the probability of finding a particle with position $\mathbf{q}$ and energy E and with momentum $\mathbf{k}$ and energy E respectively.]

Problems

(F.13.1) Prove that for ϵ real and > 0,

(a) $\dfrac{1}{\mathrm{i}\omega + \epsilon} = \displaystyle\int_{-\infty}^{\infty} \mathrm{e}^{-(\mathrm{i}\omega+\epsilon)x} \eta(x)\, \mathrm{d}x,$

(b) $\eta(x) = \dfrac{1}{2} + \dfrac{P}{2\pi\mathrm{i}} \displaystyle\int_{-\infty}^{\infty} \frac{\mathrm{e}^{\mathrm{i}\omega x}}{\omega}$

where

$$\eta(x) = 0 \ (x < 0)$$
$$= 1 \ (x > 0).$$

(F.13.2) Prove that $G^+(\mathbf{q}, \mathbf{q}', E) = G^{-*}(\mathbf{q}', \mathbf{q}, E)$, i.e. $\hat{\mathbf{G}}^+$ and $\hat{\mathbf{G}}^-$ are not hermitian operators.

(F.13.3) Another possible Green's function corresponding to standing waves is obtained by putting

$$\hat{\mathbf{G}}_0^{(1)} = \tfrac{1}{2}[\hat{\mathbf{G}}_0^+ + \hat{\mathbf{G}}_0^-].$$

Show that

$$\hat{\mathbf{G}}_0^{(1)} = P\left[\frac{1}{E - \hat{\mathbf{H}}}\right].$$

(F.13.4) Show that $\hat{\mathbf{G}}_0^{(1)}$ is a hermitian operator.

(F.13.5) Show that the Green's function (F.12.19) gives, in conjunction with equation (F.13.21) the correct density of states for the one-dimensional free electron gas.

(F.13.6) Show that the Green's function of equation (F.12.33) in conjunction with (F.13.21) gives the usual result for the density of states of a three-dimensional free electron gas.

Using the results proved in this section it is now possible to verify what was previously assumed, namely, that the eigenfunctions $[1/(2\pi)^{\frac{3}{2}}]\, \mathrm{e}^{\mathrm{i}\mathbf{k}\cdot\mathbf{q}}$ form a complete set. The completeness relation, equation (F.4.1), may be written

$$\sum_i \langle q|i\rangle\langle i|q'\rangle = \delta(q - q').$$

From equation (F.13.1) (extended to three dimensions):

$$\int_{-\infty}^{\infty} e^{i\mathbf{k}\cdot(\mathbf{q}-\mathbf{q}')}\frac{d\mathbf{k}}{(2\pi)^3} = \delta(\mathbf{q}-\mathbf{q}'), \tag{F.13.22}$$

which proves that the $[1/(2\pi)^{\frac{3}{2}}]\,e^{i\mathbf{k}\cdot\mathbf{q}}$ form a complete, orthonormalized set. Alternatively this result follows as the limit (B.7.17) of equation (B.7.7).

F.14 INTEGRAL EQUATIONS AND ITERATIVE EXPANSIONS FOR THE GREEN'S FUNCTION

The general definition of the Green's operator for the time-independent Schrödinger equation is given by equation (F.9.6); in this the boundary conditions are left unspecified. Writing this as

$$\hat{\mathbf{G}} = \frac{1}{E - \hat{\mathbf{H}}_0 - \hat{\mathbf{H}}_I} \tag{F.14.1}$$

we can now use the result of problem (F.7.8) and, putting

$$\mathbf{A} = (E - \hat{\mathbf{H}}_0)^{-1}, \tag{F.14.2}$$

$$\hat{\mathbf{B}} = \hat{\mathbf{H}}_I, \tag{F.14.3}$$

we obtain the *iterative* expansion of the Green's operator

$$\hat{\mathbf{G}} = \hat{\mathbf{G}}_0 + \hat{\mathbf{G}}_0\hat{\mathbf{H}}_I\hat{\mathbf{G}}_0 + \hat{\mathbf{G}}_0\hat{\mathbf{H}}_I\hat{\mathbf{G}}_0\hat{\mathbf{H}}_I\hat{\mathbf{G}}_0 + \cdots. \tag{F.14.4}$$

Equation (F.14.4) is quite general; the boundary conditions are built into it through whatever particular form [e.g. $\hat{\mathbf{G}}_0^+, \hat{\mathbf{G}}_0^-, \hat{\mathbf{G}}_0^{(1)}$] is assumed for $\hat{\mathbf{G}}_0$. Thus if we are interested in the particular case of outgoing wave boundary conditions.

$$\hat{\mathbf{G}}^+ = \hat{\mathbf{G}}_0^+ + \hat{\mathbf{G}}_0^+\hat{\mathbf{H}}_I\hat{\mathbf{G}}_0^+ + \hat{\mathbf{G}}_0^+\hat{\mathbf{H}}_I\hat{\mathbf{G}}_0^+\hat{\mathbf{H}}_I\hat{\mathbf{G}}_0^+ + \cdots. \tag{F.14.5}$$

The general form of equation (F.14.4) can be expressed in a particular representation by taking matrix elements and inserting appropriate sets of complete states; for example, in the position representation,

$$\begin{aligned} G(\mathbf{q},\mathbf{q}',E) &= \langle\mathbf{q}|\hat{\mathbf{G}}|\mathbf{q}'\rangle = \langle\mathbf{q}|\hat{\mathbf{G}}_0|\mathbf{q}'\rangle \\ &\quad + \sum_{\mathbf{q}_1',\mathbf{q}_2} \langle\mathbf{q}|\hat{\mathbf{G}}_0|\mathbf{q}_1\rangle\langle\mathbf{q}_1|\hat{\mathbf{H}}_I|\mathbf{q}_2'\rangle\langle\mathbf{q}_2'|\hat{\mathbf{G}}_0|\mathbf{q}'\rangle + \cdots \\ &= G_0(\mathbf{q},\mathbf{q}',E) + \int d\mathbf{q}_1 G_0(\mathbf{q},\mathbf{q}_1)H_I(\mathbf{q}_1)G_0(\mathbf{q}_1,\mathbf{q}') \\ &\quad + \int d\mathbf{q}_1\, d\mathbf{q}_2 G_0(\mathbf{q},\mathbf{q}_1)H_I(\mathbf{q}_1)G_0(\mathbf{q}_1,\mathbf{q}_2)H_I(\mathbf{q}_2)G_0(\mathbf{q}_2,\mathbf{q}') + \cdots. \end{aligned} \tag{F.14.6}$$

The result of equation (F.8.21) has been used.

Note that $\hat{\mathbf{H}}_0$ need not be $\mathbf{p}^2/2m$; it can be any unperturbed Hamiltonian we find convenient. For example $\hat{\mathbf{H}}_0$ might be the Hamiltonian of a periodic lattice and $\hat{\mathbf{H}}_I$ the interaction due to impurities.

Problems

(F.14.1) Write down the next term in the series of equation (F.14.6).

(F.14.2) Write down the first three terms in the series of equation (F.14.5) in the momentum representation.

(F.14.3) Specializing to the case of outgoing waves and the case where $\hat{\mathbf{H}}_0$ is the kinetic energy Hamiltonian, write down the first three terms of the series of equation (F.14.5) in the coordinate and momentum representation.

From equation (F.14.4), multiplying both sides by $\hat{\mathbf{H}}_I\hat{\mathbf{G}}_0$ from the right,

$$\hat{\mathbf{G}}\hat{\mathbf{H}}_I\hat{\mathbf{G}}_0 = \hat{\mathbf{G}}_0\hat{\mathbf{H}}_I\hat{\mathbf{G}}_0 + \hat{\mathbf{G}}_0\hat{\mathbf{H}}_I\hat{\mathbf{G}}_0\hat{\mathbf{H}}_I\hat{\mathbf{G}}_0 + \cdots,$$

so that, comparing with equation (F.14.4),

$$\hat{\mathbf{G}} = \hat{\mathbf{G}}\hat{\mathbf{H}}_I\hat{\mathbf{G}}_0 + \hat{\mathbf{G}}_0. \tag{F.14.7}$$

This, in general, is an operator equation for the Green's operator.

(F.14.4) Show that the equation for $\hat{\mathbf{G}}$ may also be written

$$\hat{\mathbf{G}} = \hat{\mathbf{G}}_0 + \hat{\mathbf{G}}_0\hat{\mathbf{H}}_I\hat{\mathbf{G}}.$$

In a continuous representation (F.14.7) gives rise to an integral equation; for example, in a coordinate representation,

$$G(\mathbf{q}, \mathbf{q}', E) = G_0(\mathbf{q}, \mathbf{q}', E) + \int d\mathbf{q}_1 G(\mathbf{q}, \mathbf{q}_1, E)H_I(\mathbf{q}_1)G_0(\mathbf{q}_1, \mathbf{q}', E). \tag{F.14.8}$$

(F.14.5) Write down equation (F.14.7) in the momentum representation.

F.15 THE INTEGRAL EQUATION AND ITERATIVE SERIES FOR THE WAVEFUNCTION

The Schrödinger equation $(E - \hat{\mathbf{H}})|n\rangle = 0$ can be written as an integral equation for the wave-vector. We write

$$(E - \hat{\mathbf{H}}_0 - \hat{\mathbf{H}}_I)|n\rangle = 0.$$

Thus we obtain

$$(E - \hat{\mathbf{H}}_0)|n\rangle = \hat{\mathbf{H}}_I|n\rangle,$$

i.e.

$$|n\rangle = \hat{\mathbf{G}}_0\hat{\mathbf{H}}_I|n\rangle. \tag{F.15.1}$$

In the coordinate representation this leads to an integral equation for the wavefunction:

$$\langle \mathbf{q}|n\rangle = \langle \mathbf{q}|\hat{\mathbf{G}}_0|\mathbf{q}_1\rangle \langle \mathbf{q}_1|\hat{\mathbf{H}}_I|\mathbf{q}_2\rangle \langle \mathbf{q}_2|n\rangle$$

$$= \int d\mathbf{q}_1 G_0(\mathbf{q}, \mathbf{q}_1, E) H_I(\mathbf{q}_1) \langle \mathbf{q}_1|n\rangle. \qquad \text{(F.15.2)}$$

If, corresponding to energy E, there exists a solution of the homogeneous equation, namely

$$(E - \hat{\mathbf{H}}_0)|n\rangle_0 = 0,$$

then the complete solution of (F.15.1) becomes

$$|n\rangle = |n\rangle_0 + \hat{\mathbf{G}}_0^+ \hat{\mathbf{H}}_I|n\rangle, \qquad \text{(F.15.3)}$$

which, in the coordinate representation, leads to

$$\langle \mathbf{q}|n\rangle = \langle \mathbf{q}|n\rangle_0 + \int G_0^+(\mathbf{q}, \mathbf{q}_1, E) H_I(\mathbf{q}_1) \langle \mathbf{q}_1|n\rangle \, d\mathbf{q}_1 \qquad \text{(F.15.4)}$$

Equation (F.15.1) might be used, for example, as an equation for the negative energy states of an electron in an atom, for which a solution $\langle \mathbf{q}|n\rangle_0$ does not exist. Equation (F.15.3) is the usual starting point for the calculation of wave functions in a scattering problem, for which the $\langle \mathbf{q}|n\rangle_0$ clearly exist and physically represent the waves incident on the potential. Equation (F.15.3) can also be discussed in the momentum representation[10]. The substitution $\langle \mathbf{q}|n\rangle = \langle \mathbf{q}|n\rangle_0$ under the integral sign of (F.15.4) represents the first *Born* approximation; an iterative solution may be constructed by this procedure.

Problems

(F.15.1) Construct this series.

An alternative expression for the wavefunction is obtained from equation (F.15.3) by a rearrangement of terms. We write

$$|n\rangle_{\text{total}} = |n\rangle_0 + |n\rangle_s, \qquad \text{(F.15.5)}$$

so that the l.h.s. satisfies

$$(E + i\epsilon - \hat{\mathbf{H}}_0 - \hat{\mathbf{H}}_I)|n\rangle_{\text{total}} = 0. \qquad \text{(F.15.6)}$$

Using (F.15.5) and remembering that $(E - \hat{\mathbf{H}}_0)|n\rangle_0 = 0$ we get

$$(E + i\epsilon - \hat{\mathbf{H}}_I - \hat{\mathbf{H}}_0)|n\rangle_s - \hat{\mathbf{H}}_I|n\rangle_0 = 0,$$

[10] H. A. Bethe and E. E. Salpeter, *Quantum Mechanics of One and Two Electron Atoms*, Springer, Berlin, 1957; §§8, 9.

so that

$$|n\rangle_s = (E + i\epsilon - \hat{\mathbf{H}}_I - \hat{\mathbf{H}}_0)^{-1}\hat{\mathbf{H}}_I|n\rangle_0$$

and

$$|n\rangle_{\text{total}} = |n\rangle_0 + \hat{\mathbf{G}}^+\hat{\mathbf{H}}_I|n\rangle_0. \tag{F.15.7}$$

This is another expression for the wavefunction to be compared with equation (F.15.3). This last equation is an integral equation for the wave-vector which involves the unperturbed Green's operator $\hat{\mathbf{G}}_0$; equation (F.15.7) is no longer an integral equation for the wave-vector, but requires knowledge of the Green operator $\hat{\mathbf{G}}$ associated with the total potential $\hat{\mathbf{H}} = \hat{\mathbf{H}}_0 + \hat{\mathbf{H}}_I$.

(F.15.2) Show that in the coordinate representation equation (F.15.7) may be written

$$\langle \mathbf{q}|n\rangle = \langle \mathbf{q}|n\rangle_0 + \int d\mathbf{q}' G_0^+(\mathbf{q}, \mathbf{q}', E)H_I(\mathbf{q}')\langle \mathbf{q}'|n\rangle_0. \tag{F.15.8}$$

(F.15.3) Consider equation (F.15.3) in the momentum representation, taking $\hat{\mathbf{H}}_I$ as the total potential and $\hat{\mathbf{H}}_0$ as the kinetic energy Hamiltonian. Take $|n\rangle_0$ as a plane wave and write down (F.15.3) in the momentum representation.

If $\hat{\mathbf{G}}$ in equation (F.15.7) is expanded in a series as in equation (F.14.5), we obtain once again the iterative solution which the reader was asked to construct in problem (F.15.1):

$$\begin{aligned}|n\rangle_{\text{total}} &= |n\rangle_0 + [\hat{\mathbf{G}}_0^+ + \hat{\mathbf{G}}_0^+\hat{\mathbf{H}}_I\hat{\mathbf{G}}_0^+ + \cdots]\hat{\mathbf{H}}_I|n\rangle_0 \\ &= [1 + (\hat{\mathbf{G}}_0^+ + \hat{\mathbf{G}}_0^+\hat{\mathbf{H}}_I\hat{\mathbf{G}}_0^+ + \cdots)\hat{\mathbf{H}}_I]|n\rangle_0.\end{aligned} \tag{F.15.9}$$

In this equation the operator inside the brackets may be thought of as an operator which acts upon the unperturbed wave-vector $|n\rangle_0$ to produce the 'correct' wave-vector; we can write

$$|n\rangle = \hat{\mathbf{\Omega}}|n\rangle_0. \tag{F.15.10}$$

The operator which appears in equation (F.15.10) is the Møller wave operator. This has an easy interpretation in scattering theory where $\hat{\mathbf{\Omega}}^+$ acts on the incoming wave $|n\rangle_0$ to produce the outgoing wave $|n\rangle^+$, i.e.

$$|n\rangle^+ = \hat{\mathbf{\Omega}}^+|n\rangle_0. \tag{F.15.11}$$

Comparing equations (F.15.7) and (F.15.11),

$$\hat{\mathbf{\Omega}}^+ = 1 + \hat{\mathbf{G}}^+\hat{\mathbf{H}}_I. \tag{F.15.12}$$

In the coordinate representation (F.15.9) can be given a simple diagrammatic interpretation[11]. We have

$$\begin{aligned}\langle \mathbf{q}|n\rangle = \langle \mathbf{q}|n\rangle_0 &+ \int d\mathbf{q}_1 G_0^+(\mathbf{q}, \mathbf{q}_1, E)H_I(\mathbf{q}_1)\langle \mathbf{q}_1|n\rangle_0 \\ &+ \int\int d\mathbf{q}_1\, d\mathbf{q}_2 G_0^+(\mathbf{q}, \mathbf{q}_1, E)H_I(\mathbf{q}_1)G_0^+(\mathbf{q}_1, \mathbf{q}_2, E)H_I(\mathbf{q}_2)\langle \mathbf{q}_2|n\rangle_0 + \cdots.\end{aligned} \tag{F.15.13}$$

[11] See R. P. Feynman, *Phys. Rev.*, **76**, 749 (1949).

Here the second term represents scattering of the incident wave $\langle \mathbf{q}_1|n\rangle_0$ in the volume d$\mathbf{q}$. The total contribution at $\mathbf{q}$ is the sum of all such scatterings. This is represented as in fig. F.15.1a.

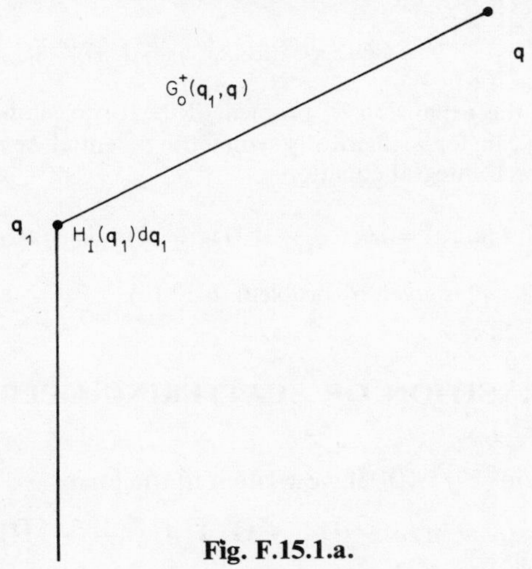

Fig. F.15.1.a.

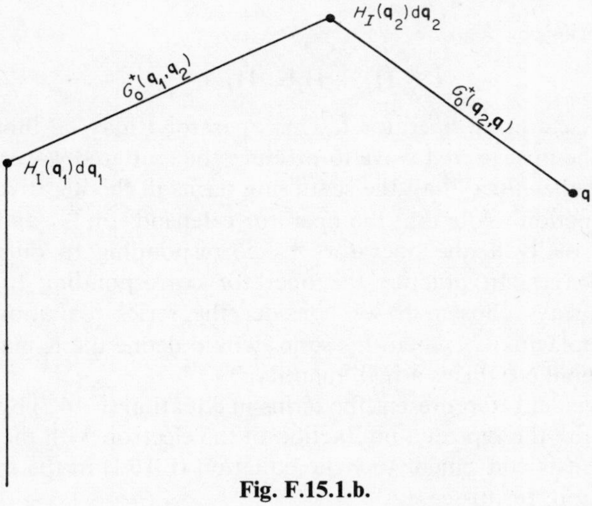

Fig. F.15.1.b.

The third term represents the scattering of the wave at the point $\mathbf{q}_1$, from which it travels to $\mathbf{q}_2$ to be scattered by the potential $H_I(\mathbf{q}_2)$. The total wavefunction at $\mathbf{q}$ can be represented as the sum of all such double scatterings and is given diagrammatically by fig. F.15.1b. The Green's function can be thought of as carrying, or propagating, the electron between scattering processes.

(F.15.4) Write down the third term in the series (F.15.13) and illustrate it by the appropriate diagram.

(F.15.5) In the expression (F.15.4) expand the wavefunction $\langle \mathbf{q}|n\rangle$ in spherical harmonics:

$$\langle \mathbf{q}|n\rangle = \sum_{l,m} A_l(k_0, q) Y_l^m(\mathbf{k}_0) Y_l^{m*}(\mathbf{q})$$

and use the expansion of problem (F.12.10) to show that this integral equation can, for a spherically symmetric potential, be reduced to the one-dimensional integral equation

$$A_l(k_0, q) = j_l(k_0, q) + \int G_l(q, q', E) A_l(k_0, q') H_I(q') q'^2 \, dq'$$

where $G_l(q, q')$ is given by problem (F.12.10).

F.16 THE TRANSITION OR SCATTERING OPERATOR (t MATRIX)

Consider equation (F.15.10). If we write it in the form

$$\begin{aligned} |n\rangle_{\text{total}} &= |n\rangle_0 + [\mathbf{G}_0^+ + \mathbf{G}_0^+ \hat{\mathbf{H}}_I \mathbf{G}_0^+ + \cdots] \hat{\mathbf{H}}_I n\rangle_0 \\ &= |n\rangle_0 + \mathbf{G}_0^+ \hat{\mathbf{t}} |n\rangle_0, \end{aligned} \tag{F.16.1}$$

then the series

$$\hat{\mathbf{t}} = \hat{\mathbf{H}}_I + \hat{\mathbf{H}}_I \hat{\mathbf{G}}_0^+ \hat{\mathbf{H}}_I + \cdots \tag{F.16.2}$$

defines the scattering operator $\hat{\mathbf{t}}$. This operator $\hat{\mathbf{t}}$ may be thought of as acting on the unscattered wave to produce the scattered wave; we incorporate in it the effect of all the scattering terms in the iterative series for the wavefunction. Note that the operator $\hat{\mathbf{t}}$ depends on $\hat{\mathbf{G}}_0$ and therefore we should really define operators $\hat{\mathbf{t}}^{\pm}$ corresponding to outgoing and incoming waves. In practice the operator corresponding to outgoing waves is always chosen. If we consider the series [equation (F.16.2)] with $\hat{\mathbf{G}}_0^{(1)}$ replacing $\hat{\mathbf{G}}_0^+$, then this series would define the $\hat{\mathbf{K}}$ matrix of the potential; unlike $\hat{\mathbf{t}}$, this is a real quantity.

It is convenient to represent the terms in equation (F.16.2) by diagrams which portray the repeated interaction of the electron with the potential. Frequently it is convenient to write equation (F.16.1) in the momentum representation. In this case

$$\langle \mathbf{k}|n\rangle = \langle \mathbf{k}|n_0\rangle + G_0^+(\mathbf{k}) \int t(\mathbf{k}, \mathbf{k}'') \psi_0(\mathbf{k}'') \, d\mathbf{k}''. \tag{F.16.3}$$

In this equation the momentum component of the incident wave $\psi_0(k'')$ [of form $e^{i\mathbf{k}\cdot\mathbf{q}}$] is fixed at the energy of the incident wave, k_0. The components of the scattered wave $\langle \mathbf{k}|n\rangle$ are also fixed at the same energy. Thus in this case $t(\mathbf{k}, \mathbf{k}'')$ has components only on the energy shell $|\mathbf{k}| = |\mathbf{k}''| = k_0$.

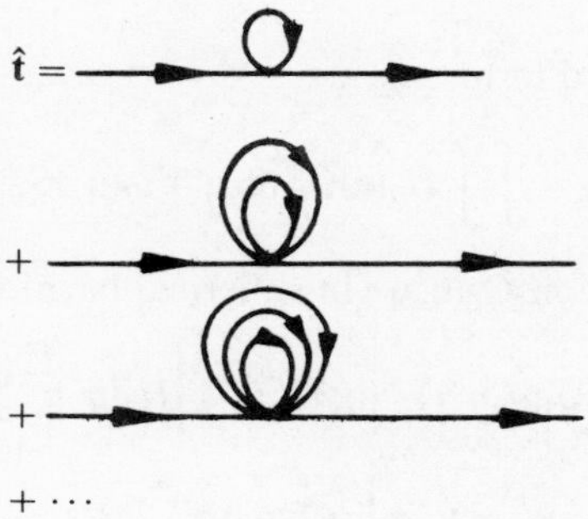

Fig. F.16.1.

In a simple scattering process the matrix elements of the t matrix in the momentum representation are confined to the energy shell. It is, however, possible to define the t matrix quite generally by the series of equation (F.16.2). In this case, off the energy shell matrix elements may be defined. Such matrix elements occur, for example, in multiple scattering processes. Note that the definition of equation (F.16.2) holds for *any* potential (not necessarily spherically symmetric).

The general calculation of the $\hat{\mathbf{t}}$ matrix is a difficult problem. Particular examples of cases in which it can be evaluated will be considered later. On the energy shell, however, matrix elements in the momentum representation may be expressed (for a spherically symmetric potential) in terms of phase shifts; this expression we shall now consider. Applying the result of problem (F.12.12) we have, quite generally,

$$t(\mathbf{q}, \mathbf{q}') = \sum_{l,m,l',m'} t_{ll'}^{mm'}(q, q') Y_l^m(\mathbf{q}) Y_{l'}^{m'*}(\mathbf{q}').$$

We shall now show that, in fact, for a spherically symmetric potential

$$t(\mathbf{q}, \mathbf{q}') = \sum_{l,m} t_l(q, q') Y_l^m(\mathbf{q}) Y_l^{m*}(\mathbf{q}'), \tag{F.16.4}$$

that is, the coefficients $t_l(q, q')$ are independent of m.

(Proof): Consider the defining series [equation (F.16.2)]. In each term expand $G_0(\mathbf{q}, \mathbf{q}')$ in spherical harmonics according to problem (F.12.10).

Then for the second term we get

$$\langle \mathbf{q}|\hat{\mathbf{H}}_I\hat{\mathbf{G}}_0\hat{\mathbf{H}}_I|\mathbf{q}'\rangle = H_I(\mathbf{q})G_0^+(\mathbf{q},\mathbf{q}')\hat{\mathbf{H}}_I(\mathbf{q}')$$
$$= H_I(\mathbf{q})\sum_{l,m} G_l(q,q)Y_l^m(\mathbf{q})Y_l^{m*}(\mathbf{q}')H_I(\mathbf{q}')$$

Now consider the third term.

$$\langle \mathbf{q}|H_IG_0^+H_IG_0^+H_I|\mathbf{q}'\rangle = \int\int H_I(\mathbf{q})G_0^+(\mathbf{q},\mathbf{q}'')H_I(\mathbf{q}'')G_0^+(\mathbf{q}'',\mathbf{q}')H_I(\mathbf{q}')\,\mathrm{d}\mathbf{q}''$$

and expand the function $G_0(\mathbf{q},\mathbf{q}')$ in spherical harmonics. We obtain

$$H_I(q)\int\left[\sum_{l,m} G_l(q,q'')Y_l^m(\mathbf{q})Y_l^{m*}(\mathbf{q}'')\right]H_I(q'')\left[\sum_{l',m'} G_l(q'',q')\right.$$
$$\left.Y_{l'}^{m'}(\mathbf{q}'')Y_{l'}^{m'*}(\mathbf{q}')\right]H_I(q')\,\mathrm{d}\Omega_{\mathbf{q}''}q''^2\,\mathrm{d}q''$$

The integration over $\mathrm{d}\Omega_{\mathbf{q}''}$ gives a factor of $\delta_{ll'}\,\delta_{mm'}$. In this way only the angular terms $Y_l^m(\mathbf{q})$, $Y_l^{m*}(\mathbf{q}')$ survive.

Finally the series may be written symbolically

$$\hat{\mathbf{t}}_l = \hat{\mathbf{H}}_I + \hat{\mathbf{H}}_I\hat{\mathbf{G}}_l\hat{\mathbf{H}}_I + \cdots \tag{F.16.5}$$

where the right-hand side (and therefore the left-hand side) depends on l only and the complete t matrix given by

$$t(\mathbf{q},\mathbf{q}') = \sum_{l,m} t_l(q,q')Y_l^m(\mathbf{q})Y_l^{m*}(\mathbf{q}'). \tag{F.16.6}$$

Now for a spherically symmetric potential the wavefunction may be expanded:

$$\psi(\mathbf{q}) = \sum_{l=0}^{\infty}\sum_{m=-l}^{m=+l} 4\pi \mathrm{i}^l A_l(k_0,q)Y_l^m(\mathbf{k}_0)Y_l^{m*}(\mathbf{q}). \tag{F.16.7}$$

The expansions of problem (F.12.10) and equation (F.16.7) may now be substituted into equation (F.15.4) to give

$$\sum_{l=0}^{\infty}\sum_{m=-l}^{l} 4\pi\mathrm{i}^l A_l(k_0,q)Y_l^m(\mathbf{k}_0)Y_l^{m*}(\mathbf{q})$$
$$= \sum_{l=0}^{\infty}\sum_{m=-l}^{+l} 4\pi\mathrm{i}^l j_l(k_0q)Y_l^m(\mathbf{k}_0)Y_l^{m*}(\mathbf{q})$$
$$+\int\sum_{l=0}^{\infty}\sum_{m=-l}^{+l} G_l(q,q')Y_l^m(\mathbf{q})Y_l^{m*}(\mathbf{q}')H_I(q')\sum_{l'=0}^{\infty}\sum_{m'=-l}^{+l}$$
$$4\pi\mathrm{i}^l A_{l'}(k_0,q)Y_{l'}^{m'}(\mathbf{k}_0)Y_{l'}^{m'}(\mathbf{q}')q'^2\,\mathrm{d}\Omega_{\mathbf{q}'}\,\mathrm{d}q'$$

The angular integration gives a factor of $\delta_{ll'}\delta_{mm'}$. Multiplying through by $Y_l^m(\mathbf{k}_0)Y_l^{m*}(\mathbf{q})$ and integrating over $d\Omega_{\mathbf{k}_0}$ and $d\Omega_{\mathbf{q}}$ gives

$$A_l(k_0, q) = j_l(k_0 q) + \int G_l(q, q')H_I(q')A_l(k_0, q')q'^2\, dq'. \quad \text{(F.16.8)}$$

On the other hand we may use equation (F.16.2), substituting the expansion of equation (F.16.6) for $t_l(q, q')$. Proceeding exactly as before we get

$$A_l(k_0, q) = j_l(k_0 q) + \int\int G_l(q, q')t_l(q', q'')j_l(k_0 q'')q'^2 q''^2\, dq'\, dq''. \quad \text{(F.16.9)}$$

Therefore, comparing equations (F.16.8) and (F.16.9),

$$\int G_l(q, q')H_I(q')A_l(k_0, q')q'^2\, dq'$$
$$= \int G_l(q, q')t_l(q', q'')j_l(k_0 q'')\, q'^2 q''^2\, dq'\, dq''. \quad \text{(F.16.10)}$$

Now it is shown in standard discussions of scattering theory that the asymptotic form of the function $A(k_0, q)$ is given by

$$A_l(k_0, q) = \frac{e^{i\delta_l}}{k_0 q}\sin\left(k_0 q - \frac{l\pi}{2} + \delta_l\right) \quad \text{(F.16.11)}$$

where the phase shifts δ_l are given by

$$\frac{1}{k_0} e^{i\delta_l}\sin\delta_l = -\int_0^\infty j_l(k_0 q)H_I(q')A_l(k_0, q')q'^2\, dq'. \quad \text{(F.16.12)}$$

This is the sought-after expression for the matrix elements of the t matrix in terms of phase shifts. The expression may be related to the expansion of the t matrix in momentum space. From equation (F.16.6),

$$t(\mathbf{k}, \mathbf{k}') = \frac{1}{(2\pi)^3}\int e^{-i\mathbf{k}\cdot\mathbf{q}}\, e^{+i\mathbf{k}'\cdot\mathbf{q}'} \sum_l t_l(q, q')Y_l^m(\mathbf{q})Y_l^{m*}(\mathbf{q}')$$
$$q^2 q'^2\, dq\, dq'\, d\Omega_{\mathbf{q}}\, d\Omega_{\mathbf{q}'}$$
$$= \frac{1}{(2\pi)^3}\int \sum_{l',m'} 4\pi i^{l'} j_{l'}(kq)Y_{l'}^{m'}(\mathbf{k})Y_{l'}^{m'}{}^*(\mathbf{q})$$
$$\sum_{l'',m''} 4\pi i^{l''} j_{l''}(k'q')Y_{l''}^{m''}(\mathbf{q}')Y_{l''}^{m''}{}^*(\mathbf{k}')\sum_l t_l(q, q')$$
$$Y_l^m(\mathbf{q})Y_l^m{}^*(\mathbf{q}')q'^2 q''^2\, d\Omega_{q'}\, d\Omega_{q''}$$
$$= \frac{2}{\pi}\sum_{l,m} Y_l(\mathbf{k})Y_l(\mathbf{k}')\int j_l(k_0 q)j_l(k'q')t_l(q, q')q'^2 q^2\, dq\, dq'$$
$$= \sum_{l,m} t_l(k, k')Y_l^m(\mathbf{k})Y_l^m{}^*(\mathbf{k}'), \quad \text{(F.16.13)}$$

This gives the coefficient $t_l(k, k')$ in the expansion

$$t(\mathbf{k}, \mathbf{k}') = \sum_{l,m} t_l(k, k')Y_l^m(\mathbf{k})Y_l^{m*}(\mathbf{k}'). \qquad \text{(F.16.14)}$$

For the particular case of $|\mathbf{k}| = |\mathbf{k}'| = k_0$, we may identify $t_l(k_0, k_0)$ by comparison with equation F.16.12. In this case,

$$t_l(k_0, k_0) = -\frac{2}{\pi k_0} e^{i\delta_l} \sin \delta_l. \qquad \text{(F.16.15)}$$

This expression will be of subsequent use in the t matrix approach to band structure.

From equations (F.16.2) and (F.14.4) we obtain the relation

$$\hat{\mathbf{G}}^+ = \hat{\mathbf{G}}_0^+ + \hat{\mathbf{G}}_0^+ \hat{\mathbf{t}} \hat{\mathbf{G}}_0^+ \qquad \text{(F.16.16)}$$

which is analogous to equation (F.16.1). Since the bound states of the system are given by the singularities of $\hat{\mathbf{G}}$ they are also given by the singularities of $\hat{\mathbf{t}}$.

An example of the foregoing theory: the one-dimensional delta function potential.

We take the potential $H_I = h\delta(x - a)$. The integral equation (F.14.7) becomes, in the coordinate representation

$$\begin{aligned} G^+(q, q') &= G_0^+(q, q') + h \int G_0^+(q, q'')\delta(q'' - a)G(q'', q')\, dq'' \\ &= G_0^+(q, q') + hG_0(q, a)G(a, q') \end{aligned}$$

Putting $q = a$ on both sides of this equation

$$G^+(a, q') = G_0(a, q') + hG_0^+(q, a)G(a, q')$$

so that

$$G_0^+(a, q') = \frac{G_0^+(a, q')}{1 - hG_0^+(a, a)}.$$

Hence

$$G_0^+(q, q') = G_0^+(q, q') + \frac{hG_0^+(q, a)G_0^+(a, q')}{1 - hG_0^+(a, a)} \qquad \text{(F.16.17)}$$

Problems

(F.16.1) Obtain the same result by summing the iterative series (F.14.5).

The poles of the Green's function are given by

$$1 - hG_0^+(a, a) = 0. \qquad \text{(F.16.18)}$$

Now for positive energies $E = k_0^2$,

$$G_0(q, q') = \frac{-\mathrm{i}}{2k_0} \mathrm{e}^{\mathrm{i}k_0|q-q'|} \tag{F.16.19}$$

and for negative energies $E = -k_0^2$,

$$G_0^+(q, q') = -\frac{1}{2k_0} \mathrm{e}^{\mathrm{i}k_0|q-q'|}. \tag{F.16.20}$$

It can be seen that equations (F.16.18) cannot be satisfied for positive energies; for negative energies it can be satisfied if the potential is negative (there are only bound states for attractive potentials).

Using equations (F.16.16) and (F.16.17) we see by comparison that

$$t(q, q') = h\frac{\delta(q - a)(q' - a)}{1 - hG_0^+(a, a)}. \tag{F.16.21}$$

(F.16.2) Obtain the same result by the summation of the series of equation (F.16.2).

Equation (F.15.1) (the homogeneous equation for bound states) is

$$\psi(q) = \int G_0^+(q, q')\psi(q')H_I(q') \tag{F.16.22}$$

where

$$H_I(q') = h\,\delta(q' - a).$$

Hence

$$\psi(q) = hG_0^+(q, a)\psi(a) \tag{F.16.23}$$

Putting $q = a$ we recover (F.16.18). [Note that equation (F.16.18) can only be satisfied for negative energies, for which $G_0(q, q')$ is real.]

For positive energies there exists a solution of the unperturbed equation of the same energy as the perturbed solution. Therefore

$$\psi(q) = \psi_0(q) + h \quad G_0^+(q, q')H_I(q')\psi(q')\,\mathrm{d}q'.$$

We take

$$\psi_0(q) = \mathrm{e}^{\mathrm{i}k_0q}.$$

Then

$$\psi(q) = \mathrm{e}^{\mathrm{i}k_0q} + hG_0^+(q, a)\psi(a)$$

Putting $q = a$,

$$\psi(a) = e^{ik_0 a} + hG_0(q, a)\psi(a)$$

and

$$\psi(a) = \frac{e^{ik_0 a}}{1 - hG_0(a, a)}.$$

Hence

$$\psi(q) = e^{ik_0 q} + \frac{hG_0(q, a)\, e^{ik_0 a}}{1 - hG_0(a, a)}.$$

For $q > a$ the wavefunction is proportional to $e^{ik_0 q}$; for $q < a$ it has two terms in $e^{ik_0 q}$ and $e^{-ik_0 q}$ respectively.

(F.16.3) Calculate the reflexion and transmission coefficients for the δ-function potential.

CHAPTER XII

Application of Green's Functions to the Theory of Electron States in Periodic Lattices

Introduction

In this chapter we shall consider methods of calculating Green's functions for lattice periodic potentials. Since the Green's operator, written in the form

$$\hat{\mathbf{G}} = \sum_n \frac{|n\rangle \langle n|}{E - E_n}$$

clearly has poles at energies $E = E_n$, it follows that in the momentum representation,

$$\langle \mathbf{k}|\hat{\mathbf{G}}|\mathbf{k}\rangle = \sum_n \frac{\langle \mathbf{k}|n\rangle \langle n|\mathbf{k}\rangle}{E - E(\mathbf{k})}$$

has poles at $E = E(\mathbf{k})$; thus the poles of the Green's function, as a function of $\mathbf{k}$, give the $E - \mathbf{k}$ relation. From equation (F.16.16) the diagonal elements of the T matrix in the momentum representation have the same property.

In §F.17 we shall consider a general formalism for multiple scattering problems which applies equally to periodic or nonperiodic lattices. In §F.18 we shall apply this formalism to the derivation of the E–$\mathbf{k}$ relation in the weak binding approximation. In §F.19 the tight-binding approximation will be considered. Finally, in §F.20 we shall consider the summation of the multiple scattering series for a periodic lattice of spherically symmetric, non-overlapping potentials. In this way we shall arrive at the well known K.K.R. relation for the allowed energy levels of such a lattice.

F.17 THE MULTIPLE SCATTERING EXPANSION OF THE T MATRIX SERIES

It is physically obvious that the wavefunction in a periodic lattice can be thought of as being gradually built up by a process of multiple scattering starting from an arbitrary wavefunction (e.g. a free electron wave). This point of view can be expressed by means of the series expansions of the Green's operator, T matrix, or wavefunction. However, if one writes down, say, the iterative expansion for the Green's function [equation (F.14.5)] for a potential of the form $V(\mathbf{q}) = \sum_{\alpha} v(\mathbf{q} - \mathbf{R}_{\alpha})$, then evaluation of each term involves in itself an infinite series. Simplification of this series can be achieved and we end up with an expansion of G in terms of the t matrices of the separate atoms rather than of the potentials. This form works equally well for disordered potentials, but only the application to ordered (i.e. periodic) lattices will be considered in this chapter.

We shall work in the $|\mathbf{k}\rangle$ representation, so that the series [equation (F.14.5)] takes the form

$$\begin{aligned}\langle \mathbf{k}|\hat{\mathbf{G}}^{+}(E)|\mathbf{k}'\rangle = \langle \mathbf{k}|\hat{\mathbf{G}}_0^{+}|\mathbf{k}'\rangle &+ \langle \mathbf{k}|\hat{\mathbf{G}}_0^{+}|\mathbf{k}_1\rangle\langle \mathbf{k}_1|\hat{\mathbf{V}}|\mathbf{k}_2\rangle\langle \mathbf{k}_2|\hat{\mathbf{G}}_0^{+}|\mathbf{k}'\rangle \\ &+ \langle \mathbf{k}|\hat{\mathbf{G}}_0^{+}|\mathbf{k}_1\rangle\langle \mathbf{k}_1|\hat{\mathbf{V}}|\mathbf{k}_2\rangle\langle \mathbf{k}_2|\hat{\mathbf{G}}_0^{+}|\mathbf{k}_3\rangle \\ &\times \langle \mathbf{k}_3|\mathbf{V}|\mathbf{k}_4\rangle\langle \mathbf{k}_4|\hat{\mathbf{G}}_0^{+}|\mathbf{k}'\rangle + \cdots. \end{aligned} \tag{F.17.1}$$

Here a particular term, say $\langle \mathbf{k}_i| V|\mathbf{k}_j\rangle$, is given by equation (F.8.29) as

$$\begin{aligned}\langle \mathbf{k}_i|\hat{\mathbf{V}}|\mathbf{k}_j\rangle &= \frac{1}{(2\pi)^3}\int \mathrm{d}\mathbf{q}\, \mathrm{e}^{\mathrm{i}(\mathbf{k}_j - \mathbf{k}_i)\cdot \mathbf{q}}\, V(\mathbf{q}) \\ &= V(\mathbf{k}_i - \mathbf{k}_j). \end{aligned} \tag{F.17.2}$$

In the rigid ion approximation

$$V(\mathbf{q}) = \sum_{\alpha} v(\mathbf{q} - \mathbf{R}_{\alpha}). \tag{F.17.3}$$

where the $\mathbf{R}_{\alpha}$ are the ionic lattice sites. The notation $v_{\alpha} = v(\mathbf{q} - \mathbf{R}_{\alpha})$ will frequently be used. Applying equation (F.17.2) we have

$$\langle \mathbf{k}_i|v(\mathbf{q} - \mathbf{R}_{\alpha})|\mathbf{k}_j\rangle = \frac{1}{(2\pi)^3}\int \mathrm{d}\mathbf{q}\, \mathrm{e}^{\mathrm{i}(\mathbf{k}_j - \mathbf{k}_i)\cdot \mathbf{q}} v(\mathbf{q} - \mathbf{R}_{\alpha}). \tag{F.17.4}$$

Making the substitution $\mathbf{Q} = \mathbf{q} - \mathbf{R}_{\alpha}$, (F.17.4) gives rise to

$$\begin{aligned}\langle \mathbf{k}_i|v(\mathbf{q} - \mathbf{R}_{\alpha})|\mathbf{k}_j\rangle &= \frac{1}{(2\pi)^3}\int \mathrm{d}\mathbf{Q}\, \mathrm{e}^{\mathrm{i}(\mathbf{k}_j - \mathbf{k}_i)\cdot(\mathbf{Q}+\mathbf{R}_{\alpha})} v(\mathbf{Q}) \\ &= \mathrm{e}^{\mathrm{i}(\mathbf{k}_j - \mathbf{k}_i)\cdot \mathbf{R}_{\alpha}} v_0(\mathbf{k}_i - \mathbf{k}_j) \end{aligned} \tag{F.17.5}$$

where the suffix zero labels the atom at the origin. In referring to the terms of equation (F.17.1) we shall adopt the convention that the term in $\hat{\mathbf{G}}_0$ is the zeroth term, the term with n V's the nth. The first term written in the $\mathbf{k}$ representation is then

$$\langle \mathbf{k}|\hat{\mathbf{G}}_0|\mathbf{k}_1\rangle \langle \mathbf{k}_1|\hat{\mathbf{V}}|\mathbf{k}_2\rangle \langle \mathbf{k}_2|\hat{\mathbf{G}}_0|\mathbf{k}'\rangle.$$

But $\hat{\mathbf{G}}_0$ is diagonal in the $\mathbf{k}$ representation. Hence this term may be written

$$G_0(\mathbf{k})V(\mathbf{k} - \mathbf{k}')G_0(\mathbf{k}').$$

If this term only were retained we would have the Born approximation to the series (F.17.1). In general the nth term of the series (F.17.1) is

$$G_0(\mathbf{k})V(\mathbf{k} - \mathbf{k}_1)\ldots V(\mathbf{k}_{n-1} - \mathbf{k}')G_0(\mathbf{k}').$$

We now consider the effect of a potential of the type (F.17.3). Using (F.17.4) we get for a typical term

$$\langle \mathbf{k}_i | V(\mathbf{q} - \mathbf{R}_\alpha) | \mathbf{k}_j \rangle = \sum_\alpha e^{i(\mathbf{k}_j - \mathbf{k}_i).\mathbf{R}_\alpha} v_0(\mathbf{k}_i - \mathbf{k}_j). \tag{F.17.6}$$

We can now write the series (F.17.1) as

$$G(\mathbf{k}, \mathbf{k}') = G_0(\mathbf{k})\delta(\mathbf{k} - \mathbf{k}') + \sum_\alpha G_0(\mathbf{k})v_\alpha(\mathbf{k} - \mathbf{k}')G_0(\mathbf{k}')$$
$$+ G_0(\mathbf{k}) \sum_\alpha v_\alpha(\mathbf{k} - \mathbf{k}_1) \sum_\beta v_\beta(\mathbf{k}_1 - \mathbf{k}')G_0(\mathbf{k}'). \tag{F.17.7}$$

Consider in this series the second term. It may be separated into the terms with $\alpha = \beta$, corresponding to scattering off the same ion, and to terms with $\alpha \neq \beta$, corresponding to scattering off different ions. These may be represented by diagrams, in which an atom scattered once is represented as follows

an atom scattered twice is represented by the diagram

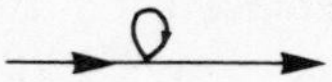

and so on. Thus the second term of equation (F.17.7) may be represented symbolically by

second term =

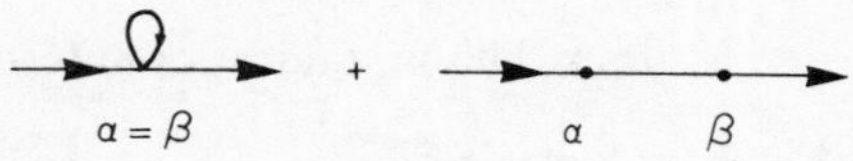

For the third term we have

third term =

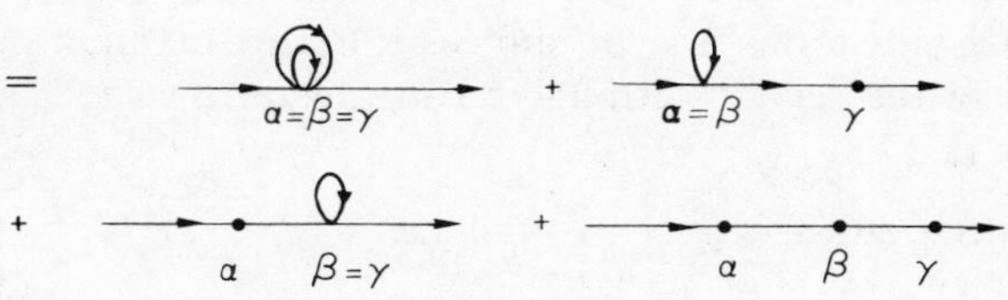

If, in the series (F.17.7), we had just one scattering centre, we could write similarly

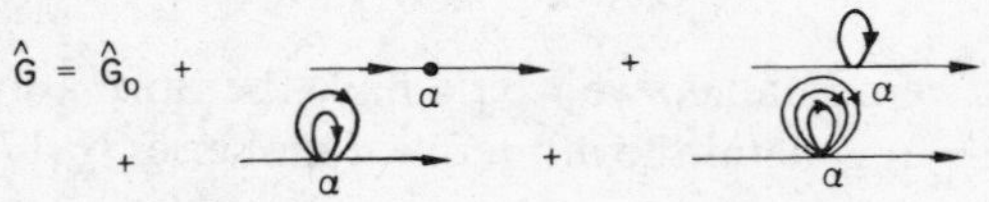

In this series, if we write $\hat{\mathbf{G}} = \hat{\mathbf{G}}_0 + \hat{\mathbf{G}}_0\hat{\mathbf{t}}_\alpha\hat{\mathbf{G}}_0$, we can equally well regard the scattering as produced by the t matrix and the series then essentially represents the t matrix expansion (F.16.2).

Consider now the effect of the terms in the series (F.17.7) which correspond to taking $\alpha = \beta$ in the second term, $\alpha = \beta = \gamma$ in the third term, and so on. We now get

$$\hat{\mathbf{G}} = \hat{\mathbf{G}}_0 + \sum_\alpha \hat{\mathbf{G}}_0\{\hat{\mathbf{v}}_\alpha + \hat{\mathbf{v}}_\alpha\hat{\mathbf{G}}_0\hat{\mathbf{v}}_\alpha + \hat{\mathbf{v}}_\alpha\hat{\mathbf{G}}_0\hat{\mathbf{v}}_\alpha\hat{\mathbf{G}}_0\hat{\mathbf{v}}_\alpha + \cdots\}\hat{\mathbf{G}}_0$$

Comparing with equation (F.16.2) the terms in brackets is just the $\hat{\mathbf{t}}$ matrix of a single atom. Therefore we can rewrite equation (F.17.8) in the form

$$\begin{aligned} G(\mathbf{k}, \mathbf{k}') = {} & G_0(\mathbf{k})\delta(\mathbf{k}, \mathbf{k}') + G_0(\mathbf{k}) \sum_\alpha t_\alpha(\mathbf{k} - \mathbf{k}')G_0(\mathbf{k}') \\ & + G_0(\mathbf{k}) \sum_\alpha v_\alpha(\mathbf{k} - \mathbf{k}_1)\, G_0(\mathbf{k}_1) \sum_{\alpha \neq \beta} v_\beta(\mathbf{k}_1 - \mathbf{k}')G_0(\mathbf{k}') \\ & + G_0(\mathbf{k}) \sum_\alpha v_\alpha(\mathbf{k} - \mathbf{k}_1)G_0(\mathbf{k}_1) \sum_{\alpha \neq \beta} v_\beta(\mathbf{k}_1 - \mathbf{k}_2) \sum_{\beta \neq \gamma} v_\gamma(\mathbf{k}_2 - \mathbf{k}') \\ & + \cdots \end{aligned} \qquad \text{(F.17.8)}$$

that is, runs of suffixes of the same kind are included in the first term. Consider next the effect of replacing $\hat{\mathbf{v}}_\alpha$ in the second term by $\hat{\mathbf{t}}_\alpha$. Using equation (F.16.2) we would get

$$\begin{aligned} & G_0(\mathbf{k})[v_\alpha(\mathbf{k} - \mathbf{k}_1) + v_\alpha(\mathbf{k} - \mathbf{k}_2)G_0(\mathbf{k}_2)v_\alpha(\mathbf{k}_2 - \mathbf{k}_1) + \cdots] \\ & \qquad \times G_0(\mathbf{k}_1) \sum_\beta v_\beta(\mathbf{k}_1 - \mathbf{k}')G_0(\mathbf{k}') \end{aligned}$$

for this particular term. The first additional term gives

$$\sum_\alpha G_0(\mathbf{k})v_\alpha(\mathbf{k} - \mathbf{k}_2)G_0(\mathbf{k}_2)v_\alpha(\mathbf{k}_2 - \mathbf{k}_1)G_0(\mathbf{k}_1) \sum_\beta v_\beta(\mathbf{k}_1 - \mathbf{k}')G_0(\mathbf{k}').$$

We can see that this is included in (F.17.8) in the third term with $\alpha = \beta \neq \gamma$. (Note that only terms with *all* suffices equal are taken into account by the $\hat{\mathbf{t}}$ matrix in the second term.) We can therefore write the third term of equation (F.17.8)

$$G_0(\mathbf{k}) \sum_\alpha v_\alpha(\mathbf{k} - \mathbf{k}_1) \sum_{\alpha \neq \beta} G_0(\mathbf{k}_1)v_\beta(\mathbf{k}_1 - \mathbf{k}_2) \sum_{\beta \neq \gamma} v_\gamma(\mathbf{k}_2 - \mathbf{k}')G_0(\mathbf{k}')$$

and replace the term $\hat{\mathbf{v}}_\alpha$ in the second term of equation (F.17.8) by $\hat{\mathbf{v}}_\alpha + \hat{\mathbf{v}}_\alpha \hat{\mathbf{G}}_0 \hat{\mathbf{v}}_\alpha$. It can similarly be seen that if we replace the term $\hat{\mathbf{v}}_\alpha$ in the second term of equation (F.17.8) by $\hat{\mathbf{v}}_\alpha + \hat{\mathbf{v}}_\alpha \hat{\mathbf{G}}_0 \hat{\mathbf{v}}_\alpha + \hat{\mathbf{v}}_\alpha \hat{\mathbf{G}}_0 \hat{\mathbf{v}}_\alpha \hat{\mathbf{G}}_0 \hat{\mathbf{v}}_\alpha$ we can cancel the last term $\hat{\mathbf{v}}_\alpha \hat{\mathbf{G}}_0 \hat{\mathbf{v}}_\alpha \hat{\mathbf{G}}_0 \hat{\mathbf{v}}_\alpha$ by writing the fourth term of equation (F.17.8)

$$\hat{\mathbf{G}}_0 \sum_\alpha \hat{\mathbf{v}}_\alpha \hat{\mathbf{G}}_0 \sum_{\alpha \neq \beta} \hat{\mathbf{v}}_\beta \hat{\mathbf{G}}_0 \sum_{\beta \neq \gamma} \hat{\mathbf{v}}_\gamma \hat{\mathbf{G}}_0 \sum_{\gamma \neq \delta} \hat{\mathbf{v}}_\delta \hat{\mathbf{G}}_0 .$$

In this way it can finally be seen that the term $\hat{\mathbf{v}}_\alpha$ in (F.17.8) in the second term can be replaced by $\hat{\mathbf{t}}_\alpha$ provided suitable adjustment of summations is made in the remaining terms of the series. This procedure can be repeated for each $\hat{\mathbf{v}}$ factor in (F.17.8) and we finally get

$$\begin{aligned} \langle \mathbf{k}|G|\mathbf{k}'\rangle &= G_0(\mathbf{k})\delta(\mathbf{k} - \mathbf{k}') + G_0(\mathbf{k}) \sum_\alpha t_\alpha(\mathbf{k} - \mathbf{k}') G_0(\mathbf{k}') \\ &\quad + G_0(\mathbf{k}) \sum_\alpha t_\alpha(\mathbf{k} - \mathbf{k}_1) G_0(\mathbf{k}_1) \sum_{\alpha \neq \beta} t_\beta(\mathbf{k}_1 - \mathbf{k}') G_0(\mathbf{k}') \\ &\quad + G_0(\mathbf{k}) \sum_\alpha t_\alpha(\mathbf{k} - \mathbf{k}_1) G_0(\mathbf{k}_1) \sum_{\alpha \neq \beta} t_\beta(\mathbf{k}_1 - \mathbf{k}_2) G_0(\mathbf{k}_2) \\ &\quad \times \sum_{\beta \neq \gamma} t_\gamma(\mathbf{k}_2 - \mathbf{k}') \mathbf{G}_0(\mathbf{k}') + \cdots \end{aligned} \tag{F.17.9}$$

Symbolically, the corresponding expansion of the T matrix [using equation (F.16.16)] is given by

$$T = \sum_\alpha \hat{\mathbf{t}}_\alpha + \sum_{\alpha \neq \beta} \hat{\mathbf{t}}_\alpha \hat{\mathbf{G}}_0 \hat{\mathbf{t}}_\beta + \sum_{\substack{\alpha \neq \beta \\ \beta \neq \gamma}} \hat{\mathbf{t}}_\alpha \hat{\mathbf{G}}_0 \hat{\mathbf{t}}_\beta \hat{\mathbf{G}}_0 \hat{\mathbf{t}}_\gamma + \cdots . \tag{F.17.10}$$

This way of writing the multiple scattering series will be used in deriving the Green's function of a periodic lattice. It has the advantage that the strength of the scattering is introduced via the $\hat{\mathbf{t}}$ matrix of the single atom; for the Born approximation we just replace the $\hat{\mathbf{t}}$ matrix of each atom by the potential.

In summary, the basic problem in calculation of the Green's function for a periodic lattice is to sum over the processes which contribute to the largest terms in the Green's function series. It is convenient to distinguish between processes in which the electron is scattered by a single ion and those in which it is scattered successively by different ions. The form of the Green's function series obtained in equation (F.17.9) provides a mathematical means of distinguishing between these two processes.

F.18 THE LOOSE BINDING APPROXIMATION

The loose binding approximation (or almost free electron approximation) is a simple way of demonstrating the band structure characteristic of a periodic lattice. It is assumed that the periodic lattice can be treated as a small perturbation on the free electrons. Using standard time-independent perturbation theory it may then be shown that the effect is to change the energy levels by a small amount except in certain regions where, it is found, 'forbidden' regions appear—that is, regions of energy levels allowed to the free electrons (whose energies form a continuum for positive energies) are forbidden when a periodic lattice is present.

The almost free electron approach is of interest particularly because of the possibility of constructing a 'pseudo-potential' which can be treated as weak even if the original potential is not. With these refinements we shall not be concerned. The Green's function approach emphasizes the assumptions which must be made for the almost free electron approximation to hold—in particular, it will be seen that in addition to the assumption that the potential of a single ion is weak the validity of the theory depends on neglecting certain multiple scattering terms. This last fact is not evident from the usual perturbation treatment and the Green's function method may therefore be said to give more physical insight into the structure of the almost free electron approximation. The use of Green's functions in the almost free electron approach is due to S. F. Edwards, who considered the one dimensional case[12] and the three dimensional case.[13] We shall depart somewhat from the presentation of both these papers in detail but the final results and the assumptions made are the same.

We start with the multiple scattering expansion of the last section—equation (F.17.9). We shall at once assume the scattering power of a single ion to be weak and replace the t matrix of each single ion by the corresponding potential. This means, of course, that we are restricting ourselves to positive energies, far away from the bound states. In this way equation (F.17.9) becomes, for the diagonal elements of $\hat{\mathbf{G}}$,

$$\begin{aligned}\langle \mathbf{k}|\hat{\mathbf{G}}|\mathbf{k}\rangle = G_0(\mathbf{k}) + G_0(\mathbf{k}) \sum_{\alpha} v_{\alpha}(\mathbf{k}-\mathbf{k}) G_0(\mathbf{k}) \\ + G_0(\mathbf{k}) \sum_{\alpha} v_{\alpha}(\mathbf{k}-\mathbf{k}_1) G_0(\mathbf{k}_1) \sum_{\alpha \neq \beta} v_{\beta}(\mathbf{k}_1-\mathbf{k}) G_0(\mathbf{k}) + \cdots\end{aligned} \tag{F.18.1}$$

It is usual in the standard perturbation approach to assume that the zero of potential can be adjusted so that the first term vanishes; it is propor-

[12] S. F. Edwards, *Phil. Mag.*, **6**, 617 (1961).

[13] S. F: Edwards, *Proc. Roy. Soc.*, **267**, 518 (1962).

tional to the integral of the potential since [cf. (B.12.1)]

$$v(\mathbf{k}-\mathbf{k}') = \frac{1}{\Omega}\int e^{i(\mathbf{k}'-\mathbf{k}).\mathbf{q}} v(\mathbf{q})\,d\mathbf{q} \tag{F.18.2}$$

and the first term involves the diagonal element $\mathbf{k} = \mathbf{k}'$.

The second term is important and turns out to be one in terms of which the sum over the whole series can be expressed. We have, taking into account normalization of the wavefunctions to volume Ω,

$$v_\alpha(\mathbf{k}-\mathbf{k}') = \frac{1}{\Omega}\int e^{i(\mathbf{k}'-\mathbf{k})\,.\,\mathbf{q}}\, v(\mathbf{q}-\mathbf{R}_\alpha)\,d\mathbf{q}$$

$$= \frac{1}{\Omega}\int e^{i(\mathbf{k}'-\mathbf{k})\,.\,(\mathbf{Q}+\mathbf{R}_\alpha)} v(\mathbf{Q})\,d\mathbf{Q}$$

where the substitution of $\mathbf{Q} = \mathbf{q} - \mathbf{R}_\alpha$ has been made.

In this way we get

$$v(\mathbf{k}-\mathbf{k}_1) = \sum_\alpha e^{i(\mathbf{k}_1-\mathbf{k})\,.\,\mathbf{R}_\alpha}\, v_0(\mathbf{k}-\mathbf{k}_1) \tag{F.18.3}$$

where $v_0(\mathbf{k})$ stands for the Fourier transform of the atom at the origin of coordinates in the direct lattice. The expression in the second term of (F.18.1) then involves

$$v(\mathbf{k}-\mathbf{k}_1)v(\mathbf{k}_1-\mathbf{k}) = \sum_\alpha\sum_\beta e^{i(\mathbf{k}_1-\mathbf{k})\,.\,(\mathbf{R}_\alpha-\mathbf{R}_\beta)}|v_0(\mathbf{k}-\mathbf{k}_1)|^2.$$

Consider the sum

$$\sum_\alpha\sum_\beta e^{i(\mathbf{k}_1-\mathbf{k})\,.\,(\mathbf{R}_\alpha-\mathbf{R}_\beta)}.$$

For fixed α we can shift the origin to $\mathbf{R}_\alpha$ and get [cf. (B.6.22) in the limit $\Omega \to \infty$]

$$\sum_\beta e^{i(\mathbf{k}-\mathbf{k}_1)\,.\,\mathbf{R}_\beta} = \sum_{\mathbf{K}} \frac{(2\pi)^3}{\tau}\delta(\mathbf{k}-\mathbf{k}_1-\mathbf{K}) \tag{F.18.4}$$

where τ is the volume of the unit cell. This happens N times (where N is the number of atoms), and the final contribution is

$$|v(\mathbf{k}-\mathbf{k}_1)|^2 = \frac{N^2}{\Omega^2}|v(\mathbf{K})|^2(2\pi)^3\delta(\mathbf{k}-\mathbf{k}_1-\mathbf{K}) \tag{F.18.5}$$

where we have put $\tau = \Omega/N$.

The second term can therefore be written

$$G_0(\mathbf{k})G_0(\mathbf{k}_1)|v(\mathbf{k}-\mathbf{k}_1)|^2G_0(\mathbf{k})$$

$$= \sum_{\mathbf{K}} \int \frac{d\mathbf{k}_1}{(2\pi)^3} G_0(\mathbf{k}_1)\delta(\mathbf{k}-\mathbf{k}_1-\mathbf{K}) \frac{N^2}{\Omega^2}(2\pi)^3|v(\mathbf{k}-\mathbf{k}_1)|^2G_0^2(\mathbf{k})$$

$$= \frac{N^2}{\Omega^2}\sum_{\mathbf{K}} \frac{G_0^2(\mathbf{k})|v(\mathbf{K})|^2}{E-(\mathbf{k}-\mathbf{K})^2}$$

$$= G_0(\mathbf{k})\,\Sigma_2(\mathbf{k})G_0(\mathbf{k}) \tag{F.18.6}$$

where

$$\Sigma_2(\mathbf{k}) = \frac{N^2}{\Omega^2}\sum_{\mathbf{K}} \frac{|v(\mathbf{K})|^2}{E-(\mathbf{k}-\mathbf{K})^2} \tag{F.18.7}$$

This expression for the second term is exact.

Consider next the fourth term in the series; this involves

$$v(\mathbf{k}-\mathbf{k}_1)v(\mathbf{k}_1-\mathbf{k}_2)v(\mathbf{k}_2-\mathbf{k}_3)v(\mathbf{k}_3-\mathbf{k}).$$

In this product we write

$$v(\mathbf{k}-\mathbf{k}_1) = \sum_{\alpha} e^{i(\mathbf{k}_1-\mathbf{k})\cdot\mathbf{R}_\alpha} v_0(\mathbf{k}-\mathbf{k}_1) \tag{F.18.8a}$$

$$v(\mathbf{k}_1-\mathbf{k}_2) = \sum_{\beta} e^{i(\mathbf{k}_2-\mathbf{k}_1)\cdot\mathbf{R}_\beta} v_0(\mathbf{k}_1-\mathbf{k}_2) \tag{F.18.8b}$$

$$v(\mathbf{k}_2-\mathbf{k}_3) = \sum_{\gamma} e^{i(\mathbf{k}_3-\mathbf{k}_2)\cdot\mathbf{R}_\gamma} v_0(\mathbf{k}_3-\mathbf{k}_2) \tag{F.18.8c}$$

$$v(\mathbf{k}-\mathbf{k}_3) = \sum_{\delta} e^{i(\mathbf{k}_3-\mathbf{k})\cdot\mathbf{R}_\delta} v_0(\mathbf{k}_3-\mathbf{k}). \tag{F.18.8d}$$

Each term may now be evaluated by equation (F.18.4) so that we get for equations (F.18.8a) and (F.18.8b) respectively the results

$$\sum_{\mathbf{K}} \frac{1}{\tau}(2\pi)^3\,\delta(\mathbf{k}-\mathbf{k}_1-\mathbf{K}) \quad \text{and} \quad \sum_{\mathbf{L}} \frac{1}{\tau}(2\pi)^3\,\delta(\mathbf{k}_1-\mathbf{k}_2-\mathbf{L}).$$

In this sum we now consider only the terms with $\mathbf{K} = -\mathbf{L}$. This gives

$$\mathbf{k}-\mathbf{k}_1 = -(\mathbf{k}_1-\mathbf{k}_2). \tag{F.18.9}$$

so that $\mathbf{k}_2 = \mathbf{k}$ and the first two terms, i.e. (F.18.8a) and (F.18.8b), now have the form of equation (F.18.5). The remaining two terms, namely (F.18.8c) and (F.18.8d) also have the same form if we assume that the second sum extends over values of the reciprocal vector which are different from those which occur in the first two potential terms. Explicitly, applying equation (F.18.4) to each term of (F.18.8), we get a product of the form

$$\sum_{\mathbf{K},\mathbf{L},\mathbf{M},\mathbf{N}} \delta(\mathbf{k} - \mathbf{k}_1 - \mathbf{K})\, \delta(\mathbf{k}_1 - \mathbf{k}_2 - \mathbf{L})\, \delta(\mathbf{k}_2 - \mathbf{k}_3 - \mathbf{M})\, \delta(\mathbf{k}_3 - \mathbf{k} - \mathbf{N}).$$

We first consider only terms in this sum with $\mathbf{K} = -\mathbf{L}$. This gives $\mathbf{k}_2 = \mathbf{k}$ and therefore $\mathbf{M} = -\mathbf{N}$. We now consider further only the cases $\mathbf{M} \neq \mathbf{K}$ so that the summations over the first two and last two potentials can be carried out independently to give a result of the form (F.18.5) for each term. In this way the fourth term can be written

$$G_0(\mathbf{k})\Sigma_2(\mathbf{k})G_0(\mathbf{k})\Sigma_2(\mathbf{k})G_0(\mathbf{k}).$$

For each even term it is clear that a similar procedure can be employed, and if we ignore the odd terms in the series we get

$$G(\mathbf{k}) = G_0(\mathbf{k}) + G_0(\mathbf{k})\Sigma_2(\mathbf{k})G_0(\mathbf{k}) + G_0(\mathbf{k})\Sigma_2(\mathbf{k})G_0(\mathbf{k})\Sigma_2(\mathbf{k})G_0(\mathbf{k}) + \cdots.$$

These terms can be summed to give

$$G(\mathbf{k}) = \frac{G_0(\mathbf{k})}{1 - G_0(\mathbf{k})\Sigma_2(\mathbf{k})} = \frac{1}{G_0(\mathbf{k})^{-1} - \Sigma_2(\mathbf{k})}. \tag{F.18.10}$$

The result (F.18.10) represents the result of two types of approximation. In the first place we have assumed the scattering power of a single scattering centre can be represented by the Born approximation, and secondly we have used approximations which only sum a subset of the total number of possible scattering processes. The second approximation is the more serious because the magnitude of the omitted terms is difficult to estimate. We are here not really concerned with the validity of the whole procedure because the almost free electron approximation is, in this form, too simple to account for the observed energy levels. The method is presented here as an example in which Green's function methods can be compared with other approaches which the reader has met elsewhere.

From equation (F.18.10) the usual results of the nearly free electron theory now follow quite easily. Equation (F.18.10) may be written in the form

$$G(\mathbf{k}) = \frac{1}{E - \mathbf{k}^2 - \dfrac{N^2}{\Omega^2} \displaystyle\sum_{\mathbf{K}} \frac{|v(\mathbf{K})|^2}{E - (\mathbf{k} - \mathbf{K})^2}}. \tag{F.18.11}$$

For energy levels not too near $E - (\mathbf{k} - \mathbf{K})^2$ the second term is small because of the assumed smallness of the potential $v(\mathbf{K})$, and $\Sigma_2(\mathbf{k})$ is then seen to be the small correction to the energy levels. If E is near

$(\mathbf{k} - \mathbf{K})^2$ then the second term is large for some particular $\mathbf{K}$—say $\mathbf{K}'$. In this case only this term need be considered in the sum and we get for the energy levels (by considering the poles of the Green's function)

$$[E - \mathbf{k}^2]^2 = \frac{N^2}{\Omega^2} |v(\mathbf{K}')|^2 \tag{F.18.12}$$

which shows that the new energy levels are given by

$$E = \mathbf{k}^2 \pm \frac{N}{\Omega} |v(\mathbf{K})|. \tag{F.18.13}$$

F.19 THE TIGHT BINDING APPROXIMATION

The states of an isolated ion can be visualized as roughly hydrogen-like with corresponding energy levels. An alternative way of looking at the band structure of the periodic lattice is to consider the effect of all the ions on a particular one as a perturbation which spreads the sharp line of the isolated level into a band. The approach is considered in almost any textbook on solid state physics. We shall again consider it from the standpoints of Green's functions, as in the previous section, not in the hope of deriving new results but simply as an illustration of the method[14].

We shall denote the atomic orbital (eigenfunction) of an isolated ion by $\varphi_n(\mathbf{q})$. For an atom at lattice site $\mathbf{R}_\alpha$ the atomic orbital is then $\varphi_n(\mathbf{q} - \mathbf{R}_\alpha)$. If $v_0(\mathbf{q})$ then denotes the potential of the isolated ion then $\varphi(\mathbf{q} - \mathbf{R}_\alpha)$ satisfies

$$\left[-\frac{\hbar^2}{2m}\nabla^2 + v_0(\mathbf{q} - \mathbf{R}_\alpha)\right]\varphi_n(\mathbf{q} - \mathbf{R}_\alpha) = E_n\varphi_n(\mathbf{q} - \mathbf{R}_\alpha). \tag{F.19.1}$$

For an isolated ion the Green's function would be

$$G(\mathbf{q}, \mathbf{q}') = \sum_n \frac{\varphi_n(\mathbf{q})\varphi_n^*(\mathbf{q}')}{E - E_n}. \tag{F.19.2}$$

If we consider energy levels E_n which are widely separated in comparison with the changes of energy levels which the interaction is likely to bring about, then for energies near E_0 an approximation for $G(\mathbf{q}, \mathbf{q}')$ would be

$$G(\mathbf{q}, \mathbf{q}') = \frac{\varphi_0(\mathbf{q})\varphi_0^*(\mathbf{q}')}{E - E_0}. \tag{F.19.3}$$

[14] For a different approach to the tight-binding formalism see J. L. Beeby and S. F. Edwards, *Proc. Roy. Soc.* **274A**, 395 1963. See also K. T. S. Somaratra, M.Sc. Dissertation (Cardiff 1967).

where the suffix 0 denotes a particular energy level near which E happens to lie.

Equation (F.19.3) does not, of course, provide a Green's function which satisfies the boundary conditions of a periodic lattice. An approximate *wavefunction* which satisfies these is

$$\psi(\mathbf{q}) = N^{-\frac{1}{2}} \sum_{\alpha} e^{i\mathbf{k}\cdot\mathbf{R}_\alpha} \varphi_0(\mathbf{q} - \mathbf{R}_\alpha). \tag{F.19.4}$$

Putting $\mathbf{q} = \mathbf{q} + \mathbf{R}_\beta$ we get

$$\psi(\mathbf{q} + \mathbf{R}_\beta) = N^{-\frac{1}{2}} \sum_{\alpha} e^{i\mathbf{k}\cdot\mathbf{R}_\alpha} \varphi_0(\mathbf{q} + \mathbf{R}_\beta - \mathbf{R}_\alpha).$$

Since the sum is over all the vectors in the direct lattice, choice of origin is immaterial and we can shift the origin by putting $\mathbf{R}_\beta - \mathbf{R}_\alpha = -\mathbf{R}_{\alpha'}$. Then

$$\begin{aligned}\psi(\mathbf{q} + \mathbf{R}_\beta) &= N^{-\frac{1}{2}} \sum_{\alpha'} e^{i\mathbf{k}\cdot(\mathbf{R}_\beta + \mathbf{R}_{\alpha'})} \varphi(\mathbf{q} - \mathbf{R}_{\alpha'}) \\ &= e^{i\mathbf{k}\cdot\mathbf{R}_\beta} \psi(\mathbf{q}).\end{aligned} \tag{F.19.5}$$

In this sense equation (F.19.4) satisfies the Bloch condition. The corresponding Green's function can be constructed:

$$G_0(\mathbf{q}, \mathbf{q}', E) = \frac{\psi(\mathbf{q})\psi^*(\mathbf{q}')}{E - E_0} \tag{F.19.6}$$

This represents the Green's function corresponding to the periodic lattice, with energy levels still at $E = E_0$. If $H_\alpha(\mathbf{q})$ denotes the Hamiltonian of an isolated atom at lattice site R_α then the perturbing Hamiltonian is

$$\sum_{\beta \neq \alpha} v_\beta(\mathbf{q}) = H_I(\mathbf{q}) = H(\mathbf{q}) - H_\alpha(\mathbf{q}).$$

We can now write

$$G(\mathbf{q}, \mathbf{q}', E) = G_0(\mathbf{q}, \mathbf{q}', E) + \int G_0(\mathbf{q}, \mathbf{q}'') H_I(\mathbf{q}'') G(\mathbf{q}'', \mathbf{q}')\, d\mathbf{q}'' \tag{F.19.7}$$

$$= \frac{\psi(\mathbf{q})\psi^*(\mathbf{q})}{E - E_0} + \int \frac{\psi(\mathbf{q})\psi^*(\mathbf{q}'')}{E - E_0} H_I(\mathbf{q}'') G(\mathbf{q}'', \mathbf{q}')\, d\mathbf{q}''$$

with $\psi(\mathbf{q})$ given by equation (F.19.4). We write this in the form

$$\begin{aligned}G(\mathbf{q}, \mathbf{q}') &= G_0(\mathbf{q}, \mathbf{q}') + \frac{\psi(\mathbf{q})}{E - E_0} \int \psi(\mathbf{q}'') H_I(\mathbf{q}'') G(\mathbf{q}'', \mathbf{q}')\, d\mathbf{q}'' \\ &= G_0(\mathbf{q}, \mathbf{q}') + \frac{\psi(\mathbf{q})}{E - E_0} f(\mathbf{q}'), \quad \text{say.}\end{aligned} \tag{F.19.8}$$

Now multiply equation (F.19.8) by $\psi^*(\mathbf{q})H_I(\mathbf{q})$ and integrate over $\mathbf{q}$.

$$\int \psi^*(\mathbf{q})H_I(\mathbf{q})G(\mathbf{q},\mathbf{q}')\,\mathrm{d}\mathbf{q}$$

$$= \int G_0(\mathbf{q},\mathbf{q}')\psi^*(\mathbf{q})H_I(\mathbf{q})\,\mathrm{d}\mathbf{q} + \left[\frac{\int \psi(\mathbf{q})H_I(\mathbf{q})\psi^*(\mathbf{q})\,\mathrm{d}\mathbf{q}}{E - E_0}\right] f(\mathbf{q}').$$

Thus

$$f(\mathbf{q}') = \int G_0(\mathbf{q},\mathbf{q}')\psi^*(\mathbf{q})H_I(\mathbf{q})\,\mathrm{d}\mathbf{q} + \left[\frac{\int \psi(\mathbf{q})H_I(\mathbf{q})\psi^*(\mathbf{q})\,\mathrm{d}\mathbf{q}}{E - E_0}\right] f(\mathbf{q}'). \tag{F.19.9}$$

Substituting for $f(\mathbf{q}')$ from equation (F.19.9) into equation (F.19.8) we get

$$G(\mathbf{q},\mathbf{q}') = G_0(\mathbf{q},\mathbf{q}') + \frac{\psi(\mathbf{q})}{E - E_0}\left[\frac{\int G_0(\mathbf{q},\mathbf{q}')\psi^*(\mathbf{q})H_I(\mathbf{q})\,\mathrm{d}\mathbf{q}}{1 - \dfrac{\int \psi(\mathbf{q})\psi^*(\mathbf{q})H_I(\mathbf{q})\,\mathrm{d}\mathbf{q}}{E - E_0}}\right] \tag{F.19.10}$$

$$= G_0(\mathbf{q},\mathbf{q}') + \frac{\psi(\mathbf{q})\int G_0(\mathbf{q},\mathbf{q}')\psi^*(\mathbf{q})\psi(\mathbf{q})\,\mathrm{d}\mathbf{q}}{E - E_0 - \int \psi(\mathbf{q})\psi^*(\mathbf{q})H_I(\mathbf{q})\,\mathrm{d}\mathbf{q}}. \tag{F.19.11}$$

Thus there arises a perturbed energy level $E - E_0 - \Gamma(\mathbf{k})$ where

$$\Gamma(\mathbf{k}) = \sum_{\substack{\alpha \\ \alpha \neq 0}} \sum_{\alpha}\sum_{\beta} \mathrm{e}^{\mathrm{i}\mathbf{k}\cdot(\mathbf{R}_\alpha - \mathbf{R}_\beta)} N^{-1} \int \varphi(\mathbf{q} - \mathbf{R}_\alpha)\varphi^*(\mathbf{q} - \mathbf{R}_\beta)v_\alpha(\mathbf{q})\,\mathrm{d}\mathbf{q} \tag{F.19.12}$$

It is usual to restrict equation (F.19.12) in various ways. The potential sum gives rise, in general, to three-centre integrals; if only two centre integrals are considered equation (F.19.12) may be approximated by

$$\Gamma(\mathbf{k}) = \sum_{\substack{\alpha \\ \alpha \neq 0}} \left[\int |\varphi(\mathbf{q})|^2 v(\mathbf{q} - \mathbf{R}_\alpha)\,\mathrm{d}\mathbf{q} + \mathrm{e}^{\mathrm{i}\mathbf{k}\cdot\mathbf{R}_\alpha}\int \varphi(\mathbf{q} - \mathbf{R}_\alpha)v(\mathbf{q} - \mathbf{R}_\alpha)\varphi^*(\mathbf{q})\mathrm{d}\mathbf{q}\right]$$

which is the usual result.

F.20 THE *T* MATRIX APPROACH TO K.K.R. THEORY

In the preceding sections equations for the diagonal elements of the Green's function for a lattice-periodic potential have been derived under two limiting approximations. In this section we shall consider the same basic problem, this time only making the assumption that the lattice consists of spherically symmetric, non-overlapping potentials. This problem was first considered from the standpoint of the integral equation for the wavefunction [equation (F.15.2)] by Korringa[15] who showed that the condition for allowed energy levels could be expressed by equating an infinite determinant to zero. This determinant consists of two distinct parts, the first characteristic only of the crystal structure of the solid, and the second determined by the phase shifts of the ions. Kohn and Rostoker[16] further developed the theory from the same standpoint and showed in a calculation of the band structure of lithium that the infinite determinant could successfully be replaced by one of small size. The method has since been applied to a number of substances, and as examples we may mention the calculation of Segall on diamond[17] and copper[18].

In this section we shall consider the theory once again from the standpoint of Green's functions, and derive the K.K.R. condition by looking for the poles of the T matrix. Such a calculation was first carried out by Beeby[19]. We shall first derive the basic integral equation of K.K.R. theory (§F.20a) and thus introduce the structural Green's function which is characteristic of periodic lattices.

In §F.20b the summation of the T matrix series for a periodic potential will be considered. In §F.20c we shall consider the derivation of some expansion formulae in spherical harmonics necessary for the summation of the T matrix series in the angular momentum representation. The summation of the T matrix series in the angular momentum representation is finally considered in §F.20d.

F.20a The integral equation for the wavefunction

In §F.15 we derived the integral equation for the wavefunction (F.15.2), which we shall here write again for a periodic potential $H_I(\mathbf{q})$:

$$\psi(\mathbf{q}) = \int G_0(\mathbf{q}, \mathbf{q}', E)\psi(\mathbf{q}')H_I(\mathbf{q}')\,d\mathbf{q}'. \tag{F.20.1}$$

This as written is a homogeneous integral equation; if a solution of the unperturbed ($H_I = 0$) equation exists *obeying the same boundary conditions*, it must be added to the right-hand side of the equation. We now consider this equation for a lattice periodic potential $H_I(\mathbf{q})$, and ask whether there exist unperturbed solutions obeying the same boundary

[15] Korringa, J., *Physica*, **13**, 392 (1947).
[16] Kohn, W., and Rostoker, N., *Phys. Rev.*, **94**, 1111 (1954).
[17] Segall, B. *J. Phys. Chem. Solids*, **8**, 371 (1959).
[18] Segall, B. *Phys. Rev.*, **125**, 109 (1962).
[19] Beeby, J. L., *Proc. Roy. Soc. A*, **279**, 82 (1964). See also J. M. Ziman, *Proc. Phys. Soc.*, **86**, 337 (1965).

conditions, that is, obeying periodic boundary conditions for only certain regions of energy. Clearly no such solutions exist. Hence, for a periodic potential, equation (F.20.1) is correct as it stands. On the other hand, if equation (F.20.1) is written with, say, G_0 chosen as the Green's function for outgoing waves, and therefore given by equation (F.12.34), then we would expect the left-hand side of (F.20.1) to obey Bloch boundary conditions. This paradox will now be resolved.

Choosing units so that $2m/\hbar^2 = 1$, we substitute equation (F.12.34) into equation (F.20.1) and obtain

$$\psi(\mathbf{q}) = -\frac{1}{4\pi}\int \mathrm{d}\mathbf{q} H_I(\mathbf{q}')\psi(\mathbf{q}')\frac{\exp[ik_0|\mathbf{q} - \mathbf{q}'|]}{|\mathbf{q} - \mathbf{q}'|} \tag{F.20.2}$$

For a periodic potential

$$H_I(\mathbf{q} + \mathbf{R}_\alpha) = H_I(\mathbf{q}) \tag{F.20.3}$$

and

$$\psi(\mathbf{q}) = -\frac{1}{4\pi}\sum_{\substack{\text{all}\\ \text{unit}\\ \text{cells}}}\int_{\text{u.c.}} \mathrm{d}\mathbf{q}' H_I(\mathbf{q}')\psi(\mathbf{q}')\frac{\exp[ik_0|\mathbf{q} - \mathbf{q}'|]}{|\mathbf{q} - \mathbf{q}'|} \tag{F.20.4}$$

We now put

$$\mathbf{q} = \mathbf{R}_\alpha + \mathbf{Q}, \qquad \mathbf{q}' = \mathbf{R}_{\alpha'} + \mathbf{Q}' \tag{F.20.5}$$

where $\mathbf{q}$ lies in the αth unit cell, and use Bloch's theorem:

$$\psi(\mathbf{Q} + \mathbf{R}_\alpha) = \mathrm{e}^{-i\mathbf{k}\cdot\mathbf{R}_\alpha}\psi(\mathbf{Q}); \qquad \psi(\mathbf{Q}' + \mathbf{R}_{\alpha'}) = \mathrm{e}^{-i\mathbf{k}\cdot\mathbf{R}_{\alpha'}}\psi(\mathbf{Q}').$$

In this way equation (F.20.4) takes the form

$$\psi(\mathbf{Q}) = -\frac{1}{4\pi}\sum_{\alpha'}\int \mathrm{d}\mathbf{Q}' H_I(\mathbf{Q}')\psi(\mathbf{Q}')\frac{\mathrm{e}^{-i\mathbf{k}\cdot(\mathbf{R}_\alpha - \mathbf{R}_{\alpha'})}}{|\mathbf{R}_\alpha - \mathbf{R}_{\alpha'} + \mathbf{Q} - \mathbf{Q}'|} \times$$
$$\mathrm{e}^{ik_0|\mathbf{R}_\alpha - \mathbf{R}_{\alpha'} + \mathbf{Q} - \mathbf{Q}'|}. \tag{F.20.6}$$

By the definitions of equation (F.20.5) the integration in (F.20.6) is over a unit cell. Hence, finally,

$$\psi(\mathbf{Q}) = \int_{\text{unit cell}} H_I(\mathbf{Q}')\psi(\mathbf{Q}')G_s(\mathbf{k}, \mathbf{Q}, \mathbf{Q}')\,\mathrm{d}\mathbf{Q}' \tag{F.20.7}$$

where

$$G_s(\mathbf{k}, \mathbf{Q}, \mathbf{Q}') = -\frac{1}{4\pi}\sum_{\alpha,\alpha'}\frac{\mathrm{e}^{-i\mathbf{k}\cdot(\mathbf{R}_\alpha - \mathbf{R}_{\alpha'})}\,\mathrm{e}^{ik_0|\mathbf{R}_\alpha - \mathbf{R}_{\alpha'} + \mathbf{Q} - \mathbf{Q}'|}}{|\mathbf{R}_\alpha - \mathbf{R}_{\alpha'} + \mathbf{Q} - \mathbf{Q}'|} \tag{F.20.8}$$

has aptly been called by Ziman the structural Green's function. In this way, by making use of Bloch's theorem and of the translational symmetry of the potential, equation (F.20.1) reduces to (F.20.7).

The structural Green's function may be written in a different way by making use of the properties of the reciprocal lattice. We recall from equation (F.12.34) that

$$\lim_{\epsilon\to 0}\frac{1}{(2\pi)^3}\int d\mathbf{k}'\,\frac{\exp[i\mathbf{k}'\cdot\mathbf{q}]}{k'^2-[k_0^2+i\epsilon]}=\frac{1}{4\pi}\,\frac{\exp[i\mathbf{k}_0|\mathbf{q}|]}{|\mathbf{q}|}.$$

Hence

$$\begin{aligned}\frac{1}{4\pi}\sum_\alpha \exp[i\mathbf{k}\cdot\mathbf{R}_\alpha]\,&\frac{\exp[ik_0|\mathbf{q}-\mathbf{q}'+\mathbf{R}_\alpha|]}{|\mathbf{q}-\mathbf{q}'+\mathbf{R}_\alpha|}\\ &=\sum_\alpha \exp[i\mathbf{k}\cdot\mathbf{R}_\alpha]\int\frac{d\mathbf{k}'}{(2\pi)^3}\,\frac{\exp[i\mathbf{k}'\cdot(\mathbf{q}-\mathbf{q}'+\mathbf{R}_\alpha)]}{\mathbf{k}'^2-k_0^2+i\epsilon}.\end{aligned} \tag{F.20.9}$$

Clearly

$$\sum_\alpha \exp[i\mathbf{R}_\alpha\cdot(\mathbf{k}-\mathbf{k}')]=\sum_\alpha\int d\mathbf{q}\,\exp[i\mathbf{q}\cdot(\mathbf{k}-\mathbf{k}')]\delta(\mathbf{q}-\mathbf{R}_\alpha).$$

However, by a standard procedure,

$$\sum_\alpha \exp[i\mathbf{R}_\alpha\cdot(\mathbf{k}-\mathbf{k}')]=\sum_n\frac{(2\pi)^3}{\tau}\delta(\mathbf{K}_n+\mathbf{k}-\mathbf{k}').$$

where $\mathbf{K}_n$ is a reciprocal lattice vector. Hence finally

$$\begin{aligned}G_s(\mathbf{k},\mathbf{q},\mathbf{q}')&=-\frac{1}{\tau}\frac{(2\pi)^3}{(2\pi)^3}\sum_n\int d\mathbf{k}'\,\frac{\exp[i\mathbf{k}'\cdot(\mathbf{q}-\mathbf{q}')]\delta(\mathbf{K}_n+\mathbf{k}-\mathbf{k}')}{\mathbf{k}'^2-(\mathbf{k}_0^2+i\epsilon)}\\ &=-\frac{1}{\tau}\sum_n\frac{\exp[i(\mathbf{q}-\mathbf{q}')\cdot(\mathbf{K}_n+\mathbf{k})]}{(\mathbf{K}_n+\mathbf{k})^2-(k_0^2+i\epsilon)}.\end{aligned} \tag{F.20.10}$$

This is another expression for the structural Green's function from which certain of its properties may be more easily derived. Ziman and Phariseau[20] have generalized equation (F.20.7) to disordered systems. Equation (F.20.7) may also be written in abstract vector space[21].

The only assumptions made in deriving equation (F.20.7) were the periodicity of the potential. If we further made the rigid ion approximation

$$H_I(\mathbf{q})=\sum_\alpha v(\mathbf{q}-\mathbf{R}_\alpha) \tag{F.20.11}$$

[20] J. M. Ziman and P. Phariseau, *Phil. Mag.*, **8**, 1487 (1963).
[21] T. Lukes and M. Roberts, *Proc. Phys. Soc.*, **91**, 211 (1967).

and assume that the potential in a unit cell is spherically symmetrical then equation (F.20.7) becomes

$$\psi(\mathbf{Q}) = \int v(\mathbf{Q})\psi(\mathbf{Q}')G_s(\mathbf{k}, \mathbf{Q}, \mathbf{Q}')\,\mathrm{d}\mathbf{Q}'. \tag{F.20.12}$$

This equation may be solved, as was shown by Kohn and Rostoker, by the variational principle, and used to derive a condition on the allowed energy levels. Instead of following their method we shall now consider the $\hat{\mathbf{T}}$ matrix approach to the problem.

F.20b The summation of the *T* matrix series for a periodic potential

The general multiple scattering expansion of the $\hat{\mathbf{T}}$ matrix series [equation (F.17.10)] applies to the periodic potential as a particular case. If, therefore, the $\hat{\mathbf{T}}$ matrix series can be summed, we can obtain the $E - \mathbf{k}$ relation by looking for the poles of the $\hat{\mathbf{T}}$ matrix. We repeat, for reference, the general equation (F.17.10), and note that in its derivation the rigid ion approximation was made from the beginning.

$$\hat{\mathbf{T}} = \sum_{\alpha} \hat{\mathbf{t}}_\alpha + \sum_{\alpha \neq \beta} \hat{\mathbf{t}}_\alpha \hat{\mathbf{G}}_0 \hat{\mathbf{t}}_\beta + \sum_{\substack{\alpha \neq \beta \\ \beta \neq \gamma}} \hat{\mathbf{t}}_\alpha \hat{\mathbf{G}}_0 \hat{\mathbf{t}}_\beta \hat{\mathbf{G}}_0 \hat{\mathbf{t}}_\gamma + \cdots \tag{F.20.13}$$

Here we remind ourselves that each term $\hat{\mathbf{t}}_\alpha$ corresponds to the t matrix of a single ion located at lattice site $\mathbf{R}_\alpha$. It is related to the potential of that site by the equation

$$t_\alpha(\mathbf{q}, \mathbf{q}') = v(\mathbf{q} - \mathbf{R}_\alpha) + \int v(\mathbf{q} - \mathbf{R}_\alpha)G_0(\mathbf{q}, \mathbf{q}'')t_\alpha(\mathbf{q}'', \mathbf{q}')\,\mathrm{d}\mathbf{q}''. \tag{F.20.14}$$

In equations (F.20.13) and (F.20.14), G_0 is the free particle Green's function. Using the fact that this depends on the difference of the arguments $\mathbf{q}$ and $\mathbf{q}''$ only, we may use equation (F.20.14) to prove that

$$t_\alpha(\mathbf{q}, \mathbf{q}') = t(\mathbf{q} - \mathbf{R}_\alpha, \mathbf{q}' - \mathbf{R}_\alpha), \tag{F.20.15}$$

which relates the $\hat{\mathbf{t}}$ matrix of the ion at lattice site $\mathbf{R}_\alpha$ to that at the origin. Equation (F.20.13) may then be written

$$T(\mathbf{q}, \mathbf{q}') = \sum_{\alpha} t(\mathbf{q} - \mathbf{R}_\alpha, \mathbf{q}' - \mathbf{R}_\alpha) + \sum_{\alpha \neq \beta} \iint t(\mathbf{q} - \mathbf{R}_\alpha, \mathbf{q}'' - \mathbf{R}_\alpha)$$
$$G_0(\mathbf{q}'', \mathbf{q}''')t(\mathbf{q}''' - \mathbf{R}_\beta, \mathbf{q}' - \mathbf{R}_\beta)\,\mathrm{d}\mathbf{q}''\,\mathrm{d}\mathbf{q}''' + \cdots. \tag{F.20.16}$$

We now write

$$\langle \mathbf{k}|T|\mathbf{k}\rangle = \langle \mathbf{k}|\mathbf{q}\rangle \langle \mathbf{q}|T|\mathbf{q}'\rangle \langle \mathbf{q}'|\mathbf{k}\rangle$$
$$= \iint \frac{\mathrm{d}\mathbf{q}\, \mathrm{d}\mathbf{q}'}{(2\pi)^3} \mathrm{e}^{-\mathrm{i}\mathbf{k}\,.\,(\mathbf{q}-\mathbf{q}')} T(\mathbf{q}, \mathbf{q}'). \tag{F.20.17}$$

Substituting for $T(\mathbf{q}, \mathbf{q}')$ from equation (F.20.16) we obtain for the first term

$$\iint \frac{\mathrm{d}\mathbf{q}\, \mathrm{d}\mathbf{q}'}{(2\pi)^3} \mathrm{e}^{\mathrm{i}\mathbf{k}\,.\,(\mathbf{q}-\mathbf{q}')} \sum_{\alpha} t(\mathbf{q} - \mathbf{R}_\alpha, \mathbf{q}' - \mathbf{R}_\alpha). \tag{F.20.18}$$

Putting $\mathbf{q} - \mathbf{R}_\alpha = \mathbf{Q}, \mathbf{q}' - \mathbf{R}_\alpha = \mathbf{Q}'$, this gives

$$\sum_{\alpha} \iint \frac{\mathrm{d}\mathbf{Q}\, \mathrm{d}\mathbf{Q}'}{(2\pi)^3} \mathrm{e}^{-\mathrm{i}\mathbf{k}\,.\,(\mathbf{Q}-\mathbf{Q}')} t(\mathbf{Q}, \mathbf{Q}') = N \iint \frac{\mathrm{d}\mathbf{Q}\, \mathrm{d}\mathbf{Q}'}{(2\pi)^3} \mathrm{e}^{-\mathrm{i}\mathbf{k}\,.\,(\mathbf{Q}-\mathbf{Q}')} t(\mathbf{Q}, \mathbf{Q}'). \tag{F.20.19}$$

For the second term we get

$$\sum_{\alpha \neq \beta} \iiiint \frac{\mathrm{d}\mathbf{q}\, \mathrm{d}\mathbf{q}'\, \mathrm{d}\mathbf{q}''\, \mathrm{d}\mathbf{q}'''}{(2\pi)^3} [\mathrm{e}^{-\mathrm{i}\mathbf{k}\,.\,(\mathbf{q}-\mathbf{q}')} t(\mathbf{q} - \mathbf{R}_\alpha, \mathbf{q}'' - \mathbf{R}_\alpha)$$
$$G_0(\mathbf{q}'', \mathbf{q}''') t(\mathbf{q}''' - \mathbf{R}_\beta, \mathbf{q}' - \mathbf{R}_\beta)]. \tag{F.20.20}$$

We now put $\mathbf{q} - \mathbf{R}_\alpha = \mathbf{Q}, \mathbf{q}'' - \mathbf{R}_\alpha = \mathbf{Q}'',\ \mathbf{q}''' - \mathbf{R}_\beta = \mathbf{Q}''', \mathbf{q}' - \mathbf{R}_\beta = \mathbf{Q}'$.

Expression (F.20.20) may then be written

$$\iiiint \frac{\mathrm{d}\mathbf{Q}\, \mathrm{d}\mathbf{Q}'\, \mathrm{d}\mathbf{Q}''\, \mathrm{d}\mathbf{Q}'''}{(2\pi)^3} \mathrm{e}^{-\mathrm{i}\mathbf{k}\,.\,(\mathbf{Q}-\mathbf{Q}')} t(\mathbf{Q}, \mathbf{Q}'')$$
$$\times \left[\sum_{\alpha \neq \beta} G_0(\mathbf{Q}'' + \mathbf{R}_\alpha, \mathbf{Q}''' + \mathbf{R}_\beta)\, \mathrm{e}^{-\mathrm{i}\mathbf{k}\cdot(\mathbf{R}_\alpha - \mathbf{R}_\beta)} \right] t(\Omega''', \mathbf{Q}'). \tag{F.20.21}$$

The middle term will be recognised as proportional to the structural Green's function $G_s(\mathbf{Q}, \mathbf{Q}')$ defined in equation (F.20.28) with the term $\mathbf{R}_\alpha = \mathbf{R}_\beta$ omitted. Defining a function $G'_{\mathbf{k}}(\mathbf{Q}, \mathbf{Q}')$ by

$$G'_k(\mathbf{Q}, \mathbf{Q}') = \frac{1}{N} G'_s(\mathbf{Q}, \mathbf{Q}'), \tag{F.20.22}$$

we may write the T matrix series symbolically as

$$T(\mathbf{k}) = \frac{N}{(2\pi)^3} \iint \mathrm{e}^{-\mathrm{i}\mathbf{k}\cdot(\mathbf{Q}-\mathbf{Q}')}\, \mathrm{d}\mathbf{Q}\, \mathrm{d}\mathbf{Q}'\, [\hat{\mathbf{t}} + \hat{\mathbf{t}}\hat{\mathbf{G}}'_k\hat{\mathbf{t}} + \hat{\mathbf{t}}\hat{\mathbf{G}}'_k\hat{\mathbf{t}}\hat{\mathbf{G}}'_k\hat{\mathbf{t}} + \cdots]$$
$$= \frac{N}{(2\pi)^3} \iint \mathrm{e}^{-\mathrm{i}\mathbf{k}\cdot(\mathbf{Q}-\mathbf{Q}')} \{\hat{\mathbf{t}}^{-1} - \hat{\mathbf{G}}'_k\}^{-1}\, \mathrm{d}\mathbf{Q}\, \mathrm{d}\mathbf{Q}'. \tag{F.20.23}$$

Formally, we may now study the $E - \mathbf{k}$ relation by looking for the poles of the matrix $[\mathbf{t}^{-1} - \mathbf{G}_k^{-1}]$. Unfortunately, in the representation we have chosen these are not easy to locate.

F.20c The structural Green's function in the angular momentum representation

It may, in fact, be shown that both in the integral equation for the wave function and in that for the Green's function the direct use of the momentum representation leads to a dispersion relation of well known form, involving the matrix element of the potential between plane waves differing by a reciprocal lattice vector. As is well known, such a relation does not converge quickly enough for practical purposes. The work of Kohn and Rostoker showed that the introduction of the angular momentum representation leads to much more rapidly convergent expressions. As a prelude to the summation of the T matrix series in this representation we consider the angular momentum representation of the structural Green's function.

Denoting the free particle Green's operator by $\hat{\mathbf{G}}_0$ we note that by a change of origin we can write

$$\langle \mathbf{q}|G'_{\mathbf{k}}|\mathbf{q}'\rangle = \sum_{\mathbf{R}_\alpha \neq 0} G_0(\mathbf{q} + \mathbf{R}_\alpha - \mathbf{q}')\, e^{-i\mathbf{k}\cdot(\mathbf{R}_\alpha)} \tag{F.20.24}$$

where the $\mathbf{R}_\alpha$ are direct lattice vectors.

In the momentum representation this can then be written

$$\begin{aligned}\langle \mathbf{k}|G'_{\mathbf{k}}|\mathbf{k}'\rangle &= \langle \mathbf{k}|\mathbf{q}\rangle \langle \mathbf{q}|G'_{\mathbf{k}}|\mathbf{q}'\rangle \langle \mathbf{q}'|\mathbf{k}'\rangle \\ &= \sum_\alpha{}' \frac{1}{(2\pi)^3} \iint d\mathbf{q}\, d\mathbf{q}'\, e^{i\mathbf{k}'\cdot\mathbf{q}'}\, e^{-i\mathbf{k}\cdot\mathbf{q}} G'_{\mathbf{k}}(\mathbf{q} + \mathbf{R}_\alpha - \mathbf{q}')\, e^{-i\mathbf{k}\cdot\mathbf{R}_\alpha}\end{aligned} \tag{F.20.25}$$

Recalling the Fourier representation of the Green's function G_0,

$$G_0(\mathbf{q}, \mathbf{q}') = \frac{1}{(2\pi)^3} \int \frac{e^{i\mathbf{j}\cdot(\mathbf{q}-\mathbf{q}')}}{E - j^2}\, d\mathbf{j}.$$

Clearly

$$G'_{\mathbf{k}}(\mathbf{q}, \mathbf{q}') = \frac{1}{(2\pi)^3} \sum_\alpha e^{-i\mathbf{k}\cdot\mathbf{R}_\alpha} \int \frac{e^{i\mathbf{j}\cdot(\mathbf{q}+\mathbf{R}_\alpha-\mathbf{q}')}}{E - j^2}\, d\mathbf{j}. \tag{F.20.26}$$

Putting (F.20.26) into (F.20.25) we get

$$G'_{\mathbf{k}}(\mathbf{k}, \mathbf{k}') = \frac{1}{(2\pi)^6} \iiint d\mathbf{q}\, d\mathbf{q}'\, d\mathbf{j}\, e^{i\mathbf{k}\cdot\mathbf{q}'}\, e^{-i\mathbf{k}\cdot\mathbf{q}} \frac{e^{i\mathbf{j}\cdot(\mathbf{q}-\mathbf{q}')}}{E - j^2} \sum_\alpha{}' e^{i\mathbf{R}_\alpha(\mathbf{j}-\mathbf{k})}$$

Using Bauer's theorem expressed in terms of real spherical harmonics this can be written

$$\frac{1}{(2\pi)^6}\int\int\int q^2\,\mathrm{d}q\,q'^2\,\mathrm{d}q'\,\mathrm{d}\Omega_{\mathbf{q}}\,\mathrm{d}\Omega_{\mathbf{q}'}\,\mathrm{d}\mathbf{j}\,\frac{\sum_\alpha \mathrm{e}^{\mathrm{i}\mathbf{j}\cdot\mathbf{R}_\alpha}}{E-j^2}\Bigg[4\pi\sum_L \mathrm{i}^l j_l(kq')Y_L(\mathbf{k})Y_L(\mathbf{q}')$$

$$4\pi\sum_{L'}\mathrm{i}^{-l'}j_{l'}(kq)Y_{L'}(\mathbf{k})Y_{L'}(\mathbf{q})4\pi\sum_{L''}\mathrm{i}^{l''}j_{l''}(j\mathbf{q})Y_{L''}(\mathbf{j})Y_{L''}(\mathbf{q})$$

$$4\pi\sum_{L'''}\mathrm{i}^{-l'''}j_{l'''}(jq')Y_{L'''}(\mathbf{j})Y_{L'''}(\mathbf{q}')\Bigg]$$

where the symbol L is an abbreviation for l, m. Making use of the orthonormality of the Y's, namely the conditions of the type

$$\int Y_L(\mathbf{q}')Y_{L'''}(\mathbf{q}')\,\mathrm{d}\Omega_{\mathbf{q}'} = \delta_{LL'''},$$

we get

$$\frac{4}{\pi^2}\int q^2\,\mathrm{d}q\,q'^2\,\mathrm{d}q'\,\mathrm{d}\mathbf{j}\sum_\alpha\frac{\mathrm{e}^{\mathrm{i}\mathbf{j}\cdot\mathbf{R}_\alpha}}{E-j^2}\sum_{L,L'}\mathrm{i}^l j_l(k'q')Y_L(\mathbf{k}')\mathrm{i}^{-l'}j_{l'}(kq)Y_{L'}(\mathbf{k})$$

$$\mathrm{i}^{l'}j_{l'}(jq)Y_{L'}(\mathbf{j})\mathrm{i}^{-l}j_l(jq')Y_L(\mathbf{j}). \qquad \text{(F.20.27)}$$

We now take $\mathrm{e}^{\mathrm{i}\mathbf{j}\cdot\mathbf{R}_\alpha}$ out of (F.20.27) and expand in the same way, obtaining

$$\mathrm{e}^{\mathrm{i}\mathbf{j}\cdot\mathbf{R}_\alpha} = 4\pi\sum_{L''}\mathrm{i}^{l''}j_{l''}(jR_\alpha)Y_{L''}(\mathbf{j})Y_{L''}(\mathbf{R}_\alpha) \qquad \text{(F.20.28)}$$

The product of three spherical harmonics cannot be eliminated by integration over angles in the same way as before. However, there is a well-known technique for dealing with such expressions in problems dealing with addition of angular momenta. We write

$$\sum_{L''} a(L, L', L'')Y_{L''}(\mathbf{j}) = Y_L(\mathbf{j})Y_{L'}(\mathbf{j}) \qquad \text{(F.20.29)}$$

where the coefficients $a(L, L', L'')$ are related to the so called Clebsch–Gordon coefficients. The coefficients $a(L, L', L'')$ have the property that they vanish unless $L + L' + L''$ is even. From F.20.29 we get

$$Y_{L'}(\mathbf{j})Y_L(\mathbf{j})Y_{L''}(\mathbf{j}) = \sum_{L'''} Y_{L''}(\mathbf{j})a(L, L', L''')Y_{L'''}(\mathbf{j}). \qquad \text{(F.20.30)}$$

Putting this back into (F.20.27) we have

$$\frac{16}{\pi}\sum_{L,L',L''}\iiint q^2\,\mathrm{d}q\,q'^2\,\mathrm{d}q'\,\mathrm{d}\mathbf{j}\left[\sum_\alpha \mathrm{i}^{l''} j_{l''}(jR_\alpha)Y_{L''}(\mathbf{R}_\alpha)\right]$$

$$[j_l(k'q')j_l(kq)j_{l'}(jq)j_l(jq')]\left[Y_{L'}(\mathbf{k})Y_L(\mathbf{k}')\sum_{L'''}Y_{L''}(\mathbf{j})\right.$$

$$\left.\times\, a(L,L',L''')Y_{L'''}(\mathbf{j})\right]. \tag{F.20.31}$$

The angular integration over Ω_j can now be carried out and gives a factor $\delta_{L''L'''}$.

We can now consider the integral

$$\int_0^\infty \mathrm{d}\mathbf{j}\frac{j_{l''}(jR_\alpha)j_{l'}(jq)j_l(jq')}{E-j^2+\mathrm{i}\epsilon}.$$

The function $j_l(x)$ has 'parity' $(-1)^l$ and the integrand is therefore, through the properties of the function $a(L, L', L'')$, an even function. We now make use of the assumption that the potentials are non-overlapping. This implies that each t matrix $t(\mathbf{q},\mathbf{q}')$ has range smaller than that of the potential, so that if R_s is the range of the potential $q, q' < R_s$. Hence it follows that $q + q' < R_\alpha$ and the behaviour of the integral will be determined by the term $j_{l''}(jR_\alpha)$. Asymptotically

$$j_l(x) = \tfrac{1}{2}[h_l^{(1)}(x)+h_l^{(2)}(x)]$$

$$\to \frac{1}{2}\left[\frac{\mathrm{e}^{\mathrm{i}x}}{x}(-\mathrm{i})^{l+1}+\frac{\mathrm{e}^{-\mathrm{i}x}}{x}(+\mathrm{i})^{l+1}\right].$$

Thus we get

$$\frac{1}{2}\int_0^\infty\left[(-i)^{l''+1}\frac{\mathrm{e}^{\mathrm{i}jR_\alpha}}{jR_\alpha}\frac{j_{l'}(jq)j_l(jq')}{E-j^2+\mathrm{i}\epsilon}+(\mathrm{i})^{l''+1}\frac{\mathrm{e}^{-\mathrm{i}jR_\alpha}}{jR_\alpha}\frac{j_{l'}(jq)j_l(jq')}{E-j^2+\mathrm{i}\epsilon}\right]\mathbf{j}^2\,\mathrm{d}j.$$

Since R_α is essentially positive we evaluate the first term by a contour in the upper half-plane and the second term by a contour in the lower half-plane. In this way we get

$$\frac{1}{2}(2\pi\mathrm{i})\left[(-\mathrm{i})^{l''+1}\frac{\mathrm{e}^{\mathrm{i}\sqrt{(E)}R_\alpha}}{R_\alpha}\frac{j_{l'}(\sqrt{E}q)j_l(\sqrt{(E)}q')}{2}\right.$$

$$\left.+(\mathrm{i})^{l''+1}\frac{\mathrm{e}^{\mathrm{i}\sqrt{(E)}R_\alpha}}{R_\alpha}\frac{j_{l'}(\sqrt{(E)}q)j_l(\sqrt{(E)}q')}{2}\right]$$

(since the contribution from the pole at the origin cancels)

$$= \frac{\pi i}{2} \frac{e^{i\sqrt{E}R_\alpha}}{R_\alpha} j_{l'}(\sqrt{(E)}q) j_l(\sqrt{(E)}q')(-i)^{l''+1}.$$

Substituting this back into (F.20.31), we are left with

$$\frac{16}{\pi} \sum_{L,L',L'',L'''} \iint q^2 \, dq \, q'^2 \, dq' \left[\frac{\pi}{2} \sum_\alpha \frac{e^{i\sqrt{E}R_\alpha}}{R_\alpha} j_{l'}(\sqrt{(E)}q) j_l(\sqrt{(E)}q') \right.$$

$$\left. j_l(k'q') j_l(kq) Y_{L'}(\mathbf{k}) Y_L(\mathbf{k}') a(L, L', L'') \right].$$

Making use of the orthonormality condition

$$\delta(\mathbf{v} - \mathbf{v}') = \frac{2}{\pi} \sum_L Y_L(\mathbf{v}) Y_L(\mathbf{v}') \int_0^\infty j_l(kv) j_l(kv') k^2 \, dk,$$

we may write

$$\langle G_k(\mathbf{k}, \mathbf{k}') \rangle = \sum_{L,L'} G_{LL'} Y_L(\mathbf{k}) Y_L(\mathbf{k}') \delta(k - k_0) \delta(k' - k_0) \quad \text{(F.20.32)}$$

where

$$k_0 = \sqrt{E}.$$

The coefficients $G_{LL'}$ involve the sum over the direct lattice vectors $\mathbf{R}_\alpha$, and therefore depend on the structure of the lattice. Explicit expansions for the structure coefficients may be obtained in a somewhat more convenient form, as shown by Kohn and Rostoker[16].

We recall from §F.11 that in the one-dimensional case a Green's function can be written in two parts, one satisfying the homogeneous equation, the second containing the inhomogeneity. The same separations can be carried out for the structural Green's function. We recall from equation (F.12.33) that a solution of the equation

$$[\nabla^2 + k_0^2] G_0(\mathbf{q}, \mathbf{q}') = \delta(\mathbf{q}, \mathbf{q}') \quad \text{(F.20.33)}$$

is given by (in units such that $2m/h^2 = 1$)

$$G_0(\mathbf{q}, \mathbf{q}') = -\frac{1}{4\pi} \frac{\exp[ik_0|\mathbf{q} - \mathbf{q}'|]}{|\mathbf{q} - \mathbf{q}'|}.$$

As we emphasized, this particular choice of Green's function corresponds to a particular boundary condition; the boundary conditions manifested themselves through the choice of a particular contour in the evaluation of the Green's function. The function

$$g_0(\mathbf{q}, \mathbf{q}') = -\frac{1}{4\pi}\frac{\cos k_0|\mathbf{q}-\mathbf{q}'|}{|\mathbf{q}-\mathbf{q}'|} \tag{F.20.34}$$

satisfies the same differential equation (F.20.33) with different boundary conditions. Now the function $G_s(\mathbf{q}, \mathbf{q}')$ also satisfies F.20.33 as follows, for example, from equation (F.20.10). Hence the difference of these functions, namely

$$D(\mathbf{q}, \mathbf{q}') = G_s(\mathbf{q}, \mathbf{q}') - g_0(\mathbf{q}, \mathbf{q}') \tag{F.20.35}$$

must satisfy the homogeneous equation, and therefore must be expressible as a product of solutions of the homogeneous equation of the form

$$D(\mathbf{q}, \mathbf{q}') = \psi(\mathbf{q})\psi(\mathbf{q}'). \tag{F.20.36}$$

Since any solutions of the homogeneous equation can be expressed as

$$\psi(\mathbf{q}) = \sum_{l,m} A_{lm} j_l(k_0 q) Y_l^m(\mathbf{q}), \tag{F.20.37}$$

we may write equation (F.20.36) as

$$D(\mathbf{q}, \mathbf{q}') = \sum_{\substack{l,m\\l',m'}} A_{lm;l'm'} j_l(k_0 q) j_{l'}(k_0 q') Y_l^m(\mathbf{q}) Y_{l'}^{m'}(\mathbf{q}'). \tag{F.20.38}$$

Now from problem (F.12.10), by equating real and imaginary parts,

$$g_0(\mathbf{q}, \mathbf{q}') = \sum_{l,m} k_0 j_l(k_0 q_<) n_l(k_0 q_>) Y_l^m(\mathbf{q}) Y_l^m(\mathbf{q}')$$

(we are now consistently using real spherical harmonics).

Hence

$$\begin{aligned} G_k(\mathbf{q}, \mathbf{q}') &= D(\mathbf{q}, \mathbf{q}') + g_0(\mathbf{q}, \mathbf{q}') \\ &= \sum_{\substack{l,m\\l',m'}} [A_{lm;l'm'} j_l(k_0 q) j_{l'}(k_0 q') + k_0 \delta_{ll'} \delta_{mm'} j_l(k_0 q_<) j_{l'}(k_0 q_>)] \times \\ &\quad Y_l^m(\mathbf{q}) Y_{l'}^{m'}(\mathbf{q}'). \end{aligned} \tag{F.20.39}$$

Again using the result of problem (F.12.10), and equating imaginary parts this time,

$$-\frac{i}{4\pi}\frac{\sin k_0|\mathbf{q}-\mathbf{q}'|}{|\mathbf{q}-\mathbf{q}'|} = \sum_{l,m} i k_0 j_l(k_0 q) j_l(k_0 q) Y_l^m(\mathbf{q}) Y_l^m(\mathbf{q}'). \tag{F.20.40}$$

Therefore

$$G'_{\mathbf{k}}(\mathbf{q}, \mathbf{q}') = G_s(\mathbf{q}, \mathbf{q}') - G_0(\mathbf{q}, \mathbf{q}')$$

$$= \sum_{\substack{l,m \\ l',m'}} [A_{lm;l'm'} j_l(k_0 q) j_{l'}(k_0 q') + k_0 \delta_{ll'} \delta_{mm'} j_l(k_0 q_<) n_l(k_0 q_>)$$

$$- k_0 \delta_{ll'} \delta_{mm'} n_l(k_0 q_>) j_l(k_0 q_<) + \mathrm{i} k_0 j_l(k_0 q_>) j_l(k_0 q_<)] Y_l^m(\mathbf{q}) Y_{l'}^{m'}(\mathbf{q}')$$

$$= \sum_{\substack{l,m \\ l',m'}} [A_{lm;l',m'} + \mathrm{i} k_0] j_l(k_0 q) j_{l'}(k_0 q') Y_l^m(\mathbf{q}) Y_{l'}^{m'}(\mathbf{q}'). \tag{F.20.41}$$

This last equation shows that $G'_s(\mathbf{q}, \mathbf{q}')$, like $D(\mathbf{q}, \mathbf{q}')$, satisfies the homogeneous equation. If we write

$$G'_{\mathbf{k}}(\mathbf{q}, \mathbf{q}') = \sum_{L,L'} G_{LL'}(\mathbf{q}, \mathbf{q}') Y_l^m(\mathbf{q}) Y_{l'}^{m'}(\mathbf{q}') \tag{F.20.42}$$

(where we have again used the symbols L, L' as abbreviations for l, m and l', m' respectively) then

$$G_{LL'}(\mathbf{q}, \mathbf{q}') = (A_{LL'} + \mathrm{i} k_0 \delta_{LL'}) j_l(k_0 q) j_{l'}(k_0 q'). \tag{F.20.43}$$

We may now transform this expression into **k**-space by double Fourier transforms and use of Bauer's theorem:

$$\langle \mathbf{k}|\mathbf{q}\rangle \langle \mathbf{q}|\hat{\mathbf{G}}|\mathbf{q}'\rangle \langle \mathbf{q}'|\mathbf{k}'\rangle = \sum_{L,L'} \frac{16\pi^2}{(2\pi)^3} \iint \mathrm{d}\mathbf{q}\, \mathrm{d}\mathbf{q}' G_{LL'}(q, q') Y_l^m(\mathbf{q})$$

$$\times\, Y_{l'}^{m'}(\mathbf{q}') \sum_{L'',L'''} (-\mathrm{i})^{l''} (+\mathrm{i})^{l'''} j_{l''}(kq) j_{l'''}(k'q') Y_{l''}^{m''}(\mathbf{k}) Y_{l''}^{m''}(\mathbf{q}) Y_{l'''}^{m'''}(\mathbf{k}') Y_{l'''}^{m'''}(\mathbf{q}') \tag{F.20.44}$$

Carrying out the integration gives a factor of $\delta(k - k_0)\, \delta(k' - k_0)$ in agreement with equation (F.20.32) and the final result

$$G'_{\mathbf{k}}(\mathbf{k}, \mathbf{k}') = \sum_{L,L'} G_{LL'} Y_l^m(\mathbf{k}) Y_{l'}^{m'}(\mathbf{k}')\, \delta(k - k_0)\, \delta(k' - k_0) \tag{F.20.45}$$

where

$$G_{LL'} = [A_{LL'} + \mathrm{i} k_0 \delta_{LL'}] \frac{\pi \mathrm{i}^{l'-l}}{2k_0^4} \tag{F.20.46}$$

We shall now obtain an explicit expression for the coefficients $A_{LL'}$ from equation F.20.10.

By a double application of Bauer's theorem,

$$\sum_n \frac{1}{\tau} \frac{e^{i(\mathbf{k}+\mathbf{K}_n)\cdot(\mathbf{q}-\mathbf{q}')}}{E - (\mathbf{k}+\mathbf{K}_n)^2}$$

$$= \left[4\pi \sum_l i^l j_l(|\mathbf{k}+\mathbf{K}_n|q) Y_l^m(\mathbf{k}+\mathbf{K}_n) Y_l^m(\mathbf{q}) \right] \times$$

$$\left[4\pi \sum_{l'} i^{-l'} j_{l'}(|\mathbf{k}+\mathbf{K}_n|q') Y_{l'}^{m'}(\mathbf{k}+\mathbf{K}_n) Y_{l'}^{m'}(\mathbf{q}') \right] \Big/ E - (\mathbf{k}+\mathbf{K}_n)^2$$

$$= \sum_{\substack{l,m \\ l',m'}} [A_{LL'} j_l(k_0 q) j_l(k_0 q') + k_0 \delta_{ll'} \delta_{mm'} j_l(k_0 q) n_l(k_0 q')][Y_l^m(\mathbf{q}) Y_{l'}^{m'}(\mathbf{q}')]$$

$$\text{(F.20.47)}$$

Hence

$$A_{LL'} = \frac{\dfrac{16\pi^2}{\tau} \sum_n [j_l(|\mathbf{k}+\mathbf{K}_n|q) j_{l'}(|\mathbf{k}+\mathbf{K}_n|q') Y_l^m(\mathbf{k}+\mathbf{K}_n) Y_{l'}^{m'}(\mathbf{k}+\mathbf{K}_n)]}{[E - (\mathbf{k}+\mathbf{K}_n)^2][j_l(k_0 q) j_l(k_0 q')]}.$$

$$\text{(F.20.48)}$$

F.20d The summation of the T matrix series in the angular momentum representation

We are now in a position to sum the series (F.20.13) in the angular momentum representation, making use of equation (F.20.45) and (F.16.15). For the first term we obtain

$$\sum_l \sum_\alpha t_l(k,k) Y_L(\mathbf{k}) Y_L(\mathbf{k}) = \sum_\alpha N t_l(k,k) Y_L(\mathbf{k}) Y_L(\mathbf{k}).$$

For the second term we get

$$\langle \mathbf{k}|\hat{t}\hat{G}_k'\hat{t}|\mathbf{k}\rangle = \langle \mathbf{k}|\hat{t}|\mathbf{k}_1\rangle \langle \mathbf{k}_1|\hat{G}_k'|\mathbf{k}_2\rangle \langle \mathbf{k}_2|\hat{t}|\mathbf{k}\rangle$$

$$= \left[\sum_L t_l(k,k_1) Y_L(\mathbf{k}) Y_L(\mathbf{k}_1) \right] \left[\sum_{L',L''} G_{L'L''}(k_1,k_2) Y_{L'}(\mathbf{k}_1) Y_{L''}(\mathbf{k}_2) \right]$$

$$\times \left[\sum_{L'''} t_{l'''}(k_2,k) Y_{L'''}(\mathbf{k}_2) Y_{L'''}(\mathbf{k}) \right]$$

Putting $d\mathbf{k}_1\, d\mathbf{k}_2 = k_1^2 k_2^2\, d\Omega_{\mathbf{k}_1}\, d\Omega_{\mathbf{k}_2}$ and integrating over the angles we make use of the orthonormality of the $Y_L(\mathbf{k}_1)$ and $Y_L(\mathbf{k}_2)$. Finally we obtain (using F.20.46)

$$\sum_{L,L'} t_l(k, k_0)k_0^4 G_{LL'}(k_0, k_0)t_{l'}(k_0, k)Y_2(\mathbf{k})Y_{L'}(\mathbf{k})$$

The other terms of these series may be summed in the same way; for the third term we get

$$\langle\mathbf{k}|\hat{\mathbf{t}}\hat{\mathbf{G}}_k\hat{\mathbf{t}}\hat{\mathbf{G}}_k\hat{\mathbf{t}}|\mathbf{k}\rangle$$
$$= \sum_{L,L',L''} t_l(k, k_0)k_0^4 G_{LL''}t_{l''}(k_0, k_0)k_0^4 G_{L''L'}t_{l'}(k_0, k)Y_L(\mathbf{k})Y_{L'}(\mathbf{k})$$

Finally the sum of the series may be written

$$\langle\mathbf{k}|T|\mathbf{k}\rangle = N\sum_{L,L'} Y_L(\mathbf{k})Y_{L'}(\mathbf{k}')\Bigg[t_l(k, k)\delta_{LL'} + t_l(k, k_0) \times \left[\frac{k_0^4\hat{\mathbf{G}}_k k_0^4}{\mathbf{I} - k_0^4\hat{\mathbf{t}}\mathbf{G}_k}\right]_{LL'} t_{l'}(k_0, k)\Bigg] \tag{F.20.49}$$

Here we note that the $\hat{\mathbf{t}}$ matrix which occurs in the geometric series is evaluated on the energy shell and may therefore be expressed in terms of the phase shifts by equation (F.16.15). Noting that we require the inverse of the matrix $\mathbf{I} - k_0^4\mathbf{t}G'_\mathbf{k}$ and that each term of this inverse will contain the determinant of the matrix in the denominator, we may write the condition for the poles as

$$\det|\delta_{LL'}t_l^{-1} - k_0^4 G_{LL'}| = 0 \tag{F.20.50}$$

From equation (F.16.15)

$$t_l^{-1} = [-k_0\cot\delta_l + ik_0]\frac{\pi}{2}. \tag{F.20.51}$$

Substituting this and equation (F.20.46) into (F.20.50) we obtain as the condition for allowed energy levels

$$\det|\delta_{ll'}k_0\cot\delta_l + A_{LL'}| = 0. \tag{F.20.52}$$

This condition is in complete agreement with that obtained from the integral equation for the wave function by Korringa and by Kohn and Rostoker. In the form in which the theory has been presented here no new results have been obtained; it has been given simply as an example of the T matrix formalism. We may mention that more recent papers have presented new developments[22].

[22] J. M. Ziman, *Proc. Phys. Soc.*, **86**, 337 (1965).

Problems

(F.20.1) Show that

$$\text{(a)} \qquad G_s(\mathbf{q} + \mathbf{R}_\alpha, \mathbf{q}', E) = e^{i\mathbf{k}\cdot\mathbf{R}_\alpha} G_s(\mathbf{q}, \mathbf{q}', E)$$

$$\text{(b)} \qquad G_s^*(\mathbf{q}, \mathbf{q}', E) = G_s(\mathbf{q}', \mathbf{q}, E)$$

$$\text{(c)} \qquad G_s(E, -\mathbf{k}) = G_s^*(E, \mathbf{k}).$$

(F.20.2) Using the previous example and equation F.20.7 show that

$$\text{(a)} \qquad \psi(E, \mathbf{k} + \mathbf{K}_n) = \psi(E, \mathbf{k})$$

$$\text{(b)} \qquad \psi(E, -\mathbf{k}) = \psi^*(E, \mathbf{k}).$$

(F.20.3) Prove the relation (F.20.15).

CHAPTER XIII

Application of Green's Functions to the Theory of Electron States in Disordered Systems

The term disordered is here used in a very general sense to describe any departure from lattice periodicity. Under this heading we might consider the influence on the electron states of phonons, of point defects, or the study of the electronic structure of liquid metals or of interacting impurities. We shall consider three simple examples of disordered systems: in §F.21 we shall consider models of localized potentials, suitable for describing impurities in metals; in §F.22 we shall consider isolated impurities with extended potentials, a model suitable for describing impurities in semiconductors; finally, in §F.23 we shall consider a simple model of interacting impurities.

F.21 LOCALIZED IMPURITIES

The interaction of an isolated spherically symmetric potential with an electron is well known. For negative energies the interaction can give rise to localized bound states for certain discrete energies; for positive energies a continuum of allowed states exists, and the wavefunction can be written as the sum of an incident wave and a scattered wave. We note that the density of states is unchanged by the interaction.

The introduction of a spherically symmetric impurity into a periodic lattice is a considerably more complicated process for a number of reasons. Only in certain regions of energy are there well-behaved solutions of the unperturbed Schrödinger equation, and therefore only for these can the wavefunction be expressed as the sum of an incident wave and a scattered wave. For such states we may write, instead of equation (F.15.3), the equation

$$|n\rangle = |n\rangle_p + \hat{\mathbf{G}}_p \hat{\mathbf{H}}_I |n\rangle$$

where the suffix p refers to the periodic lattice; the Green's operator $\hat{\mathbf{G}}_p$ is defined by

$$\hat{\mathbf{G}}_p^{\pm} = \frac{1}{E - \hat{\mathbf{H}}_p \pm i\epsilon}. \tag{F.21.1}$$

It was originally shown by Koster and Slater[23] that a successful qualitative description of localized impurities can be given by the introduction of Wannier functions together with suitable simplifying assumptions regarding the matrix elements of the periodic potential in this representation. The original treatment was based on the wave function; the introduction of the Green's function has the advantage that no distinction need be made *ab initio* between scattering states and localized states; both can be treated on the same basis. In addition, the density of states of the perturbed system can be simply obtained through equation (F.13.21).

We start with equation (F.14.7) which we write in the Wannier representation introduced in §F.8b. Making use of the result of problem (F.10.7) we obtain

$$\langle n, \alpha|\hat{\mathbf{G}}_p|n', \alpha'\rangle = \langle n, \alpha|\hat{\mathbf{G}}_p|n', \alpha'\rangle + \sum_{\alpha'', n''} \langle n, \alpha|\hat{\mathbf{G}}_p|n'', \alpha''\rangle\langle n'', \alpha''|\hat{\mathbf{H}}_I|n''', \alpha'''\rangle \times \langle n''', \alpha'''|\hat{\mathbf{G}}|n', \alpha'\rangle \tag{F.21.2}$$

Here we have made use of problem (F.10.7) in putting in explicitly the diagonal character of $\hat{\mathbf{G}}_p$ in the band indices. The Koster–Slater approximation consists of taking

$$\langle n'', \alpha''|\hat{\mathbf{H}}_I|n''', \alpha'''\rangle = H\,\delta_{\alpha''\alpha_0}\,\delta_{\alpha'''\alpha_0}\,\delta_{n''n_0}\,\delta_{n'''n_0}. \tag{F.21.3}$$

The two factors on the right-hand side correspond to two types of approximation; firstly, we take H_I to be a sharply localized potential, therefore in the first instance diagonal in the site labels α'', α; secondly, we neglect interband matrix elements and assume the effect of a single band (labelled n_0) to be predominant. Putting (F.21.3) into (F.21.2) we get

$$\langle n, \alpha|\hat{\mathbf{G}}|n', \alpha'\rangle = \langle n, \alpha|\hat{\mathbf{G}}_p|n', \alpha'\rangle\,\delta_{nn'} + H\langle n, \alpha|\hat{\mathbf{G}}_p|n'', \alpha_0\rangle\,\delta_{nn''} \times \langle n'', \alpha_0|\hat{\mathbf{G}}|n', \alpha'\rangle. \tag{F.21.4}$$

Putting $\alpha = \alpha_0$ this becomes

$$\langle n, \alpha_0|\hat{\mathbf{G}}|n', \alpha'\rangle = \langle n, \alpha_0|\hat{\mathbf{G}}_p|n', \alpha'\rangle\,\delta_{nn'} + H\langle n, \alpha_0|\hat{\mathbf{G}}_p|n_0, \alpha_0\rangle\langle n_0, \alpha_0|\hat{\mathbf{G}}|n', \alpha'\rangle. \tag{F.21.5}$$

This can be solved for $\langle n_0, \alpha_0|\hat{\mathbf{G}}|n', \alpha'\rangle$ and substituted back into (F.21.4). The result is

$$\langle n, \alpha|\hat{\mathbf{G}}|n', \alpha'\rangle = \langle n, \alpha|\hat{\mathbf{G}}|n', \alpha'\rangle\,\delta_{nn'} + \frac{H\langle n, \alpha|\hat{\mathbf{G}}_p|n_0, \alpha_0\rangle\langle n_0, \alpha_0|\hat{\mathbf{G}}_p|n', \alpha'\rangle}{1 - H\langle n_0, \alpha_0|\hat{\mathbf{G}}_p|n_0, \alpha_0\rangle}. \tag{F.21.6}$$

[23] G. F. Koster and J. C. Slater, *Phys. Rev.*, **95**, 1167 (1954).

The approximation (F.21.4) implies that the effect of a single band n_0 is predominant; we therefore only take this band into account in (F.21.6). Putting

$$\langle n_0, \alpha_0|\hat{\mathbf{G}}_p|n_0, \alpha_0\rangle = G_{00}$$

we get

$$\langle n_0, \alpha|\hat{\mathbf{G}}|n_0, \alpha'\rangle = \langle n_0, \alpha|\hat{\mathbf{G}}|n_0, \alpha'\rangle + \frac{H}{1 - HG_{00}}\langle n_0, \alpha|\hat{\mathbf{G}}_p|n_0, \alpha_0\rangle$$
$$\times \langle n_0, \alpha_0|\hat{\mathbf{G}}_p|n_0, \alpha'\rangle. \tag{F.21.7}$$

In the one-band approximation the Green's function G_p is given, by the result of problem F.10.7, by

$$\langle n_0, \alpha|\hat{\mathbf{G}}_p|n_0, \alpha'\rangle = \frac{1}{N}\sum_{\mathbf{k}} \frac{e^{i\mathbf{k}\cdot(\mathbf{R}_\alpha - \mathbf{R}_{\alpha'})}}{E - E(\mathbf{k})} \tag{F.21.8}$$

where we have introduced a normalization factor N equal to the number of atoms in the crystal.

In order to compute the density of states from equation (F.13.21) we need the trace of the Green's function. For this purpose the last term of equation (F.21.8) may be transformed as follows:

$$\sum_\alpha \langle\alpha|\hat{\mathbf{G}}_p|\alpha_0\rangle\,\langle\alpha_0|\hat{\mathbf{G}}_p|\alpha\rangle$$

$$= \frac{1}{N^2}\left[\sum_{\mathbf{k}} \frac{e^{i\mathbf{k}\cdot(\mathbf{R}_\alpha - \mathbf{R}_{\alpha_0})}}{E - E(\mathbf{k})}\right]\left[\sum_{\mathbf{k}'} \frac{e^{i\mathbf{k}'\cdot(\mathbf{R}_{\alpha_0} - \mathbf{R}_\alpha)}}{E - E(\mathbf{k}')}\right].$$

Since

$$\sum_{\alpha''} e^{i(\mathbf{k}' - \mathbf{k})\cdot\mathbf{R}_{\alpha''}} = N\,\delta(\mathbf{k} - \mathbf{k}') \tag{F.21.9}$$

we may write

$$\sum_\alpha \langle\alpha|\hat{\mathbf{G}}_p|\alpha_0\rangle\,\langle\alpha_0|\hat{\mathbf{G}}_p|\alpha\rangle = \frac{1}{N}\sum_{\mathbf{k}} \frac{1}{[E - E(\mathbf{k})]^2}$$
$$= -\frac{d}{dE}[\langle\alpha|\hat{\mathbf{G}}_p|\alpha\rangle]. \tag{F.21.10}$$

Therefore we may write, from equation (F.21.7),

$$\mathrm{Tr}[\hat{\mathbf{G}}] = \mathrm{Tr}[\hat{\mathbf{G}}_p] + \frac{d}{dE}\log_e[1 - HG_{00}]. \tag{F.21.11}$$

It is convenient to rewrite equation (F.13.21) in the alternative form

$$D(E) = \frac{1}{\pi}\,\mathrm{Tr}(\mathrm{Im}\hat{\mathbf{G}}^{-}) \tag{F.21.12}$$

and to define a density of states per atom, d(E) by

$$\mathrm{d}(E) = \frac{1}{N}\,D(E). \tag{F.21.13}$$

Putting (F.21.12), (F.21.13) and (F.21.10) into (F.21.11) we obtain

$$\mathrm{d}(E) = \mathrm{d}_0(E) + \frac{1}{\pi N}\,\mathrm{Im}\,\frac{\mathrm{d}}{\mathrm{d}E}\log_e\{1 - HG_{00}^{-}\}. \tag{F.21.14}$$

For any complex number z the logarithm is given by

$$\log_e(z) = \log_e|z| + \mathrm{i}\arg z + 2\pi n\mathrm{i} \qquad [n \text{ an integer}].$$

We choose $n = 0$ to make this function single valued.
Since

$$\begin{aligned} G_{00}^{-} &= \frac{1}{N}\sum_{\mathbf{k}}\frac{1}{E - E(\mathbf{k}) - \mathrm{i}\epsilon} \\ &= \frac{P}{N}\sum_{\mathbf{k}}\frac{1}{E - E(\mathbf{k})} + \sum_{\mathbf{k}}\frac{\pi\mathrm{i}}{N}\delta[E - E(\mathbf{k})] \qquad \text{(F.21.15)} \\ &= F_0(E) + \mathrm{i}\pi\,\mathrm{d}_0(E) \end{aligned}$$

[where the last line defines the function $F_0(E)$], we may write the last term of (F.21.14) as

$$\begin{aligned} \frac{1}{\pi N}\,\mathrm{Im}\,\frac{\mathrm{d}}{\mathrm{d}E}[\arg(1 - HG_{00}^{-})] &= \frac{1}{\pi N}\frac{\mathrm{d}}{\mathrm{d}E}\tan^{-1}\left[\frac{\pi\,\mathrm{d}_0 H}{1 - HF_0}\right] \\ &= \frac{1}{N}\left\{\frac{H^2F_0'(E)\,\mathrm{d}_0(E) + H[1 - HF_0(E)]\,\mathrm{d}_0'(E)}{[1 - HF_0(E)]^2 + \pi^2H^2\,\mathrm{d}_0^2(E)}\right\}. \qquad \text{(F.21.16)} \end{aligned}$$

Substituting this into equation (F.21.14) we obtain

$$\mathrm{d}(E) = \mathrm{d}_0(E) + \frac{1}{N}\frac{[H^2F'(E)\,\mathrm{d}_0(E) + H(1 - HF_0(E)\,\mathrm{d}_0'(E)]}{[1 - HF_0(E)]^2 + \pi^2H^2\,\mathrm{d}_0^2(E)}. \tag{F.21.17}$$

The second term has the character of a resonance in the neighbourhood of $E = E_0$, where this energy is given by

$$1 - HF_0(E) = 0. \tag{F.21.18}$$

This suggests expanding the function $F_0(E)$ in powers of $E - E_0$. We put

$$F_0(E) = \frac{1}{H} + F_0'(E)(E - E_0) \tag{F.21.19}$$

in the denominator of the second term of (F.21.17). In the numerator we simply put $F_0 H = 1$. In this way we obtain

$$\mathrm{d}(E) = \mathrm{d}_0(E) + \frac{H^2 F_0' \,\mathrm{d}(E)}{H^2 F_0'^2 (E - E_0)^2 + \pi^2 \mathrm{d}_0^2(E) H^2}. \tag{F.21.20}$$

The second term can be written in a more familiar way by defining

$$\Gamma(E) = -\frac{\pi \,\mathrm{d}_0(E)}{F_0'(E)}. \tag{F.21.21}$$

In terms of this quantity the perturbed density of states can be written

$$\mathrm{d}(E) = \mathrm{d}_0(E) + \frac{1}{\pi N} \frac{\Gamma(E)}{(E - E_0)^2 + \Gamma^2}. \tag{F.21.22}$$

This equation for the perturbed density of states applies to all energies. We may in the first instance distinguish between two cases:

(1) If the solution of equation (F.21.18) is outside the energy spectrum of the unperturbed crystal then $\mathrm{d}_0(E_0) = 0$ and from equation (F.21.21) $\Gamma(E) = 0$. In this case [making use of equation (F.13.7)] we get

$$\mathrm{d}(E) = \mathrm{d}_0(E) + \frac{1}{N} \delta(E - E_0), \tag{F.21.23}$$

that is, an additional single localized level outside the band of levels of the unperturbed crystal.

(2) If the solution of equation (F.21.18) is in the allowed band of the crystal then equation (F.21.22) predicts a sharp peak at energy $E = E_0$ in the density of states.

The conditions under which one or other of the levels arise may be investigated graphically. The function $F_0(E)$ may be expressed in terms of the density of states:

$$F_0(E) = \frac{P}{N} \sum_{\mathbf{k}} \frac{1}{E - E(\mathbf{k})} = P \int_{-\infty}^{\infty} \frac{\mathrm{d}_0(E')\,\mathrm{d}E'}{E - E'}. \tag{F.21.24}$$

A typical curve for $F_0(E)$ is shown in fig. (F.21.1), where the origin has been chosen at the centre of symmetry. Also shown is the function $\mathrm{d}_0(E)$ for a typical one band model; the points at which this function crosses the E-axis mark the limits of the energy spectrum of the unperturbed crystal.

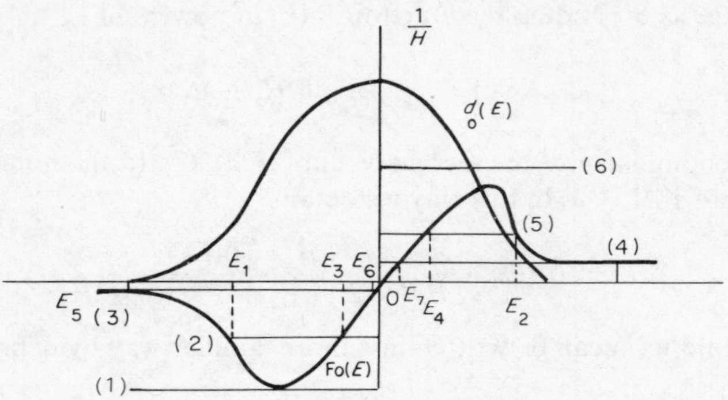

Fig. F.21.1

The graph shows six values of $1/H$. For values of $|H|$ sufficiently large—for example (1), (6),—no solutions of equation (F.21.18) exist. Cases (2) and (5) are examples in which for each value of H there are two solutions of (F.21.18) in a region of energy which is allowed for the unperturbed crystal. From equation (F.21.21) it can be seen that the sign of Γ is determined by that of $F_0'(E)$. At the points E_1 and E_2 $F_0'(E)$ is < 0, and these therefore correspond to peaks in the density of states curve. At the points E_3 and E_4, $F_0'(E)$ is > 0, and therefore these correspond to troughs in the density of states curve. However it can be seen that whereas

$$|F_0(E_1)| = |F_0(E_3)|, \qquad d(E_3) \gg d(E_1).$$

Hence the decreases in the density of states curve are much smaller in magnitude. Finally the points (3) and (4) correspond to the solution of (F.21.18) for energies outside the allowed band. Corresponding to each point there is again an anti-resonance point denoted in the diagram by E_6 and E_7 respectively.

In summary, changes in the energy spectrum can be caused by an impurity only for a sufficiently large value of $|H|$. As the value of $|H|$ increases there occurs the possibility of virtual states lying in the allowed band of the unperturbed crystal. Such states occur in pairs; corresponding to each resonant state which causes an increase in the density of states there is an anti-resonant state which causes a decrease which is, however, of much smaller magnitude. As the value of $|H|$ is increased still further localized states occur outside the energy spectrum of the unperturbed crystal; corresponding to each state there is again an 'anti-resonant' state lying within the allowed band. The density of states corresponding to localized and virtual levels is given in fig. F.21.2.

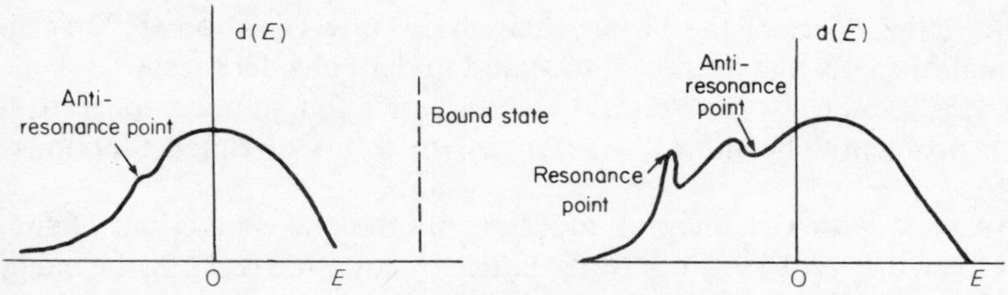

Fig. F.21.2

Problems

(F.21.1) Prove the equation before (F.21.1). [In this equation the operator G_p is related to the T matrix by equation (F.16.16) [note that G_0 in (F.16.16) is, quite generally, the unperturbed Green's function, in this case identical with G_p]. From the results of §F.20 we could therefore calculate the diagonal elements of G_p in the momentum representation. In this case, however, it is immediately clear that there is no reason why the scattered wave should, through the asymptotic behaviour of the Green's function, behave like e^{ikr}/r; in fact the problem no longer has spherical symmetry. For the forbidden region solutions may exist analogous to equation (F.15.1), whose asymptotic behaviour is again determined by the asymptotic behaviour of the Green's function. Once again, we do not have wavefunctions which behave asymptotically like decaying exponentials.]

(F.21.2) Show that for the matrix elements of the Green's function in the coordinate representation, $\langle \mathbf{q}|G_p|\mathbf{q}'\rangle$, the following relation holds:

$$G_p(\mathbf{q}, \mathbf{q}') = G_p(\mathbf{q} + \mathbf{R}, \mathbf{q}' + \mathbf{R})$$

where $\mathbf{R}$ is a direct lattice vector.

F.22 THE EFFECTIVE MASS APPROXIMATION

When a hydrogenic impurity is placed into a semiconductor the potential of the impurity is modified by the dielectric constant κ to $-C^2/\kappa r$. With a value of dielectric constant of the order of 10 the Bohr radius becomes of the order of many lattice constants. Even as an approximation it is therefore out of the question to treat the potential as localized within the unit cell. In fact, it is usual to make the opposite approximation of a slowly varying potential. Under certain conditions it is then possible to argue that for localized energy levels near a band described by a dispersion relation $E_n(k)$ the wave equation

$$[E_n(-i\nabla) + H_I(\mathbf{q})]\langle n|\mathbf{r}|\psi\rangle = \langle n, \mathbf{r}|\psi\rangle \qquad \text{(F.22.1)}$$

is valid. Equation (F.22.1) is the basic equation of effective mass theory. In practice, only parabolic dispersion relations seem to have been considered; the effect of the lattice then enters into equation (F.22.1) only through the effective masses of the band under consideration.

In this section we shall consider the Green's function approach to the same problem. The theory is discussed in terms of Bloch functions in §B.16.

We start with the integral equation for the wave function. Defining the Green operator for the periodic lattice by equation (F.21.2), the integral equation for the bound state wave function is just equation (F.15.1), with $\hat{\mathbf{H}}_0$ replaced by $\hat{\mathbf{H}}_p$:

$$|\psi\rangle = \hat{\mathbf{G}}_p \hat{\mathbf{H}}_I |\psi\rangle.$$

Writing this in the Wannier representation we obtain

$$\langle n, \alpha|\psi\rangle = \sum_{n',\alpha',\alpha''} \langle n, \alpha|\hat{\mathbf{G}}_0|n, \alpha'\rangle \langle n, \alpha'|\hat{\mathbf{H}}_I|n'', \alpha''\rangle \langle n'', \alpha''|\psi\rangle \quad \text{(F.22.2)}$$

The matrix element $\langle n, \alpha'|\hat{\mathbf{H}}_I|n'', \alpha''\rangle$ may be written more explicitly as

$$\langle n, \alpha'|\hat{\mathbf{H}}_I|n'', \alpha''\rangle = \int a_{n''}(\mathbf{q} - \mathbf{R}_{\alpha''}) H_I(\mathbf{r}) a_n^*(\mathbf{q} - \mathbf{R}_{\alpha'})\, \mathrm{d}\mathbf{q}. \quad \text{(F.22.3)}$$

If the potential $H_I(\mathbf{r})$ is sufficiently slowly varying the orthogonality of the Wannier functions will dominate the integral and give a factor of approximately $\delta_{n''n}\,\delta_{\alpha''\alpha'}$. In this way one may argue that, under these conditions,

$$\int a_{n''}(\mathbf{q} - \mathbf{R}_{\alpha''}) H_I(\mathbf{q}) a_n^*(\mathbf{q} - \mathbf{R}_{\alpha'})\, \mathrm{d}\mathbf{q} \sim H_I(\mathbf{R}_\alpha) \sim H_I(\mathbf{q}). \quad \text{(F.22.4)}$$

If this assumption is put into equation (F.22.2) the wave vector $\langle n'', \alpha''|\psi\rangle$ is replaced by $\langle n'', \mathbf{q}|\psi\rangle$. Making the same replacement on the left-hand side we arrive at the integral equation

$$\langle n'', \mathbf{q}, |\psi\rangle = \int G_p(\mathbf{q}, \mathbf{q}') H_I(\mathbf{q}') \langle n'', \mathbf{q}'|\psi\rangle\, \mathrm{d}\mathbf{q}' \quad \text{(F.22.5)}$$

where

$$G_p(\mathbf{q}, \mathbf{q}') = \frac{1}{N} \sum_{\mathbf{k}} \frac{\exp[i\mathbf{k} \,.\, (\mathbf{q} - \mathbf{q}')]}{E - E(n, \mathbf{k})}. \quad \text{(F.22.6)}$$

The approximation implied by equations (F.22.4) therefore replaces the exact Green's function given by equation (F.21.8) by the approximate form given by equation (F.22.6).

Equation (F.22.5) for the coefficients $\langle n'', \mathbf{q}, |\psi\rangle$ is exactly equivalent to the more usual differential equation. If only one band is considered and the integration over k is extended to infinite limits, equation (F.22.6) is equivalent in operator form to

$$\hat{\mathbf{G}}_p = \frac{1}{E - E_n(-\mathrm{i}\nabla)} \tag{F.22.7}$$

which implies that the Green's operator is diagonal in the momentum representation. If a parabolic form for the dispersion relation is taken,

$$E = \frac{-\hbar^2 k^2}{2m^*}, \tag{F.22.8}$$

with energy measured from the bottom of the conduction band, equation (F.22.5) for the coefficients $\langle n, \mathbf{q}|\psi\rangle$ is the same as that for the wave function $\psi(q)$ of an isolated atom, with m^* replacing m. For a hydrogen-like potential of charge Z^+ and a particle of charge Z^-, equation (F.22.5) will only have solutions for energies

$$E_n = -\frac{Z_1 Z_2 2\pi^2 m^* e^4}{n^2 h^2 \kappa} \tag{F.22.9}$$

measured relative to the bottom of the conduction band, where a dielectric constant κ has been introduced to screen the bare potential. It may be noted that equation (F.22.5) may be taken as a starting point for treating interband effects[24].

F.23 INTERACTING IMPURITIES

If the impurities in a semiconductor are sufficiently dense their wave-functions overlap and the impurity levels considered in §F.21 and §F.22 broaden into impurity bands. If the impurities are described in the effective mass approximation of the previous section, then the effects of the lattice simply appears in the replacement of m by m^*. We can then simply consider the problem of the interaction of a number of randomly placed impurity centres. Under these conditions it is no longer possible to describe the band structure by means of an $E - \mathbf{k}$ relation. We may, however, obviously consider an average density of states for the system, averaged over an ensemble of scattering centres. The Green's function approach is ideal for this problem. Making use of equation (F.13.21) we write

$$\langle\langle D(E)\rangle\rangle = -\frac{1}{\pi}\langle\langle \mathrm{Tr}[\mathscr{I}(G^+)]\rangle\rangle \tag{F.23.1}$$

[24] T. Lukes and Roberts, *Proc. Phys. Soc.*, **92**, 758 (1967).

where the brackets $\langle\langle\ \rangle\rangle$ denote an average over an ensemble of scattering centres. The problem then reduces to the computation of the average Green's function. In the rigid ion approximation, we may write for the total perturbing potential

$$H_I(\mathbf{q}) = \sum_\alpha v(\mathbf{q} - R_\alpha) \tag{F.23.2}$$

where, for randomly distributed centres, we assume

$$P(\mathbf{R}_\alpha) = \frac{\mathrm{d}\mathbf{R}_\alpha}{\Omega} \tag{F.23.3}$$

and that the various $\mathbf{R}_\alpha$'s are distributed independently of one another.

The solution of equation (F.14.7) is now an unusual problem because, in the expression

$$\hat{\mathbf{G}} = \hat{\mathbf{G}}_0 + \hat{\mathbf{G}}_0\hat{\mathbf{H}}_I\hat{\mathbf{G}} \tag{F.23.4}$$

the functions $\hat{\mathbf{G}}$ and $\hat{\mathbf{H}}_I$ both have a probability distribution. Whereas that for $\hat{\mathbf{H}}_I$ is known, that for $\hat{\mathbf{G}}$ is not, so that the average of the right-hand side cannot be taken. Instead, recourse must be made to the iterative series, which can be averaged term by term. We shall work in the momentum representation: the total potential $\hat{\mathbf{H}}_I$ is then (in rigid ion approximation)

$$H_I(\mathbf{k} - \mathbf{k}') = \sum_\alpha \mathrm{e}^{\mathrm{i}(\mathbf{k} - \mathbf{k}'_1)\cdot\mathbf{R}_\alpha} v(\mathbf{k} - \mathbf{k}') \tag{F.23.5}$$

where v is the Fourier transform of the potential of a single atom at the origin. Consider now the series

$$\hat{\mathbf{G}} = \hat{\mathbf{G}}_0 + \hat{\mathbf{G}}_0\hat{\mathbf{H}}_I\hat{\mathbf{G}}_0 + \cdots. \tag{F.23.6}$$

Various approaches to the problem of averaging the series and resuming have been made [25,26,27,28,29,30,31]. We shall consider a simple approximation based essentially on the work of Edwards[25].

The second term in equation (F.23.6) is $G_0(\mathbf{k})H_I(\mathbf{k} - \mathbf{k})G_0(\mathbf{k})$, where we consider only diagonal terms, since we are interested in the density of states. From equation (F.23.5) this is proportional to $v(\mathbf{k} - \mathbf{k})$. Now

$$v(\mathbf{k} - \mathbf{k}') = \int \mathrm{e}^{\mathrm{i}(\mathbf{k} - \mathbf{k}')\cdot\mathbf{q}} v(\mathbf{q}) \frac{\mathrm{d}\mathbf{q}}{\Omega}$$

[25] S. F. Edwards, *Phil. Mag.*, **1958**, 1020.
[26] J. R. Klauder, *Ann. Phys.*, **14**, 43 (1961).
[27] T. Matsubara and Y. Toyzawa, *Prog. Theoret. Phys.*, **26**, 739 (1961).
[28] T. Lukes, *Phil. Mag.*, **12**, 719 (1965).
[29] T. Lukes, *Phil. Mag.*, **13**, 875 (1966).
[30] M. Bastow, M.Sc. *Dissertation* (Cardiff, 1968).
[31] R. Jones and T. Lukes, *Proc. Roy. Soc.*, **309**, 457 (1969).

and so the diagonal matrix element $\mathbf{k} = \mathbf{k}_1$ is proportional to the average potential, which may be chosen to be zero. Consider now the third term, which involves

$$\int G_0(\mathbf{k}) \sum_{\alpha,\beta} e^{i(\mathbf{k}-\mathbf{k}_1)\cdot\mathbf{R}_\alpha} v(\mathbf{k}-\mathbf{k}_1) G_0(\mathbf{k}_1)\, e^{i(\mathbf{k}_1-\mathbf{k})\cdot\mathbf{R}_\beta}\, v(\mathbf{k}_1-\mathbf{k}) G_0(\mathbf{k}) \frac{d\mathbf{k}_1}{(2\pi)^3}.$$

The term $\sum_{\alpha,\beta} e^{i(\mathbf{k}-\mathbf{k}_1)\cdot(\mathbf{R}_\alpha-\mathbf{R}_\beta)}$ can be replaced by its average

$$\int \frac{d\mathbf{R}_\alpha}{\Omega} \frac{d\mathbf{R}_\beta}{\Omega} e^{i(\mathbf{k}-\mathbf{k}_1)\cdot(\mathbf{R}_\alpha-\mathbf{R}_\beta)}.$$

By the same argument as before, this vanishes unless $\alpha = \beta$, in which case it gives a contribution of N/Ω. Thus the third term contributes

$$G_0(\mathbf{k})\Sigma_2(\mathbf{k})G_0(\mathbf{k})$$

where

$$\Sigma_2(\mathbf{k}) = \frac{N}{\Omega} \int \frac{d\mathbf{k}_1}{(2\pi)^3} |v(\mathbf{k}-\mathbf{k}_1)|^2 G_0(\mathbf{k}_1). \qquad \text{(F.23.7)}$$

Consider now the third term, which involves the product

$$e^{i(\mathbf{k}-\mathbf{k}_1)\cdot\mathbf{R}_\alpha}\, e^{i(\mathbf{k}_1-\mathbf{k}_2)\cdot\mathbf{R}_\beta}\, e^{i(\mathbf{k}_2-\mathbf{k}_3)\cdot\mathbf{R}_\gamma}\, e^{i(\mathbf{k}_3-\mathbf{k})\cdot\mathbf{R}_\delta}.$$

Averaging this exactly as before, we get non-zero contributions for

$\alpha = \beta = \gamma = \delta$; $\quad \alpha = \beta, \gamma = \delta$; $\quad \alpha = \delta, \beta = \gamma$; $\quad \alpha = \gamma, \beta = \delta$.

These contributions may be represented by diagrams:

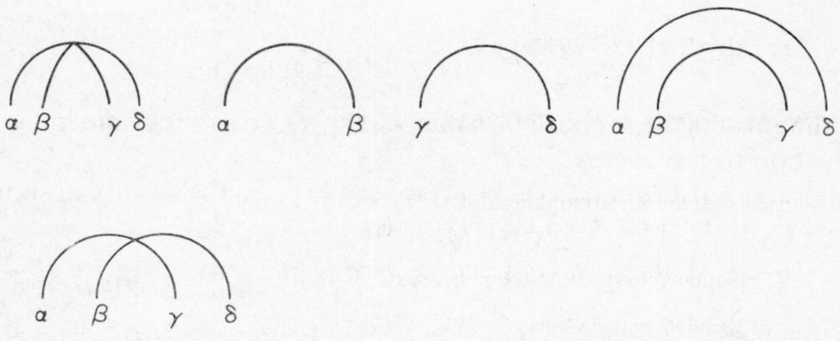

The term $\alpha = \beta = \gamma = \delta$ is, on inspection, a term which arises in the t matrix series of a single atom. It may be neglected if the scattering is treated in Born approximation. Consider now the second diagram, i.e. the

contribution $\alpha = \beta$, $\gamma = \delta$. Carrying out the averaging gives a factor $\delta(\mathbf{k} - \mathbf{k}_1 + \mathbf{k}_1 - \mathbf{k}_2)\,\delta(\mathbf{k}_3 - \mathbf{k}_4 + \mathbf{k}_4 - \mathbf{k})$. The complete term may therefore be written as

$$G_0(\mathbf{k})\Sigma_2(\mathbf{k})G_0(\mathbf{k})\Sigma_2(\mathbf{k})G_0(\mathbf{k}).$$

If only even terms in v in the series are considered, each such term can be averaged in the same way. Therefore selecting only even terms and only these diagrams, we can sum the series and obtain

$$\begin{aligned}\langle\langle G(\mathbf{k})\rangle\rangle &= \frac{G_0(\mathbf{k})}{1 - G_0(\mathbf{k})\Sigma_2(\mathbf{k})}\\ &= \frac{1}{G_0^{-1}(\mathbf{k}) - \Sigma_2(\mathbf{k})}\\ &= \frac{1}{E - k^2 - \Sigma_2(\mathbf{k})}.\end{aligned} \tag{F.23.8}$$

To interpret this expression it is useful to go back to equation (F.23.7). Making use of equation (F.13.14):

$$\begin{aligned}\Sigma_2(\mathbf{k}) &= \frac{N}{\Omega}\int\frac{d\mathbf{k}_1}{(2\pi)^3}\frac{|v(\mathbf{k} - \mathbf{k}_1)|^2}{E - k_1^2 + i\epsilon}\\ &= \frac{PN}{\Omega}\int\frac{d\mathbf{k}_1}{(2\pi)^3}\frac{|v(\mathbf{k} - \mathbf{k}_1)|^2}{E - k_1^2} - \frac{N}{\Omega}\int\frac{d\mathbf{k}_1}{(2\pi)^3}\pi i\,\delta(E - k^2)|v(\mathbf{k} - \mathbf{k}_1)|^2\\ &= A + iB, \text{ say.}\end{aligned} \tag{F.23.9}$$

This equation (F.23.8) may be written

$$\langle G(\mathbf{k})\rangle = \frac{1}{E - k^2 - [A + iB]} \tag{F.23.10}$$

and the effect of the perturbation may be expressed as a real and imaginary correction to the energy.

Making use of equation (F.13.21) we can express the density of states as

$$-\frac{1}{\pi}\mathrm{Tr}(G^+) = -\frac{1}{\pi}\int\frac{d\mathbf{k}}{(2\pi)^3}\frac{B(\mathbf{k})}{[E - k^2 - A(\mathbf{k})]^2 + B^2(\mathbf{k})}. \tag{F.23.11}$$

The imaginary correction term B is closely linked to the irreversible character of the process introduced by the averaging procedure and to the mean free path, although the exact computation of the electrical conductivity would require the average over two Green's functions. For more details the reader is referred to the original paper.

GENERAL REFERENCES

(1) Abrikosov, A. A., Gorkov, L. A. and Dzyaloshinksi, I. E., *Methods of Quantum Field Theory in Statistical Physics*, Prentice Hall, 1963.
(2) Bonch–Bruevich, V. L. and Tyablikov, S. V., *The Green's Function Method in Statistical Mechanics*, North Holland, 1962.
(3) Dirac, P. A. M., *Quantum Mechanics*, The Clarendon Press, Oxford, 1958.
(4) Goertzel, G. and Tralli, N., *Some Mathematical Methods of Physics*, McGraw-Hill, 1960.
(5) Ter Haar, D., *On the Use of Green's Functions in Statistical Mechanics*, Scottish Universities Summer School, Oliver and Boyd, 1961.
(6) Jackson, J. D., *Mathematics for Quantum Mechanics*, Benjamin, New York, 1962.
(7) Louisell, W. H., *Radiation and Noise in Quantum Electronics*, McGraw-Hill, 1964.
(8) Morse, P. M. and Feshbach, H., *Methods of Theoretical Physics*, McGraw-Hill, 1953.
(9) Powell, J. L. and Crasemann, B., *Quantum Mechanics*, Benjamin, New York, 1942.
(10) Turner, R. E. and Parry, W. E., Green's Functions in Statistical Mechanics, *Repts. Progr. Phys*, **27**, 23 (1964).
(11) Ziman, J. M., *Principles of the Theory of Solids*, Cambridge University Press, 1964.
(12) Zubarev, D. N., *Soviet Phys. Uspekhi*, **3**, 320 (1960).

Appendix

P. T. Landsberg

TIME REVERSIBILITY

1 Properties of T (classical case)

Time-reversal symmetry has been seen to be of importance in §B.11 and §C.11, since it yields the result

$$E_n(\mathbf{k}) = E_n(-\mathbf{k}) \tag{1.1}$$

for spinless particles. In this appendix a somewhat more general discussion is given.

Given a classical time-independent one-particle Hamiltonian, $H(\mathbf{r}, \mathbf{p})$, it is known that the energy of the system is invariant under time translations. This follows most generally from Nöther's theorem[1], but can be established by more elementary methods. Such a system is also invariant under the classical operation T_c of time reversal, as will now be shown. One may define T_c by the intuitive properties ($i = x, y, z$; $\dot{r}_i \equiv \mathrm{d}r_i/\mathrm{d}t$):

$$r_{\mathrm{rev}\,i}(t) \equiv T_c r_i(t) \equiv r_i(-t), \tag{1.2}$$

$$\dot{r}_{\mathrm{rev}\,i}(t) \equiv T_c \dot{r}_i(t) \equiv -\dot{r}_i(-t). \tag{1.3}$$

They state that the time-reversed motion runs through the same points in a time-reversed order and with reversed velocities. It follows that

$$p_{\mathrm{rev}\,i}(t) \equiv T_c p_i(t) = -p_i(-t), \tag{1.4}$$

$$H_{\mathrm{rev}}(\mathbf{r}, \mathbf{p}) \equiv T_c H(\mathbf{r}, \mathbf{p}) = H(\mathbf{r}, -\mathbf{p}). \tag{1.5}$$

The Hamiltonian equations of motion

$$\frac{\mathrm{d}r_i(t)}{\mathrm{d}t} = \frac{\partial H(\mathbf{r}, \mathbf{p})}{\partial p_i}, \qquad \frac{\mathrm{d}p_i(t)}{\mathrm{d}t} = -\frac{\partial H(\mathbf{r}, \mathbf{p})}{\partial r_i} \tag{1.6}$$

become under time reversal

$$\frac{\mathrm{d}r_i(-t)}{\mathrm{d}(-t)} = \frac{\partial H(\mathbf{r}, -\mathbf{p})}{\partial(-p_i)}, \qquad -\frac{\mathrm{d}p_i(-t)}{\mathrm{d}(-t)} = -\frac{\partial H(\mathbf{r}, \mathbf{p})}{\partial r_i}. \tag{1.7}$$

[1] E. Nöther, *Nachr. Kgl. Ges. Wiss. Göttingen*, **1918**, 235; E. L. Hill, *Rev. Mod. Phys.*, **23**, 253 (1951).

Writing $p_{\text{rev}\,i}$ for $-p_i$ the equations are seen to be unaltered. Thus the Hamiltonian $H(\mathbf{r}, \mathbf{p})$ cannot distinguish a direction of time.

2 Properties of T (quantum case, no spin)*

It has been seen in the problems of §B.8, equation (B.5.10) and elsewhere in this book, that changes of variable and changes of representation are achieved in quantum mechanics by transformations of the type $\mathbf{x} \to A\mathbf{x}A^{-1}$. Thus from equations (1.2) and (1.4) one requires the quantum analogue T of the classical operation T_c to have the properties

$$T\mathbf{r}T^{-1} = \mathbf{r}, \tag{2.1}$$

$$T\mathbf{p}T^{-1} = -\mathbf{p}. \tag{2.2}$$

Hence the angular momentum satisfies

$$T(\mathbf{r} \times \mathbf{p})T^{-1} = -\mathbf{r} \times \mathbf{p}. \tag{2.3}$$

The Pauli spin vector operator $\boldsymbol{\sigma} = (\sigma_x, \sigma_y, \sigma_z)$ represents an angular momentum, so that one would expect it to be inverted by T:

$$T\boldsymbol{\sigma}T^{-1} = -\boldsymbol{\sigma}. \tag{2.4}$$

To see how T changes a wavefunction, consider first the time-dependent Schrödinger equation for a spinless Hamiltonian which does not depend explicitly on the time:

$$i\hbar\frac{\partial\psi(\mathbf{r}, t)}{\partial t} = H[\mathbf{r}(t), \mathbf{p}(t)]\psi(\mathbf{r}, t). \tag{2.5}$$

On replacing t by $-t$ a negative sign is introduced on the left-hand side, and the equation becomes formally the same only if its complex conjugate is taken in addition. This yields

$$i\hbar\frac{\partial\psi^*(\mathbf{r}, -t)}{\partial t} = \{H[\mathbf{r}(-t), \mathbf{p}(-t)]\}^*\psi^*(\mathbf{r}, -t). \tag{2.6}$$

The two operations just performed suggest that one should choose

$$T\psi(\mathbf{r}, t) = \psi^*(\mathbf{r}, -t). \tag{2.7}$$

To see how T affects an operator, use it on the left-hand side of (2.5), noting from (2.7) that

$$T\mathrm{i} = -\mathrm{i}T, \qquad T\partial\phi/\partial t = -\partial T\phi/\partial t.$$

The analogue of (2.6) is then

$$\mathrm{i}\hbar\frac{\partial}{\partial t}T\psi(\mathbf{r}, t) = \{TH[\mathbf{r}(t), \mathbf{p}(t)]T^{-1}\}T\psi(\mathbf{r}, t). \tag{2.8}$$

* The author is indebted to Dr. D. J. Morgan for discussions concerning this section. In this subsection $H[\mathbf{r}(t), \mathbf{p}(t)]$ means $H(\mathbf{r}, \mathbf{p})$ and not a commutator.

Comparing (2.6) and (2.8)

$$TH[\mathbf{r}(t), \mathbf{p}(t)]T^{-1} = H[\mathbf{r}(-t), \mathbf{p}(-t)]^*. \tag{2.9}$$

This is the general result for this case. If H is a real function of its parameters, one can go one step further and find an analogue of equation (1.5):

$$TH[\mathbf{r}(t), \mathbf{p}(t)]T^{-1} = H[\mathbf{r}(-t), -\mathbf{p}(-t)]. \tag{2.10}$$

Since H does not depend explicitly on the time, one can write this more briefly as

$$TH(\mathbf{r}, \mathbf{p})T^{-1} = H(\mathbf{r}, \mathbf{p})^* = H(\mathbf{r}, -\mathbf{p}). \tag{2.11}$$

We must give more precision to the *complex conjugate of an operator* as it occurs in (2.6). One can define a real operator as one which takes a real function into a real function. One can then see that any operator P can be expressed in the form $A + \mathrm{i}B$, where A and B are real operators. Then one can define P^* as $A - \mathrm{i}B$.

If H commutes with T, one has time reversal invariance. This implies, by (2.9),

$$H(\mathbf{r}, \mathbf{p}) = H(\mathbf{r}, \mathbf{p})^*. \tag{2.12}$$

If H is a real function, this implies, by (2.11),

$$H(\mathbf{r}, \mathbf{p}) = H(\mathbf{r}, -\mathbf{p}). \tag{2.13}$$

This condition is satisfied if H involves only even powers of $\mathbf{p}$. Time reversal invariance is then satisfied. If $\mathbf{A}$ be the electromagnetic vector potential, we know already from problem (B.13.10) or equation (B.13.13) that p^2 in the Hamiltonian is replaced by a term $(\mathbf{p} + (e/c)\mathbf{A})^2$, where e is an electric charge and c is the velocity of light. This introduces linear terms in $\mathbf{p}$ and so destroys time reversal invariance.

If the Hamiltonian involves spin operators, the Schrödinger equation is no longer relevant and the argument from equation (2.5) onwards becomes inapplicable.

3 Realization of *T* (quantum case, with spin)

In order to find a realization of T, recall that the Pauli spin operators for a particle j of spin $\frac{1}{2}$ act in a two-dimensional spin space E_j and can there be represented by

$$\sigma_{jx} \equiv \begin{pmatrix} 0 & 1 \\ 1 & 0 \end{pmatrix}, \quad \sigma_{jy} \equiv \begin{pmatrix} 0 & -\mathrm{i} \\ \mathrm{i} & 0 \end{pmatrix}, \quad \sigma_{jz} \equiv \begin{pmatrix} 1 & 0 \\ 0 & -1 \end{pmatrix}, \tag{3.1}$$

whence it is easily verified as in problem (B.8.1) that

$$\sigma_{jx}\sigma_{jy} = -\sigma_{jy}\sigma_{jx} = i\sigma_{jz}, \qquad \sigma_{ji}^2 = I_j. \tag{3.2}$$

Here $j = 1, 2, \ldots, n$, where n is the number of electrons in the system, $i = x, y, z$ and I_j is the unit matrix in E_j. Let K be the complex conjugation for the Schrödinger representation in conjunction with representation (3.1). Then*

$$K\phi \equiv \phi^* \qquad (\text{all } \phi), \tag{3.3}$$

so that K^2 is the identity operator. K has the simple properties

$$K\mathbf{r} = \mathbf{r}K, \tag{3.4}$$

$$K\mathbf{p} = -\mathbf{p}K, \tag{3.5}$$

$$K\sigma_{jx} = \sigma_{jx}K, \tag{3.6}$$

$$Ki\sigma_{jy} = i\sigma_{jy}K, \tag{3.7}$$

$$K\sigma_{jz} = \sigma_{jz}K, \tag{3.8}$$

where we have used the coordinate representation $p_i = -\mathrm{i}\hbar\partial/\partial x_i$. Thus K satisfies (2.1)–(2.3), and is a possible time-reversal operator for spinless particles as discussed in §2. It is not suitable for the present purpose since it does not satisfy (2.4). To correct this, try

$$T = (\pm \mathrm{i})^n \sigma_{1y}\sigma_{2y} \cdots \sigma_{ny} K \tag{3.9}$$

which still satisfies (2.1)–(2.3). Either one of the two signs in (3.9) is admissible provided it is used consistently. Since the spin operators σ_{ji} commute for different particles,

$$T\sigma_{jx} = -\sigma_{jx}T, \tag{3.10}$$

$$T\sigma_{jy} = -\sigma_{jy}T, \tag{3.11}$$

$$T\sigma_{jz} = -\sigma_{jz}T. \tag{3.12}$$

The negative sign in (3.10) comes from (3.2), and in (3.12) it comes from the analogous relation

$$\sigma_{jy}\sigma_{jz} = -\sigma_{jz}\sigma_{jy}.$$

K plays a crucial part only in (3.11), where it causes the negative sign. Thus the *ansatz* (3.9) satisfies our conditions (2.1) to (2.4). In addition the n-particle quantum analogue of (1.5) is satisfied:

$$TH(\mathbf{r}_1, \ldots \mathbf{r}_n; \mathbf{p}_1, \ldots \mathbf{p}_n; \boldsymbol{\sigma}_1, \ldots \boldsymbol{\sigma}_n) = H(\mathbf{r}_1, \ldots; -\mathbf{p}_1, \ldots; -\boldsymbol{\sigma}_1, \ldots)T. \tag{3.13}$$

* For the effect of the choice of representation on these properties see A. Messiah, *Quantum Mechanics*, North Holland, Amsterdam, 1962. Vol. II, Chapter XV, §5.

The last result can be seen in detail by writing a general Hamiltonian symmetrical in n particles as

$$H = H^{(0)} + H^{(1)} + H^{(2)} + \cdots$$

where

$$H^{(0)} \equiv H^{0}(\mathbf{r}_1, \ldots, \mathbf{p}_n)I_1 I_2, \ldots I_n,$$

$$H^{(1)} \equiv \sum_{j=1}^{n} \left\{ [H^{(j,x)}\sigma_{jx} + H^{(j,y)}\sigma_{jy} + H^{(j,z)}\sigma_{jz}] \prod_{k(\neq j)} I_k \right\},$$

$$H^{(2)} \equiv \sum_{\substack{i,j=1 \\ (i\neq j)}}^{n} \left\{ [H^{(i,x;j,x)}\sigma_{ix}\sigma_{jx} + H^{(i,x;j,y)}\sigma_{ix}\sigma_{jy} + \cdots] \prod_{\substack{k \\ (\neq i,j)}} I_k \right\}.$$

Now K reverses the sign of all p_j and all σ_{jy}. The $i\sigma_{jy}$ factor in T reverses the sign of σ_{jx} and σ_{jz}, whence (3.13) is obtained.

4 Kramer's theorem[2]

We now adopt the operator (3.9) as the quantum-mechanical time reversal operator, and observe that it has the following properties:

$$T^2 = (-1)^n I_1 I_2 \ldots I_n, \tag{4.1}$$

$$(T\psi, T\phi) = (\phi, \psi) \qquad (\text{all } \psi, \phi). \tag{4.2}$$

To prove (4.1), note that from (3.3) and (3.7)

$$T^2 = (\mp 1)^{2n}(i\sigma_{1y})^2 \ldots (i\sigma_{ny})^2 K^2 = (-1)^n I_1 \ldots I_n.$$

To prove (4.2), note that

$$\begin{aligned}&(i^n\sigma_{1y}\ldots\sigma_{ny}K\psi, \qquad i^n\sigma_{1y}\ldots\sigma_{ny}K\phi)\\ &\qquad = (K\psi, (\sigma_{1y}\ldots\sigma_{ny})^2 K\phi) = (K\psi, K\phi) = (\psi^*, \phi^*).\end{aligned}$$

These algebraic results imply:

Theorem. Let $\psi = T\phi$, where ϕ is some wavefunction. Then

(i) $(\phi, \psi) = 0$ if n is odd; (4.3)
(ii) If H is T-invariant, then $H\phi = E\phi$ implies $H\psi = E\psi$. (4.4)

Proof. (i) With $\psi = T\phi$, (4.2) implies

$$(\phi, T\phi) = (T^2\phi, T\phi) = (-1)^n(\phi, T\phi),$$

whence the result follows.

[2] H. A. Kramer, *Proc. Amsterdam Acad.*, **33**, 959 (1930); M. J. Klein, *Am. J. Phys.*, **20**, 65 (1952).

(ii) The *T*-invariance of H implies $H = THT^{-1}$, so that $H\phi = E\phi$ yields, after operating with T on the left, $THT^{-1}T\phi = ET\phi$, i.e. $H\psi = E\psi$. Hence one sees that:

$$\left.\begin{array}{l}\text{with } \phi,\ T\phi \text{ is a degenerate eigenfunction of a } T\text{-invariant } H.\\ \text{If } n \text{ is odd the two functions are independent (even ortho-}\\ \text{gonal) so that all eigenfunctions of } H \text{ have even degeneracies.}\end{array}\right\} \tag{4.5}$$

This is Kramer's theorem, and ϕ and $T\phi$ are called **Kramers conjugates**. If $\phi_1, \ldots, \phi_l$ are degenerate they are also degenerate with $T\phi_1, \ldots, T\phi_l$, so that all degeneracies are even.

One could define a **symmetry operation A** as an operation which does not affect transition probabilities (and hence normalizations). They then satisfy

$$|(A\phi, A\psi)|^2 = |(\phi, \psi)|^2 \qquad (\text{all } \phi, \psi). \tag{4.6}$$

This leaves open two possibilities for all ϕ, ψ,

$$(A\phi, A\psi) = (\phi, \psi), \tag{4.7}$$

or

$$(A\phi, A\psi) = (\psi, \phi). \tag{4.8}$$

From (4.7) $(\phi, A^\dagger A\psi) = (\phi, \psi)$, whence if $A^\dagger$ is the Hermitian conjugate of A,

$$A^\dagger A = I.$$

Hence the symmetry operations satisfying (4.7) are **unitary**. We observe that (4.8) covers the class of symmetry operations which includes T. They are called **antiunitary**[3].

Kramers conjugates lead to opposite expectation values for the spins $\boldsymbol{\sigma}_j$, since by (4.8) and (3.13), with $\phi = \boldsymbol{\sigma}_j\psi$,

$$(\psi, \boldsymbol{\sigma}_j\psi) = (T\boldsymbol{\sigma}_j\psi, T\psi) = -(\boldsymbol{\sigma}_j T\psi, T\psi) = -(T\psi, \boldsymbol{\sigma}_j T\psi). \tag{4.9}$$

This establishes the proposition. Thus a convenient label (↑ or ↓) for a wavefunction is the expectation value of the total spin $\sum_j \boldsymbol{\sigma}_j$. If l stands for the remaining quantum numbers and the $\{\phi_{i\downarrow}\}\{\phi_{j\uparrow}\}$ form together a complete set, then the Kramers conjugate of $\phi_{l\uparrow}$ is

$$T\phi_{l\uparrow} = \sum_r a_{lr}\phi_{r\downarrow}, \tag{4.10}$$

where the a_{lr} are coefficients.

[3] For a recent discussion of the mathematical properties of antiunitary operators, see E. P. Wigner, *J. Math. Phys.*, **1**, 409 and 414 (1960).

5 The inversion in the single-particle Bloch theory

The inversion J is defined by

$$J\mathbf{r} = -\mathbf{r}J, J\mathbf{p} = -\mathbf{p}J. \tag{5.1}$$

If a lattice has a centre of inversion, the electron potential energy satisfies

$$JV(\mathbf{r}) = V(-\mathbf{r})J = V(\mathbf{r})J.$$

If the rest of the Hamiltonian also commutes with J, the Hamiltonian as a whole does. An angular momentum is invariant under J, whence the spin angular momentum σ satisfies

$$J\boldsymbol{\sigma} = \boldsymbol{\sigma}J. \tag{5.2}$$

Let $\eta_\pm$ be the two eigenfunctions of the z-component of the spin of the particle:

$$\sigma_z \eta_\pm = \pm\eta_\pm, \qquad \eta_+ \equiv \begin{pmatrix}1\\0\end{pmatrix}, \qquad \eta_- \equiv \begin{pmatrix}0\\1\end{pmatrix}.$$

Then a Bloch function can be written as

$$\psi_{n\mathbf{k}\uparrow} = e^{i\mathbf{k}\cdot\mathbf{r}}[u_{n\mathbf{k}\uparrow}^{(+)}(\mathbf{r})\eta_+ + u_{n\mathbf{k}\uparrow}^{(-)}(\mathbf{r})\eta_-] \equiv e^{i\mathbf{k}\cdot\mathbf{r}}u_{n\mathbf{k}\uparrow}(\mathbf{r}). \tag{5.3}$$

Note that it is not in general in a pure spin state η_+ or η_-, the arrow having the meaning given to it in (4.10). Equation (5.3) specifies what may be called a Bloch spinor.

The equation for the Bloch function (5.3) can be put in the usual form [see equation (B.15.11)]:

$$H_\mathbf{k}u_{n\mathbf{k}\uparrow}(\mathbf{r}) = E_n(\mathbf{k})_\uparrow u_{n\mathbf{k}\uparrow}(\mathbf{r}), \qquad H_\mathbf{k} \equiv e^{-i\mathbf{k}\cdot\mathbf{r}}He^{i\mathbf{k}\cdot\mathbf{r}}. \tag{5.4}$$

The basic assumption of this section is to suppose that the Hamiltonian is invariant under inversion, whence

$$JH_\mathbf{k} = H_{-\mathbf{k}}J. \tag{5.5}$$

Replacing $\mathbf{k}$ by $-\mathbf{k}$ in (5.4) and then applying J,

$$H_{-\mathbf{k}}u_{n,-\mathbf{k}\uparrow}(\mathbf{r}) = E_n(-\mathbf{k})_\uparrow u_{n,-\mathbf{k}\uparrow}(\mathbf{r}),$$

$$H_\mathbf{k}u_{n,-\mathbf{k}\uparrow}(-\mathbf{r}) = E_n(-\mathbf{k})_\uparrow u_{n,-\mathbf{k}\uparrow}(-\mathbf{r}). \tag{5.6}$$

In acting with J on u in (5.6) the spin arrow is not changed by an argument analogous to (4.9), using (5.2). Since the operator $H_\mathbf{k}$ is the same in (5.4) and (5.6), it follows that

$$\text{if } JH = HJ, \qquad \text{then } Ju_{n\mathbf{k}}(\mathbf{r}) = u_{n\mathbf{k}}(-\mathbf{r}) = u_{n,-\mathbf{k}}(\mathbf{r}) \tag{5.7}$$

and

$$E_n(\mathbf{k})_\uparrow = E_n(-\mathbf{k})_\uparrow. \tag{5.8}$$

It has been assumed here that the **k**-dependent and spin-dependent phase factors which may occur in

$$J\psi_{n\mathbf{k}\uparrow}(-\mathbf{r}) = \psi_{n,\,-\mathbf{k}\uparrow}(\mathbf{r})\,\mathrm{e}^{i\theta(\mathbf{k},s)} \tag{5.9}$$

have been removed by an appropriate choice of functions.

Equation (5.8) is a form of equation (1.1) and the condition in (5.7) will be fulfilled if the crystal has a centre of inversion. Even without this, however, a form of equation (1.1) is valid, as will now be shown.

6 The Kramers conjugates in a single particle Bloch theory

In a one-electron theory the theorem (4.5) concerning even degeneracies applies, and it is therefore desirable to locate the Kramers degeneracy in **k**-space. To this end, apply T to the Bloch function (5.3). The exponent has its sign reversed because K is a factor of T. The exponential is found to be multiplied by a lattice periodic spinor whose spin arrow is turned downwards by (4.9). Thus one obtains another Bloch spinor:

$$T\psi_{n\mathbf{k}\uparrow}(\mathbf{r}) = \psi_{n,-\mathbf{k},\downarrow}(\mathbf{r})\,\mathrm{e}^{\mathrm{i}\theta(\mathbf{k},s)}. \tag{6.1}$$

The phase factor will be neglected as before.

The main assumption of this section is to suppose that the Hamiltonian is invariant under T. Applying this operation to the Schrödinger equation

$$H\psi_{n\mathbf{k}\uparrow}(\mathbf{r}) = E_n(\mathbf{k})_\uparrow\psi_{n\mathbf{k}\uparrow}(\mathbf{r}),$$

one then finds

$$H\psi_{n,-\mathbf{k},\downarrow}(\mathbf{r}) = E_n(\mathbf{k})_\uparrow\psi_{n,-\mathbf{k},\downarrow}(\mathbf{r}).$$

But this equation has the eigenvalue $E_n(-\mathbf{k})_\downarrow$, whence

$$\text{if } TH = HT, \qquad \text{then } E_n(\mathbf{k})_\uparrow = E_n(-\mathbf{k})_\downarrow. \tag{6.2}$$

This is another form of (1.1).

One observes that if one starts with a Bloch theory without spin, then addition of spin doubles the degeneracies. If one then adds spin–orbit interaction, this can lift only those degeneracies which exceed two. One can interpret (6.1) and (6.2) in a theory without spin by saying that in such a theory time reversal symmetry leads to

$$\psi^*_{n\mathbf{k}}(\mathbf{r}) = \psi_{n,-\mathbf{k}}(\mathbf{r}) \quad \text{and} \quad E_n(\mathbf{k}) = E_n(-\mathbf{k}). \tag{6.3}$$

By (6.3) one can make $E_n(\mathbf{k})$ periodic in $\mathbf{k}$ by extending a typical range $-\pi/a < k_i \leqslant \pi/a$ for a general component of $\mathbf{k}$ to the lower limit and beyond, as one pleases. One then has, at a general point,

$$\frac{\partial E_n(\mathbf{k})}{\partial k_i} = -\frac{\partial E_n(-\mathbf{k})}{\partial k_i}. \tag{6.4}$$

But these slopes at the limiting points on the zone boundary must be equal, so that it follows from (6.4) that they must vanish.

Suppose now that both symmetry elements J and T are present. Operating on (5.9) with T,

$$TJ\psi_{n\mathbf{k}\uparrow}(\mathbf{r}) = T\psi_{n,-\mathbf{k}\uparrow}(\mathbf{r}) = \psi_{n\mathbf{k}\downarrow}(\mathbf{r}). \tag{6.5}$$

Operating on (6.1) with J,

$$JT\psi_{n\mathbf{k}\uparrow}(\mathbf{r}) = J\psi_{n,-\mathbf{k}\downarrow}(\mathbf{r}) = \psi_{n\mathbf{k}\downarrow}(\mathbf{r}). \tag{6.6}$$

Hence the so-called **conjugation** C satisfies

$$C \equiv JT = TJ, \tag{6.7}$$

and (5.7) and (6.2) yield (see fig. 1):

$$\text{if } JH = HJ \quad \text{and} \quad TH = HT, \qquad \text{then } E_n(\mathbf{k})_\uparrow = E_n(\mathbf{k})_\downarrow. \tag{6.8}$$

The inversion can be defined from (5.1) for an n-particle system. Then, since $J^2 = 1$, it follows from (6.7) that the conjugation satisfies, by (4.1),

$$C^2 = T^2 = (-1)^n I_1 \ldots I_n. \tag{6.9}$$

Also

$$(C\psi, C\phi) = (TJ\psi, TJ\phi) = (J\phi, J\psi) = (\phi, \psi). \tag{6.10}$$

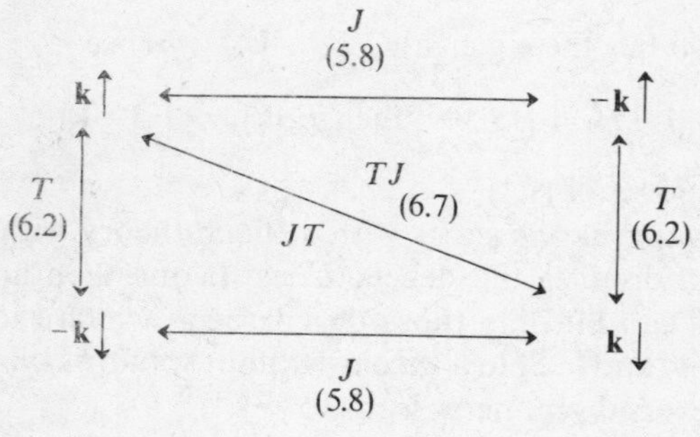

Fig. 1. Degeneracies of $E_n(\mathbf{k})_\uparrow$.

Thus the conjugation shares the basic properties (4.1) and (4.2) with the time reversal operator. These are valid for a system of n electrons. Note that by (6.9)

$$C = (-1)^n C^{-1}, \qquad T = (-1)^n T^{-1}. \tag{6.11}$$

7 Theorems for matrix elements

We now introduce an abstract specification of a class of symmetry operations Y for an n-electron system. All we know of Y is that it has the properties

$$Y^2 = (-1)^n I_1 I_2 \ldots I_n, \tag{7.1}$$

$$(Y\psi, Y\phi) = (\phi, \psi) \qquad (\text{all } \phi, \psi). \tag{7.2}$$

By (4.1) and (4.2) they are satisfied by time reversal, and by (6.9) and (6.10) they are also satisfied by the conjugation.

Theorem 1. If an operator A satisfies

$$YA = \mathrm{d}A^\dagger Y, \tag{7.3}$$

(where d is a number, d^* its complex conjugate, and $A^\dagger$ the hermitian conjugate of A) then

(i) $$(\psi, A\phi) = d^*(Y\phi, AY\psi), \tag{7.4}$$

(ii) $$(\psi, AY\phi) = (-1)^n d^*(\phi, AY\psi). \tag{7.5}$$

Proof. (i) By (7.2) and (7.3),

$$(\psi, A\phi) = (YA\phi, Y\psi) = d^*(A^\dagger Y\phi, Y\psi) = d^*(Y\phi, AY\psi).$$

(ii) Interpret, in (7.4), ϕ as $Y^{-1}\phi'$ where ϕ' is another wavefunction. Then, using (6.11),

$$\phi = (-1)^n Y\phi',$$

and (7.5) results when the primes are dropped again.

Theorem 2[4]. In a one-fermion Bloch theory in which the Hamiltonian is invariant under Y, which satisfies (7.1)–(7.3),

(i) $$(\psi_{\mathbf{nk}\uparrow}, A\psi_{\mathbf{nk}\uparrow}) = d^*(Y\psi_{\mathbf{nk}\uparrow}, AY\psi_{\mathbf{nk}\uparrow}),$$

(ii) $$(\psi_{\mathbf{nk}\uparrow}, AY\psi_{\mathbf{nk}\uparrow}) = -d^*(\psi_{\mathbf{nk}\uparrow}, AY\psi_{\mathbf{nk}\uparrow}).$$

This is simply the special case $\phi = \psi = \psi_{\mathbf{nk}\uparrow}$ of theorem 1. Its uses are illustrated in the table below. The results derived in this Appendix are of use in band theory and in many of its applications. In particular, it is clear that equation (6.3) can often be used to simplify theoretical deductions.

[4] Generalization of theorems given by Y. Yafet in *Solid State Physics*, **14**, 14 (1963). See also C. Kittel, *Quantum Theory of Solids*, Wiley, New York, 1963, Chapter 9.

Some Interpretations of Y, d, and A

Y	d	A	No. of theorem	Use of theorem
T	1	A function of coordinates and/or of products of even numbers of momentum and/or spin components. Simple cases: $A = \mathbf{r}$, $A = p_x p_y$, $A = \boldsymbol{\sigma} \times \text{grad } \mathbf{V} \cdot \mathbf{p}$ (spin–orbit interaction).	2(i) 2(ii)	$(\psi_{n\mathbf{k}\uparrow}, \mathbf{r}\psi_{n\mathbf{k}\uparrow}) = (\psi_{n,-\mathbf{k}\downarrow}, \mathbf{r}\psi_{n,-\mathbf{k},\downarrow})$ $(\psi_{n\mathbf{k}\uparrow}, \mathbf{r}\psi_{n,-\mathbf{k}\downarrow}) = 0$
T	−1	A function of products of an odd number of momentum and/or spin components. Simple cases: $A = \mathbf{p}$, $A = \sigma$.	2(i)	$(\psi_{n\mathbf{k}\uparrow}, \mathbf{p}\psi_{n\mathbf{k}\uparrow}) = -(\psi_{n,-\mathbf{k}\downarrow}, \mathbf{p}\psi_{n,-\mathbf{k}\downarrow})$
C	1	A function of momentum components, and of products of an even number of coordinates and/or spin components. Simple cases: $A = \mathbf{p}$, $A = \boldsymbol{\sigma} \times \text{grad } \mathbf{V} \cdot \mathbf{p}$.	2(i) 2(ii)	$(\psi_{n\mathbf{k}\uparrow}, \mathbf{p}\psi_{n\mathbf{k}\uparrow}) = (\psi_{n\mathbf{k}\downarrow}, \mathbf{p}\psi_{n\mathbf{k}\downarrow})$ $(\psi_{n\mathbf{k}\uparrow}, \mathbf{p}\psi_{n\mathbf{k}\downarrow}) = 0$
C	−1	A function of products of an odd number of coordinates and/or spin components. Simple cases: $A = \mathbf{r}$, $A = \boldsymbol{\sigma}$, $A = \mathbf{r} \times \mathbf{p}$.	2(i)	$(\psi_{n\mathbf{k}\uparrow}, \mathbf{r}\psi_{n\mathbf{k}\uparrow}) = -(\psi_{n\mathbf{k}\downarrow}, \mathbf{r}\psi_{n\mathbf{k}\downarrow})$

Author Index

Subject Index